Obtaining and Characterization of New Materials, Volume III

Obtaining and Characterization of New Materials, Volume III

Editor

Andrei Victor Sandu

Basel • Beijing • Wuhan • Barcelona • Belgrade • Novi Sad • Cluj • Manchester

Editor
Andrei Victor Sandu
Faculty of Materials Science
and Engineering
Gheorghe Asachi Technical
University of Iasi
Iasi
Romania

Editorial Office
MDPI
St. Alban-Anlage 66
4052 Basel, Switzerland

This is a reprint of articles from the Special Issue published online in the open access journal *Materials* (ISSN 1996-1944) (available at: www.mdpi.com/journal/materials/special_issues/obtain_charaterization_mater_volumeIII).

For citation purposes, cite each article independently as indicated on the article page online and as indicated below:

Lastname, A.A.; Lastname, B.B. Article Title. *Journal Name* **Year**, *Volume Number*, Page Range.

ISBN 978-3-0365-8671-7 (Hbk)
ISBN 978-3-0365-8670-0 (PDF)
doi.org/10.3390/books978-3-0365-8670-0

© 2023 by the authors. Articles in this book are Open Access and distributed under the Creative Commons Attribution (CC BY) license. The book as a whole is distributed by MDPI under the terms and conditions of the Creative Commons Attribution-NonCommercial-NoDerivs (CC BY-NC-ND) license.

Contents

About the Editor vii

Andreea Hegyi, Adrian-Victor Lăzărescu, Adrian Alexandru Ciobanu, Brăduţ Alexandru Ionescu, Elvira Grebenişan and Mihail Chira et al.
Study on the Possibilities of Developing Cementitious or Geopolymer Composite Materials with Specific Performances by Exploiting the Photocatalytic Properties of TiO_2 Nanoparticles
Reprinted from: *Materials* 2023, 16, 3741, doi:10.3390/ma16103741 1

Liliana Maria Nicula, Daniela Lucia Manea, Dorina Simedru, Oana Cadar, Anca Becze and Mihai Liviu Dragomir
The Influence of Blast Furnace Slag on Cement Concrete Road by Microstructure Characterization and Assessment of Physical-Mechanical Resistances at 150/480 Days
Reprinted from: *Materials* 2023, 16, 3332, doi:10.3390/ma16093332 35

Aimi Noorliyana Hashim, Mohd Arif Anuar Mohd Salleh, Muhammad Mahyiddin Ramli, Mohd Mustafa Al Bakri Abdullah, Andrei Victor Sandu and Petrica Vizureanu et al.
Effect of Isothermal Annealing on Sn Whisker Growth Behavior of Sn0.7Cu0.05Ni Solder Joint
Reprinted from: *Materials* 2023, 16, 1852, doi:10.3390/ma16051852 56

Normah Kassim, Shayfull Zamree Abd Rahim, Wan Abd Rahman Assyahid Wan Ibrahim, Norshah Afizi Shuaib, Irfan Abd Rahim and Norizah Abd Karim et al.
Sustainable Packaging Design for Molded Expanded Polystyrene Cushion
Reprinted from: *Materials* 2023, 16, 1723, doi:10.3390/ma16041723 70

Ismail Luhar, Salmabanu Luhar, Mohd Mustafa Al Bakri Abdullah, Andrei Victor Sandu, Petrica Vizureanu and Rafiza Abdul Razak et al.
Solidification/Stabilization Technology for Radioactive Wastes Using Cement: An Appraisal
Reprinted from: *Materials* 2023, 16, 954, doi:10.3390/ma16030954 89

Chirawat Wattanapanich, Thanongsak Imjai, Reyes Garcia, Nur Liza Rahim, Mohd Mustafa Al Bakri Abdullah and Andrei Victor Sandu et al.
Computer Simulations of End-Tapering Anchorages of EBR FRP-Strengthened Prestressed Concrete Slabs at Service Conditions
Reprinted from: *Materials* 2023, 16, 851, doi:10.3390/ma16020851 118

Yuri P. Piryatinski, Markiian B. Malynovskyi, Maryna M. Sevryukova, Anatoli B. Verbitsky, Olga A. Kapush and Aleksey G. Rozhin et al.
Mixing of Excitons in Nanostructures Based on a Perylene Dye with CdTe Quantum Dots
Reprinted from: *Materials* 2023, 16, 552, doi:10.3390/ma16020552 132

Zahrah Ramadlan Mubarokah, Norsuria Mahmed, Mohd Natashah Norizan, Ili Salwani Mohamad, Mohd Mustafa Al Bakri Abdullah and Katarzyna Błoch et al.
Near-Infrared (NIR) Silver Sulfide (Ag_2S) Semiconductor Photocatalyst Film for Degradation of Methylene Blue Solution
Reprinted from: *Materials* 2023, 16, 437, doi:10.3390/ma16010437 149

Siti Faqihah Roduan, Juyana A. Wahab, Mohd Arif Anuar Mohd Salleh, Nurul Aida Husna Mohd Mahayuddin, Mohd Mustafa Al Bakri Abdullah and Aiman Bin Mohd Halil et al.
Effectiveness of Dimple Microtextured Copper Substrate on Performance of Sn-0.7Cu Solder Alloy
Reprinted from: *Materials* 2022, 16, 96, doi:10.3390/ma16010096 170

Md Azree Othuman Mydin, Mohd Mustafa Al Bakri Abdullah, Mohd Nasrun Mohd Nawi, Zarina Yahya, Liyana Ahmad Sofri and Madalina Simona Baltatu et al.
Influence of Polyformaldehyde Monofilament Fiber on the Engineering Properties of Foamed Concrete
Reprinted from: *Materials* **2022**, *15*, 8984, doi:10.3390/ma15248984 . . . . **184**

Li Liu, Leixin Liu, Zhaohui Liu, Chengcheng Yang, Boyang Pan and Wenbo Li
Study on the Effect of Ultraviolet Absorber UV-531 on the Performance of SBS-Modified Asphalt
Reprinted from: *Materials* **2022**, *15*, 8110, doi:10.3390/ma15228110 . . . . **204**

Brăduț Alexandru Ionescu, Mihail Chira, Horațiu Vermeșan, Andreea Hegyi, Adrian-Victor Lăzărescu and Gyorgy Thalmaier et al.
Influence of Fe_2O_3, MgO and Molarity of NaOH Solution on the Mechanical Properties of Fly Ash-Based Geopolymers
Reprinted from: *Materials* **2022**, *15*, 6965, doi:10.3390/ma15196965 . . . . **221**

Mariusz Tryznowski, Tomasz Gołofit, Selim Gürgen, Patrycja Kręcisz and Marcin Chmielewski
Unexpected Method of High-Viscosity Shear Thickening Fluids Based on Polypropylene Glycols Development via Thermal Treatment
Reprinted from: *Materials* **2022**, *15*, 5818, doi:10.3390/ma15175818 . . . . **239**

Chiemela Victor Amaechi, Emmanuel Folarin Adefuye, Irish Mpho Kgosiemang, Bo Huang and Ebube Charles Amaechi
Scientometric Review for Research Patterns on Additive Manufacturing of Lattice Structures
Reprinted from: *Materials* **2022**, *15*, 5323, doi:10.3390/ma15155323 . . . . **249**

Laura Madalina Cursaru, Miruna Iota, Roxana Mioara Piticescu, Daniela Tarnita, Sorin Vasile Savu and Ionel Dănuț Savu et al.
Hydroxyapatite from Natural Sources for Medical Applications
Reprinted from: *Materials* **2022**, *15*, 5091, doi:10.3390/ma15155091 . . . . **283**

Srijita Nundy, Aritra Ghosh, Abdelhakim Mesloub, Emad Noaime and Mabrouk Touahmia
Comfort Analysis of Hafnium (Hf) Doped ZnO Coated Self-Cleaning Glazing for Energy-Efficient Fenestration Application
Reprinted from: *Materials* **2022**, *15*, 4934, doi:10.3390/ma15144934 . . . . **302**

About the Editor

Andrei Victor Sandu

Dr.Eng. Andrei Victor SANDU is Associate Professor at Faculty of Materials Science and Engineering, Technical University "Gheorghe Asachi" of Iași. He was awarded his PhD in Materials Engineering in 2012 with summa cum laudae. He has published over 450 scientific articles, over 400 indexed by SCOPUS and more than 300 indexed by ISI Web of Science. His H-index is 28. He is the co-owner of 40 patents, with another 10 patent applications pending (in Romania, R. Moldova, and Malaysia). He has published 10 books, 3 of them in the USA. He is Publishing Wditor for *International Journal of Conservation Science* (indexed on Web of Science and Scopus) and the *European Journal of Materials Science and Engineering*; he is also a reviewer for more than 20 journals which are indexed Web of Science. He is visiting professor at Universiti Malaysia Perlis and is President of the Romanian Inventors Forum. Based on his expertise, he is also Senior Researcher for the National Institute for Research and Development for Environmental Protection INCDPM and the Representative for Romania at IFIA (International Federation of Inventors' Associations) and WIIPA (World Invention Intellectual Property Associations). His main field is materials science, with involvement in environmental issues, as well as the advanced characterization and synthesis of geopolymers and biomaterials.

Review

Study on the Possibilities of Developing Cementitious or Geopolymer Composite Materials with Specific Performances by Exploiting the Photocatalytic Properties of TiO$_2$ Nanoparticles

Andreea Hegyi [1], Adrian-Victor Lăzărescu [1,*], Adrian Alexandru Ciobanu [2,*], Brăduț Alexandru Ionescu [1], Elvira Grebenișan [1], Mihail Chira [1], Carmen Florean [1,2], Horațiu Vermeșan [3] and Vlad Stoian [4]

[1] NIRD URBAN-INCERC Cluj-Napoca Branch, 117 Calea Floresti, 400524 Cluj-Napoca, Romania; andreea.hegyi@incerc-cluj.ro (A.H.); bradut.ionescu@incerc-cluj.ro (B.A.I.); elvira.grebenisan@incerc-cluj.ro (E.G.); mihail.chira@incerc-cluj.ro (M.C.); carmen.florean@incerc-cluj.ro (C.F.)
[2] NIRD URBAN-INCERC Iași Branch, 6 Anton Șesan Street, 700048 Iași, Romania
[3] Faculty of Materials and Environmental Engineering, Technical University of Cluj-Napoca, 103–105 Muncii Boulevard, 400641 Cluj-Napoca, Romania; horatiu.vermesan@imadd.utcluj.ro
[4] Department of Microbiology, Facutly of Agriculture, University of Agricultural Sciences and Veterinary Medicine Cluj-Napoca, 3–5 Calea Mănăștur, 400372 Cluj-Napoca, Romania; vlad.stoian@usamvcluj.ro
* Correspondence: adrian.lazarescu@incerc-cluj.ro (A.-V.L.); adrian.ciobanu@incd.ro (A.A.C.)

Citation: Hegyi, A.; Lăzărescu, A.-V.; Ciobanu, A.A.; Ionescu, B.A.; Grebenișan, E.; Chira, M.; Florean, C.; Vermeșan, H.; Stoian, V. Study on the Possibilities of Developing Cementitious or Geopolymer Composite Materials with Specific Performances by Exploiting the Photocatalytic Properties of TiO$_2$ Nanoparticles. *Materials* **2023**, *16*, 3741. https://doi.org/10.3390/ma16103741

Academic Editor: Claudio Ferone

Received: 20 March 2023
Revised: 6 May 2023
Accepted: 11 May 2023
Published: 15 May 2023

Copyright: © 2023 by the authors. Licensee MDPI, Basel, Switzerland. This article is an open access article distributed under the terms and conditions of the Creative Commons Attribution (CC BY) license (https://creativecommons.org/licenses/by/4.0/).

Abstract: Starting from the context of the principles of Sustainable Development and Circular Economy concepts, the paper presents a synthesis of research in the field of the development of materials of interest, such as cementitious composites or alkali-activated geopolymers. Based on the reviewed literature, the influence of compositional or technological factors on the physical-mechanical performance, self-healing capacity and biocidal capacity obtained was analyzed. The inclusion of TiO$_2$ nanoparticles in the matrix increase the performances of cementitious composites, producing a self-cleaning capacity and an anti-microbial biocidal mechanism. As an alternative, the self-cleaning capacity can be achieved through geopolymerization, which provides a similar biocidal mechanism. The results of the research carried out indicate the real and growing interest for the development of these materials but also the existence of some elements still controversial or insufficiently analyzed, therefore concluding the need for further research in these areas. The scientific contribution of this study consists of bringing together two apparently distinct research directions in order to identify convergent points, to create a favorable framework for the development of an area of research little addressed so far, namely, the development of innovative building materials by combining improved performance with the possibility of reducing environmental impact, awareness and implementation of the concept of a Circular Economy.

Keywords: self-cleaning cementitious composites; geopolymer; TiO$_2$ nanoparticles; physical-mechanical performance; microorganism resistance

1. Introduction

Worldwide, in line with the principles of Sustainable Development and Circular Economy, there is a strong orientation towards reducing the consumption of non-renewable raw materials, increasing sustainability, reducing soil, water and air pollution and, consequently, reducing the volume of waste or identifying possibilities for its recovery. In the construction sector, there is a huge consumption of cement, which is the main raw material in many technological processes specific to this sector [1]. Each ton of cement produced requires between 60 and 130 kg of liquid fuel or its equivalent, depending on the type of cement and the manufacturing process used, and about 110 kWh of electricity [2]. One ton of manufactured cement also releases between 0.8 and 1.1 tons of CO$_2$ into the atmosphere as a consequence of fuel combustion and limestone calcination [3]. In 2008 alone, world cement production was about 2.9 billion tons and in 2014, about 4.2 billion tons [4,5]. Producing such a large volume of cement/concrete is directly associated with environmental problems—cement production is responsible for about 5–8% of total carbon dioxide emissions. A "green" alternative to cement-intensive concrete can be the so-called geopolymer concrete [6–8].

New binder materials, known as "geopolymers" were introduced in 1987 by Davidovits to describe a group of mineral binders, with a chemical composition similar to natural zeolitic materials, but with an amorphous microstructure. He stated that "binders could be produced by a polymeric reaction of alkali liquids with silicate and aluminium, with source materials of geological origin or with waste materials such as thermal power plant ash" [9]. Geopolymer cements are under development, with research being driven mainly by the need to reduce global CO_2 emissions. With excellent mechanical properties and strengths in aggressive environments, these materials represent an opportunity for both the environment and engineering, an alternative to traditional technology [10,11]. The concept of Alkali Activated Geopolymer Materials (AAGM) as an alternative to Portland cement has been known since the 1980s. The durability of materials produced by this process has also been demonstrated over the years in Belgium, Finland, Russia, China and recently Australia and the UK. Since the 1990s, most research in the field of alkali-activated geopolymeric materials has focused on the microstructure of these materials and much less on studies on predictions of the engineering properties of these materials, their bearing capacities under different types of stresses, durability, etc., and their actual serviceability. On the other hand, the possibility of producing cementitious composite materials with self-cleaning properties is currently reported worldwide, due to the photocatalytic properties of TiO_2 nanoparticles used as an addition or as a substitute for certain amount of cement. The first tests were carried out using TiO_2-enriched white cement-based mixtures. In 1996 the first relevant results were reported, and in 2003, the first large-scale construction of this kind, the Dives in Misericordia church in Rome, Italy, was put into service.

Overall, an analysis of the trend of evolution of research in the field indicates that, if at the beginning, this topic was approached more from the point of view of the effects, respective of the influence of NT on the physical-mechanical performances, at the macroscopic level, nowadays, the emphasis is put on deepening the study of the phenomenon at the micro-structural level.

Although there are numerous references in the literature, various research groups have approached the problem from different points of view (physico-mechanical aspect, chemical aspect, microstructural aspect, durability or micro-organism resistance aspects, etc.), due to its complexity. It is therefore not possible to identify a pattern on which to build a research programme, given the lack of flexibility and the possibility of recalibration along the way, depending on the intermediate results obtained. This is a direct consequence of the specific character of the locally/regionally available raw materials.

Therefore, the motivation of the research approach aims to contribute to the implementation of the Circular Economy concept by analyzing the possibilities of exploiting waste resources and industrial by-products potentially supplying alumino-silicates, while exploiting the specific performance of TiO_2 nanoparticles (NT), with the direct destination in the construction industry.

The aim of this work is to carry out a study on the specific performances of NT-containing cementitious composite materials, respectively, alkali-activated geopolymer (GP) materials, as a review of the available literature, thus contributing to the creation of the context favorable to the development of future experimental research.

The scientific contribution of this study consists of bringing together two apparently distinct research directions in order to identify convergent points, to create a favorable framework for the development of an area of research little addressed so far, namely, the development of innovative building materials by combining improved performance with the possibility of reducing environmental impact, awareness and implementation of the concept of the Circular Economy.

The paper is a synthesis analysis of some of the papers published in the last 25 years and is structured on the principle of "mirroring" the positioning of two innovative building materials: cementitious composites with self-cleaning capacity vs. geopolymer composites.

On the one hand, based on references that indicate a better possibility of documentation due to the earlier approach of the field, Section 2 presents a synthesis of research in the field on the possibility of inducing self-healing capacity and increasing resistance to the action of biological agents in cementitious composites by exploiting the specific properties of NT. Other influences of NT use on cementitious composites are also presented, i.e., effects on the physical-mechanical performance of both fresh and hardened cementitious composites. Finally, for a good argumentation of the specific innovative element characteristic of these cementitious composites, aspects related to the NT specific photoactivation phenomenon and the specific mechanisms leading to the self-healing and biocidal performance are presented.

On the other hand, in Section 3, aspects related to geopolymer materials are addressed, touching both specific elements of the geopolymerization reaction and specific elements of the physical-mechanical performances. The field of self-healing geopolymers is still poorly represented at the level of reports in the literature, there is still a strong interest in understanding the mechanism and kinetics of the geopolymerization phenomenon in general, the influencing factors and, only subsequently, the impact that the introduction of nano-TiO_2 has on the performance of this type of material.

Therefore, the challenge of carrying out such a synthesis study included identifying as many bibliographical sources as possible for the two directions addressed. If in the case of cementitious composites, the mechanism of hydration-hydrolysis reactions of cement can be said to be, at present, amply documented. In the case of geopolymers, there are still many aspects that need clarification. Moreover, in the case of the influence of NT on the physical-mechanical performance, durability, self-healing and biocidal capacity, for cementitious composites there is a wealth of reported studies. There is already a history, in contrast to the situation of geopolymer materials which can be said to be still in their infancy from this point of view. However, it should be specified that the purpose of this work and the way of documentation was aimed to analyze strictly on the field of cementitious composites produced with Portland cement vs. geopolymer binders, with NT content, without analyzing other material, e.g., non-Portland cements, lime mortars, lime and cement mortars and hydraulic lime mortars modified with TiO_2.

2. Cementitious Composites with Self-Cleaning Capacity

Numerous studies indicate that the introduction of nanoparticles, including TiO_2 nanoparticles (NT) in the mixture of a cementitious matrix, has beneficial effects on the physico-mechanical properties and durability and induces some specific performances resulting from photoactivation [12–18]. From the outset, the literature indicates two ways of exploiting the self-cleaning and biocidal effects of NT-specific performance: by applying an NT-containing film to the surface, or by developing NT-containing composite material in mass. In this work the study focused on the documentation and analysis of these NT-containing composite materials in the matrix.

In general, for use, NT is found as a mixture of rutile and anatase (both being crystallographic forms of TiO_2). There is also the situation where the crystallographic form anatase is predominant (over 90%), and the literature recommends that TiO_2 nanoparticles be added dry, by direct mixing with cement powder, followed by the addition of hydration water. Water does not chemically react with any crystallographic form of titanium dioxide, nor does a chemical reaction occur between the photosensitive nanoparticles and the hydration-hydrolysis phases of the cement, therefore the hydration-hydrolysis reactions are not chemically influenced [19], but, as will be seen in the following, their kinetics will be influenced.

However, the size and particle size distribution of the TiO$_2$ powder used influences the cementitious composite material, so research has been carried out at the University of Milan, Italy, using TiO$_2$ powder with micrometric size, m-TiO$_2$, and nanometric size, n-TiO$_2$, respectively. The results showed that there are advantages and disadvantages in both cases [19]. Thus, the dispersion and distribution in the matrix is more convenient when using micrometer-sized granules, while nanometer-sized ones tend to agglomerate, thus reducing the overall reactive surface available for the initiation of photocatalytic reactions. On the other hand, the use of nanometric granules is advantageous and preferable, although more difficult, because, under conditions of homogeneous, satisfactory dispersion, the composite will have a better capacity for adsorption of pollutant oxides (NOx, SOx, dust, exhaust fumes), which can easily penetrate the agglomerations of nanoparticles, and the hydrophilicity effect is more intense, but the research carried out so far has not shown a definite influence of particle size on the efficiency of TiO$_2$-enriched cement matrix in terms of the decomposition capacity of pollutant molecules and dirt particles [20]. Research to date has explored both micrometer- and nanometer-sized TiO$_2$. However, although the photoactivation phenomenon of nano-TiO$_2$ is more obvious in terms of the properties induced in the cementitious matrix in which it is embedded, there is also the possibility of using micro-TiO$_2$, each of which has advantages and disadvantages. On the one hand, nanoparticles raise the problem of homogeneous dispersion in the mass, which is easier to achieve if the grain size is within micro limits. On the other hand, the physico-mechanical characteristics, the hydrophilicity of the surface, the decomposition capacity of organic/dirt particles and the biocidal effect are lower when using micro-sized TiO$_2$ granules. A possible solution to improve the dispersion rate of TiO$_2$ particles would be to use dispersing agents and superplasticizers.

However, the influence of the qualitative and quantitative parameters of the photoactivation light should not be neglected either, as research has shown that there are differences, even when keeping all the designed preparation parameters, whether the composite preparation takes place in artificial light, natural light or darkness, or in the same type of light but with different intensities [19,21].

The unanimously accepted conclusion is that, for each case of a particular cementitious composite, it is necessary to determine the optimal amount of TiO$_2$ nanoparticles used in the composite, as an additional amount is not economically justified and very often may negatively influence some parameters of the cementitious composite matrix enriched with TiO$_2$ nanoparticles.

2.1. Influence on the Physical-Mechanical Characteristics of Fresh Cementitious Composites

Given the complex structure of cementitious composites and the hydration-hydrolysis mechanisms underlying the formation of this structure, i.e., the C-S-H gel formation phase, the introduction of NT involves a number of difficulties, especially in terms of their tendency to agglomerate, which reduces the potential to obtain a homogeneous dispersion in the cementitious mass. Therefore, this aspect induces some modifications in the C-S-H gel and, implicitly, on the physico-mechanical properties of the whole composite [22]. According to reported research, one of the biggest challenges is the uniform and homogeneous dispersion of NT in the composite matrix [22–24]. This problem has been analyzed by applying various dispersion methods, i.e., the success of water preparation using plasticizers and superplasticizers [25,26] or the use of ultrasonic waves [27]. Thus, Perez-Nicolas et al. [28] have shown the possibility of obtaining a homogeneous dispersion using polycarboxylate-based superplasticizers that effectively prevented NT agglomeration. Similarly, Zhao et al. [29] report results of research on engineered cementitious composites (ECC) with NT, but also with polyvinyl alcohol fibers (PVA), in the preparation of which polycarboxylate-based superplasticizers were also used, an additive which again proved effective. As reported by Perez-Nicolas et al. [28], plasticizers and superplasticizers are frequently used in the field of cementitious composites due to their mechanism of reducing water requirements and facilitating homogeneity through a mechanism based on the induction of these molecules which exhibit a dispersing action between cement particles due their electrostatic repulsion and/or steric hindrance effects [30,31]. Their research [28], however, focused on coatings made with water dispersions of different nanoparticles of photocatalytic additives (titania and titania doped with iron and vanadium), analyzing the efficacy of three types of polycarboxylate-based polymers, synthesized in the authors' research, demonstrated the effect of preventing agglomeration of nanoparticles, improving aqueous dispersions, unlike a fourth, commercial superplasticizer, which led to the formation of NT agglomerates. Therefore, not only the use of superplasticizers is important for the production of homogeneously dispersed NT cementitious composites and therefore with uniform and homogeneous properties, but also the type of superplasticizer, especially as this technique disperses nanoparticles in the preparation water, as a raw material preparation phase prior to the preparation of the cementitious composite, is also frequently encountered when considering the production of these composites with nanoparticles dispersed in the mass, the homogeneous dispersion problem also being encountered in the case of SiO_2 nanoparticles [32]. Other research has addressed this issue in a different way, e.g., the possibility of coating the NT granule with nano-SiO_2 [26] was investigated, leading to a reactive powder which, following microstructural analysis, was found to be uniformly dispersed in the composite matrix.

Regarding the fresh self-cleaning cementitious composite, following the embedding of TiO_2 nanoparticles (NT), the following were observed: increased water requirement to reach standard consistency, decreased workability, decreased initial and final setting time, accelerated hydration-hydrolysis processes, increased heat of hydration and cement hydration rate, reduced porosity of the cement paste due to changes in pore size and distribution, altered size and orientation of cement hydration product crystals and formation of a larger amount of C-H-S calcium hydrosilicate gel [33–35]. Thus, Nazari et al. showed that partial replacement of cement, or addition of nano-TiO_2, contributes to the reduction in setting time and intensification of cement hydration-hydrolysis processes, findings that correlate with those reported by Lee et al. and Pimenta Teixeira [12,15–17,36–39]. The phenomena can be explained both because TiO_2 nanoparticles function as possible concentrating nuclei for the hydration products and on the catalytic effect they have on the cement hydration reaction. The first effect observed, as early as the preparation phase of cement-based mortars and concrete, is an increase in the water required to reach the standard consistency [33].

In terms of workability, it can be said that it decreases with an increasing percentage of TiO_2 nanoparticles, introduced into the cement paste either as an addition or as a substitution of a part of OPC [40–47], a similar issue being reported regarding the flowability of the composite mixture [48]. These phenomena are reported in the literature for a percentage amount of NT used, varying in the range of 0.5–10% relative to the amount of OPC. At the same time, the literature indicates that the workability of nano-TiO_2-containing cementitious composites does not vary in proportion to the amount of nanoparticles used and, moreover, different reductions in workability are identified even for the same amount of nano-TiO_2 [48]. This phenomenon can be attributed to differences in terms of the type of cement used, its oxide composition, water/cement ratio, the type of additive used and in terms of preparation conditions.

Thus, Salemi et al. [41] indicate a reduction of up to 50% in the flowability of the composite when using a maximum of 2% NT; Nazari et al. [42] indicate a reduction in workability of about 25% when replacing 2% OPC with NT for a water/cement ratio of 0.4 and Li et al. [43,44] report workability reductions of over 50% when replacing a maximum of 3% OPC with NT and even over 70% when using 5% NT. In contrast, Meng et al. [47] report a workability reduction of only 21% and 40%, respectively, under the conditions of substitution of OPC with 5% and 10% NT, respectively, for a water/cement ratio of 0.5. Other research [34] indicated that substitution of OPC with max. 1% NT, mass ratio, does not significantly influence the flowability of cement-based mortar. Therefore, the existence of this controversy, the impossibility of a quantifiable prediction of how the workability of the fresh composite will vary depending only on the amount of NT used, induces the need for further experimental research and preliminary case-by-case customized analyses specific to the raw materials used before designing any nano-TiO_2-containing cementitious composite mixtures.

In terms of setting time, research has shown that from 0.5% to 10% NT content, the initial and final setting times decrease with increasing NT content, but at the same time, the importance of NT type on these parameters is also indicated, i.e., rutile and anatase concentration, both of which are crystallization forms of TiO_2 [49–51]. Further research has shown increased heat of hydration and increased hydration rate of OPC as well as change in the structural orientation of CH crystals, i.e., their size [52]. Lee and Kurtis [53] showed that the intensification of hydration processes, manifested by increased heat, hydration rate and degree of hydration of OPC granules occurs during more than 3 days after preparation of the cementitious composite. In agreement with them, Chen et al. [37] indicated the intensification of the hydration process of OPC granule in the present NT, with an increase in the hydration heat especially in the first 30 h after preparation, i.e., 22 h after preparation as indicated by Senff et al. [54].

Additionally, Nazari and Riahi [55], Jalal et al. [56], Jayapalan et al. [57,58], Kurikara and Maruyama [59], Baoguo et al. [60], Sakthivel et al. [61], Khataee et al. [62] and Hamidi et al. [24] confirm the acceleration of OPC hydration processes in the presence of NT, for amounts of up to 5% NT as a substitute of OPC used, the general result being a gain in terms of mechanical strength at early ages, a phenomenon also influenced by the rutile–anatase composition of NT, respectively, the grain size distribution of NT.

The decrease in workability and grip type, concomitant with the increase in hydration heat and hydration rate of OPC leads to a possible explanation: due to the significant increase in the specific surface area of the particles in the mixture used for the preparation of the composite, a large, specific surface area of NT induces higher and faster water consumption but also an increased reactivity overall, NT functioning as a catalytic effect of the hydration-hydrolysis processes and functioning as potential accumulation nuclei of hydration products.

2.2. Influence on the Physical-Mechanical Characteristics of Cementitious Composites in the Hardened State

Nazari et al. showed that partial replacement of OPC or addition into the composite cement matrix of NT contributes to densification of the composite material, reduction and resizing of porosity and increase in mechanical strengths, findings that correlate with those reported by Lee et al., Pimenta Teixeira and Feng et al. [12,15–17,36–39]. Up to a certain dosage of TiO_2 nanoparticles, the dosage limit above which a reduction in these parameters occurs, also in agreement with those reported by Zhang and Li. [43,44].

Thus, as presented by Rashad [55], referring to Nazari et al. [15–17,53], shows that partial replacement of OPC with 0.5–2% NT, for a water/cement ratio of 0.4, induces an increase in mechanical strength at 28 days of specimens, by 15.91–25.00% for flexural strength, by 5.55–66.67% for splitting tensile strength and by 6.79–17.93% for compressive strength. However, it should be noted that these variations are not linearly increasing. The variation curves showing a maximum for the case of OPC substitution with 1% NT, after which the slope of increase in mechanical performance follows a downward trend. This behavior could be explained precisely based on the phenomena that are proven to occur in the preparation and early phase of the cementitious composite, namely, the consumption of a quantity of the mixing water by the NT and the acceleration of the hydration-hydrolysis processes of the cement granule in the first approximately 72 h after casting. It could thus be said that substitution of OPC with an amount of more than 1% NT could have better results if additional water was added to the mix.

In agreement with Nazari et al., Li et al. [24] report an increase in flexural strength of 10.27% and 2.93%, respectively, for cementitious composites at 28 days of age for replacement of OPC with 1% and 3% NT under the condition of maintaining the water/cement ratio of 0.42. In terms of compressive strength, Li et al. [24] but also Zhang and Li [63], under the same conditions of water/cement ratio of 0.42 and testing at the age of 28 days, indicate an increase of the parameter by 18.03% for 1% NT, 12.76% for 3% NT and 1.55% for 5% NT. Agreement with the optimal NT percentage indicated by Nazari et al. is observed, i.e., 1% mass percentage, an amount of NT that substitutes OPC without inducing an increase in the preparation water requirement and that provides maximum benefit.

In contrast to general trends, Behfarnia et al. [64], Sandu et al. [65] and Meng et al. [47] report reductions in compressive strength with the introduction of NT into the cementitious composite. A series of reported results on the variation of compressive strength as a function of the amount of NT substituting OPC is presented in Figure 1. Similarly, the variation of the tensile strength is shown in Figure 2.

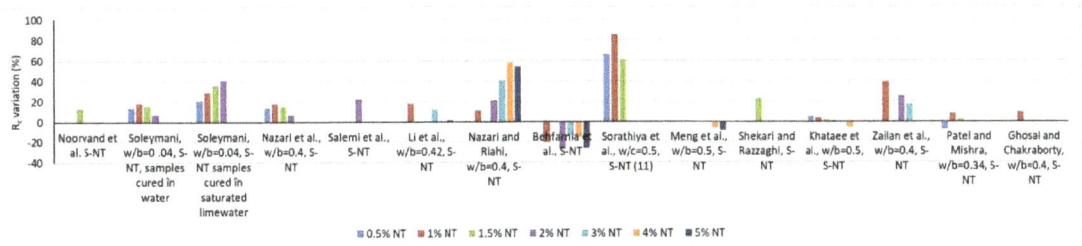

Figure 1. Variation of the compressive strength based on several research [15–18,41,43,44,47,48,62,64,66–71].

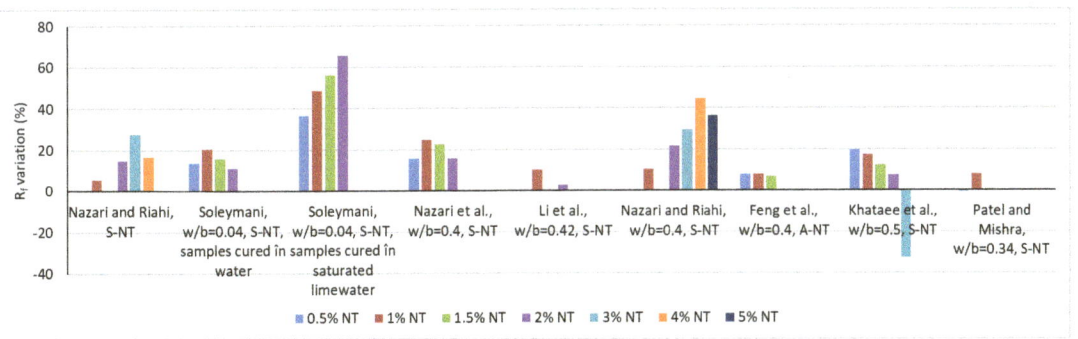

Figure 2. Variation of the flexural strength based on several research [15–17,24,25,43,44,62,66,67,70].

In terms of internal structure, NT causes a reduction in open porosity, which will also lead to a reduction in water absorption. Thus, Salemi et al. [41], in agreement with Nazari [72], indicate that the introduction of 2% NT, respectively, 0.5%, 1%, 1.5% and 2%, into the cementitious composite for specimens at 28 days of age, causes a reduction in water absorption by 22%, 59.1%, 54.1%, 51.1% and 47.5%, as shown in Figure 3. Regarding the resistance to chloride ion penetration, Hi and Shi [73], in agreement with Li et al. [74], indicate a reduction in the penetration rate as a result of OPC substitution, i.e., the addition of 1% NT, a phenomenon that can also be explained by a reduction in the porosity of the composite. If a simple extension of these improvements in physical-mechanical performance is made, it can be appreciated that a foreseeable consequence of the introduction of NT in cementitious composites will also be an improvement in wear resistance, frost-thaw resistance, essential elements of the general durability of these materials, as in fact indicated by Li et al. [44], Hassan et al. [75] or Farzadnia et al. [76].

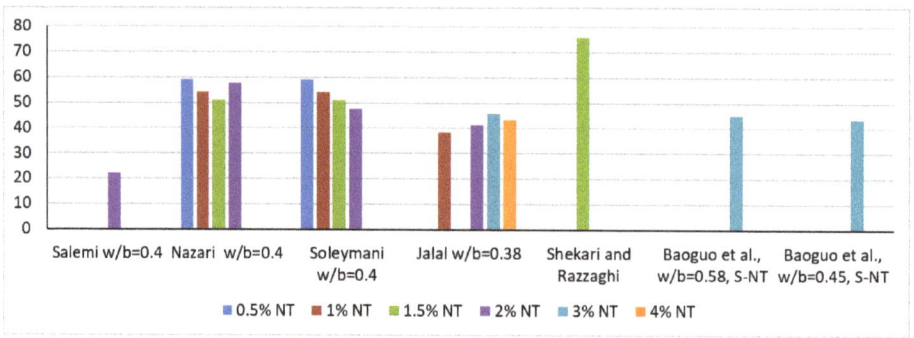

Figure 3. Water absorption of the samples according to several research [15–17,40,41,69].

In terms of influence on physical-mechanical performance, in accordance with the literature, the identification of the optimal dosage of TiO_2 nanoparticles introduced into the cementitious composite matrix is still a controversial parameter, influenced by several parameters: characteristics of the other raw materials, water/cement ratio, conditioning conditions and test age, temperature, mode of introduction (mixing in the dry or wet phase, introduction as an addition or substitute of OPC) to the preparation, knowing that the photocatalytic activity, expected in these TiO_2 cementitious materials, is critical for this choice. Too much photocatalyst often leads to electron-hole coupling which reduces the efficiency. Furthermore, Feng et al. [77] identify structural changes present in the cement paste that could explain the evolution of the other physico-mechanical properties, in particular, mechanical strengths, density and porosity. The authors, starting from what is known, namely, that in Portland cement paste (OPC), through hydration reactions of the cement granule, the formation of hydrated calcium silicate (C-S-H) takes place which, in the hardened phase, becomes C_3S with a Ca/Si ratio in the range 1.25–2.1. Consider, in accordance with the literature, that in the hardened cement matrix this C-S-H divides into two types of formations, one type where the silicate anions are entirely monomeric and the other where a linear silicate chain is present in the tobermorite. With the introduction of 1% (mass percentage) TiO_2 nanoparticles and ensuring a good dispersion in the matrix, a series of micro and nanostructural changes are achieved: reduction in microcracks and internal defects in the cement paste, obtaining a densification of the material, reduction in nanoroughness concomitant with the formation of nanoprecipitated, needle-shaped hydration products that will function in the reinforced composite matrix as a dispersed nano-reinforcement. Comparative analysis of the influence of NT both in cementitious-based matrices (hydraulic binders) vs. lime-based composites (non-hydraulic binders) showed on the one hand that the photocatalytic activity of TiO_2 is not influenced by the type of binder, as both materials are porous and NT acts by filling these pores, and on the other hand that this positioning of NT in the composite matrix will lead to improved physico-mechanical performance [22]. TiO_2 will act as a filler for these pores and make it possible to absorb NT in the case of hydration products and, at the same time, reduce the available surface area of the photocatalyst. Furthermore, the literature shows that an increase in electron-hole recombination can occur for adsorbed species [78,79]. In addition, as the material ages, an increase in volume of almost 10% is observed, due to carbonation. This phenomenon leads to a decrease in capillary adsorption and precipitation of calcium carbonate. These precipitates will act as a barrier and decrease the TiO_2 capacity [80,81].

2.3. Influence of the Introduction of TiO_2 Nanoparticles into Cementitious Composites on Their Resistance to the Action of Microorganisms—Self-Cleaning Capacity and Biocidal Mechanism

For the first time, in 1985, Matsunaga et al. [82] demonstrated the photocatalytic cell killing mechanism of *Saccharomyces cerevisiae* (yeast), *Lactobacillus acidophilus* and

Escherichia coli (bacteria) and *Chlorella vulgaris* (green algae) in water treated with TiO_2 nanoparticles [82].

It is now known that the biocidal mechanism and self-cleaning capacity of composite surfaces containing TiO_2 nanoparticles is the result of two mechanisms, that of superhydrophilicity and that of degradation, destruction of molecules of an organic nature, therefore, implicitly of the cells of microorganisms these having a structure of an organic nature [18].

If in the case of inorganic substances adhering to the composite surface, the superhydrophilicity mechanism is sufficient for the manifestation of the self-cleaning capacity (the water film picks up and washes the inorganic particles from the composite surface). In the case of organic substances, large-volume organic molecules and microorganism cells, it is necessary that this superhydrophilicity is complemented by the degradation capacity of the organic molecules and/or cells. The schematic representation of the self-cleaning phenomenon of composite cementitious surfaces containing TiO_2 nanoparticles is presented in Figure 4.

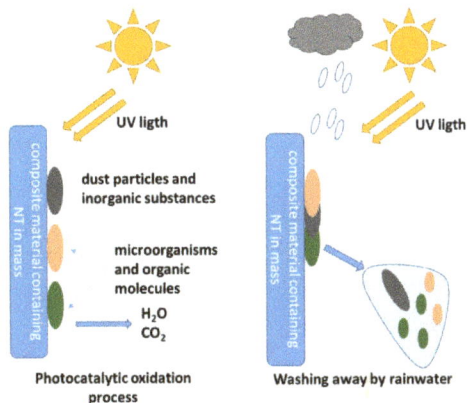

Figure 4. Schematic representation of the self-cleaning phenomenon of composite cementitious surfaces containing TiO_2 nanoparticles.

An experimental test method based on a simple principle but demonstrating the accumulation of the two mechanisms is the decolorization method for *Rhodamine B* (RhB), an organic substance, N-9-(2-Carboxyphenyl)-6-(diethylamino)-N,N-diethyl-3H-xanthen-3-iminium chloride, which is characterized by an intense pink-magenta coloring. The principle of the method consists of the fact that, by applying a spot of RhB solution on a cementitious composite surface with TiO_2 nanoparticles, subsequently, under the influence of UV radiation, with or without the accumulation of washing steps, discoloration is rapidly observed, as a result of the degradation of the organic molecule (oxido-reduction reactions) and, in the case of the existence of the washing step, the uptake of the reaction products by the water film developed on the cement composite surface which has become superhydrophilic [83]. A similar method has been identified in the literature as using *Methylene blue*, Methylthioninium chloride with the formula $C_{16}H_{18}ClN_3S$ as the organic staining substance [84], with the mention that the advantage of the RhB staining method is primarily represented by the more intense highlighting of the stain and the bleaching effect.

The mechanism underlying the increase in hydrophilicity of TiO_2 nanoparticle-containing cementitious composite surfaces under UV exposure can be explained based on the increase in hydroxyl (OH^-) groups identified by X-ray photoemission spectroscopy (XPS), Fourier transform infrared spectroscopy (FTIR) or nuclear magnetic resonance (NMR) [85–90]. The surface transition, under the influence of UV radiation, into a thermodynamically metastable state is the result of the coexistence of two molecular forms of water: molecular water and dissociated water. In general, under the action of UV rays, titanium dioxide being a semiconductor with a band gap of about 3.0 eV, by absorbing energy, generates electrons (e^-) and voids (h^+). Electrons tend to reduce Ti(IV) cations to Ti(III) ions, and voids oxidize O^{2-} anions. This process will release oxygen, creating vacancies on the surface of the titanium dioxide, vacancies that allow the water molecules to bind with the release of hydroxyl groups (OH^-). Additionally, in the case of TiO_2-containing cementitious composite surfaces, the literature also indicates [85–87,91] that photogenerated (h^+) vacancies (voids) cause the bond length within the TiO_2 network to increase, bringing the surface into a metastable state allowing adsorption of molecular water, simultaneously with the formation of new hydroxyl groups and the release of a proton (Equations (1)–(3), Figure 5). Research has shown that these generated hydroxyl groups are less thermodynamically stable, therefore, the surface will allow flattening of the water droplet to cover a larger area for stabilization purposes [85–87,91–94], thus achieving the effect of superhydrophilicity, i.e., the water that will arrive on the photoactivated composite surface will no longer form individual droplets, but lamellae that favor the uptake of impurities (Figure 5).

$$TiO_2 + h\upsilon \rightarrow e^- + h^+ \quad (1)$$

$$e^- + Ti^{4+} \rightarrow Ti^{3+} \quad (2)$$

$$4h^+ + 4O^{2-} \rightarrow 2O_2 \quad (3)$$

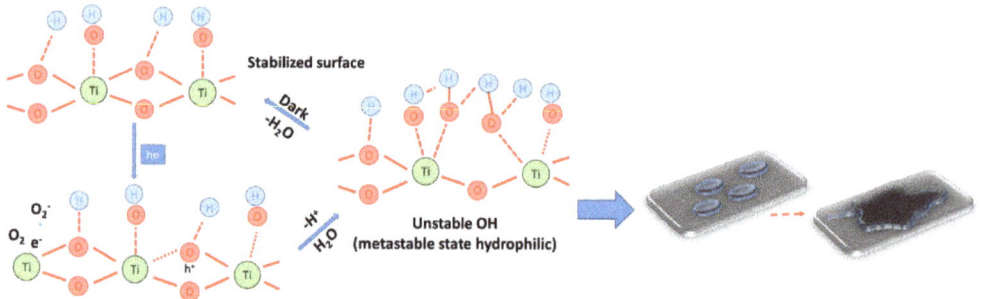

Figure 5. Schematic representation of the superhydrophilicity mechanism.

In conjunction with the development of superhydrophilicity, using the energy provided by UV radiation, which is greater than the valence band gap of TiO_2, pairs of electrons (e^-) and voids (h^+) are generated, which react with O_2 and H_2O to form anionic radicals (O^{2-}) and (OH). These oxidative species (h^+, $\cdot(O^{2-})$ and (OH)) are all highly reactive, contributing to the destruction of microorganism cells [95] according to a sequence of chemical reactions shown in Equations (4)–(9). This mechanism of organic cell destruction is coupled with the superhydrophilicity mechanism specific to the surfaces of TiO_2 nanoparticle-enriched cementitious composites, so that after the initiation of destruction at the cellular level, the organic residue, not yet completely degraded to CO_2 and H_2O, can be more quickly and easily removed from the surface by simple washing (under rainy conditions), leading to increased performance in terms of surface safety and hygiene.

$$TiO_2 + h\upsilon \rightarrow TiO_2\ (e^-{}_{cb} + h^+{}_{vb}) \quad (4)$$

$$O_2 + e^-{}_{cb} \rightarrow O^{2-} \quad (5)$$

$$H_2O + h^+_{vb} \rightarrow \bullet OH + H^+ \quad (6)$$

$$\bullet OH + \bullet OH \rightarrow H_2O_2 \quad (7)$$

$$O^{2-\bullet} + H_2O_2 \rightarrow \bullet OH + OH^- + O_2 \quad (8)$$

$$\bullet OH + Organic + O_2 \rightarrow CO_2 + H_2O \quad (9)$$

Subsequent research has examined hypotheses regarding the biocidal effect indicating that the cell membrane is photocatalytically destroyed in *Escherichia coli*, as reported by Sunada et al. [96]. The same results have been also supported by reports by Oguma et al. [97], who propose a destruction mechanism explained by both cell wall destruction and induction of cellular disruption following contact of the microorganism with TiO_2, while Saito et al. [98] propose a hypothesis of a destruction mechanism explained by inhibition of bacterial cell respiration function once in contact with TiO_2. Research indicates a wavelength of incident TiO_2 photoactivating radiation in the range 320–400 nm that produces strong destruction at the cellular level [99].

Gogniat et al. [100] showed that the adsorption capacity of TiO_2 in contact with the cell wall is positively correlated with the biocidal effect and adsorption was consistently associated with a reduction in bacterial membrane integrity. The authors indicate the hypothesis that adsorption of cells onto photocatalyzed TiO_2 is followed by loss of membrane integrity, which is key to the biocidal effect.

Mazurkova et al. [101], analyzing the effect of photocatalyzed TiO_2 nanoparticles on the *Influenza* virus, showed that after 15 min of incubation, nanoparticles adhered to the outer surface of the virus, the surface spinules of the virus were glued together, and the outer membrane, lipoprotein in nature, was ruptured. The degree of destruction in the influenza virus increased after 30 min of testing, and after 1–5 h of incubation, the virus was destroyed. The researchers indicate that the biocidal effect is dependent on the existence or not of photoactivation, the type of incident photoactivating radiation (natural light, UV rays) the duration of exposure/incubation, the viral concentration and the concentration of nano-TiO_2, but in all cases, this effect was recorded sooner or later.

In support of the biocidal effect argument, tests by Adams et al. [102] showed that the concentration of *Bacillus subtilis* and *Escherichia coli* was reduced 3.7 times, respectively, 2 times, upon contact with a nano-TiO_2 suspension under photoactivation conditions under natural light. A similar effect on *Bacillus subtilis* was also reported by Armelao et al. [103].

Research by Dedkova et al. [104] on samples of kaolinic composite material enriched with TiO_2 nanoparticles also confirmed the biocidal effect for *Escherichia coli*, *Enterococus faecalis* and *Pseudomonas aeruginosa* after 2 days of exposure to artificial light, reporting also in agreement with Gurr [105], who estimated that the antibacterial effect of TiO_2 composites manifests itself in the presence of natural light, without necessarily requiring additional photoactivation with incident radiation with wavelength exclusively in the UV spectrum.

As for cementitious composites enriched with 3% and 5% TiO_2 nanoparticles, it has been shown that the viability rate of *E-coli* bacteria is reduced after 24 h by 60% to 70%, compared to the situation without TiO_2 nanoparticles, according to the study reported by Hamdany [106]. The results of this research agree with reports by Davidson et al. [107], Lorenzeti et al. [108] and Peng et al. [109]. The explanation for this ability would be in line also with reports by Daly et al. [110], Carre et al. [111] and Kubacka et al. [112], which confirm that by the formation of free radicals and strongly oxidized anions by photoactivation of nano-TiO_2 (OH• and O^{2-}) at the cellular level, the cell membrane is ruptured and plasma components such as DNA, RNA, lipids and proteins are destroyed.

In support of these hypotheses are also the results of pilot research by Huang et al. [113], who showed that metals such as Ag and Cu enriched with TiO_2 have the ability to reduce the survival rate of *Aspergillius Niger* spores under conditions of lack of light radiation and even more so under UV irradiation, respectively, UV irradiation and presence of ozone. Yadav et al. [114] summarized the reports of research results in the field, pointing out that microorganisms are destroyed when they reach the photocatalytic surface, the effect being demonstrated on both Gram-positive and Gram-negative bacteria, endospores, fungi, algae and viruses.

Similar results supporting the antimicrobial character imparted by TiO_2 nanoparticles to composite cementitious surfaces have been presented by numerous studies [90,115–118], showing that cementitious composites with 10 wt% TiO_2 powder reduce algal growth on the surface by 66% compared to cementitious surfaces without nanoparticle content.

Regarding the ability of different building materials to resist microorganism attack, the most convenient method is the one adapted from the antibiogram method used in medicine and known as the halo inhibition method [119] or the Kirby–Bauer method, currently standardized according to AATCC TM147 and AATCC TM30.

The need for uniform evaluation of the behavior of various materials to the action of micro-organisms has led to the development of dedicated method standards as well as non-standardized but frequently used methods. Thus, method standards for qualitative assessment and method standards for quantitative assessment are known [120]:

- ASTM E2149—presents a quantitative method for evaluating the behavior of irregular surfaces to bacterial action. The principle of the method is to immerse the material in a suspension with a known concentration of bacteria and to follow the evolution of this concentration over time. The antimicrobial activity of the material is considered positive when the concentration of the bacterial suspension is significantly reduced [121];
- ASTM E2180—presents a quantitative method for evaluating the behavior of hydrophobic surfaces to bacterial action. The principle of the method consists of making a pseudo-film of nutrient medium on the surface of the material, on which bacteria are inoculated in a suspension of known concentration and monitoring the evolution of the concentration compared to a control [122];
- ISO 22196—presents a quantitative method. The principle of this method is also to follow the variation of the concentration of the bacterial suspension inoculated on a nutrient medium [123];
- ASTM E1428—presents a qualitative method featuring the so-called "pink spot test", in which an inoculation with *Streptoverticullium* reticulum is used and the appearance of pink spots on the surface of the tested material is observed [124];
- STAS 12718/1989 offers the possibility of semi-quantitative quantification of the microbiological load of the system, providing a quantification grid as follows: 0(−) no growth (sterile); 1(+) 1–10 colonies of microorganisms; 2(++) more than 10 colonies of microorganisms; 3(+++) areas with confluent colonies; 4(++++) growth over the whole surface area [125];
- ISO 27447—presents a method for evaluating the antibacterial activity of semiconducting photocatalytic materials, can be applied for the analysis of some ceramic, photocatalytic materials but not for permeable or rough materials [126].

Although the potential benefits of TiO_2 nanoparticles have been known since 1921 [127], the ability to destroy microorganisms was not documented until over 60 years ago, starting with Matsunaga et al., in 1985 [128–130]. Currently, it is known that the anti-algal and antifungal activity of nano-TiO_2 is evident in 11 genera of filamentous fungi, 3 yeasts, 2 amoebae, 1 apicomplexan, 1 diplomonad, 1 ciliate and 7 algae, including 1 diatom, with fungal spores being generally more resistant than vegetative forms [131–141]. Sassolini et al. [142] considered the realization of surfaces from TiO_2 nanoparticle-enriched cementitious materials as a passive form of safety technology to increase safety in biological, radiological and nuclear (CBRN) accidents.

3. Geopolymer Composites with Self-Cleaning Capability

By geopolymer, it is meant that type of amorphous, alumino-silicate cementitious material which can be synthesized by the polycondensation reaction between a geopolymeric material and alkali polysilicates. This process is called geopolymerization [143]. This innovative technology allows for the transformation of alumino-silicate materials into products called geopolymers or inorganic polymers. Geopolymers, therefore, represent a material developed as an environmentally friendly alternative for the construction industry, but also as a solution for exploitation, reintroduction into the economic circuit of some industrial wastes and by-products, the most common being fly ash, slag kaolin and metakaolin mostly activated with alkaline solutions based on Na_2SiO_3 in combination with NaOH [8,144–147].

In 1979, Davidovits described this new family of materials, with a matrix based on a Si-O-Al-O structure, by alternating tetrahedra of SiO_4 and AlO_4, joined together in three directions with all oxygen atoms, calling them geopolymer materials [148]. In a simplified method, it can be stated that geopolymers can be synthesized by alkaline activation of materials that are rich in SiO_2 and Al_2O_3 [149]. Although the whole process is still lacking consistent data, there is concrete information that in geopolymerization mechanisms, the dissolution of aluminium (Al) and silicon (Si) in an alkaline medium occurs, followed by the transport (orientation) of the dissolved elements and then the polycondensation phenomenon, as a result of which a 3D network of alumino-silicate structures is formed [150]. In 1999, Palomo proposed that pozzolanic materials (blast furnace slag, thermal power plant ash) can be activated "using alkaline liquids to form a binder and totally replace the use of Portland cement in concrete production" [151]. When the two components of the geopolymer material (reactive solids and alkaline solution) react, an alumino-silicate network is formed, resulting in a hard, water-resistant product, the geopolymerization process can be expressed according to the following sequence of reactions (Equation (10), Figure 6 and Equation (11)) [152]:

$$2SiO_2 \cdot Al_2O_3 + 3OH^- + 3H_2O \rightarrow 2[Al(OH)_4]^- + [SiO_2(OH)_2]^{2-} \tag{10}$$

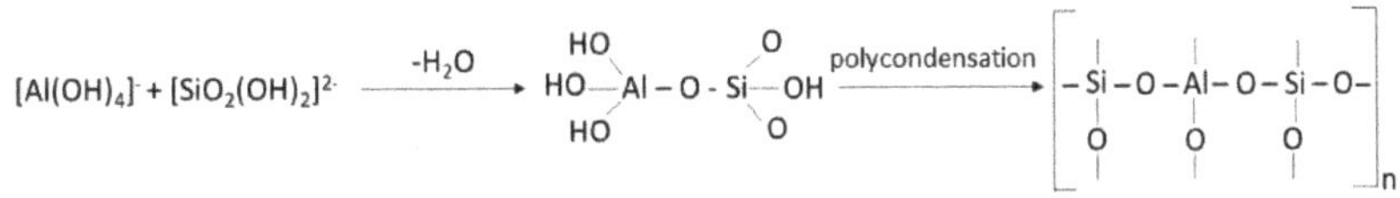

Figure 6. Geopolymer polycondensation process [152].

The empirical formula for the whole polymerization product is shown in Equation (11) [153]:

$$Mn\{-(SiO_2)z-AlO_2\}n, wH_2O \tag{11}$$

where M—represents the alkali element (which can be: K (potassium); Na (sodium); Ca (calcium)), the symbol "-" indicates the presence of a bond; n—represents the degree of polycondensation (or polymerization); z—is the Si/Al ratio which can be −1, 2, 3 or higher, up to 32.

Wallah [154] presented a more detailed equation showing the process by which the powder particles dissolve in alkali to produce the reactant product. In this equation, the reaction of aluminosilicate, hydroxide (Na, K), water and silicate in gel form is shown, followed by the dehydroxylation process to form a network of geopolymers. The reaction is presented in Figure 7.

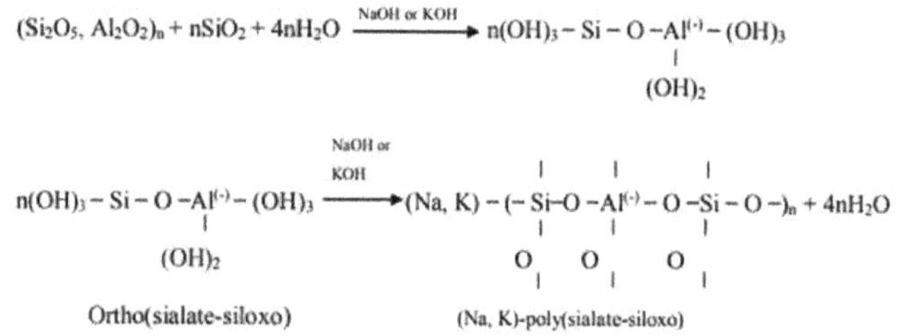

$$(Si_2O_5, Al_2O_2)_n + nSiO_2 + 4nH_2O \xrightarrow{NaOH \text{ or } KOH} n(OH)_3 - Si - O - Al^{(-)} - (OH)_3$$
$$| \quad (OH)_2$$

$$n(OH)_3 - Si - O - Al^{(-)} - (OH)_3 \xrightarrow{NaOH \text{ or } KOH} (Na, K) - (-Si-O-Al^{(-)}-O-Si-O-)_n + 4nH_2O$$

Ortho(sialate-siloxo) \qquad (Na, K)-poly(sialate-siloxo)

Figure 7. Geopolymerization process according to Wallah [154].

According to studies, it is known that the reaction of alumino-silicate materials in a strongly alkaline environment first causes the Si-O-Si bonds to break, subsequently, new phases are produced, and the formation mechanism appears to be synthesis via solution. The most important part of this process is the penetration of Al atoms into the original Si-O-Si structure and the formation of alumino-silicate gels. The composition of these gels can be characterized by Figure 7. The C-S-H and C-A-H phases may also originate from their direct dependence on the chemical composition of the materials used and the conditions of reaction production [155]. Additionally, the concentration of solid matter plays an important role in the alkali-activation process [156].

Fly ash contains high percentages of aluminium and amorphous silicon, making it suitable to produce geopolymers [157]. The steps of the chemical process producing geopolymers by alkaline activation of fly ash, according to Buchwald et al. [158], can be defined by the chemical reactions shown in Equations (12)–(17), specific to each constituent oxide of the individual fly ash, and the general chemical reaction according to Equation (18).

- hydration process for vitreous silica with a pH > 12:

$$SiO_2 + 2OH^- \rightarrow SiO_3^{3-} + H_2O \qquad (12)$$

- hydration of Al_2O_3:

$$Al_2O_3 + 2OH^- \rightarrow 2AlO_2^{2-} + H_2O \qquad (13)$$

- reaction of CaO and MgO:

$$CaO + H_2O \rightarrow Ca^{2+} + 2OH^- \qquad (14)$$

- reaction of Na_2O and K_2O:

$$Na_2O + H_2O \rightarrow 2Na^+ + 2OH^- \qquad (15)$$

- the Fe_2O_3 reaction:

$$Fe_2O_3 + 3H_2O \rightarrow 2Fe^{3+} + 6OH^- \qquad (16)$$

- hydration of TiO_2:

$$TiO_2 + OH^- \rightarrow HTiO_3^{-} \qquad (17)$$

$$SiO_2 \cdot \alpha Al_2O_3 \cdot \beta CaO \cdot \gamma Na_2O \cdot \delta Fe_2O_3 \cdot \varepsilon TiO_2 + (\beta + \gamma + 3\delta)H_2O + (2 + 2\alpha + \varepsilon)OH \rightarrow SiO_3^{2-} + 2\alpha AlO_2^{2-} + \beta Ca^{2+} + 2\gamma Na^+ + 2\delta Fe^{3+} + \varepsilon HTiO_3^{-} + (1 + \alpha)H_2O + 2(\beta + \gamma + 3\delta) \qquad (18)$$

3.1. Physical-Mechanical Characteristics of Geopolymer Composites

The Si/Al ratio is a significant factor affecting the degree of crystallization and reaction with alkali activators [159], forming amorphous to semi-crystalline phases. Both polysialate-siloxo (Si/Al = 2) and polysialate-diloxo (Si/Al = 3) provided good strength of geopolymers. Polysialate-siloxo (Si/Al = 2) appears to be formed faster and has lower compressive strength than polysialate-diloxo (Si/Al = 3). The monomeric group of $[SiO(OH)_3]^-$, $[SiO_2(OH)_2]^{2-}$ and $[Al(OH)_4]^-$ normally forms later than Si and Al species, because small aluminium silicate oligomers can enhance geopolymer formation [160]. Research tends to indicate that metakaolin-based geopolymers exhibit satisfactory compressive strength with a Si/Al ratio of 1.9 ÷ 3, while the appropriate ratio for fly-ash-based geopolymers is about 2 ÷ 4 [161,162]. Due to the large availability of fly ash, and the need to reintroduce it into the economic circuit, in line with the principles of the Circular Economy, much research has been directed towards making these geopolymers using fly ash. In this context, the issue of integrating affordable eco-efficient solutions into the value chains is a challenge, while combating climate change and reducing the negative impact on the environment by reducing waste and by-products from industry, reducing cement consumption and thus reducing greenhouse gas emissions, etc., while ensuring the development of advanced innovative materials. Class F fly ash has been identified as the most suitable raw material for geopolymer materials due to its reactivity and availability. The mass ratio of SiO_2 and Al_2O_3 in Class F fly ash is between 1.7–4.0, while the amorphous content is generally higher than 50% [163–165]. Class C fly ash has a calcium content of 15–40%. It offers a different geopolymer structure compared to class F fly ash due to the increased calcium content. Temuujin et al. (2013) [165] reported that class C fly ash has self-cementing properties, which, through alkaline activation, allow it to harden at room temperature. Having a low calcium content, Class F fly ash, according to the literature studied, was preferred for producing geopolymer binders because the high amount of calcium may impact the polymerization process which may lead to changes in the microstructure of the final product [165–169].

During geopolymerization, fly ash reacts with the alkaline medium and specifically with aqueous solutions of polyisalates, leading to the formation of geopolymer materials comprising alumino-silicate-hydrate (A-S-H) gel [167]. Fly-ash-based geopolymers have demonstrated good mechanical strength and enhanced durability [167,168,170].

Several additions introduced at preparation in the geopolymer composite can have beneficial effects, improving the characteristics of the material in the fresh state but especially in the hardened state. Thus, literature indicates beneficial effects of nano-SiO_2 in small amounts (max. 10 wt% relative to the amount of FA) on workability, polymerization reaction, mechanical strength and durability [171–176], and the use of MgO powder as an addition can improve aspects related to compressive strength, workability, drying shrinkage and porosity [177]. However, all these should be analyzed in relation to the oxide composition and specific characteristics of fly ash used as the main raw material.

The concentration of the alkali activator has a significant effect on the compressive strength of the geopolymer. The test age and the temperature at which curing (heat treatment) of the geopolymer takes place are other variables that influence the compressive strength of the geopolymer. However, with all the still existing uncertainties, research has shown that a sufficient concentration of activator is required during geopolymerization, as the NaOH concentration has a greater influence on the compressive strength values than on the curing time [178–180]. A common alkaline solution used for making fly-ash-based geopolymers consists of sodium silicate and sodium or potassium hydroxide, respectively, a solution with molar concentrations of sodium hydroxide between 7 and 10 M. The combination of NaOH and sodium silicate (Na_2SiO_3) is the most suitable for the alkaline activator because sodium silicate contains partially polymerized and dissolved silicon, which reacts more easily. Research shows that geopolymers with higher strength, for the same type of fly ash, are obtained when the ratio (Na_2SiO_3/NaOH) is between

0.67 and 1.00 [181], with a molar concentration of sodium hydroxide (NaOH) of at least 8–10 M [166,182].

In this direction, Al Bakri et al. [183] studied the effect of $Na_2SiO_3/NaOH$ ratio on the compressive strength of fly-ash-based geopolymer, showing that by increasing the $Na_2SiO_3/NaOH$ ratio from 0.6 to 1.00, the compressive strength increased, with the maximum compressive strength obtained for $Na_2SiO_3/NaOH$ ratio equal to 1.00 [184,185]. Morsy et al. (2014) [186] studied the effect of $Na_2SiO_3/NaOH$ ratios of 0.5, 1.0, 1.5, 2.0 and 2.5 on the strength of fly-ash-based geopolymer mortar. The highest value of compressive strength was obtained when the ratio was equal to 1.0 and increasing the $Na_2SiO_3/NaOH$ ratio caused a decrease in compressive strength. Álvarez-Ayuso et al. (2008) [187] showed that when the NaOH concentration is higher, geopolymerization can be achieved even without soluble sodium silicate.

Rangan et al. [188] proposed $0.2 < Na_2O/SiO_2 < 0.28$ and $15 < H_2O/Na_2O < 17.5$ as optimum oxide ratios in the activation solution to achieve improved performance in geopolymer concrete. Provis et al. [189] stated that if the activator to binder mass ratio is between $0.6 \div 0.7$ and the activator has a SiO_2/Na_2O ratio in the range of $1 \div 1.5$, the resulting geopolymer binder imparts better mechanical properties. Another study showed that as the ratio of sodium hydroxide to sodium silicate increased from 1 to 2.5 and the molar concentration of NaOH increased from 8 to 16 M, the compressive strength of fly-ash-based geopolymers increased, i.e., the highest compressive strength was obtained for NaOH (16 M) concentration and sodium hydroxide/sodium silicate ratio in the range of 1.5 to 2.0 [190].

Vora and Dave [191] concluded that an increase in the sodium hydroxide/sodium silicate ratio and a higher molar concentration of sodium hydroxide result in higher compressive strength. However, Sindhunata et al. [192] reported that a soluble silicate content ($SiO_2/M_2O > 2$, where M is alkali ion) may reduce the reactivity in alkali-activated mixtures, as the concentration of cyclic silicate species may inhibit further condensation of aluminium ions. Therefore, higher concentration is judged to give rise to stronger ion pair formation and provide a more complete and faster polycondensation process of the particle interface [184,193], improving the dissolution of silicon and aluminium-containing materials in the presence of activators [194], but too high concentration could lead to an increase in the coagulated structure [195], causing lower workability with rapid curing behavior [196].

The mass ratio of alkaline solution to raw material is widely used in geopolymer synthesis to define both alkaline dosage and water content. In most cases, fly ash was used and the ratio was called alkali activator to fly ash ratio (AA/FA). Barbosa et al. [197] analyzed the effect of AA/FA ratio on strength development using 10 M NaOH solution as alkaline solution with AA/FA ratio = $0.34 \div 0.46$. It was observed that the compressive strength increased when the AA/FA ratio increased to 0.40. Too high of an AA/FA ratio could lead to a decrease in concentration as more sodium carbonate was formed and obstructed the geopolymerization process [198]. It was found that, depending on the type of alumina-silicate source materials, the recommended AA/FA ratio could be between $0.35 \div 0.50$ to have good compressive strength as well as good workability [199,200].

In terms of heat treatment temperature, research conducted on geopolymer paste and mortars for a temperature in the range (30 °C and 90 °C) demonstrated an increase in the overall geopolymerization reaction, leading to an increase in compressive strength from an early age [201]. However, exposure of the material to a treatment temperature above 90 °C will result in a geopolymer with a porous structure due to the rapid loss of water from the mixture, which will result in a possible decrease in mechanical performance [202]. Therefore, heat treatment is considered to improve mechanical properties, but the optimum temperature that improves the geopolymerization process and leads to the development of a suitable geopolymer microstructure is in the range $60 \div 75$ °C [203].

It is very difficult to accurately identify all the factors influencing the properties of alkali-activated geopolymer materials and to quantify the influence of each. According to the literature, the main factors influencing the properties of geopolymer materials are: type of raw materials and their Si and Al percentage content; Si and Al ratio; type of alkaline activator used; molarity of NaOH or KOH solution used; $Na_2SiO_3/NaOH$ mass ratio; duration and temperature of heat treatment; test age of the geopolymer material [204–214].

In the case of geopolymers, temperature and curing time play a significant role in the kinetics of the unfolding of chemical reactions: at low temperatures the evaporation of geopolymer precursors and water molecules occur simultaneously, preventing the formation of voids and cracks within the material, thus increasing the compressive strength [215,216]. This suggests that a longer curing time at low temperatures is preferable for the synthesis of geopolymer with higher compressive strength.

3.2. Influence of TiO_2 Adding Nanoparticles in Geopolymer Composites

It should be noted that, as reported by several research [217–222], a problem in producing these composites is the way of incorporation of NT, as this is relatively difficult to ensure a homogeneous dispersion, NT tending to agglomerate. Another controversy is in terms of self-cleaning capacity, with Guerrero et al. [220] indicating that 1 wt% NT is sufficient and Yang et al. [221] indicating an optimum requirement of 10 wt% NT.

Geopolymer concrete has been used in various applications such as waste management [205], civil engineering [222,223], cements and concretes [224,225], building retrofit [226] and as an alternative binder to replace OPC [226–228]. Applications have been extended to pavements [229–232] and building facades [223,229,232] and special function coatings [233]. Aguirre-Guerrero et al. [234] explored the potential of hybrid geopolymer coatings as a protective coating for reinforced concrete structures subjected to a marine environment. Sikora et al. [217] studied the applicability of geopolymer mortar as a coating to protect concrete from chemical attack and corrosion. Self-cleaning facades are gaining momentum in the construction industry. The applicability of self-cleaning facades is due to improved performance in brightness and reflectivity without the need for frequent maintenance [218,219].

Research reports on the influence of NT introduction in the geopolymer matrix on micro- and macro-structural characteristics, physical-mechanical and durability performances are found in a limited number of literatures. The literature indicates possibilities to induce self-healing capacity on geopolymeric materials using ZnO, TiO_2, WO_3 or Fe_2O_3 nanoparticles, with preference for NT and ZnO due to their high stability and low toxicity [9,235–242]. However, although they represent an environmentally friendly alternative in the construction industry, the properties of geopolymer materials with self-healing capacity are not yet sufficiently investigated so that they can be exploited.

3.2.1. NT Influence on Geopolymer Paste Properties

Experimental research conducted by Ambikakumari Sanalkumar and Yang [243] for geopolymer produced using metakaolin and containing 1–10% nano-TiO_2 (mass percentages reported to the mass of metakaolin) showed that the addition of nano-TiO_2 induces reduction in the flowability of the geopolymer paste. Such behavior could be associated with the high surface area of nano-TiO_2 particles, which creates the water demand of the mixtures [244]. In addition, the existence of very fine particles brings stronger cohesive van der Waals forces inside the particles, resulting in flocculated and agglomerated particles, which also makes the fluidity of the geopolymer paste in its fresh state become lower. Specifically, the fluidity of the fresh geopolymer paste decreases, relative to the fluidity of the control fresh geopolymer paste (without nanoTiO_2) by 3.22%; 5.65%; 25%, and by 37% when 1%, 2%, 5% and 10% nanoTiO_2 are introduced into the geopolymer paste (mass percentages relative to the amount of metakaolin).

Research by Duan et al. showed that in the case of replacing 1%, 3% or 5% fly ash with nano-TiO_2 (massive percentages relative to the amount of ash), the effects are also significant. Thus, a reduction in the workability of the nano-TiO_2-containing geopolymer is reported, the workability decreasing by 7.9% for the case of 1% nanoTiO, and by up to 20% if 5% fly ash is substituted with nano-TiO$_2$.

3.2.2. Influence of NT on the Properties of Geopolymer in Hardened State

Mechanical properties: Several studies have reported that the inclusion of TiO$_2$ in geopolymer binders induces certain effects on the mechanical properties of the resulting binder. Duan et al. [245] conducted a study to examine the impact of nano-TiO$_2$ inclusion on the physical and mechanical properties of a fly-ash-based geopolymer. The study showed that the incorporation of nano-TiO$_2$ particles increases compressive strength and carbonation resistance, reduces the drying shrinkage of the geopolymer [245]. The addition of nano-TiO$_2$ densifies the microstructure of the geopolymer matrix, in contradiction to a study [244] that reported that the inclusion of micro TiO$_2$ particles does not improve the mechanical properties of the geopolymer.

In terms of density, experimental research [244] reported an increase of up to 12% in density and a reduction of up to 41% in the volume of pore sizes ranging from 2 nm to 5 μm, concomitant with an increase in nano-TiO$_2$ content. These manifestations are thought to be due to the filling of fine pores by TiO$_2$ nanoparticles, thus inducing changes in the geopolymer at the microstructural level. Because of the densification of the material, but not only, as expected, the compressive strength at 7 days was improved by up to 41% (compared to the compressive strength of the control sample without nano-TiO$_2$) as the amount of nano-TiO$_2$ used was increased. Samples stored and tested at longer than 7 days (14 days, 28 days) also showed increases in compressive strength but at lower levels. This behavior shows that, unlike cementitious composites, where the increase in compressive strength occurs as a result of hydration-hydrolysis processes taking place over time (28 days), in the case of geopolymer binders the strength performance is obtained as a result of aluminosilicate dissolution processes and formation of the specific three-dimensional structure, and a densification of the material induces beneficial effects.

Research by Duan et al. showed that in the case of replacing 1%, 3% or 5% fly ash with nano-TiO$_2$ (massive percentages relative to the amount of ash), in terms of drying shrinkage, the presence of nano-TiO$_2$ in the geopolymer paste has a beneficial effect in reducing this indicator. This behavior is attributed to the possibility of filling the pores of the geopolymer paste with nanoparticles, which leads to densification of the material.

In terms of the performance of the hardened composite, there is an increase in the compressive strength at 7 days after peening, compared to the control sample, by more than 4% in the case of substitution of 1% fly ash with nanoTiO$_2$ and by more than 17% in the case of 5% nano-TiO$_2$. This increase in compressive strength is also evident when testing specimens at lower or higher ages, i.e., 1 day, 3 days after pouring or 28, 56, even 90 days after pouring. However, in agreement with other reports in the literature, the intensity of the effect on the compressive strength at early ages is noted, with the highest increases in compressive strength recorded compared to the control sample 24 h after casting, respectively, 7.1% for 1% nano-TiO$_2$ and 51% for 5% nano-TiO$_2$, a sign that these particles act as a densification spinner and induce microstructural changes.

According to Zulkifli et al. [246], the geopolymer made from metakaolin with NT content shows a much more homogeneous, compact microstructure with low porosity compared to the control sample made without NT. Syamsidar et al. [247] present results on a geopolymer material made by heat treatment at 50 °C based on class C fly ash from Bosowa Power Plant Jeneponto, Suth Sulawesi alkali activated, $SiO_2/Al_2O_3 = 3$; $Na_2O/SiO_2 = 2$ and $H_2O/Na_2O = 10$, in which 5%, 10% or 15% NT (mass percentage relative to the amount of fly ash) were introduced. Tests on specimens matured up to 7 days showed a much more compact and smooth surface appearance, without apparent porosity or surface defects, with apparent density increasing slightly with increasing %NT (2.85%, 3.1% and 4.76% increase in apparent density for 5%, 10% and 15% NT samples, compared to the apparent density of the control—0% NT). The compressive strength did not show a continuous increase as %NT increased, with a maximum recorded for the

10% NT composite, i.e., an increase of 12.36% for the 5%NT sample, 53.74% for the 10%NT sample and 45.64% for the 15%NT sample, suggesting the need to identify the optimal NT content interval in the geopolymer matrix in order to obtain the best compressive strength. XRD diffractograms performed for the NT geopolymer showed that, at microstructural level and compared to the control geopolymer (0%NT), no new crystallization phases are identified in the structure, therefore NT added during the preparation does not react with the constituents of the geopolymer, like the behavior of NT in cementitious matrices. Supporting the results obtained in the compressive strength evaluation, SEM analysis of the specimens indicates numerical reduction and decrease in microcracks opening for the case of NT specimens compared to the control geopolymer. Additionally, SEM analysis indicates a good adherence of the NT in the geopolymer matrix. In terms of resistance to H_2SO_4 action (1M, immersion 3 days), the strong reaction with CaO in fly ash is indicated with the formation of gypsum crystals ($CaSO_4 \cdot 2H_2O$), which is favored by the existence of NT in the geopolymer matrix. The formation of these crystals will induce internal stresses in the composite matrix which will favor the degradation process of the material, therefore, the use of a CaO-rich fly ash is not favorable for the use of a geopolymer composite with NT if it is intended for use in an acidic environment. Regarding the self-healing capacity, it was analyzed on specimens immersed in red clay solution showing that after removal from the solution, the surface of the specimens remained clean, without adherent red clay particles.

Guzman-Aponte et al. [248] showed that the inclusion of up to 10 wt.% NT did not influence the development of calcium silicate hydrate gel but did not indicate in detail the effect of NT addition on the physico-mechanical properties of GP.

Duan et al. [245], analyzing the influence of NT on the physico-mechanical characteristics of geopolymer paste prepared by alkaline activation of fly ash (alkaline activator prepared based on Na_2SiO_3 and NaOH), showed that 5 wt%, NT relative to the amount of fly ash, compared to the geopolymer control sample without NT, contributes to an increase in the compressive strength both at early ages and 28 days after casting, to obtain a more compact microstructure with less microcracks and improves the carbonation resistance of the composite. Duan et al. contribute by their research, and given that, reports on the carbonation resistance of nano-TiO_2-containing geopolymer are rare. Thus, it is shown that, as a result of these microstructurally induced changes, the depth of carbonation decreases as the number of nanoparticles substituting fly ash increases, the effect being even more evident compared to the control geopolymer, the longer the duration of exposure to carbonation. These results agree with the results reported by Sastry et al. [249], which indicate the increase in compressive strength of alkaline-activated fly-ash-based geopolymer for 2.5 wt% NT and Yang et al. respectively, [248] who, for the case of geopolymer made by alkaline activation of slag indicate the role of NT on several factors, namely, on geopolymer formation reactions, reduction in microcracks, cracks and improvement of compaction at the microstructural level.

Subaer et al. [250] analyze the thermo-mechanical properties of geopolymer composites made from alkali-activated metakaolin with soil. Na_2SiO_3 + NaOH, dispersive reinforced with 1–2 wt% carbon fibres and with NT applied as surface coating. The results showed that the geopolymer represents a good adhesion incorporating NT, but since NT are not included in the composite matrix and represent only a surface coating, their influence on the thermo-mechanical performance of the composite is not noticeable, carbon fibers having a more significant influence.

Mohamed et al., analysing the effect of TiO_2 on the performance of alkaline activated meta-haloiste based geopolymers with potassium hydroxide and potassium silicate-based activator, indicate that additions of 2.5%, 5%, 7.5% or 10% nanoparticles (mass percentage) reduce the total pore size (total porosity) by up to 49% compared to the control geopolymer, proportional to the amount of NT used. An increase in tensile strength of up to 78% is also reported, proportional to the amount of nano-TiO_2 used. All this leads, in agreement with other reports in the literature, to the conclusion that NT has a beneficial densification effect at the microstructural level, but it is noted that rutile TiO has a stronger effect than anatase TiO.

SEM-EDS analyses reported by Bonilla et al. [251] indicate the possibility of obtaining more homogeneous, smooth, compact surfaces with a reduced number of cracks compared to the control sample. In terms of physical-mechanical properties, a slight increase in density was observed, but, probably due to the heterogeneous distribution and agglomeration of nanoparticles both in the NT and nano-ZnO cases, the compressive strength decreased significantly by more than 2.5 times compared to the control.

Self-cleaning capacity and biocidal capacity: The self-cleaning performance of nano-TiO_2 modified geopolymer as a potential building material has been rarely reported in the literature [248]. However, as with cementitious composites, this self-cleaning ability is the sum of two main mechanisms: the ability to modify the surface's hydrophilicity and the ability to decompose organic molecules and even microorganism cells through redox reactions. The most common methods of evaluation from this point of view are oriented towards the evaluation of surface hydrophilicity (induction of superhydrophilicity of the surface), the evaluation of the decolorization capacity of rhodamine B or methylene blue, respectively, the evaluation of resistance in the presence of an environment contaminated with microorganisms. Experimental research by Ambikakumari Sanalkumar and Yang [243] showed an improvement of up to 15% in total solar reflectance (TSR) for nanoTiO_2-containing geopolymers compared to the control sample. Additionally, an improvement in hydrophilicity and surface self-healing ability is indicated and obtained on the one hand, as a result of the photoactivation of nano-TiO_2 and, on the other hand, as a result of the use of NaOH in the alkaline activation of the raw materials to obtain the geopolymer, since it is known that NaOH has a strong decomposition effect on organic molecules [252,253].

In agreement with Loh et al. [254], it is estimated that incorporation of NT into fly ash or kaolin-based composites has the effect of increasing the photocatalytic activity of NT. Moreover, even in the absence of light, based on the MB decolorization test and tests using microbiological techniques, antifungal properties were demonstrated.

Zailan et al. [255] report a review on the induction of self-cleaning ability by introducing 2.5%, 5% and 7.5% NT into the geopolymer matrix and evaluating/demonstrating this performance using *rhodamine B* (RhB) and *methylene blue* (MB) staining tests, respectively. They also analyze the influence of ZnO nanoparticles on the performance of geopolymer prepared based on fly ash (FA) class F from CIMA plant Perlis, Malaysia and alkaline activator based on $Na_2SiO:NaOH$, 12 M = 2.5:1.0 in which 2.5; 5; 7.5 and 10, wt%, ZnO nanoparticles were introduced. The results of the research on specimens matured up to 28 days showed a reduction of the compressive strength by approx. 29–54% compared to the control sample (0% ZnO nanoparticles), depending on the amount of nano-ZnO used, a behavior also supported by microstructural analysis by XRD and Sem which reveals changes in the crystallization phases. In terms of surface self-healing ability, based on the methylene blue (MB) staining test, a continuously increasing stain discoloration is recorded over time (over the evaluation time of 150 min UV exposure) and in relation to the amount of nano-ZnO used. This phenomenon is explained by a mechanism similar to the one presented for NT and confirms the possibility of inducing photocatalytic character on geopolymer not only using NT but also using nano-ZnO.

In consensus, Min et al. [256], Gasca-Tirado et al. [257], Zhang and Liu [258] and Luhar et al. [259], Kaya et al. [260], Chen et al. [261] indicate good results in terms of self-healing ability by photocatalytically activated degradation of methylene blue (MB) stains—Zhang and Liu indicating a 93% reduction of methylene blue stain staining for the case of photoactivated alkaline activated fly-ash-based NT-containing geopolymer surface within the first 6 h. However, research shows that part of the reduction in the degree of staining is due to the absorption of the dye into the geopolymer mass, and the rest is due to the degradation of the dye under the action of photoactivated NT [258,259]. Additionally, some research even indicates that GP itself has its own antimicrobial, antifungal and decolorizing capacity of staining substances, but the addition of NT has a strong increasing effect on these properties [254,260,262,263].

Yang et al. [264] report the effects of introducing 10 wt% NT on the performance of a geopolymer matrix made from fly ash from Shenhua Junggar Energy Corporation in Junggar, Inner Mongolia, China. Experimental XRD, SEM, BET analysis and photocatalytic activity results indicated the possibility of uniform distribution of NT in the geopolymer matrix and the influence of NT distribution mode on the specific geopolymer–NT surface area and photoactivity, i.e., the decolorization capacity of MB, which collapsed with increasing distribution homogeneity.

Alouani et al. [265] investigated the ability of geopolymer material produced by alkaline activation of metakaolin as an adsorbent to remove methylene blue. Strini et al. [266] showed that for 3 wt% NT added in geopolymer paste made from fly ash and metakaolin, geopolymer binders can be effective matrices to support photocatalytic activation of NT and induce specific material properties. Relating also to self-cleaning properties, Wang et al. [267] show that 5 wt% NT would be the optimum percentage for maximum MB decolorization effect, but research is insufficient because GP performance is influenced by several factors, as shown in the head above. Finally, and in terms of the influence on the physical-mechanical properties of GP, it is still controversial how much NT is introduced into the GP paste to achieve optimal performance.

Qin et al. [268] indicate the possibility of making superhydrophobic geopolymer surfaces by alkaline activation of blast furnace slag (also Si and Al oxide supplier waste) with reported good results in terms of hydrophobicity and therefore durability, including the fouling resistance capability of the material. In a similar direction of research development, Chindaprasirt et al. [269] indicate the possibility of inducing superhydrophobicity and self-cleaning capability of the surface of a geopolymer made based on alkali-activated fly ash, $Na_2SiO_3/NaOH = 2$, but in this case which benefited from a polymeric surface coating. Permatasari et al. [270] indicate the possibility of inducing self-healing performance for the surface of a geopolymer made from Gowa Regency soil deposit laterite, to which a thin film of NT solution coating was sprayed. This method allows for inducing a self-cleaning character, without influencing the flexural strength, with the mention that, once this film is destroyed, the self-cleaning capacity is lost.

In terms of bactericidal effect on *K. pneumoniae* and *P. aeruginosa*, Bonilla et al. [271] analyze a geopolymer composite based on powder material consisting of alumino-silicate precursors (85%) and Portland cement (15%), alkaline activated and containing 5 wt%, 10 wt% NT, 5 wt%, 10 wt% nano-ZnO, respectively. Research results demonstrate the development of inhibition halos and bactericidal effect such as a gentamicin antibiotic for both composite paste and composite mortar. In the same paper, the authors also indicate the self-cleaning effect developed by photocativation, an effect that causes rhodamine B, RhB, to decolorize 76.4% after 24 h for NT and over 98% for nano-ZnO.

4. Future Perspectives

At present, possibilities for increasing the specific effect induced by NT in both cementitious and alkali-activated geopolymer composites are reported in the literature. Thus, due to the specificity of NT being a wide bandgap semiconductor, i.e., 3.2 eV vs. the normal hydrogen electrode (NHE), it is indicated that through photoactivation, electrons jump from the valence band to the conduction band but a rapid charge recombination, i.e., a return of electrons to the valence band, occurs. This behavior is not beneficial from the point of view of the electron lifetime in the conduction band and, therefore, the performance of the composite by photoactivation. In order to improve these performances of the composite, acquired by photoactivation, in terms of increasing the charge separation efficiency (production of electrons and holes) and implicitly, the duration of manifestation of the self-cleaning effect as well as the biocidal effect, after the light source is removed, the literature indicates the possibility of introducing carbon-based materials (graphene, graphene oxide, carbon particles, fullerenes, which function as absorbers, acceptors and electron carriers) [271–281]. Therefore, the composites will show an increased absorption of light also in the visible spectrum, this technique becoming a possibility to improve the performance of the composites, i.e., to induce the possibility of photoactivation also with incident rays with wavelength in the visible spectrum. Hamidi and Aslani [22,282–287] indicated that the efficiency of the NT photoactivation process is influenced by a number of factors, the most important of which are: efficient absorption of sunlight, separation of products from the photocatalyst surface, rapid charge separation after light absorption to prevent electron-hole recombination, compatibility of the redox potentials of the valence band hole and conduction band electron with those of the donor and acceptor species, and long-term stability of the photocatalyst. Furthermore, research has shown [22,287–289] that the photocatalytic properties of TiO_2 are influenced by particle size, surface area, pore volume, surface hydroxyl content and degree of crystallinity. Crystallinity is an important factor contributing to the high photoactivity, as the presence of an amorphous phase would facilitate the recombination of photoexcited electrons and holes. Therefore, on a case-by-case basis, the optimal dosage of NT in cementitious composites or geopolymer materials is strongly influenced by a combination of all these factors.

This paper addresses a current research topic, with the aim of contributing to the creation of a favorable framework for the development and validation of materials with specific performances of the "smart-innovative" concept (the ability of a composite intended for the construction sector to harden in the absence of cement, while at the same exploiting the self-cleaning capacity), thus opening up new opportunities for the immediate exploitation and valorization of waste and by-products in accordance with current international environmental and sustainable development policies.

The multi- and transdisciplinary as well as eco-innovative character derives from the strongly applicative-exploratory approach of a research field involving on the one hand aspects of sustainable development of the built environment, and on the other hand specific aspects responding to the European requirements of environmental impact and population health. At the same time, a cross-cutting connection is made between the industries generating waste/industrial by-products (energy, metallurgy, natural resource processing) and the construction industry, creating a favorable framework for the implementation of emerging technologies with a high degree of novelty and the possibility of obtaining innovative intelligent products with the potential to advance in global value-added chains. Moreover, the exploitation of the specific performance of nanomaterials provides a link with other areas (safety, hygiene, health) by contributing to the production of knowledge needed to develop technologies that induce a high degree of safety in terms of surface hygiene, with high resistance to the development of biological films of microorganisms (moulds, lichens, algae, bacteria). Consequently, a synergistic connection is achieved between applied research activities aimed at (1) developing innovative composite materials for the sustainable development of the built environment; (2) developing materials with low environmental impact; (3) expanding the range of possibilities for introducing industrial wastes and by-products into the economic circuit; (4) developing innovative intelligent materials with self-cleaning capacity and increased resistance to biological agents; and (5) developing effective solutions for increasing safety in terms of public health.

Therefore, the contribution of this review responds to the need to connect current research to the European and global scientific frontier, in line with societal challenges related to responsible consumption and production; combating climate change and reducing negative environmental impacts.

5. Conclusions

The aim of this study was to summarize, as far as possible in parallel, the current state of research into the development of high-performance, self-healing composites with low environmental impact, or which allow the reduction in environmental impact by using industrial waste or by-products as raw materials. Thus, an analysis focused on the field of cementitious composites based on Portland cement vs. geopolymer composites based on alkali-activated fly ash, with NT content in mass, without analysing other materials, (non-Portland cements, lime mortars, lime and cement mortars, hydraulic lime mortars modified with TiO_2).

Based on the above presented, in general, the following can be said:
- in the current context, where the need to identify sustainable development solutions is imperative, the development of innovative materials contributes to the creation of a favorable framework for increasing the implementation of the principles of the Circular Economy, reducing environmental impact and increasing sustainability in the construction sector;
- innovative directions in this respect, still niche, are the development of cementitious composites or geopolymer composites that include nanoparticles in the matrix, the most used being NT;
- in parallel, the development of geopolymer materials allows for the reuse of waste or industrial by-products which contributes to reducing the environmental impact of other industries.
- In terms of cementitious composites, studies conducted to date have shown that:
- inducing exceptional properties by exploiting specific features of the nanoparticles embedded in the composite matrix has already proven to be a possible way forward;
- to date, although the research carried out is encouraging, there are several controversies and uncertainties, which point to further research;
- the introduction of NT into the cementitious matrix has consequences on the physical-mechanical performance, durability or resistance performance to the action of microorganisms, improving them;
- the results of the research carried out at microstructural level, corroborated and reflected by the results of the research carried out at macrostructural level, indicate the need for in-depth analysis so as to gain a thorough knowledge of the mechanisms underlying the phenomenon and to be able to determine more easily and more precisely the optimal quantity of nanoparticles and the way in which they are introduced during preparation, so as to achieve good performance in terms of physical-mechanical properties, self-healing capacity and increased surface hygiene.
- In terms of geopolymer materials, studies conducted so far have shown that:
- the field of self-healing geopolymers is an area of interest, but one that has been addressed only in recent years;
- so far, a number of results are reported, but there is still some controversy about the mechanisms of the geopolymerization reaction, the influence of a significant number of factors (e.g., type and oxidative composition of the raw material, characteristics of the alkaline activator, existence or not of nanoparticles or the type of nanoparticles used, working temperature, etc.) on the physical-mechanical performances, which have been studied more intensively, but also on some performances of durability, self-healing capacity, resistance to the action of microorganisms, etc.

The scientific contribution of the paper derives from the cumulative approach, in parallel, presenting possibilities of integrating smart functions within the eco-friendly function for the development of technologies and materials for the construction sector, in the context of applying and integrating the principles of the Circular Economy as a tool for Sustainable Development. Therefore, the general objective of the work has been set in order to respond to the need to connect current research to the European and global scientific frontier, in line with societal challenges related to responsible consumption and production through the efficient use of resources and raw materials, having as basic benchmarks the elements of scientific novelty in the field of smart-eco-innovative composite materials.

Author Contributions: Conceptualization, A.H., A.-V.L., A.A.C., B.A.I., E.G. and M.C.; methodology, A.H., A.-V.L., H.V. and V.S.; formal analysis, B.A.I., E.G., M.C. and C.F.; investigation, A.H., A.-V.L., A.A.C., B.A.I., E.G., M.C., C.F., H.V. and V.S.; data curation, A.H., C.F., H.V. and V.S.; writing—original draft preparation, A.H. and A.-V.L.; writing—review and editing, A.-V.L., A.A.C., H.V. and

V.S.; project administration, A.-V.L. All authors have read and agreed to the published version of the manuscript.

Funding: This research was funded by the Romanian Government Ministry of Research Innovation and Digitization, project No. PN 23 35 05 01 "Innovative sustainable solutions to implement emerging technologies with cross-cutting impact on local industries and the environment, and to facilitate technology transfer through the development of advanced, eco-smart composite materials in the context of sustainable development of the built environment".

Institutional Review Board Statement: Not applicable.

Informed Consent Statement: Not applicable.

Data Availability Statement: Not applicable.

Conflicts of Interest: The authors declare no conflict of interest.

References

1. Zailan, S.N.; Mahmed, N.; Al Bakri Abdullah, M.M.; Sandu, A.V. Self-cleaning geopolymer concrete—A review. *IOP Conf. Ser. Mater. Sci. Eng.* **2016**, *133*, 012026. [CrossRef]
2. Malhotra, V.M. Making Concrete "Greener" With Fly Ash. *ACI Conc. Int.* **1999**, *21*, 61–66.
3. Cement Technology Roadmap: Carbon Emissions Reductions up to 2050. Available online: https://www.iea.org/reports/cement-technology-roadmap-carbon-emissions-reductions-up-to-2050 (accessed on 10 March 2023).
4. US Geological Survey. Mineral Commodity Summaries: Cement. 2012. Available online: https://d9-wret.s3.us-west-2.amazonaws.com/assets/palladium/production/mineral-pubs/mcs/mcs2012.pdf (accessed on 28 February 2023).
5. Cembureau. Available online: https://www.cembureau.eu/library/reports/2050-carbon-neutrality-roadmap/ (accessed on 21 February 2023).
6. Aitcin, P.-C. Cements of yesterday and today; Concrete of tomorrow. *Cem. Concr. Res.* **2000**, *30*, 1349–1359. [CrossRef]
7. Sandu, A.V. Obtaining and Characterization of New Materials. *Materials* **2021**, *14*, 6606. [CrossRef]
8. Jamaludin, L.; Razak, R.A.; Abdullah, M.M.A.B.; Vizureanu, P.; Bras, A.; Imjai, T.; Sandu, A.V.; Abd Rahim, S.Z.; Yong, H.C. The Suitability of Photocatalyst Precursor Materials in Geopolymer Coating Applications: A Review. *Coatings* **2022**, *12*, 1348. [CrossRef]
9. Davidovits, J. Chemistry of Geopolymeric Systems Terminology. In Proceedings of the Geopolymer International Conference, Saint-Quentin, France, 30 June–2 July 1999.
10. Wazien, A.Z.W.; Mustafa, M.; Abdullah, A.B.; Razak, R.A.; Rozainy, M.M.A.Z.R.; Faheem, M.; Tahir, M.; Faris, M.A.; Hamzah, H.N. Review on Potential of Geopolymer for Concrete Repair and Rehabilitation. *MATEC Web Conf.* **2016**, *78*, 01065. [CrossRef]
11. Lloyd, N.A.; Rangan, B.V. Geopolymer concrete with fly ash. In Proceedings of the Second International Conference on Sustainable Materials and Technologies, Ancona, Italy, 28 June 2010; Zachar, J., Claisse, P., Naik, T.R., Ganjian, E., Eds.; Università Politecnica delle Marche: Ancona, Italy, 2010.
12. Wang, L.; Zhang, H.; Gao, Y. Effect of TiO_2 Nanoparticles on Physical and Mechanical Properties of Cement at Low Temperatures. *Adv. Mater. Sci. Eng.* **2018**, *2018*, 8934689. [CrossRef]
13. Falah, M.; Mackenzie, K.J.D. Photocatalytic Nanocomposite Materials Based on Inorganic Polymers (Geopolymers): A Review. *Catalysts* **2020**, *10*, 1158. [CrossRef]
14. Mohseni, E.; Miyandehi, B.M.; Yang, J.; Yazdi, M.A. Single and combined effects of nano-SiO_2, nano-Al_2O_3 and nano-TiO_2 on the mechanical, rheological and durability properties of self-compacting mortar containing fly ash. *Constr. Build. Mater.* **2015**, *84*, 331–340. [CrossRef]
15. Nazari, A.; Riahi, S. The effects of TiO_2 nanoparticles on physical, thermal and mechanical properties of concrete using ground granulated blast furnace slag as binder. *Mater. Sci. Eng. A* **2011**, *528*, 2085–2092. [CrossRef]
16. Nazari, A.; Riahi, S. The effects of TiO_2 nanoparticles on properties of binary blended concrete. *J. Compos. Mater.* **2011**, *45*, 1181–1188. [CrossRef]
17. Nazari, A.; Riahi, S. The effects of TiO_2 nanoparticles on flexural damage of self-compacting concrete. *Int. J. Damage Mech.* **2011**, *20*, 1049–1072. [CrossRef]
18. Zailan, S.N.; Mahmed, N.; Al Bakri Abdullah, M.M.; Sandu, A.V.; Shahedan, N.F. Review on Characterization and Mechanical Performance of Self-cleaning Concrete. *MATEC Web Conf.* **2017**, *97*, 01022. [CrossRef]
19. Quagliarini, E.; Bondioli, F.; Goffredo, G.B.; Cordoni, C.; Munafò, P. Self-cleaning and de-polluting stone surfaces: TiO_2 nanoparticles for limestone. *Constr. Build. Mater.* **2012**, *37*, 51–57. [CrossRef]
20. Zhang, S.M.-H.; Tanadi, D.; Li, W. Effect of photocatalyst TiO_2 on workability, strength, and self-cleaning efficiency of mortars for applications in tropical environment. In Proceedings of the 35th Conference on Our World in Concrete and Structures, Singapore, 25–27 August 2010.

21. Cassar, L. Nanotechnology and photocatalysis in cementitous materials. In Proceedings of the NICOM 2: 2nd International Symposium on Nanotechnology in Construction, Bilbao, Spain, 13–16 November 2005; de Miguel, Y., Porro, A., Bartos, P.J.M., Eds.; RILEM Publications SARL: Champs-sur-Marne, France, 2006; pp. 277–284.
22. Hamidi, F.; Aslani, F. TiO_2-based Photocatalytic Cementitious Composites: Materials, Properties, Influential Parameters, and Assessment Techniques. *Nanomaterials* **2019**, *9*, 1444. [CrossRef] [PubMed]
23. Reches, Y. Nanoparticles as concrete additives: Review and perspectives. *Constr. Build. Mater.* **2018**, *175*, 483–495. [CrossRef]
24. Li, H.; Xiao, H.-G.; Yuan, J.; Ou, J. Microstructure of cement mortar with nano-particles. *Compos. Part B Eng.* **2004**, *35*, 185–189. [CrossRef]
25. Li, Z.; Han, B.; Yu, X.; Dong, S.; Zhang, L.; Dong, X.; Ou, J. Effect of nano-titanium dioxide on mechanical and electrical properties and microstructure of reactive powder concrete. *Mater. Res. Express* **2017**, *4*, 095008. [CrossRef]
26. Han, B.; Li, Z.; Zhang, L.; Zeng, S.; Yu, X.; Han, B.; Ou, J. Reactive powder concrete reinforced with nano SiO_2-coated TiO_2. *Constr. Build. Mater.* **2017**, *148*, 104–112. [CrossRef]
27. Yang, L.; Jia, Z.; Zhang, Y.; Dai, J. Effects of nano-TiO_2 on strength, shrinkage and microstructure of alkali activated slag pastes. *Cem. Concr. Compos.* **2015**, *57*, 1–7. [CrossRef]
28. Pérez-Nicolás, M.; Plank, J.; Ruiz-Izuriaga, D.; Navarro-Blasco, I.; Fernandez, J.; Alvarez, J.I. Photocatalytically active coatings for cement and air lime mortars: Enhancement of the activity by incorporation of superplasticizers. *Constr. Build. Mater.* **2018**, *162*, 628–648. [CrossRef]
29. Zhao, A.; Yang, J.; Yang, E.-H. Self-cleaning engineered cementitious composites. *Cem. Concr. Res.* **2015**, *64*, 74–83. [CrossRef]
30. Li, C.-Z.; Feng, N.-Q.; Li, Y.-D.; Chen, R.-J. Effects of polyethylene oxide chains on the performance of polycarboxylate-type water reducers. *Cem. Concr. Res.* **2005**, *35*, 867–873. [CrossRef]
31. Navarro-Blasco, I.; Pérez-Nicolás, M.; Fernández, J.M.; Duran, A.; Sirera, R.; Alvarez, J.I. Assessment of the interaction of polycarboxylate superplasticizers in hydrated lime pastes modified with nanosilica or metakaolin as pozzolanic reactives. *Constr. Build. Mater.* **2014**, *73*, 1–12. [CrossRef]
32. Zapata, L.E.; Portela, G.; Suarez, O.M.; Carrasquillo, O. Rheological performance and compressive strength of superplasticized cementitious mixtures with micro/nano-SiO_2 additions. *Constr. Build. Mater.* **2013**, *41*, 708–716. [CrossRef]
33. Ranjit, K.; Odedra; Parmar, K.A.; Arora, N.K. Photocatalytic Self cleaning Concrete. *IJSRD Int. J. Sci. Res. Dev.* **2014**, *1*, 2521–2523.
34. Lazăr, M.; Fiat, D.; Hubcă, G. The influence of TiO_2 nanometric photocatalytic pigment on the proprieties of the film forming products based on organic binders in aqueous dispersion. *Rom. J. Mater.* **2015**, *40*, 178–187.
35. Aslanidou, D.; Karapanagiotis, I.; Lampakis, D. Waterborne Superhydrophobic Coatings for the Protection of Marble and Sandstone. *Materials* **2018**, *1*, 585. [CrossRef] [PubMed]
36. Nazari, A.; Riahi, S.; Riahi, S.; Shamekhi, S.F.; Khademno, A. Improvement the mechanical properties of the cementitious composite by using TiO_2 nanoparticles. *Am. J. Sci.* **2010**, *6*, 98–101.
37. Midtdal, K.; Jelle, B.P. Self-cleaning glazing products: A state-of-the-art review and future research pathways. *Sol. Energy Mater. Sol. Cells* **2013**, *109*, 126–141. [CrossRef]
38. Lee, B.Y.; Jayapalan, A.R.; Kurtis, K.E. Effects of nanoTiO_2 on properties of cement-based materials. *Mag. Concr. Res.* **2013**, *65*, 1293–1302. [CrossRef]
39. Pimenta Teixeira, K.; Perdigão Rocha, I.; De S'a Carneiro, L.; Flores, J.; Dauer, E.A.; Ghahremaninezhad, A. The effect of curing temperature on the properties of cement pastes modified with TiO_2 nanoparticles. *Materials* **2016**, *9*, 952. [CrossRef] [PubMed]
40. Jalal, M.; Fathi, M.; Farzad, M. Effects of fly ash and TiO_2 nanoparticles on rheological, mechanical, microstructural and thermal properties of high strength self compacting concrete. *Mech. Mater.* **2013**, *61*, 11–27. [CrossRef]
41. Salemi, N.; Behfamia, K.; Zaree, S.A. Effect of nanoparticles on frost durability of concrete. *Asian J. Civ. Eng. (BHRC)* **2014**, *15*, 411–420.
42. Nazari, A.; Riahi, S.; Shamekhi, S.F.; Khademno, A. Assessment of the Effects of the Cement Paste Composite in Presence TiO_2 Nanoparticles. *Am. J. Sci.* **2010**, *6*, 43–46.
43. Li, H.; Zhang, M.H.; Ou, J.P. Flexural fatigue performance of concrete containing nanoparticles for pavement. *Int. J. Fatigue* **2007**, *29*, 1292–1301. [CrossRef]
44. Li, H.; Zhang, M.; Ou, J. Abrasion resistance of concrete containing nano-particles for pavement. *Wear* **2006**, *260*, 1262–1266. [CrossRef]
45. Smits, M.; Chan, C.K.; Tytgat, T.; Craeye, B.; Costarramone, N.; Lacombe, S.; Lenaerts, S. Photocatalytic degradation of soot deposition: Self-cleaning effect on titanium dioxide coated cementitious materials. *Chem. Eng. J.* **2013**, *222*, 411–418. [CrossRef]
46. Jalal, M.; Ramezanianpour, A.A.; Pool, M.K. Effects of titanium dioxide nanopowder on rheological properties of self compacting concrete. *Am. J. Sci.* **2012**, *8*, 285–288.
47. Meng, T.; Yu, Y.; Qian, X.; Zhan, S.; Qian, K. Effect of nano-TiO_2 on the mechanical properties of cement mortar. *Constr. Build. Mater.* **2012**, *29*, 241–245. [CrossRef]
48. Noorvand, H.; Ali, A.A.A.; Demirboga, R.; Farzadnia, N.; Noorvand, H. Incorporation of nano TiO_2 in black rice husk ash mortars. *Constr. Build. Mater.* **2013**, *47*, 1350–1361. [CrossRef]
49. Lee, B.Y. Effect of Titanium Dioxide Nanoparticles on Early Age and Long Term Properties of Cementitious Materials. Ph.D. Thesis, School of Civil & Environmental Engineering, Georgia Institute of Technology, Atlanta, GA, USA, August 2012.

50. Chen, J.; Kou, S.; Poon, C. Hydration and properties of nano-TiO$_2$ blended cement composites. *Cem. Concr. Comp.* **2012**, *34*, 642–649. [CrossRef]
51. Essaway, A.A.; El Aleem, S.A. Physico-mechanical properties, potent adsorptive and photocatalytic efficacies of sulfate resisting cement blends containing micro silica and nano-TiO$_2$. *Constr. Build. Mater.* **2014**, *52*, 1–8. [CrossRef]
52. Rashad, A.M. A synopsis about the effect of nano-titanium dioxide on some properties of cementitious materials—A short guide for civil engineer. *Rev. Adv. Mater. Sci.* **2015**, *40*, 72–88.
53. Lee, B.Y.; Kurtis, K.E. Influence of TiO$_2$ Nanoparticles on Early C$_3$S Hydration. *J. Am. Ceram. Soc.* **2010**, *93*, 3399–3405. [CrossRef]
54. Sneff, L.; Hotza, D.; Lucas, S.; Ferreira, V.M.; Labrinca, J.A. Effect of nano-SiO$_2$ and nano-TiO$_2$ addition on the rheological behavior and the hardened properties of cement mortars. *Mater. Sci. Eng. A* **2012**, *532*, 354–361. [CrossRef]
55. Nazari, A.; Riahi, S. The effect of TiO$_2$ nanoparticles on water permeability and thermal and mechanical properties of high strength self-compacting concrete. *Mater. Sci. Eng. A* **2010**, *582*, 756–763. [CrossRef]
56. Jalal, M.; Mortazavi Ali, A.; Nemat, H. Thermal properties of TiO$_2$ nanoparticles binary blended cementitious composites. *Am. J. Sci.* **2012**, *8*, 391–394.
57. Jayapalan, A.R.; Lee, B.Y.; Kurtis, K.E. Effect of Nano-sized Titanium Dioxide on Early Age Hydration of Portland Cement. In *Nanotechnology in Construction 3*; Bittnar, Z., Bartos, P.J.M., Němeček, J., Šmilauer, V., Zeman, J., Eds.; Springer: Berlin/Heidelberg, Germany, 2009.
58. Jayapalan, A.R.; Lee, B.Y.; Kurtis, K.E. Can nanotechnology be 'green'? Comparing efficacy of nano and microparticles in cementitious materials. *Cem. Concr. Comp.* **2013**, *36*, 16–24. [CrossRef]
59. Kurihara, R.; Maruyama, I. Influences of Nano-TiO$_2$ particles on alteration of microstructure of cement. *JCI* **2016**, *38*, 219–224.
60. Baoguo, M.A.; Hainan, L.I.; Junpeng, M.E.I.; Pei, O. Effect of Nano-TiO$_2$ Addition on the Hydration and Hardening Process of Sulphoaluminate Cement. *J. Wuhan Univ. Technol. Mater. Sci. Ed.* **2015**, *30*, 768–773.
61. Sakthivel, R.; Arun, K.T.; Dhanabal, M.; Aravindan, V.; Aravindh, S. Experimental study of photocatalytic concrete using titanium dioxide. *Int. J. Innov. Res. Sci. Technol.* **2018**, *4*, 117–123.
62. Khataee, R.; Heydari, V.; Moradkhannejhad, L.; Safarpour, M.; Joo, S.W. Self-cleaning and mechanical properties of modified white cement with nanostructured TiO$_2$. *J. Nanosci. Nanotechnol.* **2013**, *13*, 5109–5114. [CrossRef]
63. Zhang, M.; Li, H. Pore structure and chloride permeability of concrete containing nano-particles for pavement. *Constr. Build. Mater.* **2011**, *25*, 608–616. [CrossRef]
64. Behfarnia, K.; Azarkeivan, A.; Keivan, A. The effects of TiO$_2$ and ZnO nanoparticles on physical and mechanical properties of normal concrete. *Asian J. Civ. Eng. (Bhrc)* **2013**, *14*, 517–531.
65. Zailan, S.N.; Mahmed, N.; Abdullah, M.M.A.B.; Rahim, S.Z.A.; Halin, D.S.C.; Sandu, A.V.; Vizureanu, P.; Yahya, Z. Potential Applications of Geopolymer Cement-Based Composite as Self-Cleaning Coating: A Review. *Coatings* **2022**, *12*, 133. [CrossRef]
66. Soleymani, F. Assessments of the effects of limewater on water permeability of TiO$_2$ nanoparticles binary blended limestone aggregate-based concrete. *J. Am. Sci.* **2011**, *7*, 7–12.
67. Soleymani, F. The filler effects TiO$_2$ nanoparticles on increasing compressive strength of palm oil clinker aggregate-based concrete. *J. Am. Sci.* **2012**, *8*, 21–24.
68. Sorathiya, J.; Shah, S.; Kacha, S. Effect on Addition of Nano "Titanium Dioxide"(TiO$_2$) on Compressive Strength of Cementitious Concrete. *Kalpa Publ. Civil Eng.* **2017**, *1*, 219–225.
69. Shekari, A.H.; Razzaghi, M.S. Influence of Nano Particles on Durability and Mechanical Properties of High Performance Concrete. *Proc. Eng.* **2011**, *14*, 3036–3041. [CrossRef]
70. Patel, N.; Mishra, C.B. Laboratory Investigation of nano titanium dioxide (TiO$_2$) in concrete for pavement. *Int. Res. J. Eng. Technol. (IRJET)* **2018**, *5*, 1634–1638.
71. Ghosal, M.; Chakraborty, A.K. A comparative assessment of nano-SiO$_2$ and nano-TiO$_2$ insertion in concrete. *Eur. J. Adv. Eng. Technol.* **2015**, *2*, 44–48.
72. Nazari, A. The effects of curing medium on flexural strength and water permeability of concrete incorporating TiO$_2$ nanoparticles. *Mater. Struct.* **2011**, *44*, 773–786. [CrossRef]
73. He, X.; Shi, X. Chloride Permeability and Microstructure of Portland Cement Mortars Incorporating Nanomaterials. *Transp. Res. Rec.* **2008**, *2070*, 13–21. [CrossRef]
74. Li, H.; Xiao, H.; Guan, X.; Wang, Z.; Yu, L. Chloride diffusion in concrete containing nano-TiO$_2$ under coupled effect of scouring. *Compos. B. Eng.* **2014**, *56*, 698–704. [CrossRef]
75. Hassan, M.M.; Dylla, H.; Mohammad, L.N.; Rupnow, T. Methods for the application of titanium dioxide coatings to concrete pavement. *Int. J. Pavement Res. Technol.* **2012**, *5*, 12–20.
76. Farzadnia, N.; Ali, A.A.A.; Demirboga, R.; Parvez, M. Characterization of high strength mortars with nano Titania at elevated temperatures. *Contr. Build. Mater.* **2013**, *43*, 469–479. [CrossRef]
77. Feng, D.; Xie, N.; Gong, C. Portland cement paste modified by TiO$_2$ nanoparticles: A microstructure perspective. *Ind. Eng. Chem. Res.* **2013**, *52*, 11575–11582. [CrossRef]
78. Lackhoff, M.; Prieto, X.; Nestle, N.; Dehn, F.; Niessner, R. Photocatalytic activity of semiconductor-modified cement—Influence of semiconductor type and cement ageing. *Appl. Catal. B Environ.* **2003**, *43*, 205–216. [CrossRef]
79. Kwon, J.M.; Kim, Y.H.; Song, B.K.; Yeom, S.H.; Kim, B.S.; Im, J.B. Novel immobilization of titanium dioxide (TiO$_2$) on the fluidizing carrier and its application to the degradation of azo-dye. *J. Hazard. Mater.* **2006**, *134*, 230–236. [CrossRef]

80. De Ceukelaire, L.; Van Nieuwenburg, D. Accelerated carbonation of a blast-furnace cement concrete. *Cem. Concr. Res.* **1993**, *23*, 442–452. [CrossRef]
81. Castellote, M.; Fernandez, L.; Andrade, C.; Alonso, C. Chemical changes and phase analysis of OPC pastes carbonated at different CO_2 concentrations. *Mater. Struct.* **2009**, *42*, 515–525. [CrossRef]
82. Matsunaga, T.; Tomoda, R.; Nakajima, T.; Nakamura, N.; Komine, T. Continuous-sterilization system that uses photosemiconductor powders. *Appl. Environ. Microbiol.* **1988**, *54*, 1330–1333. [CrossRef] [PubMed]
83. Folli, A.; Jakobsen, U.H.; Guerrini, G.L.; Macphee, D.E. Rhodamine B discolouration on TiO_2 in the cement environment: A look at fundamental aspects of the self-cleaning effect in concretes. *J. Adv. Oxid. Technol.* **2009**, *12*, 126–133. [CrossRef]
84. Jimenez-Relinque, E.; Castellote, M. Quantification of hydroxyl radicals on cementitious materials by fluorescence spectrophotometry as a method to assess the photocatalytic activity. *Cem. Concr. Res.* **2015**, *74*, 108–115. [CrossRef]
85. Dharma, H.N.C.; Jaafar, J.; Widiastuti, N.; Matsuyama, H.; Rajabsadeh, S.; Othman, M.H.D.; Rahman, M.A.; Jafri, N.N.M.; Suhaimin, N.S.; Nasir, A.M.; et al. A Review of Titanium Dioxide (TiO_2)-Based Photocatalyst for Oilfield-Produced Water Treatment. *Membranes* **2022**, *12*, 345. [CrossRef]
86. Anucha, C.B.; Altin, I.; Bacaksiz, E.; Stathopoulos, V.N. Titanium Dioxide (TiO_2)-Based Photocatalyst Materials Activity Enhancement for Contaminants of Emerging Concern (CECs) Degradation: In the Light of Modification Strategies. *Chem. Eng. J. Adv.* **2022**, *10*, 100262. [CrossRef]
87. Sharifi, T.; Crmaric, D.; Kovacic, M.; Popovic, M.; Rokovic, M.K.; Kusic, H.; Jozić, D.; Ambrožić, G.; Kralj, D.; Kontrec, J.; et al. Tailored $BiVO_4$ for Enhanced Visible-Light Photocatalytic Performance. *J. Environ. Chem. Eng.* **2021**, *9*, 106025. [CrossRef]
88. Baxter, D.M.; Perkins, J.L.; McGhee, C.R.; Seltzer, J.M. A Regional Comparison of Mold Spore Concentrations Outdoors and Inside "Clean" and "Mold Contaminated" Southern California Buildings. *J. Occup. Environ. Hyg.* **2005**, *2*, 8–18. [CrossRef] [PubMed]
89. Hui Chen, P.E.; Song Deng, P.E.; Homer Bruner, C.E.M., Jr.; Garcia, J. Roots of Mold Problems and Humidity Control Measures in Institutional Buildings with Pre-Existing Mold Condition. In Proceedings of the Fourteenth Symposium on Improving Building Systems in Hot and Humid Climates, Richardson, TX, USA, 17–20 May 2004.
90. Wang, L.; Hu, C.; Shao, L. The antimicrobial activity of nanoparticles: Present situation and prospects for the future. *Int. J. Nanomed.* **2017**, *12*, 1227–1249. [CrossRef] [PubMed]
91. Kühn, K.P.; Chaberny, I.F.; Massholder, K.; Stickler, M.; Benz, V.W.; Sonntag, H.G.; Erdinger, L. Disinfection of surfaces by photocatalytic oxidation with titanium dioxide and UVA light. *Chemosphere* **2003**, *53*, 71–77. [CrossRef] [PubMed]
92. Machida, M.; Norimoto, K.; Kimura, T. Antibacterial Activity of Photocatalytic Titanium Dioxide Thin Films with Photodeposited Silver on the Surface of Sanitary Ware. *J. Am. Ceram. Soc.* **2005**, *88*, 95–100. [CrossRef]
93. Watts, R.J.; Kong, S.; Orr, M.P.; Miller, G.C.; Henry, B.E. Photocatalytic inactivation of coliform bacteria and viruses in secondary wastewater effluent. *Water Res.* **1995**, *29*, 95–100. [CrossRef]
94. Vohra, A.; Goswami, D.Y.; Deshpande, D.A.; Block, S.S. Enhanced photocatalytic inactivation of bacterial spores on surfaces in air. *J. Ind. Microbiol. Biotechnol.* **2005**, *32*, 364–370. [CrossRef] [PubMed]
95. Haleem Khan, A.A.; Mohan Karuppayil, S. Fungal pollution of indoor environments and its management. *Saudi J. Biol. Sci.* **2012**, *19*, 405–426. [CrossRef] [PubMed]
96. Sunada, K.; Watanabe, T.; Hashimoto, K. Bactericidal Activity of Copper-Deposited TiO_2 Thin Film under Weak UV Light Illumination. *Environ. Sci. Technol.* **2003**, *37*, 4785–4789. [CrossRef] [PubMed]
97. Oguma, K.; Katayama, H.; Ohgaki, S. Photoreactivation of Escherichia coli after Low- or Medium-Pressure UV Disinfection Determined by an Endonuclease Sensitive Site Assay. *Appl. Environ. Microbiol.* **2002**, *68*, 6029–6035. [CrossRef] [PubMed]
98. Saito, T.; Iwase, T.; Horie, J.; Morioka, T. Mode of photocatalytic bactericidal action of powdered semiconductor TiO_2 on mutans streptococci. *J. Photochem. Photobiol. B* **1992**, *14*, 369–379. [CrossRef] [PubMed]
99. Linkous, C.A.; Carter, G.J.; Locuson, D.B.; Ouellette, A.J.; Slattery, D.K.; Smitha, L.A. Photocatalytic inhibition of algae growth using TiO_2, WO_3, and cocatalyst modifications. *Environ. Sci. Technol.* **2000**, *34*, 4754–4758. [CrossRef]
100. Gogniat, G.; Thyssen, M.; Denis, M.; Pulgarin, C.; Dukan, S. The bactericidal effect of TiO_2 photocatalysis involves adsorption onto catalyst and the loss of membrane integrity. *FEMS Microbiol. Lett.* **2006**, *258*, 18–24. [CrossRef]
101. Mazurkova, N.A.; Spitsyna, Y.E.; Shikina, N.V.; Ismagilov, Z.R.; Zagrebel'Nyi, S.N.; Ryabchikova, E.I. Interaction of titanium dioxide nanoparticles with influenza virus. *Nanotechnol. Russ.* **2010**, *5*, 417–420. [CrossRef]
102. Adams, L.K.; Lyon, D.Y.; McIntosh, A.; Alvarez, P.J.J. Comparative toxicity of nano-scale TiO_2, SiO_2 and ZnO water suspensions. *Water Sci. Technol.* **2006**, *54*, 327–334. [CrossRef] [PubMed]
103. Armelao, L.; Barreca, D.; Bottaro, G.; Gasparotto, A.; Maccato, C.; Maragno, C.; Tondello, E.; Stangar, U.L.; Bergant, M.; Mahne, D. Photocatalytic and antibacterial activity of TiO_2 and Au/TiO_2 nanosystems. *Nanotechnology* **2007**, *18*, 375709. [CrossRef]
104. Dědková, K.; Matějová, K.; Lang, J.; Peikertová, P.; Kutláková, K.M.; Neuwirthová, L.; Frydrýšek, K.; Kukutschová, J. Antibacterial activity of kaolinite/nanoTiO_2 composites in relation to irradiation time. *J. Photochem. Photobiol. B* **2014**, *135*, 17–22. [CrossRef] [PubMed]
105. Gurr, J.R.; Wang, A.S.S.; Chen, C.H. Ultrafine titanium dioxide particles in the absence of photoactivation can induce oxidative damage to human bronchial epithelial cells. *Toxicology* **2005**, *213*, 66–73. [CrossRef] [PubMed]
106. Shaaban, I.G.; El-Sayad, H.; El-Ghaly, A.E.; Moussa, S. Effect of micro TiO_2 on cement mortar. *EJME* **2020**, *5*, 58–68. [CrossRef]
107. Davidson, H.; Poon, M.; Saunders, R.; Shapiro, I.M.; Hickok, N.J.; Adams, C.S. Tetracycline tethered to titanium inhibits colonization by Gram-negative bacteria. *J. Biomed. Mater. Res. B Appl. Biomater.* **2015**, *103*, 1381–1389. [CrossRef]

108. Lorenzetti, M.; Dogša, I.; Stošicki, T.; Stopar, D.; Kalin, M.; Kobe, S.; Novak, S. The Influence of Surface Modification on Bacterial Adhesion to Titanium-Based Substrates. *ACS Appl. Mater. Interfaces* **2015**, *7*, 1644–1651. [CrossRef] [PubMed]
109. Peng, Z.; Ni, J.; Zheng, K.; Shen, Y.; Wang, X.; He, G.; Jin, S.; Tang, T. Dual effects and mechanism of TiO_2 nanotube arrays in reducing bacterial colonization and enhancing C3H10T1/2 cell adhesion. *Int. J. Nanomed.* **2013**, *8*, 3093–3105.
110. Daly, M.J.; Gaidamakova, E.K.; Matrosova, V.Y.; Vasilenko, A.; Zhai, M.; Leapman, R.D.; Lai, B.; Ravel, B.; Li, S.-M.W.; Kemner, K.M.; et al. Protein Oxidation Implicated as the Primary Determinant of Bacterial Radioresistance. *PLoS Biol.* **2007**, *5*, 92. [CrossRef] [PubMed]
111. Carre, G.; Estner, M.; Gies, J.-P.; Andre, P.; Hamon, E.; Ennahar, S.; Keller, V.; Keller, N.; Lett, M.-C.; Horvatovich, P. TiO_2 Photocatalysis Damages Lipids and Proteins in Escherichia coli. *Appl. Environ. Microbiol.* **2014**, *80*, 2573–2581. [CrossRef]
112. Kubacka, A.; Diez, M.S.; Rojo, D.; Bargiela, R.; Ciordia, S.; Zapico, I.; Albar, J.P.; Barbas, C.; Martins dos Santos, V.A.P.; Fernández-garcía, M.; et al. Understanding the antimicrobial mechanism of TiO_2-based nanocomposite films in a pathogenic bacterium. *Sci. Rep.* **2014**, *4*, 4134. [CrossRef] [PubMed]
113. Huang, Y.-T.; Yu, K.-P.; Yang, K.-R.; Yang, S.-C.; Chen, Y.-L. Evaluation the antifungal effects of nano-metals loaded titanium dioxide on fungal spore. In Proceedings of the 10th International Conference on Healthy Buildings 2012, Brisbane, Australia, 8–12 July 2012.
114. Yadav, H.M.; Kim, J.S.; Pawar, S.H. Developments in photocatalytic antibacterial activity of nano TiO_2: A review. *Korean J. Chem. Eng.* **2016**, *33*, 1989–1998. [CrossRef]
115. Chen, J.; Poon, C.-S. Photocatalytic construction and building materials: From fundamentals to applications. *Build. Environ.* **2009**, *44*, 1899–1906. [CrossRef]
116. Dubosc, A.; Escadeillas, G.; Blanc, P. Characterization of biological stains on external concrete walls and influence of concrete as underlying material. *Cem. Concr. Res.* **2001**, *31*, 1613–1617. [CrossRef]
117. Kurth, J.C.; Giannantonio, D.J.; Allain, F.; Sobecky, P.A.; Kurtis, K.E. Mitigating biofilm growth through the modification of concrete design and practice. In Proceedings of the International RILEM Symposium on Photocatalysis, Environment and Construction Materials, Florence, Italy, 8–9 October 2007; pp. 8–9.
118. Maury, A.; De Belie, N. State of the art of TiO_2 containing cementitious materials: Self-cleaning properties. *Mater. Constr.* **2010**, *60*, 33–50. [CrossRef]
119. Mejía, J.M.; Mendoza, J.D.; Yucuma, J.; Mejía de Gutiérrez, R.; Mejía, D.E.; Astudillo, M. Mechanical, in-vitro biological and antimicrobial characterization as an evaluation protocol of a ceramic material based on alkaline activated metakaolin. *Appl. Clay Sci.* **2019**, *178*, 105141. [CrossRef]
120. Damian, L.; Patachia, S. Method for testing the antimicrobial character of the materials and their fitting to the scope. *Bull. Transilv. Univ. Bras.* **2014**, *7*, 37–44.
121. ASTM E2149; Standard Test Method for Determining the Antimicrobial Activity of Antimicrobial Agents under Dynamic Contact Conditions. ASTM International: West Conshohocken, PA, USA, 2020.
122. ASTM E2180; Test for Hydrophobic Antimicrobial Surfaces. ASTM International: West Conshohocken, PA, USA, 2018.
123. ISO 22196:2011; Measurement of Antibacterial Activity on Plastics and Other Non-Porous Surfaces. International Organization for Standardization: Geneva, Switzerland, 2020.
124. ASTM E1428; Antimicrobial Pink Stain Test. ASTM International: West Conshohocken, PA, USA, 2015.
125. STAS 12718/1989; Lacquers and Paints. Determination of the Sterility or Degree of Contamination with Micro-Organisms of Film-Forming Products. Romanian Institute for Standardization: Bucharest, Romania, 1989. (In Romanian)
126. ISO 27447; Fine Ceramics (Advanced Ceramics, Advanced Technical Ceramics)—Test Method for Antibacterial Activity of Semiconducting Photocatalytic Materials. International Organization for Standardization: Geneva, Switzerland, 2019.
127. Renz, C. Lichtreaktionen der Oxyde des Titans, Cers und der Erdsauren. *Helv. Chim. Acta* **1921**, *4*, 961–968. [CrossRef]
128. Matsunaga, T. Sterilization with particulate photosemiconductor. *J. Antibact. Antifung. Agents* **1985**, *13*, 211–220.
129. Matsunaga, T.; Tomoda, R.; Nakajima, T.; Wake, H. Photoelectrochemical sterilization of microbial cells by semiconductor powders. *FEMS Microbiol. Lett.* **1985**, *29*, 211–214. [CrossRef]
130. Aissa, A.H.; Puzenat, E.; Plassais, A.; Herrmann, J.-M.; Haehnel, C.; Guillard, C. Characterization and photocatalytic performance in air of cementitious materials containing TiO_2. Case study of formaldehyde removal. *Appl. Catal. B Environ.* **2011**, *107*, 1–8. [CrossRef]
131. Giannantonio, D.J.; Kurth, J.C.; Kurtis, K.E.; Sobecky, P.A. Effects of concrete properties and nutrients on fungal colonization and fouling. *Int. Biodeterior. Biodegrad.* **2009**, *63*, 252–259. [CrossRef]
132. Sökmen, M.; Candan, F.; Sümer, Z. Disinfection of E. coli by the Ag–TiO_2/UV system: Lipidperoxidation. *J. Photochem. Photobiol. A* **2001**, *143*, 241–244. [CrossRef]
133. Sökmen, M.; Degerli, S.; Aslan, A. Photocatalytic disinfection of Giardia intestinalis and Acanthamoeba castellani cysts in water. *Exp. Parasitol.* **2008**, *119*, 44–48. [CrossRef]
134. Kikuchi, Y.; Sunada, K.; Iyoda, T.; Hashimoto, K.; Fujishima, A. Photocatalytic bactericidal effect of TiO_2 thin films: Dynamic view of the active oxygen species responsible for the effect. *J. Photochem. Photobiol. A* **1997**, *106*, 51–56. [CrossRef]
135. Sunada, K.; Watanabe, T.; Hashimoto, K. Studies on photokilling of bacteria on TiO_2 thin film. *J. Photochem. Photobiol. A* **2003**, *156*, 227–233. [CrossRef]

136. Ditta, I.B.; Steele, A.; Liptrot, C.; Tobin, J.; Tyler, H.; Yates, H.M.; Sheel, D.W.; Foster, H.A. Photocatalytic antimicrobial activity of thin surface films of TiO_2, CuO and TiO_2/CuO dual layers on Escherichia coli and bacteriophage T4. *Appl. Microbiol. Biotechnol.* **2008**, *79*, 127–133. [CrossRef]
137. Brook, L.A.; Evans, P.; Foster, H.A.; Pemble, M.E.; Steele, A.; Sheel, D.W.; Yates, H.M. Highly bioactive silver and silver/titania composite films grown by chemical vapour deposition. *J. Photochem. Photobiol.* **2007**, *187*, 53–63. [CrossRef]
138. Yates, H.M.; Brook, L.A.; Ditta, I.B.; Evans, P.; Foster, H.A.; Sheel, D.W.; Steele, A. Photo-induced self-cleaning and biocidal behaviour of titania and copper oxide multilayers. *J. Photochem. Photobiol. A* **2008**, *195*, 197–205. [CrossRef]
139. Yates, H.M.; Brook, L.A.; Sheel, D.W.; Ditta, I.B.; Steele, A.; Foster, H.A. The growth of copper oxides on glass by flame assisted chemical vapour deposition. *Thin Solid Films* **2008**, *517*, 517–521. [CrossRef]
140. Sahana, R. Setting Time, Compressive Strength and Microstructure of Geopolymer Paste. *Int. J. Innov. Res. Sci. Eng. Technol.* **2013**, *2*, 311–316.
141. Yahya, Z.; Abdullah, M.M.A.B.; Ramli, N.M.; Burduhos-Nergis, D.D.; Abd Razak, R. Influence of Kaolin in Fly Ash Based Geopolymer Concrete: Destructive and Non-Destructive Testing. *IOP Conf. Ser. Mater. Sci. Eng.* **2018**, *374*, 012068. [CrossRef]
142. Sassolini, A.; Malizia, A.; D'Amico, F.; Carestia, M.; Di Giovanni, D.; Cenciarelli, O.; Bellecci, C.; Gaudio, P. Evaluation of the effectiveness of titanium dioxide (TiO_2) self-cleaning coating for increased protection against cbrn incidents in critical infrastructures. *Def. S&T Tech. Bull.* **2014**, *7*, 9–17.
143. Al Bakri Abdullah, A.M.; Kamarudin, H.; Binhussain, M.; Nizar, K.; Mastura, W.I.W. Mechanism and Chemical Reaction of Fly Ash Geopolymer Cement—A Review. *Asian J. Sci. Res.* **2011**, *1*, 247–253.
144. Adewuyi, Y.G. Recent Advances in Fly-Ash-Based Geopolymers: Potential on the Utilization for Sustainable Environmental Remediation. *ACS Omega* **2021**, *6*, 15532–15542. [CrossRef]
145. Abbas, R.; Khereby, M.A.; Ghorab, H.Y.; Elkhoshkhany, N. Preparation of Geopolymer Concrete Using Egyptian Kaolin Clay and the Study of Its Environmental Effects and Economic Cost. *Clean Technol. Environ. Policy* **2020**, *22*, 669–687. [CrossRef]
146. Albidah, A.; Alghannam, M.; Abbas, H.; Almusallam, T.; Al-Salloum, Y. Characteristics of Metakaolin-Based Geopolymer Concrete for Different Mix Design Parameters. *J. Mater. Res. Technol.* **2021**, *10*, 84–98. [CrossRef]
147. Ionescu, B.A.; Lăzărescu, A.-V.; Hegyi, A. The Possibility of Using Slag for the Production of Geopolymer Materials and Its Influence on Mechanical Performances—A Review. *Proceedings* **2020**, *63*, 30.
148. Davidovits, J. Synthesis of new high-temperature Geopolymers for reinforced plastics and composites. In Proceedings of the PACTEC'79 Society of Plastics Engineers, Costa Mesa, CA, USA, 1 January–2 February 1979; pp. 151–154.
149. Assi, L.; Ghahari, S.; Deaver, E.; Leaphart, D.; Ziehl, P. Improvement of the early and final compressive strength of fly ash-based geopolymer concrete at ambient conditions. *Constr. Build. Mater.* **2016**, *123*, 806–813. [CrossRef]
150. De Silva, P.; Sagoe-Crenstil, K.; Sirivivatnanon, V. Kinetics of geopolymerisation: Role of Al_2O_3 and SiO_2. *Cem. Concr. Res.* **2007**, *37*, 512–518. [CrossRef]
151. Palomo, A.; Grutzeck, M.W.; Blanco, M.T. Alkali-Activated Fly Ashes. A Cement for the Future. *Cem. Concr. Res.* **1999**, *29*, 1323–1329. [CrossRef]
152. Duxson, P.; Lukey, G.C.; Van Deventer, J.S.J. Physical evolution of Na-geopolymer derived from metakaolin up to 1000 °C. *J. Mater. Sci.* **2007**, *42*, 3044–3054. [CrossRef]
153. Davidovits, J. Geopolymers of the first generation: SILIFACE-Process. In Proceedings of the Geopolymer'88, First European Conference on Soft Mineralogy, Compiegne, France, 1–3 June 1988; pp. 49–67.
154. Wallah, S. Drying shrinkage of heat-cured fly ash-based geopolymer concrete. *Mod. Appl. Sci.* **2009**, *3*, 14–21. [CrossRef]
155. Al Bakri Abdullah, A.M.; Hussin, K.; Bnuhussain, M.; Ismail, K.N.; Ahmad, M.I. Chemical Reactions in the Geopolymerisation Process Using Fly Ash-Based Geopolymer: A review. *AJBAS* **2011**, *5*, 1199–1203.
156. Skvara, F. *Alkali Activated Material—Geopolymer*; Department of Glass and Ceramics, Faculty of Chemical Technology, ICT Prague: Prague, Czech Republic, 2007; pp. 661–676.
157. Amran, M.; Fediuk, R.; Murali, G.; Avudaiappan, S.; Ozbakkaloglu, T.; Vatin, N.; Karelina, M.; Klyuev, S.; Gholampour, A. Fly Ash-Based Eco-Efficient Concretes: A Comprehensive Review of the Short-Term Properties. *Materials* **2021**, *14*, 4264. [CrossRef]
158. Buchwald, A. What are geopolymers? Current State of Research and Technology, The Opportunities They Offer, and Their Significance for The Precast Industry, Concrete Precasting Plant and Technology. *Betonw. Fert.-Tech.* **2006**, *72*, 42.
159. Xu, H.; van Deventer, J. The effect of alkali metals on the formation of geopolymeric gels from alkali-feldspats. *Colloids Surf. A Physicochem. Eng. Asp.* **2013**, *216*, 27–44. [CrossRef]
160. Weng, L.; Sagoe-Crentsil, K. Dissolution processes, hydrolysis and condensation reactions during geopolymer synthesis: Part I—Low Si/Al ratio systems. *J. Mater. Sci.* **2007**, *42*, 2997–3006. [CrossRef]
161. Andini, S.; Cioffi, R.; Colangelo, F.; Grieco, T.; Montagnaro, F.; Santoro, L. Coal fly ash as raw material for the manufacture of geopolymer-based product. *J. Waste Manag.* **2008**, *28*, 416–423. [CrossRef] [PubMed]
162. Duxon, P.; Fernande-Jimenez, A.; Provis, J.L.; Lukey, G.C.; Palomo, A.; van Deventer, J.S.J. Geopolymer technology: The current state of the art. *J. Mater. Sci.* **2007**, *42*, 2917–2933. [CrossRef]
163. Moreno, N.; Querol, X.; Andrés, J.M.; Stanton, K.; Towler, M.; Jurcovicova, M.; Jones, R. Physico-chemical characteristics of European pulverized coal combustion fly ashes. *Fuel* **2005**, *84*, 1351–1563. [CrossRef]
164. Chen-Tan, N.W.; Van Riessen, A.; Ly, C.V.; Southam, D. Determining the reactivity of a fly ash for production of geopolymer. *J. Am. Ceram. Soc.* **2009**, *92*, 881–887. [CrossRef]

165. Temuujin, J.; van Riessen, A.; Williams, R. Influence of calcium compounds on the mechanical properties of fly ash geopolymer pastes. *J. Hazard. Mater.* **2009**, *167*, 82–88. [CrossRef] [PubMed]
166. Nath, P.; Sarker, P.K.; Rangan, V.B. Early Age Properties of Low-calcium Fly Ash Geopolymer Concrete Suitable for Ambient Curing. *Procedia Eng.* **2015**, *125*, 601–607. [CrossRef]
167. Swanepoel, J.C.; Strydom, C.A. Utilisation of fly ash in a geopolymeric material. *Appl. Geochem.* **2002**, *17*, 114–148. [CrossRef]
168. Goretta, K.C.; Gutierrez-Mora, F.; Singh, D.; Routbort, J.L.; Lukey, G.C.; van Deventer, J.S.J. Erosion of geopolymers made from industrial waste. *J. Mater. Sci.* **2007**, *42*, 3066–3072. [CrossRef]
169. Puertas, F.; Martinez-Ramirez, S.; Alonso, S.; Vazquez, T. Alkali activated fly ash/slag cements: Strength behavior and hydration products. *Cem. Concr. Res.* **2000**, *30*, 1625–1632. [CrossRef]
170. Farhana, Z.; Kamarudin, H.; Rahmat, A.; Al Bakri, A.M. The Relationship between Water Absorption and Porosity for Geopolymer Paste. *Mater. Sci. Forum* **2014**, *803*, 166–172. [CrossRef]
171. Aly, M.; Hashmi, M.S.; Olabi, A.G.; Messeiry, M. Effect of colloidal nano-silica on the mechanical and physical behavior of waste-glass cement mortar. *Mater. Des.* **2012**, *33*, 127135. [CrossRef]
172. Khater, M.H. Effect of nano-silica on microstructure formation of low-cost geopolymer binder. *Nanocomposites* **2016**, *2*, 84–97. [CrossRef]
173. Khater, M.H. Physicomechanical properties of nano-silica effect on geopolymer composites. *J. Build. Mater. Struct.* **2016**, *3*, 1–14. [CrossRef]
174. Assaedi, H.; Shaikh, F.U.; Low, I.M. Effect of nanoclay on durability and mechanical properties of flax fabric reinforced geopolymer composites. *J. Asian Ceram. Soc.* **2017**, *5*, 62–70. [CrossRef]
175. Adak, D.; Sarkar, M.; Mandal, S. Effect of nano-silica on strength and durability of fly ash based geopolymer mortar. *Construct. Build. Mater.* **2014**, *70*, 453–459. [CrossRef]
176. Shaikh, F.U.; Supit, S.W.; Sarker, P.K. A study on the effect of nano silica on compressive strength of high volume fly ash mortars and concretes. *Mater. Des.* **2014**, *60*, 433–442. [CrossRef]
177. Zhaoheng, L.; Wei, Z.; Ruilan, W.; Fangzhu, C.; Xichun, J.; Peitong, G. Effects of Reactive MgO on the Reaction Process of Geopolymer. *Materials* **2019**, *12*, 526.
178. Hu, W.; Zhu, X.; Long, F. Alkali-activated fly ash-based geopolymers with zeolite or bentonite as additives. *Cem. Concr. Compos.* **2009**, *31*, 762–768. [CrossRef]
179. Bakharev, T. Geopolymeric materials prepared using Class F fly ash elevated temperature curing. *Cem. Concr. Res.* **2005**, *35*, 1224–1232. [CrossRef]
180. Atis, C.D.; Görür, E.B.; Karahan, O.; Bilim, C.; Ilkentapar, S.; Luga, E. Very high strength (120 MPa) Class F fly ash geopolymer mortar activated at different NaOH amount, heat curing temperature and heat curing duration. *Constr Build. Mater.* **2015**, *96*, 673–678. [CrossRef]
181. Al Bakri, M.M.; Mohammed, H.; Kamarudin, H.; Niza, K.; Zarina, Y. Review of Fly Ash-Based Geopolymer Concrete Without Portland Cement. *J. Eng. Technol.* **2011**, *3*, 1–4.
182. Hardjito, D.; Rangan, B.V. *Development and Properties of Low-Calcium Fly Ash-Based Geopolymer Concrete*; Technical Report GC1; Civil Engineering Faculty, Technical University: Perth, Australia, 2005.
183. Al Bakri Mustafa, A.M.; Kamarudin, H.; Binhussain, M.; Niza, I.K. The effect of curing temperature on physical and chemical properties of geopolymers. *Phys. Procedia* **2011**, *22*, 286–291.
184. Al Bakri Mustafa, A.M.; Kamarudin, H.; Bnhussain, M.; Nizar, I.K.; Rafiza, A.R.; Zarina, Y. The processing, characterization, and properties of fly ash based geopolymer concrete. *Rev. Adv. Mater. Sci.* **2012**, *30*, 90–97.
185. Chindaprasirt, P.; Chareerat, T.; Sirivivatnano, V. Workability and strength of coarse high calcium fly ash geopolymer. *Cem. Conc. Comp.* **2007**, *29*, 224–229. [CrossRef]
186. Morsy, M.S.; Alsaye, S.H.; Al-Salloum, Y.; Almusallam, T. Effect of sodium silicate to sodium hydroxide ratios on strength and microstructure of fly ash geopolymer binder. *Arab. J. Sci. Eng.* **2014**, *39*, 4333–4339. [CrossRef]
187. Álvarez-Ayuso, E.; Querol, X.; Plana, F.; Alastuey, A.; Moreno, N.; Izquierdo, M.; Font, O.; Moreno, T.; Diez, S.; Vasquez, K.; et al. Environmental, physical and structural characterisation of geopolymer matrixes synthesised from coal (co-)combustion fly ashes. *J. Hazard. Mater.* **2008**, *154*, 175–183. [CrossRef] [PubMed]
188. Hardjito, D.; Rangan, B.V. *Development and Properties of Low-Calcium Fly Ash-Based Geopolymer Concrete*; Technical Report GC2; Civil Engineering Faculty, Technical University: Perth, Australia, 2005.
189. Provis, J.L.; Yong, C.Z.; Duxson, P.; van Deventer, J. Correlating mechanical and thermal properties of sodium silicate-fly ash geopolymers. *Colloids Surf. A Physicochem. Eng.* **2009**, *336*, 57–63. [CrossRef]
190. Sumajouw, D.; Hardjito, D.; Wallah, S.; Rangan, B. Fly ash-based geopolymer concrete: Study of slender reinforced columns. *J. Mater. Sci.* **2007**, *42*, 3124–3130. [CrossRef]
191. Vora, P.; Dave, U. Parametric Studies on Compressive Strength of Geopolymer Concrete. *Procedia Eng.* **2013**, *51*, 210–219. [CrossRef]
192. Sindhunata; van Deventer, J.S.J.; Lukey, G.C.; Xu, H. Effect of Curing Temperature and Silicate Concentration on Fly-Ash-Based Geopolymerization. *Ind. Eng. Chem. Res.* **2006**, *45*, 3559–3569. [CrossRef]
193. Raijiwala, D.; Patil, H. Geopolymer concrete: A green concrete. In Proceedings of the 2nd International Conference on Chemical, Biological and Environmental Engineering (ICBEE 2010), Cairo, Egypt, 2–4 November 2010.

194. Mishra, A.; Choudhary, D.; Jain, N.; Kumar, M.; Sharda, N.; Dutta, D. Effect of concentration of alkaline liquid and curing time on strength and water absorption of geopolymer concrete. *J. Eng. Appl. Sci.* **2008**, *3*, 14–18.
195. Alonso, S.; Palomo, A. Alkaline activation of metakaolin and calcium hydroxide mixtures: Influence of temperature, activator concentration and solids ratio. *Mater. Lett.* **2001**, *47*, 55–62. [CrossRef]
196. Memon, F.; Nuruddin, M.F.; Khan, S.H.; Shafiq, N.R. Effect of sodium hydroxide concentration on fresh properties and compressive strength of self-compacting geopolymer concrete. *J. Eng. Sci. Technol.* **2013**, *8*, 44–56.
197. Barbosa, V.; Mackenzie, K.; Thaumaturgo, C. Synthesis and characterisation of sodium polysialate inorganic polymer based on alumina and silica. In Proceedings of the Geopolymer'99 International Conference, Saint-Quentin, Geopolymer Institute, Saint-Quentin, France, 30 June–2 July 1999.
198. Luhar, S.; Dave, U. Investigations on mechanical properties of fly ash and slag based geopolymer concrete. *Ind. Concr. J.* **2016**, 34–41.
199. Ma, Y.; Hu, J.; Ye, G. The effect of activating solution on the mechanical strength, reaction rate, mineralogy, and microstructure of alkali-activated fly ash. *J. Mater. Sci.* **2012**, *47*, 4568–4578. [CrossRef]
200. Xie, J.; Yin, J.; Chen, J.; Xu, J. Study on the geopolymer based on fly ash and slag. *Energy Environ.* **2009**, *3*, 578–581.
201. van Jaarsveld, J.; van Deventer, J.; Lukey, G. The effect of composition and temperature on the properties of fly ash-and kaolinite-based geopolymers. *J. Chem. Eng.* **2002**, *89*, 63–73. [CrossRef]
202. Rovnaník, P. Effect of curing temperature on the development of hard structure of metakaolin-based geopolymer. *Constr. Build. Mater.* **2010**, *24*, 1176–1183. [CrossRef]
203. Chindaprasirt, P.; Chareerat, T.; Hatanaka, S.; Cao, T. High-strength geopolymer using fine high-calcium fly ash. *Mater. Civ. Eng.* **2010**, *23*, 264–270. [CrossRef]
204. Fernández-Jiménez, A.; Garcia-Lodeiro, I.; Palomo, A. Durability of alkali-activated fly ash cementitious materials. *J. Mater. Sci.* **2007**, *42*, 3055–3065. [CrossRef]
205. Phoo-ngernkham, T.; Sinsiri, T. Workability and compressive strength of geopolymer mortar from fly ash containing diatomite. *Eng. J.* **2011**, *38*, 11–26.
206. Kong, D.; Sanjayan, J. Effect of elevated temperatures on geopolymer paste, mortar and concrete. *Cem. Concr. Res.* **2010**, *40*, 334–339. [CrossRef]
207. Guo, X.; Shi, H. Self-solidification/stabilization of heavy metal wastes of class C fly ash-based geopolymers. *J. Mater. Civ. Eng.* **2012**, *25*, 491–496. [CrossRef]
208. Rashad, A.; Zeedan, S. The effect of activator concentration on the residual strength of alkali-activated fly ash pastes subjected to thermal load. *Constr. Build. Mater.* **2011**, *25*, 3098–3107. [CrossRef]
209. Sukmak, P.; Horpibulsuk, S.; Shen, S. Strength development in clay-fly ash geopolymer. *Contr. Build. Mater.* **2013**, *40*, 566–574. [CrossRef]
210. Taebuanhuad, S.; Rattanasak, U.; Jenjirapanya, S. Strength behavior of fly ash geopolymer with microwave pre-radiation curing. *J. Ind. Technol.* **2012**, *8*, 1–8.
211. Lăzărescu, A.V.; Szilagyi, H.; Baeră, C.; Ioani, A. Parameters Affecting the Mechanical Properties of Fly Ash-Based Geopolymer Binders–Experimental Results. *IOP Conf. Ser. Mater. Sci. Eng.* **2018**, *374*, 012035. [CrossRef]
212. Hardjito, D.; Rangan, B.V. *Development and Properties of Low-Calcium Fly Ash-Based Geopolymer Concrete*; Technical Report GC3; Civil Engineering Faculty, Technical University: Perth, Australia, 2006.
213. Fernandez-Jimenez, A.; Palomo, A. Composition and microstructure of alkali activated fly ash binder: Effect of the activator. *Cem. Concr. Res.* **2005**, *35*, 1984–1992. [CrossRef]
214. Omar, O.M.; Heniegal, A.M.; Abd Elhameed, G.D.; Mohamadien, H.A. Effect of Local Steel Slag as a Coarse Aggregate on Properties of Fly Ash Based-Geopolymer Concrete. *Int. J. Civ. Environ.* **2015**, *3*, 1452–1460.
215. Perera, D.S.; Uchida, O.; Vance, E.R.; Finnie, K.S. Influence of curing schedule on the integrity of geopolymers. *J. Mater. Sci.* **2007**, *42*, 3099–3106. [CrossRef]
216. Davidovits, J. Properties of geopolymer cements. In Proceedings of the First International Conference on Alkaline Cements and Concretes, Kiev, Ukraine, 11–14 October 1994; SRIBM, Kiev State Technical University: Kiev, Ukraine, 1994; pp. 131–149.
217. Sikora, S.; Gapys, E.; Michalowski, B.; Horbanowicz, T.; Hynowski, M. Geopolymer coating as protection of concrete against chemical attack and corrosion. *E3S Web Conf.* **2018**, *49*, 00101. [CrossRef]
218. Cassar, L.; Pepe, C.; Tognon, G.; Guerrini, G.L.; Amadelli, R. White Cement for Architectural Concrete Possessing Photocatalytic Properties. In Proceedings of the 11th International Congress on the Chemistry of Cement, Durban, South Africa, 11–16 May 2003.
219. Andaloro, A.; Mazzucchelli, E.S.; Lucchini, A.; Pedeferri, M.P. Photocatalytic self-cleaning coating for building façade maintainance. Performance analysis through a case-study application. *J. Façade Des. Eng.* **2016**, *4*, 115–129.
220. Guerrero, L.E.; Gómez-Zamorano, L.; Jiménez-Relinque, E. Effect of the addition of TiO_2 nanoparticles in alkali-activated materials. *Constr. Build. Mater.* **2020**, *245*, 118370. [CrossRef]
221. Yang, X.; Liu, Y.; Yan, C.; Peng, R.; Wang, H. Geopolymer-TiO_2 Nanocomposites for Photocsatalysis: Synthesis by One-Step Adding Treatment Versus Two-Step Acidification Calcination. *Minerals* **2019**, *9*, 658. [CrossRef]
222. Singh, B.; Ishwaraya, G.; Gupta, M.; Bhattacharyya, S.K. Review: Geopolymer concrete: A review of some recent developments. *Constr. Build. Mater.* **2015**, *85*, 78–90. [CrossRef]

223. Provis, J.L.; Van Deventer, J.S.J. *Geopolymers: Structure, Processing, Properties and Industrial Applications*; Woodhead Publishing Limited: Cambridge, UK, 2009.
224. Ma, C.K.; Awang, A.Z.; Omar, W. Structural and material perfromance of geopolymer concrete: A review. *Constr. Build Mater.* **2018**, *186*, 90–102. [CrossRef]
225. Davidovits, J. Geopolymers: Inorganic polymeric new materials. *J. Therm. Anal.* **1991**, *37*, 1633–1656. [CrossRef]
226. Abdel-Gawwad, H.A.; Khalil, K.A. Application of thermal treatment on cement kiln dust and feldspar to create one-part geopolymer cement. *Constr. Build. Mater.* **2018**, *187*, 231–237. [CrossRef]
227. Fernandez-Jimenez, A.; Palomo, A. Characterization of fly ashes. Potential reactivity as alkaline cements. *Fuel* **2003**, *82*, 2259–2265. [CrossRef]
228. Malek, R.I.A.; Roy, D.M. Structure and properties of alkaline activated cementitious materials. In Proceedings of the 97th Annual Meeting and the 1995 Fall Meetings of the Materials & Equipment and Whitewares Divisions, Cincinnati, OH, USA, 30 April–3 May 1995.
229. Provis, J.L.; Brice, D.G.; Buchwald, A.; Duxson, P.; Kavalerova, E.; Krivenko, P.V.; Shi, C.; van Deventer, J.S.J.; Wiercx, J.A. *Demonstration Projects in Building and Civil Infrastructure*; Alkali-Activated Materials: State-of-the-Art Report; RILEM TC 224-AAM; Springer: Dordrecht, The Netherlands, 2014; pp. 309–338.
230. Hoy, M.; Horpibulsuk, S.; Arulrajah, A. Strength development of recycled asphalt pavement—Fly ash geopolymer as a road construction material. *Constr. Build. Mater.* **2016**, *117*, 209–219. [CrossRef]
231. Murgod, G.; Shetty, K.; Raja, A. Self-consolidating paving grade geopolymer concrete. *IOP Conf. Ser. Mater. Sci. Eng.* **2018**, *431*, 92006.
232. Moutinho, S.; Costa, C.; Cerqueira, A.; Rocha, F.; Velosa, A. Geopolymers and polymers in the conservatiopn of tile facades. *Constr. Build. Mater.* **2019**, *197*, 175. [CrossRef]
233. Salwa, M.; Al-Bakri Mustafa, M.M.; Kamarudin, H.; Ruzaidi, C.; Binhussain, M.; Syed Zuber, S.Z. Review on current geopolymer as a coating material. *Aust. J. Basic Appl. Sci.* **2013**, *7*, 246–257.
234. Aguirre-Guerreo, A.M.; Robayo-Salazar, R.A.; de Gutierrez, R.M. A novel geopolymer application: Coatings to protect reinforced concrete against corrosion. *Appl Clay Sci.* **2017**, *135*, 437–446. [CrossRef]
235. Zhang, J.; Tian, B.; Wang, L.; Xing, M.; Lei, J. Mechanism of Photocatalysis. In *Photocatalysis. Lecture Notes in Chemistry*; Springer: Singapore, 2018; Volume 100.
236. Rabajczyk, A.; Zielecka, M.; Klapsa, W.; Dziechciarz, A. Self-Cleaning Coatings and Surfaces of Modern Building Materials for the Removal of Some Air Pollutants. *Materials* **2021**, *14*, 2161. [CrossRef] [PubMed]
237. Joshi, N.C.; Gururani, P.; Gairola, S.P. Metal Oxide Nanoparticles and Their Nanocomposite-Based Materials as Photocatalysts in the Degradation of Dyes. *Biointerface Res. Appl. Chem.* **2022**, *12*, 6557–6579.
238. Burduhos Nergis, D.D.; Vizureanu, P.; Ardelean, I.; Sandu, A.V.; Corbu, O.C.; Matei, E. Revealing the Influence of Microparticles on Geopolymers' Synthesis and Porosity. *Materials* **2020**, *13*, 3211. [CrossRef] [PubMed]
239. Meor Ahmad Tajudin, M.A.F.; Abdullah, M.M.A.B.; Sandu, A.V.; Nizar, K.; Moga, L.; Neculai, O.; Muniandy, R. Assessment of Alkali Activated Geopolymer Binders as an Alternative of Portland Cement. *Mater. Plastice.* **2017**, *54*, 145–154.
240. Vizureanu, P.; Samoila, C.; Cotfas, D. Materials Processing using Solar Energy. *Environ. Eng. Manag. J.* **2009**, *8*, 301–306. [CrossRef]
241. Azimi, E.A.; Abdullah, M.M.A.B.; Vizureanu, P.; Salleh, M.A.A.M.; Sandu, A.V.; Chaiprapa, J.; Yoriya, S.; Hussin, K.; Aziz, I.H. Strength Development and Elemental Distribution of Dolomite/Fly Ash Geopolymer Composite under Elevated Temperature. *Materials* **2020**, *13*, 1015. [CrossRef] [PubMed]
242. Burduhos Nergis, D.D.; Vizureanu, P.; Corbu, O. Synthesis and Characteristics of Local Fly Ash Based Geopolymers Mixed with Natural Aggregates. *Rev. De Chim.* **2019**, *70*, 1262–1267. [CrossRef]
243. Krishnan, U.; Sanalkumar, A.; Yahg, E.-H. Self-cleaning performance of nano-TiO_2 modified metakaolin-based geopolymers. *Cem. Concr. Res.* **2021**, *115*, 103847.
244. *ASTM C1437-07*; Standard Test Metod for Flow of Hydrauloc Cement Mortar. ASTM International: West Conshohocken, PA, USA, 2007.
245. Duan, P.; Yan, C.; Luo, W.; Zhou, W. Effects of adding nano-TiO_2 on compressive strength, drying shrinkage, carbonation and microstructure of fluidized bed fly ash based geopolymer paste. *Constr. Build. Mater.* **2016**, *106*, 115–125. [CrossRef]
246. Zulkifly, K.; Heah, C.Y.; Liew, Y.M.; Abdullah, M.M.A.B.; Abdullah, S.F.A. The Synergetic Compressive Strength and Microstructure of Fly Ash and Metakaolin Blend Geopolymer Pastes. *AIP Conf. Proc.* **2018**, *2045*, 020100.
247. Syamsidar, D. The properties of nano TiO_2-geopolymer composite as a material for functional surface application. In *MATEC Web of Conferences*; EDP Sciences: Les Ulis, France, 2017; Volume 97, p. 01013.
248. Guzmán-Aponte, L.A.; de Gutiérrez, R.M.; Maury-Ramírez, A. Metakaolin-Based Geopolymer with Added TiO_2 Particles: Physicomechanical Characteristics. *Coatings* **2017**, *7*, 233. [CrossRef]
249. Sastry, K.G.K.; Sahitya, P.; Raviteha, A. Influence of nano TiO_2 on strength and durability properties of geopolymer concrete. *Mater. Today Proc.* **2020**, *45*, 1017–1025. [CrossRef]
250. Subaer; Haris, A.; Noor Afifah, K.; Akifah, N.; Zulwiyati, R. Thermo-Mechanical Properties of Geopolymer/Carbon Fiber/TiO_2 Nanoparticles (NPs) Composite. *Mater. Sci. Forum* **2019**, *967*, 267–273. [CrossRef]
251. Bonilla, A.; Villaquirán-Caicedo, M.A.; Mejía de Gutiérrez, R. Novel Alkali-Activated Materials with Photocatalytic and Bactericidal Properties Based on Ceramic Tile Waste. *Coatings* **2022**, *12*, 35. [CrossRef]

252. Sakai, N.; Fujishima, A.; Watanabe, T.; Hashimoto, K. Enhancement of the photoinduced hydrofilic conversion rate of TiO_2 film electrode surfaces by anodic polarization. *J. Phys. Chem. B.* **2001**, *105*, 3023–3026. [CrossRef]
253. Takeuchi, M.; Sakamoto, K.; Martra, G.; Coluccia, S.; Anpo, M. Mechanism of photoinduced superhydrophilicity on the TiO_2 photocatalyst surface. *J. Phys. Chem. B.* **2005**, *109*, 15422–15428. [CrossRef] [PubMed]
254. Loh, K.; Gaylarde, C.C.; Shirakawa, M.A. Photocatalytic Activity of ZnO and TiO_2 'Nanoparticles' for Use in Cement Mixes. *Constr. Build. Mater.* **2018**, *167*, 853–859. [CrossRef]
255. Zailan, S.N.; Bouaissi, A.; Mahmed, N. Influence of ZnO Nanoparticles on Mechanical Properties and Photocatalytic Activity of Self-cleaning ZnO-Based Geopolymer Paste. *J. Inorg. Organomet. Polym.* **2020**, *30*, 2007–2016. [CrossRef]
256. Min Li, C.; He, Y.; Tang, Q.; Tuo Wang, K.; Min Cui, X.; Min Li, C.; He, Y.; Tang, Q.; Tuo Wang, K.; Min Cui, X. Study of the preparation of CdS on the surface of geopolymer spheres and photocatalyst performance. *Mater. Chem. Phys.* **2016**, *178*, 204–210.
257. Gasca-Tirado, J.R.; Manzano-Ramırez, A.; Villasenor-Mora, C.; Muniz-Villarreal, M.S.; Zaldivar-Cadena, A.A.; Rubio-Avalos, J.C.; Borras, V.A.; Mendoza, R.N. Incorporation of photoactive TiO_2 in an aluminosilicate inorganic polymer by ion exchange. *Microporous Mesoporous Mater.* **2012**, *153*, 282–287. [CrossRef]
258. Zhang, Y.; Liu, L. Fly ash-based geopolymer as a novel photocatalyst for degradation of dye from wastewater. *Particuology* **2013**, *11*, 353–358. [CrossRef]
259. Luhar, I.; Luhar, S.; Abdullah, M.M.A.B.; Razak, R.A.; Vizureanu, P.; Sandu, A.V.; Matasaru, P.-D. A State-of-the-Art Review on Innovative Geopolymer Composites Designed for Water and Wastewater Treatment. *Materials* **2021**, *14*, 7456. [CrossRef] [PubMed]
260. Kaya-Özkiper, K.; Uzun, A.; Soyer-Uzun, S. Red Mud- and Metakaolin-Based Geopolymers for Adsorption and Photocatalytic Degradation of Methylene Blue: Towards Self-Cleaning Construction Materials. *J. Clean. Prod.* **2021**, *288*, 125120. [CrossRef]
261. Chen, L.; Zheng, K.; Liu, Y. Geopolymer-supported photocatalytic TiO_2 film: Preparation and characterization. *Constr. Build. Mater.* **2017**, *151*, 63–70. [CrossRef]
262. Saufi, H.; el Alouani, M.; Alehyen, S.; el Achouri, M.; Aride, J.; Taibi, M. Photocatalytic Degradation of Methylene Blue from Aqueous Medium onto Perlite-Based Geopolymer. *Int. J. Chem. Eng.* **2020**, *2020*, 9498349. [CrossRef]
263. Jdm, K.; Yusoff, M.M.; Aqilah, N.S. Degradation of Methylene Blue via Geopolymer Composite Photocatalyst. *Solid State Sci. Technol.* **2013**, *21*, 23–30.
264. Yang, J.; Wang, F.; Du, D.; Liu, P.; Zhang, W.; Hu, S. Enhanced photocatalytic efficiency and long-term performance of TiO_2 in cementitious materials by activated zeolite fly ash bead carrier. *Constr. Build. Mater.* **2016**, *126*, 886–893. [CrossRef]
265. El Alouani, M.; Alehyen, S.; El Achouri, M.; Taibi, M. Preparation, Characterization, and Application of Metakaolin-Based Geopolymer for Removal of Methylene Blue from Aqueous Solution. *J. Chem.* **2019**, *2019*, 4212901. [CrossRef]
266. Strini, A.; Roviello, G.; Ricciotti, L.; Ferone, C.; Messina, F.; Schiavi, L.; Corsaro, D.; Cioffi, R. TiO_2-Based Photocatalytic Geopolymers for Nitric Oxide Degradation. *Materials* **2016**, *9*, 513. [CrossRef] [PubMed]
267. Wang, L.; Geddes, D.A.; Walkley, B.; Provis, J.L.; Mechtcherine, V.; Tsang, D.C.W. The Role of Zinc in Metakaolin-Based Geopolymers. *Cem. Concr. Res.* **2020**, *136*, 106194. [CrossRef]
268. Qin, Y.; Fang, Z.; Chai, X.; Cui, X. A Superhydrophobic Alkali Activated Materials Coating by Facile Preparation. *Coatings* **2022**, *12*, 864. [CrossRef]
269. Chindaprasirt, P.; Jitsangiam, P.; Pachana, P.K. Self-cleaning superhydrophobic fly ash geopolymer. *Sci. Rep.* **2023**, *13*, 44. [CrossRef]
270. Permatasari, A.D.; Fahira, N.; Husna Muslimin, N.; Subaer. Development of Photoactive Nano TiO_2 Thin Film-Geopolymer Based on Laterite Soils Deposit Gowa Regency as Self-Cleaning Materia. *Mater. Sci. Forum* **2019**, *967*, 274–280. [CrossRef]
271. Vancea, D.P.C.; Kamer-Ainur, A.; Simion, L.; Vanghele, D. Export expansion policies. An analysis of Romanian exports between 2005–2020 using Principal Component Analysis method and short recommendations for increasing this activity. *Transform. Bus. Econ.* **2021**, *20*, 614–634.
272. Kamer-Ainur, A.; Munteanu, I.F.; Stan, M.-I.; Chiriac, A. A Multivariate Analysis on the Links Between Transport Noncompliance and Financial Uncertainty in Times of COVID-19 Pandemics and War. *Sustainability* **2022**, *14*, 10040.
273. Batrancea, L.M.; Pop, M.C.; Rathnaswamy, M.M.; Batrancea, I.; Rus, M.-I. An Empirical Investigationon the Transition Process toward a Green Economy. *Sustainability* **2021**, *13*, 13151. [CrossRef]
274. Batrancea, L.; Rathnaswamy, M.M.; Rus, M.I.; Tulai., H. Determinants of Economic Growth for the Last Half Century: A Panel Data Analysis on 50 Countries. *J. Knowl. Econ.* **2022**, 1–25. [CrossRef]
275. Yao, Q.; Jahanshahi, H.; Batrancea, L.M.; Alotaibi, N.D.; Rus, M.-I. Fixed-Time Output-Constrained Synchronization of Unknown Chaotic Financial Systems Using Neural Learning. *Mathematics* **2022**, *10*, 3682. [CrossRef]
276. Cheng, G.; Xu, F.; Xiong, J.; Tian, F.; Ding, J.; Stadler, F.J.; Chen, R. Enhanced adsorption and photocatalysis capability of generally synthesized TiO_2-carbon materials hybrids. *Adv. Powder Technol.* **2016**, *27*, 1949–1962. [CrossRef]
277. Tian, H.; Shen, K.; Hu, X.; Qiao, L.; Zheng, W. N, S co-doped graphene quantum dots-graphene-TiO_2 nanotubes composite with enhanced photocatalytic activity. *J. Alloys Compd.* **2017**, *691*, 369–377. [CrossRef]
278. Razzaq, A.; Grimes, C.A.; In, S. Il Facile fabrication of a noble metal-free photocatalyst: TiO_2 nanotube arrays covered with reduced graphene oxide. *Carbon* **2016**, *98*, 537–544. [CrossRef]

279. Lettieri, S.; Gargiulo, V.; Pallotti, D.K.; Vitiello, G.; Maddalena, P.; Alfè, M.; Marotta, R. Evidencing opposite charge-transfer processes at TiO_2/graphene-related materials interface through combined EPR, photoluminescence and photocatalysis assessment. *Catal. Today* **2018**, *315*, 19–30. [CrossRef]
280. Andreozzi, M.; Álvarez, M.G.; Contreras, S.; Medina, F.; Clarizia, L.; Vitiello, G.; Llorca, J.; Marotta, R. Treatment of saline produced water through photocatalysis using rGO-TiO_2 nanocomposites. *Catal. Today* **2018**, *315*, 194–204. [CrossRef]
281. Khalid, N.R.; Majid, A.; Tahir, M.B.; Niaz, N.A.; Khalid, S. Carbonaceous-TiO_2 nanomaterials for photocatalytic degradation of pollutants: A review. *Ceram. Int.* **2017**, *43*, 14552–14571. [CrossRef]
282. Hanus, M.J.; Harris, A.T. Nanotechnology innovations for the construction industry. *Prog. Mater. Sci.* **2013**, *58*, 1056–1102. [CrossRef]
283. Tsai, S.-J.; Cheng, S. Effect of TiO_2 crystalline structure in photocatalytic degradation of phenolic contaminants. *Catal. Today* **1997**, *33*, 227–237. [CrossRef]
284. Folli, A.; Macphee, D. Photocatalytic Concretes–The interface between photocatalysis and cement chemistry. In Proceedings of the 33rd Cement and Concrete Science Conference, Portsmouth, UK, 2–3 September 2013.
285. Cassar, L. Photocatalysis of cementitious materials: Clean buildings and clean air. *Mrs Bull.* **2004**, *29*, 328–331. [CrossRef]
286. Bellardita, M.; Di Paola, A.; Megna, B.; Palmisano, L. Determination of the crystallinity of TiO_2 photocatalysts. *J. Photochem. Photobiol. A Chem.* **2018**, *367*, 312–320. [CrossRef]
287. Jimenez-Relinque, E.; Rodriguez-Garcia, J.R.; Castillo, A.; Castellote, M. Characteristics and efficiency of photocatalytic cementitious materials: Type of binder, roughness and microstructure. *Cem. Concr. Res.* **2015**, *71*, 124–131. [CrossRef]
288. Addamo, M.; Augugliaro, V.; Bellardita, M.; Di Paola, A.; Loddo, V.; Palmisano, G.; Palmisano, L.; Yurdakal, S. Environmentally friendly photocatalytic oxidation of aromatic alcohol to aldehyde in aqueous suspension of brookite TiO_2. *Catal. Lett.* **2008**, *126*, 58–62. [CrossRef]
289. Jang, H.D.; Kim, S.-K.; Kim, S.-J. Effect of particle size and phase composition of titanium dioxide nanoparticles on the photocatalytic properties. *J. Nanopart. Res.* **2001**, *3*, 141–147. [CrossRef]

Disclaimer/Publisher's Note: The statements, opinions and data contained in all publications are solely those of the individual author(s) and contributor(s) and not of MDPI and/or the editor(s). MDPI and/or the editor(s) disclaim responsibility for any injury to people or property resulting from any ideas, methods, instructions or products referred to in the content.

Article

The Influence of Blast Furnace Slag on Cement Concrete Road by Microstructure Characterization and Assessment of Physical-Mechanical Resistances at 150/480 Days

Liliana Maria Nicula [1,2,*], Daniela Lucia Manea [1,*], Dorina Simedru [3], Oana Cadar [3], Anca Becze [3] and Mihai Liviu Dragomir [1]

1. Faculty of Civil Engineering, Technical University of Cluj-Napoca, 28, Memorandumului Street, 400114 Cluj-Napoca, Romania; mihai.dragomir@cfdp.utcluj.ro
2. Faculty of Construction, Cadastre and Architecture, University of Oradea, 4, B.S. Delavrancea Street, 410058 Oradea, Romania
3. INCDO-INOE2000, Subsidiary Research Institute for Analytical Instrumentation Cluj-Napoca, 67 Donath Street, 400293 Cluj-Napoca, Romania; dorina.simedru@icia.ro (D.S.); oana.cadar@icia.ro (O.C.); anca.naghiu@icia.ro (A.B.)
* Correspondence: liliana.nicula@infra.utcluj.ro (L.M.N.); daniela.manea@ccm.utcluj.ro (D.L.M.)

Abstract: The results presented in this paper on the appropriateness of using of blast furnace slag (BFS) in the composition of roads make an original contribution to the development of sustainable materials with the aim to reduce the carbon footprint and the consumption of natural resources. The novelty of this work consists of determining the optimal percentage of BSF in road concrete, in order to: increase mechanical resistances, reduce contractions in the hardening process, and ensure increased corrosion resistances, even superior to classic cement-based mixtures. Thus, the physical-mechanical characteristics and the microstructure of some road concretes were studied in the laboratory for three different recipes. We kept the same amount of ground granulated blast furnace slag (GGBS) as a substitute for Portland cement, respectively three percentages of 20%, 40%, 60% air-cooled blast furnace slag (ACBFS) and crushed as sand substitute from now on called S54/20, S54/40, S54/60. Drying shrinkage, mechanical resistances, carbonation-induced corrosion, microstructure characterization of hardened concretes, and degree of crystallinity by SEM and XRD measurements were analyzed after a longer curing period of 150/480 days. The obtained results on the three BSF mixtures indicated a reduction of drying shrinkage and implicitly increased the tensile resistance by bending to 150 days well above the level of the blank composition. The degree of crystallinity and the content of the majority phases of the mineralogical compounds, albites, quartz, and tobermorite out of the three BSF samples justifies the increase in the compressive strengths at the age of 480 days in comparison with the test samples. Scanning electron microscope (SEM) and X-ray diffraction measurements showed the highest compactness and lowest portlandite crystal content for the S54/20 slag composite. Future research concerns are the realization of experimental sections in situ, the study of the influence of BFS on the elasticity module of road concrete, and the opportunity to use other green materials that can contribute to the reduction of the carbon footprint, keeping the physical and mechanical properties of road concrete at a high level.

Keywords: concrete shrinkage; mechanical resistances; carbonation; concrete microstructure; ground granulated blast furnace slag (GGBS); air cooled blast furnace slag (ACBFS); cement concrete roads

Citation: Nicula, L.M.; Manea, D.L.; Simedru, D.; Cadar, O.; Becze, A.; Dragomir, M.L. The Influence of Blast Furnace Slag on Cement Concrete Road by Microstructure Characterization and Assessment of Physical-Mechanical Resistances at 150/480 Days. *Materials* **2023**, *16*, 3332. https://doi.org/10.3390/ma16093332

Academic Editor: Baoguo Han

Received: 9 March 2023
Revised: 18 April 2023
Accepted: 21 April 2023
Published: 24 April 2023

Copyright: © 2023 by the authors. Licensee MDPI, Basel, Switzerland. This article is an open access article distributed under the terms and conditions of the Creative Commons Attribution (CC BY) license (https://creativecommons.org/licenses/by/4.0/).

1. Introduction

The construction industry has developed numerous measures to reduce the greenhouse gas emissions associated with cement production [1,2]. Moreover, the hydraulic cement manufacturing process is responsible for approximately 7–9% of global carbon dioxide emissions, on par with emissions from fuel combustion [3–5]. There are several materials resulting as byproducts or industrial waste that can be used as multi-component binders in cement mixtures to minimize the carbon footprint [3,6–14]. Using industrial byproducts such as GGBS from blast furnace iron ore extraction as a substitute for cement can reduce greenhouse gas (GHG) emissions by 47.5% [15,16]. The global environmental emission factor for producing of one ton of GGBS is 0.143 t CO_2-e/ton, well below the value of 0.91 t CO_2-e/ton for cement, which includes the cement transporter to the plants for the concrete mixture preparation [17]. Globally, slag production (GGBS) is almost 530 million tons, of which only 65% is absorbed by the construction industry [18,19]. Research has been carried out on the influence of GGBS on the performance of some types of concrete and mortar [3,20,21]. The research done in the study [22] shows that the use of an optimal combination of microsilica (MS) and GGBS can improve the resistance characteristics of concretes compared to the individual use of these additional cementitious materials. The experimental investigations within the paper [23] show that geopolymer concrete with GGBS as the main binder and substitution of slag powder with 20% micronized biomass silica (MBS) made from rice husk without Portland cement achieved optimal resistance and durability performance. The effects on the durability characteristics of waste glass-derived nanopowder (WGBNP) with the inclusion of fly ash (FA) and ground blast furnace slag (GBFS), evaluated on alkali-activated mortars (AAM), led to improved durability performances through reduced drying shrinkage and increased resistance to sulfuric acid, wear, and freeze-thaw cycles [24]. According to other research, the optimal dosage for replacing cement with GGBS is limited to max. 20%, because the resistances decreased significantly above this level compared to the reference concrete, due to low workability and increased porosity [15,25].

The impact of GGBS mixed with cement develops properties in fresh and hardened concrete, such as workability, reduced bleeding of fresh concrete, and hydration heat, increases long-term resistance, and increased resistance to corrosion, porosity, and low permeability [26–29]. The study [3] shows an increase in the workability of concrete up to a 40% substitution level with GGBS. Increased consistency is due to better particle dispersion (GGBS) [30]. More cement paste minimizes internal friction between concrete components by filling the micro-spaces in the concrete aggregate, resulting in more workable concrete [25]. Developing tensile strength, the main characteristic of road concrete, requires the careful establishment of design parameters for any structural element that requires crack control [31]. An important clue for determining the cracking resistance of concrete results from the evolution of shrinkage and the maximum level of shrinkage. The drier the air and the higher the temperature, the stronger the shrinkage [32]. Materials such as cement, water, and aggregates significantly influence concrete shrinkage. Cement leads to increased drying shrinkage by increasing dosage, tricalcium aluminate C3A component, gel component, alkali content of cement, deficiency, and excess of gypsum influence shrinkage curing [33]. The volume contraction of the cement paste represents approx. 1% of the absolute volume of dry cement [33,34]. Currently, the shrinkage has a value below 0.6 mm/m; this can be exceeded in concretes rich in binders [35]. Some authors appreciate the evolution of concrete shrinkage as follows: after ½ month 5% is recorded, after 3 months 60%, and after a year 75% of its maximum value [32]. The increase in the A/C ratio leads to an increase in concrete shrinkage, because the number of concrete pores also increases. Increasing the amount of aggregates reduces shrinkage due to their nature, rigidity, and granularity [36]. An increased concentration of CO_2 and a humidity greater than 50% causes an increase in shrinkage to a value more than double. Carbon dioxide decalcifies and dehydrates hydrosilicates of the C_2S type [35]. Increasing CO_2 concentrations in the external environment also increases the carbonation rate for permeable concretes [37]. The carbonation phenomenon lowers the pH value from the typical values (12–13) to less than 9 in the pore solution and destroys the passivity of the reinforcing bars embedded in the concrete, triggering the corrosion process [38–40].

Another significant impact on the environment is the consumption of aggregates since they represent the largest share of the mass and volume of concrete. It is estimated that worldwide demand for construction aggregates is over 10 billion tons annually [41,42]. Much research explores the durability of concrete containing recycled aggregates (RAC) from building demolition [43,44], from road asphalt pavements (RAP) [45–47], quarry sand (QS) [48], or ecological mortars in which natural aggregates have been replaced by glass waste [49].

A sustainable source for substituting natural aggregates in road concrete is air-cooled crushed blast furnace slag (ACBFS) [50,51]. There are research studies related to the use of blast furnace slag as a substitute for natural aggregates in asphalt mixtures leading to these conclusions [52]. The Japanese Guide using blast furnace slag aggregates for concrete structures recommends using fine aggregates mixed with natural fine aggregates at a ratio between 20 and 60% [53]. In our country, since 2003, the characteristics of blast furnace slag aggregates used for concrete have been covered by the SR EN 12620 standard [54]. These byproducts from the steel industry can be used in road concrete mixes, but require proper evaluation to ensure that their properties do not adversely affect fresh and hardened concrete [55]. Applying reusable artificial materials and new technologies will positively impact on the environment by protecting non-renewable materials and reducing production costs [56,57].

The objective of this study is to continue investigations into the physical and mechanical characteristics of road concrete compositions with blast furnace slag made in previous works [58,59], to establish the optimal percentages of BSF to improve the quality of durability in concrete road production. It is known that cements containing blast furnace slag are characterized by a slower hydration rate, lower hydration heat, higher resistance after longer hardening periods, and greater resistance to chemical aggression [60]. The curing mechanism specific to concretes containing SCMs, such as GGBS, makes it possible to measure reference mechanical resistances after the age of 56 days [32]. Most concrete research with GGBS monitors curing times at 56 days [25] and 90 days [3] and fewer up to 360 days [61]. The relative humidity of the environment has a great influence on the size of the contraction. Thus, for a relative humidity of 100%, contraction decreases with age, while for a relative humidity of 50–70% (specific to road concretes that are kept in the air), the contraction can increase up to 20 years [32]. Variations in drying contraction influence the development over time of mechanical resistances (stretch and compression) and durability such as carbonation depth. This study aims to examine the influence of powdered blast furnace slag (GGBS) on the road concretes, analyzing the evolution of contraction and long-term resistance, 150/480 days. More than most previous research, road concrete compositions added fine aggregates (ACBFS) made of blast furnace slag. Consequently, tests were carried out on the samples made in the laboratory to evaluate some physical and mechanical properties (shrinkage on drying, tensile strength by bending at 150 days, and compressive strength at 480 days), durability (corrosion from carbonation at 150 and 480 days old), and microstructural properties (X-ray diffraction and SEM scanning electron microscopy at 480 days old). Mechanical resistance values obtained were reported at the reference age of 28 days.

2. Materials and Methods

2.1. Materials

As binder, CEM I 42.5R cement acquired from Holcim Romania was used according to SR EN 197-1 [62] and GGBS from local sources, Galati Steel Mill (in Romanian). The 28-day activity index of granulated slag, amounting to 0.95, was taken from the manufacturer's tests. By grinding BSF to a size smaller than 63 µm, slag activation by grinding fineness was pursued [63–65]. The GGBS powder recorded after grinding a specific surface area of 3775 cm^2/g lower than that of Portland cement of 4385 cm^2/g. The GGBS characteristics were analyzed according to the system (CaO-SiO_2-Al_2O_3-MgO), in compliance with SR EN 15167:1 [66], from the XRF spectral analysis by the manufacturer. The requirements of this standard require that the sum of the masses of CaO + MgO + SiO_2 be greater than 2/3 and the mass ratio $(CaO + MgO)/(SiO_2) > 1$. The values recorded in Table 1 show that these requirements are met.

Table 1. Oxidative analysis of granulated slag (GGBS).

GGBS	SiO_2	Al_2O_3	MnO	MgO	CaO	Fe_2O_3	Na_2O	K_2O
(%)	36.70	9.50	0.23	8.70	42.00	0.55	0.28	0.53

The aggregate (ACBFS) crushed to a size of 0/4 mm was produced by the Galati Steel Mill. XRD measurements performed on the aggregates (ACBFS) in the paper [58] indicated the percentage of 100% crystalline phase, the danger of disaggregation being removed [35]. The fineness modulus (Mf), evaluated in Table 2, resulting as the sum of the total percentages retained on the site series, placed the aggregates (ACBFS) in the category of large-grained sands with values between 2.4 and 4.0 and fine natural aggregate (NA) in the category of sand with medium grains having Mf between 1.5 and 2.8, in compliance with SR EN 12620 [54]. The water absorption coefficient WA_{24} for natural fine aggregates and ACBFS brought to the condition SS with the saturated surface and to the condition SSD with the dry saturated surface recorded the values in Table 2.

Table 2. Fineness modulus (Mf) and water absorption coefficient (WA_{24}) of aggregates of size 0/4 mm [59].

Aggregate Mixture	Mf	WA_{24} (SS)	WA_{24} (SSD)
100% (NA)	2.72	20% (water saturated)	2% (after 4 days air cooled)
100% (ACBFS)	3.15	30% (water saturated)	2% (after 8 days air cooled)

The selection of natural aggregates and the granulometric curve of the total mixture was in accordance with the requirements of the national standard NE 014 [67]. River sand (0/4), crushed river gravel (4/8), and crushed quarry screening were used for sorting (8/16) and (16/25) acquired from local sources (Balastiera Beclănuț and Bologa quarry). When preparing the concrete, the superplasticizer additive Master-Glenium SKY 527, (SP MG-SKY 527) and the air trainer additive Master Air 9060 (MA 9060) were added, having characteristics in compliance with SR EN 934-2 [68]. The additives (Ad) used were acquired from the Master Builders Solutions Romania group. The water was taken from the supply system of the city of Cluj-Napoca, the characteristics being in compliance with SR EN 1008 [69].

2.2. Concrete Mixtures

Five mixtures were made, of which the first two compositions were made with conventional materials, with Portland cement, CEM I 42.5R, and natural aggregates. In three mixtures, GGBS and the ACBFS aggregates were added. A quantity of 54 kg/m^3 of slag (GGBS) was used as an addition in the percentage of 15% of the mass of cement in the control composition S360, as a substitute in the percentage of 13% of the mass of cement in the control composition S414. The sand (NA) was substituted in percentages of 20%, 40%, and 60% with crushed aggregates (ACBFS) with a size of 0/4 mm.

The amounts of materials per m^3 are summarized in Table 3, and the abbreviations for mixtures are symbolically noted as follows.

In the preparation of slag concretes, the added water content and additives were helped to achieve consistency within the range 20–40 mm, workability being an important feature of road concretes.

Table 3. Quantities of materials per m^3.

Mixtures (kg/m^3)	S360	S414	S54/20	S54/40	S54/60
Cement	360	414	360	360	360
(GGBS)	-	-	54	54	54
Total binder (l)	360	414	414	414	414
Water (w)	166.47	174.39	172.91	181.23	167.64
w/l	0.46	0.42	0.42	0.44	0.41
(NA_0/4 mm)	607	594	478	355	240
(ACBFS_0/4 mm)	-	-	119	237	359
(CA_4/25 mm)	1290	1261	1269	1256	1275
(SP MG-SKY 527)	3.60	4.14	4.39	4.55	4.97
(MA 9060),	1.80	2.07	2.07	2.07	2.07

The design parameters of road concrete classified in the BcR 5.0 class followed the requirements set out in NE 014 [67], presented in Table 4, and the obtained results are presented in detail in the paper [59].

Table 4. The design parameters of road concrete composites in compliance with NE 014.

Cement Dosage	(w/l)	Consistency	Occluded Air Content	fcm 28 Days	fcfm 28 Days
min. 360 kg/m^3	max. 0.45	(30 ± 10) mm	(3.5 ± 0.5)%	min. 50 MPa	min. 5.5 MPa

2.3. Methods

2.3.1. The Drying Shrinkage of Road Concretes with Blast Furnace Slag

The shrinkage of hardened concrete was measured in compliance with SR 2833 [70], with the help of the Huggenberger deformer, the landmarks being located at a distance of 250.50 mm. The shrinkage measurement was carried out on prisms with dimensions of 150 × 150 × 600 mm^3. The samples were kept in water until the age of 7 days, then in air until the age of 150 days at a humidity of 65 ± 5% and a temperature of 20 ± 2 °C, as in Figure 1a,b. The initial reading was taken at 7 days, followed by further readings at 14, 28, 42, 56, 90, and 150 days. At each test age, the shrinkage was calculated as the arithmetic mean of the values obtained on three samples on three test tubes, applying (Equation (1)):

$$\varepsilon_{ci} = \frac{\delta_0 - \delta_i}{l} \left(\frac{mm}{m}\right), \tag{1}$$

In which:

ε_{ci}—shrinkage of the hardened concrete, in mm/m;
δ_0—initial reading at 7 days old (standard) with deformer, in mm;
δ_i—reading at the age of i days, with deformer, in mm;
l—distance between landmarks, in mm.

2.3.2. Tensile Strengths by Bending, Compression, and Carbonatation

The determination of mechanical resistances was carried out on three prism test tubes, with dimensions of 150 × 150 × 600 mm^3, respectively on three cubes with a side of 150 mm for each composite. The prismatic samples for determining the bending tensile strength in compliance with SR EN 12390-5 [71] were the same as those monitored for drying shrinkage evaluation up to the age of 150 days, images in Figure 1c. The compressive strengths were determined in compliance with SR EN 12390-3 [72], after a longer curing period, at the age of 480 days, images Figure 1d. The evolution of the mechanical resistances was compared to those at the reference age of 28 days, found in the paper [59]. The determination of the resistance to carbonation at 150/480 days was performed on the freshly crushed faces of the three cubes and prisms remaining after the mechanical tests from each composite. According to the methodology of SR CR 12793 [73], they were sprayed with 1% phenolphthalein solution to measure the depth of the carbonation layer dk (mm).

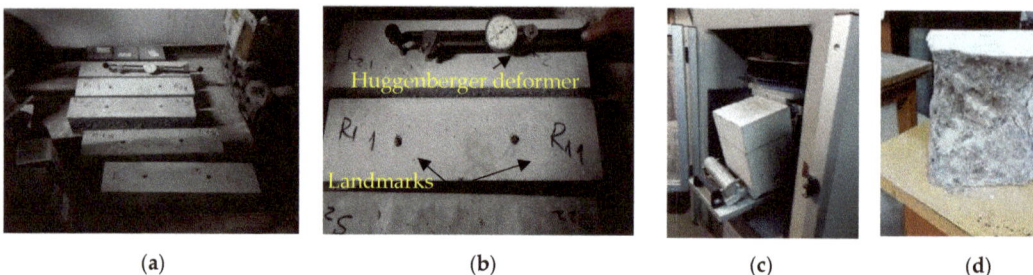

Figure 1. (**a**,**b**) Pictures of drying shrinkage measurement; (**c**) Images after the bending tensile test of the 150 × 150 × 600 mm prism; (**d**) Images after the compression test on the 150 mm cube.

2.3.3. Characterization of the Microstructure of Road Concrete

The XRD patterns were recorded using a D8 Advance diffractometer (Bruker, Karlsruhe, Germany) with Ni-filtered CuK α1 radiation of λ = 1.54060 Å wavelength, operating at 40 kV and 40 mA, at room temperature. The degree of crystallinity was determined as the ratio between the area of diffraction peaks and the total area of diffraction peaks and halos.

SEM-EDX scanning electron microscope measurements were performed on small samples at the age of 480 days. The SEM-EDX analysis was performed at room temperature using a scanning electronic microscope (VEGAS 3 SBU, Tescan, Brno-Kohoutovice, Czech Republic) with a Quantax EDX XFlash (Bruker, Karlsruhe, Germany) detector. Samples of ~4 mm^2 were mounted with carbon tape on an SEM stub. For each composite, XRD-SEM-EDX measurements were performed on a single sample.

3. Results

3.1. The Drying Shrinkage of Road Concretes with Blast Furnace Slag

In the same conservation conditions (7 days in the humid environment and the rest in the air with controlled humidity), Table 5 shows the values of the contraction upon curing (ε) up to the age of 150 days.

Table 5. Curing shrinkage of road concretes up to the age of 480 days.

Curing Shrinkage	S 360	S 414	S 54/20	S 54/40	S 54/60
ε (mm/m)—14 days	0.039	0.057	0.040	0.057	0.051
ε (mm/m)—28 days	0.065	0.112	0.064	0.104	0.076
ε (mm/m)—42 days	0.083	0.134	0.083	0.125	0.092
ε (mm/m)—56 days	0.097	0.150	0.098	0.141	0.105
ε (mm/m)—90 days	0.104	0.161	0.106	0.154	0.114
ε (mm/m)—120 days	0.110	0.172	0.113	0.165	0.122
ε (mm/m)—150 days	0.116	0.178	0.121	0.174	0.129

3.2. Tensile Strengths by Bending and Compression

Tensile flexural strengths at 150 days, compression at 480 days, standard deviation (SD), and coefficient of variation (CoV) of mechanical strengths are given in Table 6.

Table 6. Tensile flexural strengths at 150 days, compression at 480 days, standard deviation (SD), and coefficient of variation (CoV) of mechanical strengths.

Mixture	S 360	S 414	S 54/20	S 54/40	S 54/60
f_{cfm} 150 days (MPa)	6.06	5.77	6.57	5.78	6.31
SD-f_{cfm} (MPa)	0.33	0.35	0.19	0.28	0.29
CoV-f_{cfm} (%)	0.05	0.06	0.03	0.05	0.05
f_{cm} 480 days (MPa)	72.5	80.97	83.44	81.56	87.00
SD-f_{cm} (MPa)	2.94	4.65	4.27	1.10	4.48
CoV-f_{cm} (%)	0.04	0.06	0.05	0.01	0.05

3.3. Corrosion Resistances from Carbonation

Figure 2a presents photo images of fragments from prisms tested at the age of 150 days. In Figure 2b, fragments from cubes were tested at the age of 480 days after one hour of spraying with phenolphthalein solution in a concentration of 1%.

Figure 2. Samples after one hour of spraying with 1% phenolphthalein solution; (**a**) Fragments from prisms tested at the age of 150 days; (**b**) Fragments from cubes tested at the age of 480 days.

3.4. Characterization of the Microstructure of Road Concrete

3.4.1. X-ray Diffraction

The XRD diffraction patterns of the samples S360, S414, S54/20, S54/40, and S54/60 are presented in Figure 3.

XRD patterns showed the existence of quartz (SiO_2), portlandite ($Ca(OH)_2$), ettringite ($Ca_6Al_2(SO_4)_3(OH)_{12} \cdot 26H_2O$), calcium silicate hydrate ($CaSiO_3 \cdot H_2O$), and albite ($NaAlSi_3O_8$). The RIR (Reference Intensity Ratio) method was used for the quantitative phase analysis of the samples investigated at 480 days (Table 7).

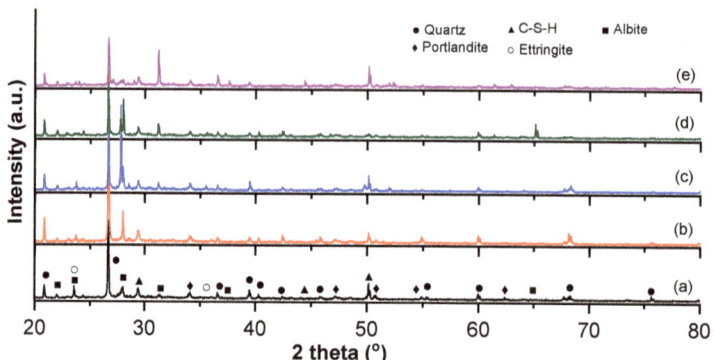

Figure 3. Diffraction patterns of the samples (**a**) S360; (**b**) S414; (**c**) S54/20; (**d**) S54/40; (**e**) S54/60.

Table 7. Quantitative analysis results obtained by the RIR method (%) of the samples investigated at 480 days.

Sample	S 360	S 414	S 54/20	S 54/40	S 54/60
Degree of crystallinity (%)	77	75	69	74	71
Amorphous phase (%)	23	25	31	26	29
Albite (Ab)	++	++	+++	+++	+++
Quartz	+++	+++	+++	++	+++
Tobermorites (C-S-H)	++	++	+++	+	+++
Portlandite (CH)	+	+	+	+	+
Ettringite (C-A-S-H)	+	+	+	+	+

+++ major phase (>20%), ++ minor phase (5–10%), + phases in traces (<5%).

3.4.2. SEM-EDX Scanning Electron Microscopy Measurements

SEM investigations, at sizes from 20 μm to 500 μm, resulted in the surface topography for the compositions S360, S414, S54/20, S54/40 and S54/60, aged 480 days. Table 8 shows the pore size measured on the studied samples.

Table 8. Pore size was measured on the studied samples from the compositions S360, S414, S54/20, S54/40, and S54/60.

Sample	Pore Identification Code	Pore Radius (μm)	Pore Diameter (μm)	Distance Identification Code Qi (Ci-Ci+n)	Distance (μm)
S360	C1	2.61	5.22	Q1 (C1–C2)	7.95
	C2	1.75	3.51		
S414	C1	13.69	23.79		
	C2	11.05	22.10	Q1 (C1–C3)	288.16
	C3	22.76	45.51	Q2 (C2–C4)	44.47
	C4	6.58	13.17	Q3 (C1–C2)	77.69
	C5	10.11	20.21	Q4 (C5–C6)	203.85
	C6	6.74	13.48		
S54/20	C1	8.09	16.19	Q1 (C1–C2)	380.79
	C2	6.44	12.89		
S54/40	C1	31.41	62.83	-	-
S54/60	C1	6.60	13.20	Q1 (C1–C2)	76.22
	C2	4.67	9.34	Q2 (C1–C3)	28.77
	C3	6.52	13.05		

EDX was used to map the surface of the measured samples, the results for the concentration of the identified elements are presented in Table 9.

Table 9. Element concentrations (%) in S360, S414, S54/20, S54/40, and S54/60 obtained by mapping the sample surface.

Mixture	O	Ca	Si	Al	Ca/Si
S360	53.30	22.37	21.92	2.41	0.42
S414	50.83	4.45	30.44	9.07	0.15
S54/20	54.18	28.54	14.54	2.74	1.96
S54/40	55.57	21.68	19.31	3.44	1.12
S54/60	68.79	27.02	2.10	2.10	12.86

4. Discussion

4.1. The Drying Shrinkage of Road Concretes with Blast Furnace Slag

The highest shrinkage value was recorded for the S414 composite, 0.178 mm/m, which represents 63.33% of the maximum shrinkage value, respectively 0.283 mm/m, below the allowed value of 0.6 mm/m according to the principle presented in the paper [32,35], Figure 4a,b. The lowest shrinkage corresponds to the composite S360 in the amount of 0.116 mm/m. This result is justified by a reduced dosage of cement 360 kg/m^3 compared to the dosage of 414 kg/m^3 used for the rest of the composites. It is observed, for all composites, that the shrinkage value decreases with the decrease of the water content in the mixtures, recorded in Table 3.

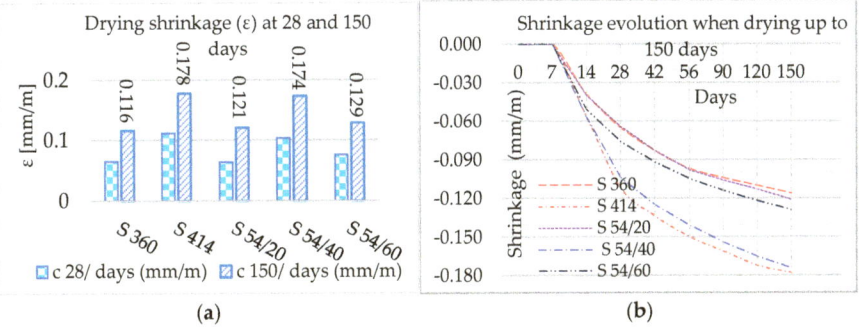

Figure 4. (a) Drying shrinkage at 28 and 150 days; (b) Evolution of drying shrinkage up to 150 days.

Reducing the amount of cement by substitution with GGBS reduced shrinkage in slag composites. Moreover, from the pozzolanic reaction of GGBS with calcium hydroxide (CH), a lower drying shrinkage value resulted in samples with GGBS compared to the control mixture with the same binder dosage [12]. Increasing the amount of water while reducing the volume of aggregates in the S54/40 composite increased the shrinkage value compared to the S54/20 and S54/60 composites. The shrinkage mitigation of S54/20, S54/40, and S54/60 composites compared to S414 can be justified by the grinding fineness of GGBS with the specific surface (3775 cm^2/g) lower than the specific surface of Portland cement (4385 cm^2/g). Finer ground cements react more energetically with water and develop greater shrinkage after setting through stronger hydration and greater increase in the amount of gels [32].

The slag aggregates (ACBFS) also influenced the concrete shrinkage through grain size, porosity, and angularity greater than fine sand (NA) [74]. The ACBFS aggregates have larger grains than (NA) and higher porosity. After 24 h of immersion in water, the absorption coefficient was 10% higher in ACBFS aggregates compared to NA, the results in Table 2. Due to the crushing process, the angularity is higher in ACBFS aggregates than in NA. The larger grain size, shape, and porosity of ACBFS aggregates compared to NA lead to higher water absorption, in agreement with NA [75]. To avoid early shrinkage and reduction of freeze-thaw resistance, the slag aggregates must be brought to the SSD, a saturated state with the dry surface at the time of concrete preparation, as well as the use of a water-reducing admixture and an air-entraining admixture [76], conditions met in the laboratory for the compositions in this experiment.

4.2. Tensile Strengths by Bending and Compression

The tensile strength results show acceptable values between 3 and 6% for CoV and deviations between 0.19 and 0.35 MPa, of max. 6.06% for SD. For the compressive strengths, the coefficient of variation (CoV) of 4.65 MPa and the standard deviation (SD) of 0.6% with the highest value was recorded for the control composite S414. It can be appreciated that the dispersion of the results in Table 6 has a reasonable quality in a range from 1 to 6%, below the allowed limit of 15%, indicated in the paper [77].

Figure 5a presents the diagram of the tensile strengths at the age of 150 days and the evolution coefficients related to the reference resistances at 28 days. There is a more pronounced growth in the control composite S360 (1.23) and the composite with the furnace slag S54/20 (1.19) compared to the rest of the composites, in which the ratio was within the range 1.10–1.11. The decrease in tensile strength of the composite S414 (5.77 MPa) and S54/40 (5.78 MPa) has been largely influenced by the increase of shrinkage, caused by the superficial microcracking that develops especially at high cement dosages [32]. In Figure 5b, it is observed in the composites with high dosage of cement 414 kg/m^3 a decrease in tensile strength with the increase of contraction, the function of the power between the two features registered a very good correlation coefficient ($R^2 = 0.98$), confirming results also found in the paper [78]. Substitution (NA) with aggregate ACBFS led to increased tensile strength in the composite S54/20 (6.57 MPa) and S54/60 (6.31 MPa). The use of aggregate (ACBFS) increased adhesion to cement paste and tensile strength, because the asperities of the crushed aggregate surface are greater than those of NA [76].

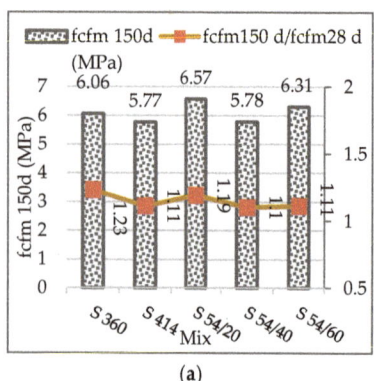
(a)

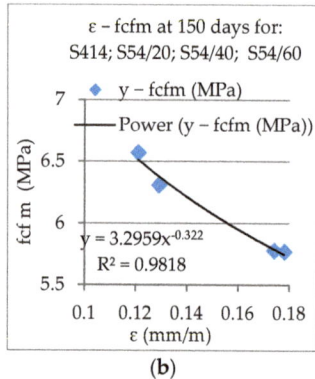
(b)

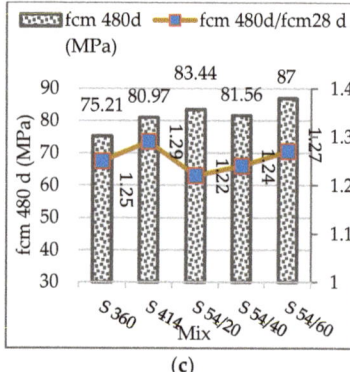
(c)

Figure 5. (a) The tensile strengths at 150 days and evolution coefficients from 28 to 150 days; (b) the relationship between contraction and tensile strength at 150 days; (c) compression strength at 480 days and evolution coefficients from 28 to 480 days.

Figure 5c presents the compression resistance diagram and evolution coefficients related to the reference resistances at 28 days. The highest compression resistance is observed for composite S54/60, closer for S54/20, and lower for S54/40, but the values registered are above the level of the control composites S360 and S414, at the age of 480 days. As expected, the evolution coefficients of mechanical resistances at the age of 480 are higher than those at 150 days compared to the reference attempts at 28 days. According to other studies, compression resistances have improved considerably after the age of 56 and 90 days of hardening for the concrete that contained GGBS. The improvement of compression resistances is attributed to the pozzolanic reaction of GGBS, which continues gradually compared to hydration (OPC) [79–81].

4.3. Corrosion Resistances in Carbonation

The photo images in Figure 2a,b show that the samples were not affected by the carbonation up to the age of 480 days, as the color of the indicator solution (red-purple) remained uniform throughout the surface of the cement stone, from the inner region to the outer edges of the samples. These results suggest that the diffusion of carbon dioxide has not occurred in the cement matrix, a phenomenon prevented by the compactness of the concrete cement stone [37,82].

4.4. Characterization of the Microstructure of Road Concrete

4.4.1. X-ray Diffraction

Albite (Ab) is a component of the plagioclase feldspar family, which is often present in the siliceous mineral aggregates that make up concrete [83] and in cement-based mortars containing recycled fine aggregates [84]. Ab is the major crystalline component for composites with blast furnace slag S54/20, S54/40, and S54/60, while for the control samples S360 and S414, quartz (SiO_2) becomes the main crystalline phase. The C-S-H product, which influences the increase of mechanical resistance and impermeability [76,85], registers a high frequency for the S54/20 and S54/60 composites, a reduced frequency for the control (S360 and S414) compositions, and a low frequency for the S54/40 composite (Table 7). This evolution is consistent with the mechanical resistance results presented above. The CH product does not influence the mechanical resistances, but the addition of SCM in the cement paste can form additional C-S-H [86]. The ettringite (C-A-S-H) affects concrete structures due to the formation of secondary ettringite, which is expansive within the material [87]. After a longer curing period, the products (CH and C-A-S-H) display lower frequency. It is known that the presence of the amorphous phase leads to the development of a high pozzolanic activity [88]. The amorphous phase presented as the difference from the crystalline phase suggests the highest pozzolanic activity in the composites S54/20 and S54/60, followed by S54/40, all being above the level of control samples S414, a possible explanation being the addition of GGBS (Figure 6a). The slag (GGBS) reacts slowly in the presence of water but becomes reactive in the presence of calcium hydroxide (CH) in the pore solution of hydrated Portland cement [76]. For the same cement dosage, the compressive strengths increase with the decrease of the crystalline phase according to a second-order polynomial relationship, having a very good correlation coefficient ($R^2 = 0.96$) (Figure 6b).

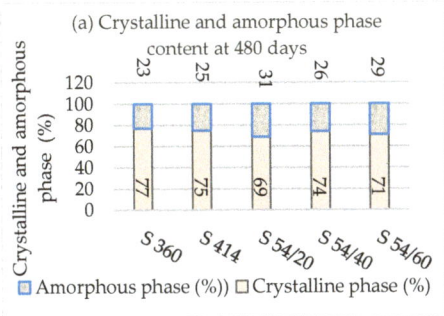
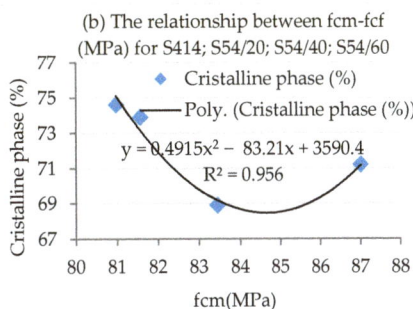

Figure 6. (a) The crystalline and amorphous phase content at 480 days of age: (b) Relationship between compressive strength and crystalline phase of S414, S54/20, S54/40, S54/60 samples at 480 days.

4.4.2. Measurements with Electronic Microscopy with SEM-EDX Scanning

The topography, pore size, and morphology of S360, S414, S54/20, S54/40, and S54/60 samples at advanced age (480 days) were studied by SEM-EDX and are shown in Figure 7. For a more complex characterization of the pore structure, some authors use methods such as mercury intrusion porosimetry (MIP) and fractal dimensions [89–91]. Still, for this work, the SEM technique was used because it provides information about the surface structure and several information about the porosity of the sample up to the minimum size at which one can be identified. As can be observed, their surfaces are irregular and inhomogeneous with large pores characteristic of air holes (>several µm) [92]. The presence of several characteristic mineral phases can be observed. Figure 7a–e show higher magnification images of examined samples, giving more information about the structure on the surface of the sample. S360 (Figure 7a) has large pores filled with ettringite needles and calcium hydroxide platelets [93,94]. With increasing Portland cement content, S414, the presence of well-defined portlandite crystals [94] can be observed at high magnification. In the case of S54/20 and S54/40, the addition of slag leads to a decrease of portlandite crystals, on the surface of which only small crystals can be observed. For S54/60, the dendritic growth of $CaCO_3$ [94] can be observed in addition to the calcium hydroxide platelets. SEM analysis is consistent with XRD patterns showing maximum peaks for portlandite at S360 and S54/60.

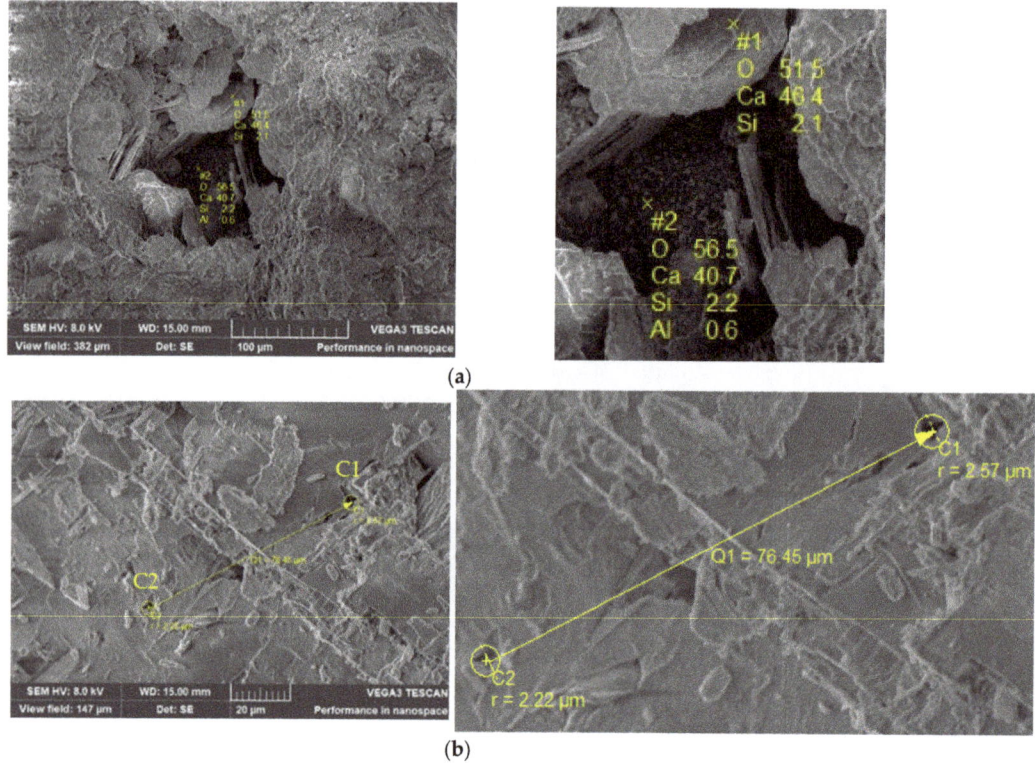

Figure 7. *Cont.*

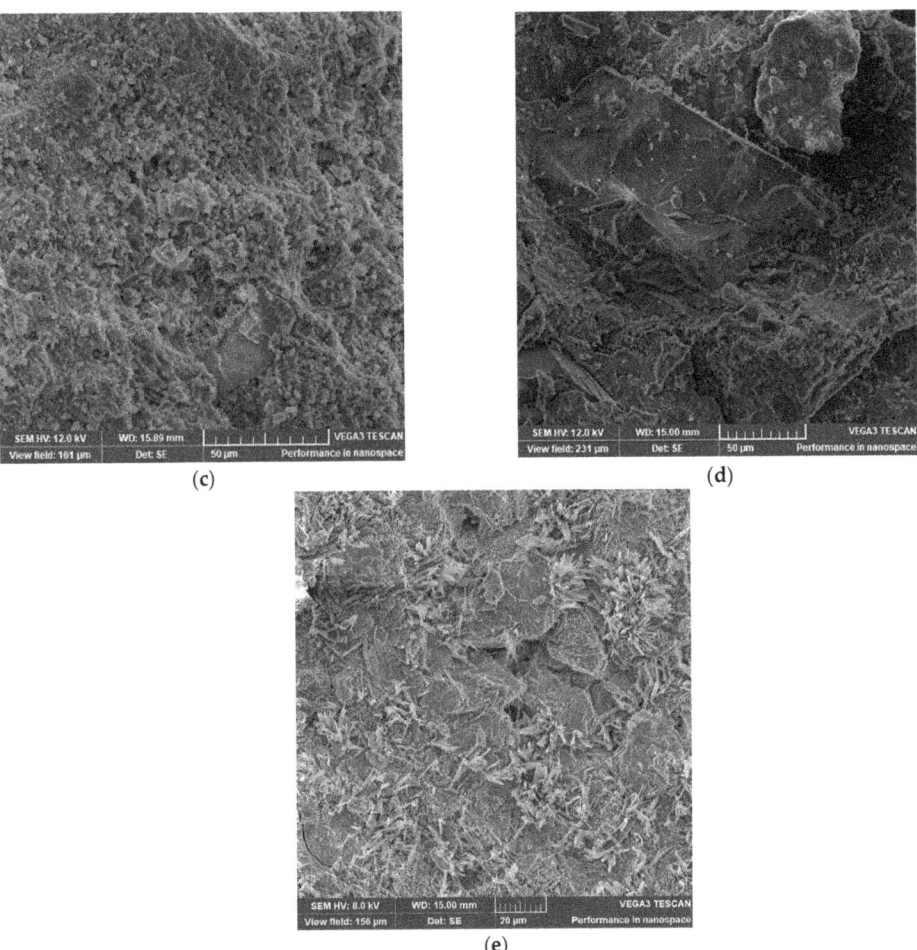

Figure 7. SEM images of (**a**) S360; (**b**) S414; (**c**) S54/20; (**d**) S54/40; (**e**) S54/60.

Knowing that the pore radius and spacing demonstrate the samples' compactness, a comparison was made using the data in Table 8 and Figure 8a–e. The smaller pore radius (characteristic of air holes) is for samples S360 and S54/60 which show the highest content of portlandite crystals. The highest spacing between pores was obtained for S54/20, which shows the lowest content of portlandite crystals. The results indicate a correlation between the samples' compactness and the portlandite crystals' content. Portlandite being the most soluble constituent, it is the easiest to wash with water, it affects the cement stone through corrosion. The presence of a higher portlandite content in S360 and S54/60 results in smaller pores located at smaller a distance between them and thus a higher pore density. A lower quantity of portlandite content is observed in S54/20 which has higher pores at higher distance between them.

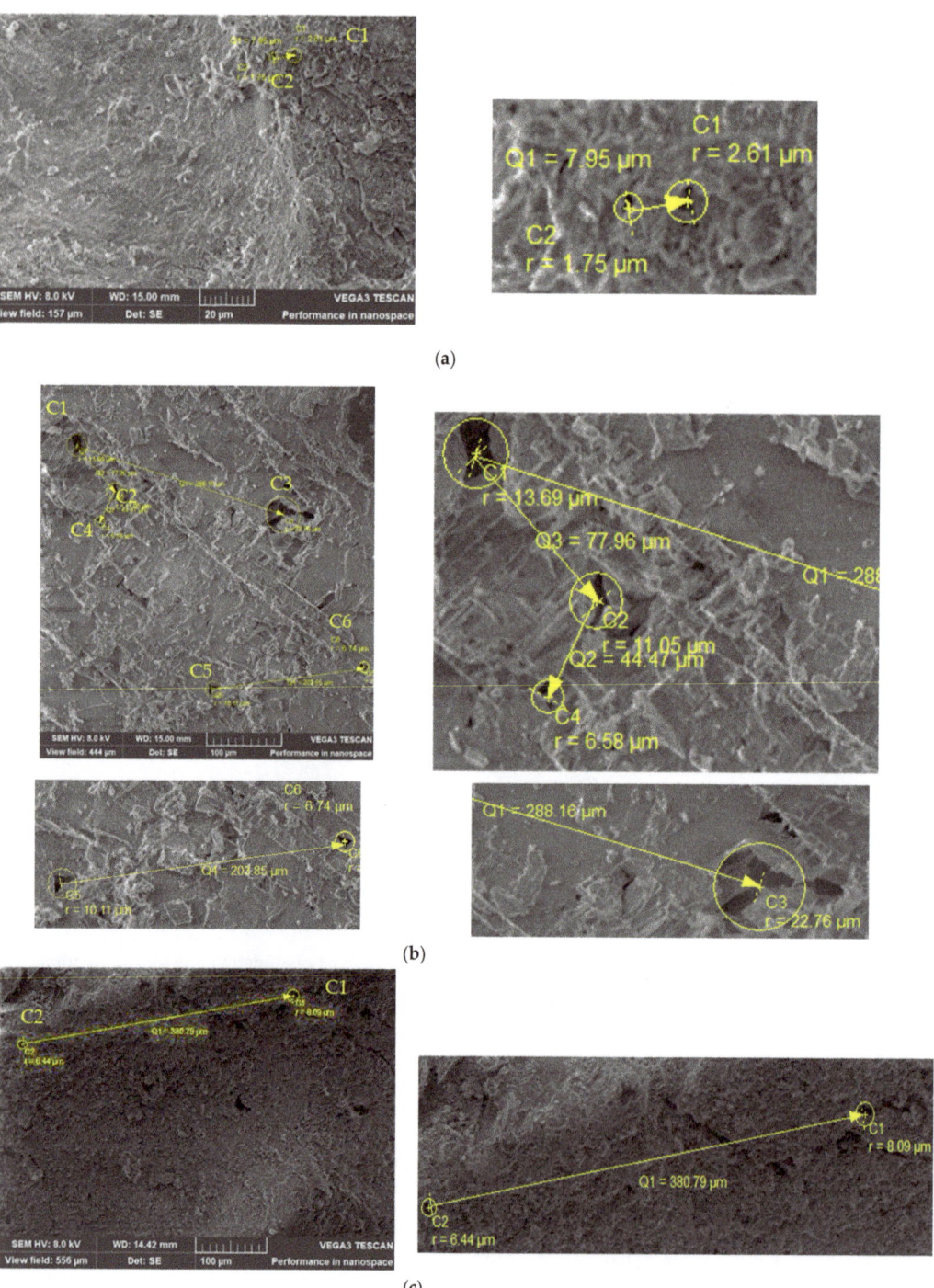

Figure 8. *Cont.*

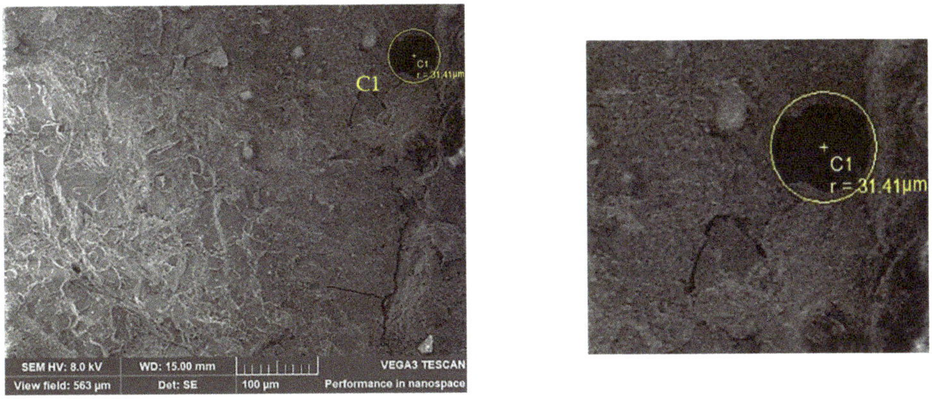

(d)

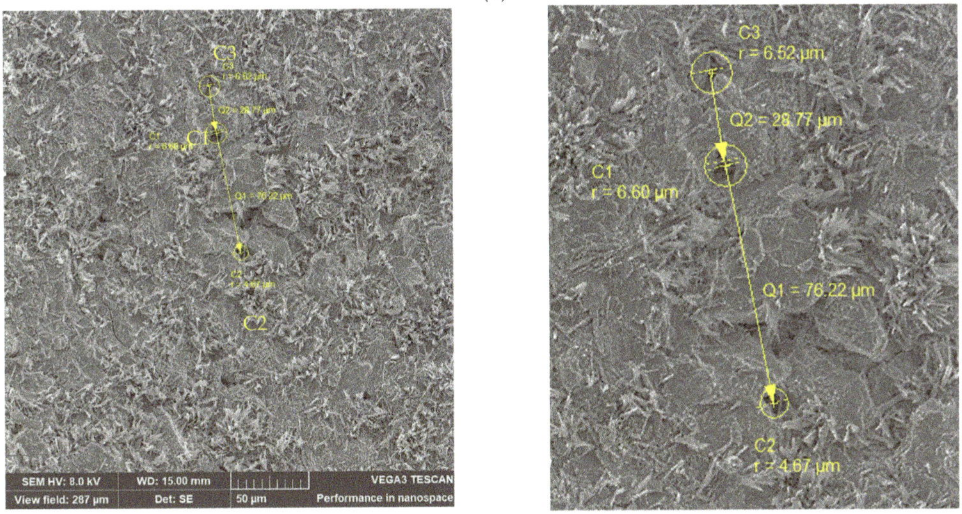

(e)

Figure 8. Pore structure of (**a**) S360; (**b**) S414; (**c**) S54/20; (**d**) S54/40; (**e**) S54/60 samples.

Figure 9a–j show the map surface and spectra of S360, S414, S54/20, S54/40, and S54/60 samples and the results obtained are presented in Table 9.

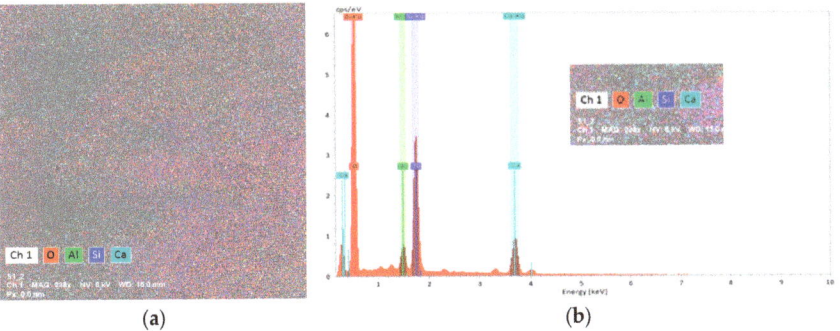

(a)   (b)

Figure 9. *Cont.*

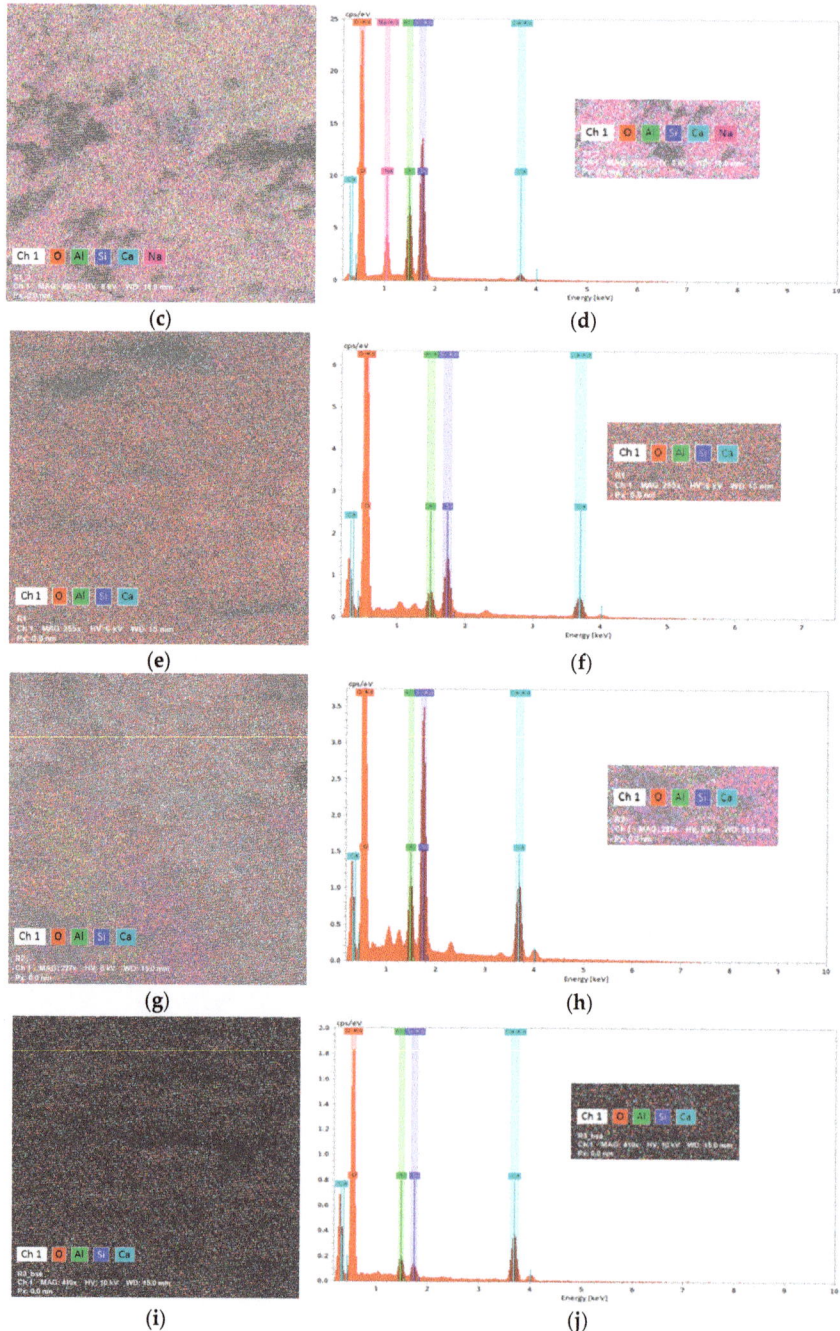

Figure 9. Surface mapping and EDX spectrum of (**a**,**b**) S360; (**c**,**d**) S414; (**e**,**f**) S54/20; (**g**,**h**) S54/40; (**i**,**j**) S54/60 samples at high magnification.

For all samples except S414, the predominant element besides oxygen is calcium. The results obtained for S414 indicate an inhomogeneity of the sample, with areas with many portlandite crystals and areas with small amounts of calcium, with the formation of silicon structures.

The ratio Ca/Si shows different variation:
- for S360 and S414, the values are below a value of 0.66 [95,96]. This indicates the formation of other silicate phases in the samples.
- for S54/20 and S54/40, the values are in the interval of values indicating the presence of different forms of C-S-H in the samples [95,96];
- for S54/60, the value of 12.86 indicates the presence of a majority of calcium-based structures.

5. Conclusions

In this paper, the performance of road concretes with GGBS and ACBFS were evaluated in the long term up to the age of 150 and 480 days. The results obtained from the evaluation of the contraction and the tensile strength at the age of 150 days, of the compression, carbonation, and microstructure characterization by the degree of crystallinity and the mineralogical composition at the age of 480 days lead to the following conclusions.

Up to the age of 480 days, the road concrete composites were not affected by carbonation corrosion.

XRD-SEM-EDX analysis suggests the lowest content of portlandite crystals for the S54/20 composite.

Very good compactness and low portlandite content result in the best corrosion resistance for the S54/20 blend.

The compressive strengths recorded the highest values for the compositions S54/20, S54/40, and S54/60.

The reaction products of the tobermorite group (C-S-H) as well as albite and quartz content were the main phases for the S54/20 and S54/60 composites with obtained the best mechanical strengths.

The reduction of the water-to-binding ratio as well as the grinding fineness of the GGBS powder diminishes the contractions in the slag mixtures compared to the control samples.

The reduction of hardening contractions positively influenced the tensile strength by bending to the composites S54/20 and S54/60 compared to the control composite S414.

Using crushed aggregate (ACBFS) with higher asperities than sand led to increased tensile strengths in S54/20 and S54/60 composites due to higher adhesion to the cement paste.

For the S54/20 and S360 composites, due to the decrease of contractions in the hardening process, there were the largest increases in bending tensile strengths at 150 days compared to 28 days.

For the same dosage of binder, the value of shrinkage during hardening decreased by 32.02% in the S54/20 composite and by 27.53% in the S54/60 composite compared to the control mixture S414 at the age of 150 days.

For the control composites, the main phase was quartz, while for the S54/40 composite it was albite. The addition of GGBS indicated a change in the morphology of the cement paste, favoring the formation of hydration product (C-S-H) for S54/20, S54/40, and calcium-based structures for S54/60.

The results presented in this paper contribute favorably to the opportunity and necessity of using road concrete containing blast furnace slag.

Future research should focus on the realizing of experimental concrete sectors with blast furnace slag subject to monitoring under the combined effect of temperature variations, traffic and chemical actions.

Last but not least, it is necessary to study the variation of the elasticity module in the case of slag road concretes as well as the opportunity to use other green materials that can reduce the carbon footprint, while maintaining the physical and mechanical properties at a high level.

Author Contributions: Conceptualization, L.M.N., D.S., D.L.M. and M.L.D.; methodology, L.M.N., D.S., O.C. and A.B.; investigation, L.M.N., D.S., O.C., A.B. and M.L.D.; writing—original draft preparation, L.M.N.; writing—review and editing, L.M.N., D.S., O.C., A.B., D.L.M. and M.L.D.; supervision, D.L.M., D.S. and M.L.D. All authors have read and agreed to the published version of the manuscript.

Funding: This research was funded by Project "Network of excellence in applied research and innovation for doctoral and postdoctoral programs/InoHubDoc", project co-funded by the European Social Fund financing agreement no. POCU/993/6/13/153437.

Institutional Review Board Statement: Not applicable.

Informed Consent Statement: Not applicable.

Data Availability Statement: All the required data that support the finding are presented in the manuscript.

Acknowledgments: This paper was financially supported by the Project "Network of excellence in applied research and innovation for doctoral and postdoctoral programs/InoHubDoc", project co-funded by the European Social Fund financing agreement no. POCU/993/6/13/153437. D.S., O.C. and A.B. acknowledge the financial support by the Ministry of Research, Innovation, and Digitization through Program 1—Development of the national research & development system, Subprogram 1.2—Institutional performance—Projects that finance the RDI excellence, Contract no. 18PFE / 30.12.2021.

Conflicts of Interest: The authors declare no conflict of interest.

References

1. Prakash, R.; Thenmozhi, R.; Raman, S.N. Mechanical characterisation and flexural performance of eco-friendly concrete produced with fly ash as cement replacement and coconut shell coarse aggregate. *Int. J. Environ. Sustain. Dev.* **2019**, *18*, 131–148. [CrossRef]
2. Prakash, R.; Thenmozhi, R.; Raman, S.N.; Subramanian, C. Characterization of eco-friendly steel fiber-reinforced concrete containing waste coconut shell as coarse aggregates and fly ash as partial cement replacement. *Struct. Concr.* **2020**, *21*, 437–447. [CrossRef]
3. Ahmad, J.; Kontoleon, K.J.; Majdi, A.; Naqash, M.T.; Deifalla, A.F.; Ben Kahla, N.; Isleem, H.F.; Qaidi, S.M.A. A Comprehensive Review on the Ground Granulated Blast Furnace Slag (GGBS) in Concrete Production. *Sustainability* **2022**, *14*, 8783. [CrossRef]
4. Monteiro, P.J.M.; Miller, S.A.; Horvath, A. Towards sustainable concrete. *Nat. Mater.* **2017**, *16*, 698–699. [CrossRef] [PubMed]
5. Olivier, J.G.J.; Peters, J.A.H. *Trends in Global CO_2 and Total Greenhouse Gas Emission Report*; PBL Netherlands Environmental Assessment Agency: The Hague, The Netherlands, 2020; pp. 1–85.
6. Smirnova, O. Compatibility of shungisite microfillers with polycarboxylate admixtures in cement compositions. *ARPN J. Eng. Appl. Sci.* **2019**, *14*, 600–610.
7. Bakharev, T. Geopolymeric materials prepared using Class F fly ash and elevated temperature curing. *Cem. Concr. Res.* **2005**, *35*, 1224–1232. [CrossRef]
8. Mwiti, M.J.; Karanja, T.J.; Muthengia, W.J. Thermal Resistivity of Chemically Activated Calcined Clays-Based Cements. In *Calcined Clays for Sustainable Concrete*; Martirena, F., Favier, A., Scrivener, K., Eds.; Springer: Dordrecht, The Netherlands, 2018; pp. 327–333.
9. Smirnova, O.M.; de Navascués, I.M.P.; Mikhailevskii, V.R.; Kolosov, O.I.; Skolota, N.S. Sound-absorbing composites with rubber crumb from used tires. *Appl. Sci.* **2021**, *11*, 7347. [CrossRef]
10. De Domenico, D.; Faleschini, F.; Pellegrino, C.; Ricciardi, G. Structural behavior of RC beams containing EAF slag as recycled aggregate: Numerical versus experimental results. *Constr. Build. Mater.* **2018**, *171*, 321–337. [CrossRef]
11. Brooks, J.J.; Megat Johari, M.A. Effect of metakaolin on creep and shrinkage of concrete. *Cem. Concr. Compos.* **2001**, *23*, 495–502. [CrossRef]
12. Bheel, N.; Abbasi, S.A.; Awoyera, P.; Olalusi, O.B.; Sohu, S.; Rondon, C.; Mar, A. Fresh and Hardened Properties of Concrete Incorporating Binary Blend of Metakaolin and Ground Granulated Blast Furnace Slag as Supplementary Cementitious Material. *Adv. Civ. Eng.* **2020**, *2020*, 8851030. [CrossRef]
13. Khassaf, S.I.; Jasim, A.T.; Mahdi, F.K. Investigation The Properties Of Concrete Containing Rice Husk Ash To Reduction The Seepage In Canals. *Int. J. Sci. Technol. Res.* **2014**, *3*, 348–354.
14. Burduhos Nergis, D.D.; Vizureanu, P.; Corbu, O. Synthesis and characteristics of local fly ash based geopolymers mixed with natural aggregates. *Rev. Chim.* **2019**, *70*, 1262–1267. [CrossRef]
15. Ganesh, P.; Murthy, A.R. Tensile behaviour and durability aspects of sustainable ultra-high performance concrete incorporated with GGBS as cementitious material. *Constr. Build. Mater.* **2019**, *197*, 667–680. [CrossRef]

16. Crossin, E. The greenhouse gas implications of using ground granulated blast furnace slag as a cement substitute. *J. Clean. Prod.* **2015**, *95*, 101–108. [CrossRef]
17. Şanal, İ. Fresh-state performance design of green concrete mixes with reduced carbon dioxide emissions. *Greenh. Gases Sci. Technol.* **2018**, *8*, 1134–1145. [CrossRef]
18. Gholampour, A.; Ozbakkaloglu, T. Performance of sustainable concretes containing very high volume Class-F fly ash and ground granulated blast furnace slag. *J. Clean. Prod.* **2017**, *162*, 1407–1417. [CrossRef]
19. Zhao, H.; Sun, W.; Wu, X.; Gao, B. The properties of the self-compacting concrete with fly ash and ground granulated blast furnace slag mineral admixtures. *J. Clean. Prod.* **2015**, *95*, 66–74. [CrossRef]
20. Abbass, M.; Singh, D.; Singh, G. Properties of hybrid geopolymer concrete prepared using rice husk ash, fly ash and GGBS with coconut fiber. *Mater. Today Proc.* **2021**, *45*, 4964–4970. [CrossRef]
21. Prakash, S.; Kumar, S.; Biswas, R.; Rai, B. Influence of silica fume and ground granulated blast furnace slag on the engineering properties of ultra-high-performance concrete. *Innov. Infrastruct. Solut.* **2021**, *7*, 117. [CrossRef]
22. Suda, V.B.R.; Srinivasa Rao, P. Experimental investigation on optimum usage of Micro silica and GGBS for the strength characteristics of concrete. *Mater. Today Proc.* **2020**, *27*, 805–811. [CrossRef]
23. Vediyappan, S.; Chinnaraj, P.K.; Hanumantraya, B.B.; Subramanian, S.K. An Experimental Investigation on Geopolymer Concrete Utilising Micronized Biomass Silica and GGBS. *KSCE J. Civ. Eng.* **2021**, *25*, 2134–2142. [CrossRef]
24. Hamzah, H.K.; Huseien, G.F.; Asaad, M.A.; Georgescu, D.P.; Ghoshal, S.K.; Alrshoudi, F. Effect of waste glass bottles-derived nanopowder as slag replacement on mortars with alkali activation: Durability characteristics. *Case Stud. Constr. Mater.* **2021**, *15*, e00775. [CrossRef]
25. Ahmad, J.; Martínez-García, R.; Szelag, M.; De-Prado-gil, J.; Marzouki, R.; Alqurashi, M.; Hussein, E.E. Effects of steel fibers (Sf) and ground granulated blast furnace slag (ggbs) on recycled aggregate concrete. *Materials* **2021**, *14*, 7497. [CrossRef] [PubMed]
26. Yu, R.; Spiesz, P.; Brouwers, H.J.H. Development of an eco-friendly Ultra-High Performance Concrete (UHPC) with efficient cement and mineral admixtures uses. *Cem. Concr. Compos.* **2015**, *55*, 383–394. [CrossRef]
27. Pal, S.C.; Mukherjee, A.; Pathak, S.R. Investigation of hydraulic activity of ground granulated blast furnace slag in concrete. *Cem. Concr. Res.* **2003**, *33*, 1481–1486. [CrossRef]
28. Yeau, K.Y.; Kim, E.K. An experimental study on corrosion resistance of concrete with ground granulate blast-furnace slag. *Cem. Concr. Res.* **2005**, *35*, 1391–1399. [CrossRef]
29. Teng, S.; Lim, T.Y.D.; Sabet Divsholi, B. Durability and mechanical properties of high strength concrete incorporating ultra fine Ground Granulated Blast-furnace Slag. *Constr. Build. Mater.* **2013**, *40*, 875–881. [CrossRef]
30. Lenka, B.P.; Majhi, R.K.; Singh, S.; Nayak, A.N. Eco-friendly and cost-effective concrete utilizing high-volume blast furnace slag and demolition waste with lime. *Eur. J. Environ. Civ. Eng.* **2022**, *26*, 5351–5373. [CrossRef]
31. Constantinescu, H.; Gherman, O.; Negrutiu, C.; Ioan, S.P. Mechanical Properties of Hardened High Strength Concrete. *Procedia Technol.* **2016**, *22*, 219–226. [CrossRef]
32. Jercan, S. *Concrete Roads*; Corvin Publishing House: Deva, Romania, 2002.
33. Nicolescu, L. *Hydrotechnical Concretes for Land Improvement Works*; Publishing house CERES: Bucharest, Romania, 1997.
34. Neville, A.M. *Properties of Concrete*; Bucharest Technical Publishing House: Bucharest, Romania, 1979.
35. Teoreanu, I. *Technology of Binders and Concretes*; Didactic and Pedagogical Publishing House: Bucharest, Romania, 1967.
36. Iureș, L. Research report: Concretes with reduced shrinkage made with special additives. *J. Sci. Policy Sci.* **2005**, 1582–1218.
37. Medeiros, R.A.; Lima, M.G.; Yazigi, R.; Medeiros, M.H.F. Carbonation depth in 57 years old concrete structure. *Steel Compos. Struct.* **2015**, *19*, 953–966. [CrossRef]
38. Kim, G.; Kim, J.-Y.; Kurtis, K.E.; Jacobs, L.J.; Pape, Y.L.; Guimaraes, M. Quantitative evaluation of carbonation in concrete using nonlinear ultrasound. *Mater. Struct.* **2016**, *49*, 399–409. [CrossRef]
39. Papadakis, V.G.; Vayenas, C.G.; Fardis, M.N. Fundamental modeling and experimental investigation of concrete carbonation. *ACI Mater. J.* **1991**, *88*, 363–373.
40. Papadakis, V.G.; Vayenas, C.G. Experimental investigation and mathematica modeling of the concrete carbonation problem. *Chem. Eng. Sci.* **1991**, *46*, 1333–1338. [CrossRef]
41. Pacheco-Torgal, F.; Ding, Y.; Jalali, S. Properties and durability of concrete containing polymeric wastes (tyre rubber and polyethylene terephthalate bottles). *Constr. Build. Mater.* **2012**, *30*, 714–724. [CrossRef]
42. José, D.R.; Adelardo, V.; Leiva, C.; la Concha, M.-D.; Antonio, M.; Cifuentes, H. Analysis of the utilization of air-cooled blast furnace slag as industrial waste aggregates in self-compacting concrete. *Sustainability* **2019**, *11*, 1702. [CrossRef]
43. Corbu, O.; Puskas, A.; Sandu, A.V.; Ioani, A.M.; Hussin, K.; Sandu, I.G. New Concrete with Recycled Aggregates from Leftover Concrete. *Appl. Mech. Mater.* **2015**, *754–755*, 389–394. [CrossRef]
44. Majhi, R.K.; Nayak, A.N.; Mukharjee, B.B. Characterization of lime activated recycled aggregate concrete with high-volume ground granulated blast furnace slag. *Constr. Build. Mater.* **2020**, *259*, 119882. [CrossRef]
45. Forton, A.; Mangiafico, S.; Sauzéat, C.; Di Benedetto, H.; Marc, P. Behaviour of binder blends: Experimental results and modelling from LVE properties of pure binder, RAP binder and rejuvenator. *Road Mater. Pavement Des.* **2021**, *22*, S197–S213. [CrossRef]
46. Cadar, R.D.; Boitor, R.M.; Dragomir, M.L. An Analysis of Reclaimed Asphalt Pavement from a Single Source—Case Study: A Secondary Road in Romania. *Sustainability* **2022**, *14*, 7057. [CrossRef]

47. Forton, A.; Mangiafico, S.; Sauzéat, C.; Di Benedetto, H.; Marc, P. Properties of blends of fresh and RAP binders with rejuvenator: Experimental and estimated results. *Constr. Build. Mater.* **2020**, *236*, 117555. [CrossRef]
48. Soni, Y.; Gupta, N. Experimental Investigation on Workability of Concrete with Partial Replacement of Cement by Ground Granulated Blast Furnace and Sand by Quarry Dust. *IJIRST-Int. J. Innov. Res. Sci. Technol.* **2016**, *3*, 312–316.
49. Molnar, L.M.; Manea, D.L.; Aciu, C.; Jumate, E. Innovative Plastering Mortars Based on Recycled Waste Glass. *Procedia Technol.* **2015**, *19*, 299–306. [CrossRef]
50. Ozbakkaloglu, T.; Gu, L.; Fallah Pour, A. Normal- and high-strength concretes incorporating air-cooled blast furnace slag coarse aggregates: Effect of slag size and content on the behavior. *Constr. Build. Mater.* **2016**, *126*, 138–146. [CrossRef]
51. Cao, Q.; Nawaz, U.; Jiang, X.; Zhang, L.; Ansari, W.S. Effect of air-cooled blast furnace slag aggregate on mechanical properties of ultra-high-performance concrete. *Case Stud. Constr. Mater.* **2022**, *16*, e01027. [CrossRef]
52. Popescu, D.; Burlacu, A. Considerations on the Benefits of Using Recyclable Materials for Road Construction. *Rom. J. Transp. Infrastruct.* **2017**, *6*, 43–53. [CrossRef]
53. Concrete Ubrary of JSCE. Guidelines for Construction Using Blast-Furnace Slag Aggregate Concrete. Available online: https://www.jsce.or.jp/committee/concrete/e/web/pdf/22-2.pdf (accessed on 16 February 2023).
54. SR EN 12620+A1; Aggregates for Concrete. ASRO: Bucharest, Romania, 2008.
55. ASA 36pp Guide—(Iron & Steel) Slag Association. A Guide to the Use of Slag in Roads. *Revision 2*. Available online: https://www.yumpu.com/en/document/view/20477792/asa-36pp-guide-iron-steel-slag-association (accessed on 16 February 2023).
56. Burlacu, A.; Racanel, C. Reducing Cost of Infrastructure Works Using New Technologies. Available online: https://www.oecd-ilibrary.org/development/road-and-rail-infrastructure-in-asia_9789264302563-en (accessed on 27 February 2023).
57. Dimulescu, C.; Burlacu, A. Industrial Waste Materials as Alternative Fillers in Asphalt Mixtures. *Sustainability* **2021**, *13*, 8068. [CrossRef]
58. Nicula, L.M.; Corbu, O.; Iliescu, M.; Sandu, A.V.; Hegyi, A. Study on the Durability of Road Concrete with Blast Furnace Slag Affected by the Corrosion Initiated by Chloride. *Adv. Civ. Eng.* **2021**, *2021*, 8851005. [CrossRef]
59. Nicula, L.M.; Corbu, O.; Iliescu, M.; Dumitraș, D.G. Using the blast furnace slag as alternative source in mixtures for the road concrete for a more sustainable and a cleaner environment. *Rom. J. Mater.* **2020**, *50*, 545–555.
60. Król, A.; Giergiczny, Z.; Kuterasińska-Warwas, J. Properties of Concrete Made with Low-Emission Cements CEM II/C-M and CEM VI. *Materials* **2020**, *13*, 2257. [CrossRef]
61. Bazaldúa-Medellín, M.E.; Fuentes, A.F.; Gorokhovsky, A.; Escalante-García, J.I. Early and late hydration of supersulphated cements of blast furnace slag with fluorgypsum. *Mater. Constr.* **2015**, *65*, e043. [CrossRef]
62. SR EN 197; Cement Part 1: Composition, Specification, and Conformity Criteria Common Cements. ASRO: Bucharest, Romania, 2011.
63. Kazberuk, M.K. Surface Scaling Resistance of Concrete with Fly Ash from Co-Combustion of Coal and Biomass. *Procedia Eng.* **2013**, *57*, 605–613. [CrossRef]
64. Papadakis, V.G.; Tsimas, S. Supplementary cementing materials in concrete. Part I: Efficiency and design. *Cem. Concr. Res.* **2002**, *32*, 1525–1532. [CrossRef]
65. Sinsiri, T.; Kroehong, W.; Jaturapitakkul, C.; Chindaprasirt, P. Assessing the effect of biomass ashes with different finenesses on the compressive strength of blended cement paste. *Mater. Des.* **2012**, *42*, 424–433. [CrossRef]
66. SR EN 15167; Ground Granulated Blast Furnace Slag for Use in Concrete, Mortar and Grout Part 1: Definitions, Specifications and Conformity Criteria. ASRO: Bucharest, Romania, 2007.
67. NE 014:2002; The Norm for the Execution of Cement Concrete Road Pavements in a Fixed and Sliding Formwork System. Matrix ROM: Bucharest, Romania, 2007; ISBN 978-973-755-185-6.
68. SR EN 934-2+A1; Concrete Additives. ASRO: Bucharest, Romania, 2002.
69. SR EN 1008; Mixing Water for Concrete. ASRO: Bucharest, Romania, 2003.
70. SR 2833; Tests on Concrete. Determination of the Axial Contraction of Hardened Concrete. ASRO: Bucharest, Romania, 2009.
71. SR EN 12390; Test on Hardened Concrete. Part 5: Bending Tensile Strength of Specimens. ASRO: Bucharest, Romania, 2019.
72. SR EN 12390; Standard for Test-Hardened Concrete—Part 3: Compressive Strength of Test Specimens. ASRO: Bucharest, Romania, 2019.
73. SR CR 12793; Determination of the Depth of the Carbonation Layer of Hardened Concrete. ASRO: Bucharest, Romania, 2002.
74. Chesner, W.H.; Collins, R.J.; MacKay, M.H.; Emery, J. User Guidelines for Waste and By-Product Materials in Pavement Construction (No. FHWA-RD-97-148, Guideline Manual, Rept No. 480017). Available online: https://rosap.ntl.bts.gov/view/dot/38365 (accessed on 23 February 2023).
75. Smith, K.D.; Morian, D.A.; Van Dam, T.J. *Use of Air-Cooled Blast Furnace Slag as Coarse Aggregate in Concrete Pavements—A Guide to Best Practice*; Report No. FHWA-HIF-12-009; Federal Highway Administration: Philadelphia, PA, USA, 2012.
76. Taylor, G.P.; Van Dam, T.; Sutter, L.; Fick, G. Integrated Materials and Construction Practices for Concrete Pavement. Available online: https://issuu.com/ich_mkt/docs/2019_cptech_-_integrated_materials_and_constructio (accessed on 17 February 2023).
77. Badr, A.; Ashour, A.F.; Platten, A.K. Statistical variations in impact resistance of polypropylene fibre-reinforced concrete. *Int. J. Impact Eng.* **2006**, *32*, 1907–1920. [CrossRef]
78. Alexanderson, J. Relations between structure and mechanical properties of autoclaved aerated concrete. *Cem. Concr. Res.* **1979**, *9*, 507–514. [CrossRef]

79. Ahmad, J.; Martínez-García, R.; De-Prado-gil, J.; Irshad, K.; El-Shorbagy, M.A.; Fediuk, R.; Vatin, N.I. Concrete with Partial Substitution of Waste Glass and Recycled Concrete Aggregate. *Materials* **2022**, *15*, 430. [CrossRef]
80. Ahmad, J.; Aslam, F.; Martinez-Garcia, R.; de-Prado-Gil, J.; Qaidi, S.M.A.; Brahmia, A. Effects of waste glass and waste marble on mechanical and durability performance of concrete. *Sci. Rep.* **2021**, *11*, 21525. [CrossRef]
81. Ahmad, J.; Tufail, R.F.; Aslam, F.; Mosavi, A.; Alyousef, R.; Javed, M.F.; Zaid, O.; Khan Niazi, M.S. A step towards sustainable self-compacting concrete by using partial substitution of wheat straw ash and bentonite clay instead of cement. *Sustainability* **2021**, *13*, 824. [CrossRef]
82. Pimienta, P.; Albert, B.; Huetb, B.; Dierkens, M.; Francisco, P.; Rougeaud, P. Durability performance assessment of non-standard cementitious materials for buildings: A general method applied to the French context, Fact sheet 1—Risk of steel corrosion induced by carbonation. *RILEM Tech. Lett.* **2016**, *1*, 102–108. [CrossRef]
83. Hsiao, Y.-H.; La Plante, E.C.; Anoop Krishnan, N.M.; Le Pape, Y.; Neithalath, N.; Bauchy, M.; Sant, G. Effects of Irradiation on Albite's Chemical Durability. *J. Phys. Chem. A* **2017**, *121*, 7835–7845. [CrossRef] [PubMed]
84. Saiz-Martínez, P.; González-Cortina, M.; Fernández-Martínez, F. Characterization and influence of fine recycled aggregates on masonry mortars properties. *Mater. Construc.* **2015**, *65*, e058. [CrossRef]
85. Raki, L.; Beaudoin, J.; Alizadeh, R.; Makar, J.; Sato, T. Cement and concrete nanoscience and nanotechnology. *Materials* **2010**, *3*, 918–942. [CrossRef]
86. Kosmatka, S.H.; Panarese, W.C.; Kerkhoff, B. *Design and Control of Concrete Mixtures Fourteenth Edition*; Portland Cement Association: Washington, DC, USA, 2011; ISBN 0893122173.
87. Silva, A.S.; Soares, D.; Matos, L.; Salta, M.; Divet, L.; Pavoine, A.; Candeias, A.; Mirão, J. Influence of Mineral Additions in the Inhibition of Delayed Ettringite Formation in Cement based Materials—A Microstructural Characterization. *Mater. Sci. Forum Vols.* **2010**, *636–637*, 1272–1279.
88. Alam, S.; Kumar Das, S.; Hanumantha Rao, B. Strength and durability characteristic of alkali activated GGBS stabilized red mud as geo-material. *Constr. Build. Mater.* **2019**, *211*, 932–942. [CrossRef]
89. Huang, J.; Li, W.; Huang, D.; Wang, L.; Chen, E.; Wu, C.; Wang, B.; Deng, H.; Tang, S.; Shi, Y.; et al. Fractal Analysis on Pore Structure and Hydration of Magnesium Oxysulfate Cements by First Principle, Thermodynamic and Microstructure-Based Methods. *Fractal Fract.* **2021**, *5*, 164. [CrossRef]
90. Peng, Y.; Tang, S.; Huang, J.; Tang, C.; Wang, L.; Liu, Y. Fractal Analysis on Pore Structure and Modeling of Hydration of Magnesium Phosphate Cement Paste. *Fractal Fract.* **2022**, *6*, 337. [CrossRef]
91. Zeng, Q.; Li, K.; Fen-chong, T.; Dangla, P. Pore structure characterization of cement pastes blended with high-volume fly-ash. *Cem. Concr. Res.* **2012**, *42*, 194–204. [CrossRef]
92. Song, Y.; Zhou, J.; Bian, Z.; Dai, G. Pore Structure Characterization of Hardened Cement Paste by Multiple Methods. *Adv. Mater. Sci. Eng.* **2019**, *2019*, 3726953. [CrossRef]
93. Portlandite—Basis for a Very Important Building Material. Available online: https://crystalsymmetry.wordpress.com/ (accessed on 28 February 2023).
94. Galan, I.; Glasser, F.P.; Baza, D.; Andrade, C. Assessment of the protective effect of carbonation on portlandite crystals. *Cem. Concr. Res.* **2015**, *74*, 68–77. [CrossRef]
95. Richardson, I.G. The nature of C-S-H in hardened cements. *Cem. Concr. Res.* **1999**, *29*, 1131–1147. [CrossRef]
96. Chu, D.C.; Kleib, J.; Amar, M.; Benzerzour, M.; Abriak, N.-E. Determination of the degree of hydration of Portland cement using three different approaches: Scanning electron microscopy (SEM-BSE) and Thermogravimetric analysis (TGA). *Case Stud. Constr. Mater.* **2021**, *15*, e00754. [CrossRef]

Disclaimer/Publisher's Note: The statements, opinions and data contained in all publications are solely those of the individual author(s) and contributor(s) and not of MDPI and/or the editor(s). MDPI and/or the editor(s) disclaim responsibility for any injury to people or property resulting from any ideas, methods, instructions or products referred to in the content.

Article

Effect of Isothermal Annealing on Sn Whisker Growth Behavior of Sn0.7Cu0.05Ni Solder Joint

Aimi Noorliyana Hashim [1,2], Mohd Arif Anuar Mohd Salleh [1,2,*], Muhammad Mahyiddin Ramli [3], Mohd Mustafa Al Bakri Abdullah [1,2], Andrei Victor Sandu [4,5,6,*], Petrica Vizureanu [4,7] and Ioan Gabriel Sandu [4]

1. Centre of Excellence Geopolymer and Green Technology, Universiti Malaysia Perlis (UniMAP), Taman Muhibah, Jejawi, Arau 02600, Perlis, Malaysia
2. Faculty of Chemical Engineering and Technology, Universiti Malaysia Perlis (UniMAP), Arau 02600, Perlis, Malaysia
3. School of Microelectronic Engineering, Pauh Putra Campus, Universiti Malaysia Perlis (UniMAP), Arau 02600, Perlis, Malaysia
4. Faculty of Materials Science and Engineering, Gheorghe Asachi Technical University of Iasi, Blvd. D. Mangeron 71, 700050 Iasi, Romania
5. Romanian Inventors Forum, Str. Sf. P. Movila 3, 700089 Iasi, Romania
6. National Institute for Research and Development in Environmental Protection INCDPM, Splaiul Independentei 294, 060031 Bucharest, Romania
7. Technical Sciences Academy of Romania, Dacia Blvd 26, 030167 Bucharest, Romania
* Correspondence: arifanuar@unimap.edu.my (M.A.A.M.S.); sav@tuiasi.ro (A.V.S.)

Abstract: This paper presents an assessment of the effect of isothermal annealing of Sn whisker growth behavior on the surface of Sn0.7Cu0.05Ni solder joints using the hot-dip soldering technique. Sn0.7Cu and Sn0.7Cu0.05Ni solder joints with a similar solder coating thickness was aged up to 600 h in room temperature and annealed under 50 °C and 105 °C conditions. Through the observations, the significant outcome was the suppressing effect of Sn0.7Cu0.05Ni on Sn whisker growth in terms of density and length reduction. The fast atomic diffusion of isothermal annealing consequently reduced the stress gradient of Sn whisker growth on the Sn0.7Cu0.05Ni solder joint. It was also established that the smaller $(Cu,Ni)_6Sn_5$ grain size and stability characteristic of hexagonal η-Cu_6Sn_5 considerably contribute to the residual stress diminished in the $(Cu,Ni)_6Sn_5$ IMC interfacial layer and are able to suppress the growth of Sn whiskers on the Sn0.7Cu0.05Ni solder joint. The findings of this study provide environmental acceptance with the aim of suppressing Sn whisker growth and upsurging the reliability of the Sn0.7Cu0.05Ni solder joint at the electronic-device-operation temperature.

Keywords: Sn whisker; Sn0.7Cu0.05Ni; IMC interfacial; $(Cu,Ni)_6Sn_5$; solder joint; annealing; suppression; mitigation

1. Introduction

The main impetus of Sn whisker nucleation and growth is residual stress [1–3] by soldering. As a response over time, stresses release mechanism by diffusion [2,4,5] and compressive residual stress, and Sn whiskers are initiated and protrude out spontaneously at their base of substrate that may contribute immense apprehension to the electronics industry [1,6,7]. The major source of compressive stress in lead-free solder joints is intermetallic compound (IMC) interfacial evolution as per the continuous reaction of a Sn-rich solder with a copper (Cu) substrate forming intermetallic compounds within the grain boundaries [2,5,7–11]. The formation of the Cu_6Sn_5 layer by the IMC growth–diffusion gradient between 109 °C and 220 °C is an adequate and continuously controlled deliberate reaction process [1]. Illés et al. established the ability of spontaneous Sn whiskers to initiate and grow on surface of a solder joint with an average thickness of solder coating of 400 nm as deposited. They also provided evidence that the main stimulus of the whiskering intensity on the solder coating surface is induced by high compressive stress of IMC interfacial layer [2].

The properties of the lead-free solder joint between the Sn-rich solder and the Cu substrate affect the Sn whisker growth behavior of the surface. Horvath et al. concluded that the thickness of the solder coating and the grain size are the main properties accountable for Sn whisker behavior [12,13]. A thicker solder coating provides better stress relaxation capacity [2,14] and exhibits a lower inclination to whisker nucleation and growth. A thicker solder coating requires more incubation time for the interfacial IMC to diffuse and move up the grain boundaries to develop the occupied compressive stress cell [13]. Horváth et al. endorsed that the thickness of the solder coating for electronic devices should be 8 μm at minimum in order to suppress the vulnerability of Sn whisker growth, as well to prolong incubation time [13]. The effect of solder grain size also correspondingly contributes stress that is induced in solder coating. To correlate the atomic diffusion of the grain boundary and the formation of the IMC interfacial layer, it is accepted that more stress is produced during grain boundary diffusion of smaller grains. However, Sn whiskers proceed with lower stress for the purpose of grain boundary sliding for nucleation and growth in small grain boundary diffusion. As an outcome, the resistance of smaller grains is improved against the low residual stress gradient on behalf of the stress relaxation of the solder coating layer [13]. In a previous study, Hashim et al. found that Sn0.7Cu0.05Ni with a smaller grain size of $(Cu,Ni)_6Sn_5$ had a lower inclination to Sn whisker propensity compared to the Sn0.7Cu solder joint [14].

Numerous studies of mitigation methods have been suggested to suppress Sn whisker formation. The conformal coating or the added-nickel (Ni) under-layer between the solder and the Cu substrate is a well-known method used to mitigate Sn whisker formation [10], which has comparable outcomes to avoiding large grain growth of the IMC interfacial layer between a Sn-rich solder and a Cu substrate. Distinctly, these approaches are significant in terms of temporary effect as, in many cases, Sn whiskers are able to penetrate conformal coatings. The findings of these reference studies are accepted in terms of delayed incubation time for Sn whisker growth [3,10]. Isothermal annealing is also one of the common methods used to mitigate the Sn whisker by developing a fine and uniform IMC interfacial layer between the Sn-rich solder and the Cu substrate. This IMC interfacial layer inhibits the formation of a large grain in the IMC layer by promoting fast diffusion of Cu grain boundaries in the Sn-rich layer and yielding the relief of internal stresses in the solder joint [8,13].

In the view of the constrained temperature range for the nucleation and growth of Sn whiskers, there has been limited significant research on how isothermal annealing affects Sn whisker behavior [15]. When the temperature is too low, slow atomic diffusion, takes place, owing to the insufficient kinetics. Furthermore, when the temperature is too high, fast atomic diffusion occurs; therefore, there is not enough of a driving gradient because of stress release [8]. As stated in previous research, isothermal annealing at 150 °C and aged for 30 min is a representative technique used for the inhibition of Sn whisker growth at ambient temperatures. Fukuda et al. conducted a reduction assessment of whisker density periodically over 8 months with isothermal annealing at 50 °C [7]. They also detected a reduction in maximum Sn whisker length with isothermal annealing at 150 °C and ageing for 1 h. In addition, Kim et al. also asserts that annealing at 125 °C is useful to the inhibition of Sn whisker growth without the growth of Cu_3Sn [13].

The purpose of this study was to analyze the effects of isothermal annealing on Sn whisker growth behavior of the Sn0.7Cu0.05Ni solder joint using the hot-dip soldering technique. Isothermal annealing at 50 °C and 105 °C with an ageing time of up to 600 h pointedly represents an accelerated test to obtain life extrapolation at the electronic-device-operation temperature. To evaluate the practicality of the mitigation strategies of Sn whiskers, the Sn0.7Cu solder joint was used as a point of reference.

2. Materials and Methods

2.1. Materials and Sample Preparation

The lead-free solder alloys of Sn-0.7Cu and Sn-0.7Cu-0.05Ni were provided by Nihon Superior Co., Ltd. in Osaka, Japan. For sample preparation, the high-purity Cu substrate (1.5 cm × 1.5 cm × 0.1 cm) was cleaned with a 5% hydrochloric acid solution and deionized water for 3 min to remove surface oxides, then it was rinsed with acetone followed by distilled water, and subsequently air-dried to dispose of oil impurity. Then, the Cu substrate was dipped in a standard halogen rosin-activated flux solution of the Japanese Industrial Standard, JIS Z3198-4, in order to seal out air and improve soldering wetting characteristics.

2.2. Hot-Dip Soldering

The Sn-0.7Cu and Sn-0.7Cu-0.05Ni solder joint was synthesized using the hot-dip soldering technique. This technique provides well-controllable lead-free solder coating thickness parameters. During the hot-dip soldering process, the molten solder pot was heated up to 265 °C and the Cu substrate was placed between a pair of blower air knives. The Cu substrate was immersed in molten solder for 2 s and withdrawn at a speed of 10 mm/s. During withdraw, the air knives system wiped down the excessive solder (tinning process) adhered on both sides of the substrate surface by impinging the pressured hot air as shown in Figure 1. The controllable hot air of the knives was used to control the thickness of the total solder coating and the uniformity of the coating, particularly near the substrate edge. The pressure of hot air comprised the range 0.1 MPa to 1.0 MPa to obtain the condition of comparable solder coating thickness for Sn whisker growth evaluation. The samples were then washed with acetone and distilled water after cooling for 30 min. In order to saturate the internal residue stress induced by Cu_6Sn_5 IMC formation, samples were aged for 48 h at room temperature.

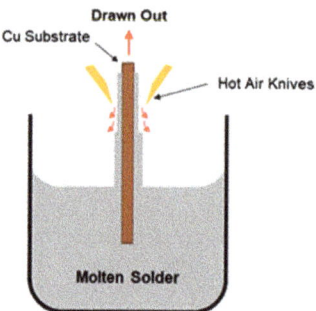

Figure 1. Schematic diagram of a hot-dip soldering process.

2.3. Testing and Characterization

The samples of the Sn0.7Cu and Sn0.7Cu-0.05Ni solder joint was investigated by performing isothermal annealing at 50 °C and 105 °C for ageing times 0, 100, 200, 300, 400, 500, and 600 h. The samples were annealed in a mechanical-convection heating oven provided by Thermo Scientific. For reference assessment, the samples were observed at room temperature with the same ageing time (0–600 h).

The metallography surface and cross-section of the solder joint morphology of Sn whisker behavior were found using the secondary and backscattered electron imaging approaches of a scanning electron microscope, SEM (JEOL-JSM-6010LA), JEOL Ltd., Tokyo, Japan. The samples were deep-etched using 2% 2-nitrophenol, 5% sodium hydroxide solution, and 93% distilled water for a top view of the Cu_6Sn_5 IMC interfacial layer. Figure 2 presents the illustrative cross-sectional schematic of the solder joint. A measured average value of five interpretations of the thickness value of the solder coating, the IMC interfacial layer, and particle size analyses were examined using Image-J programmed software (1.8.0 open source software) as specified in previous research [14].

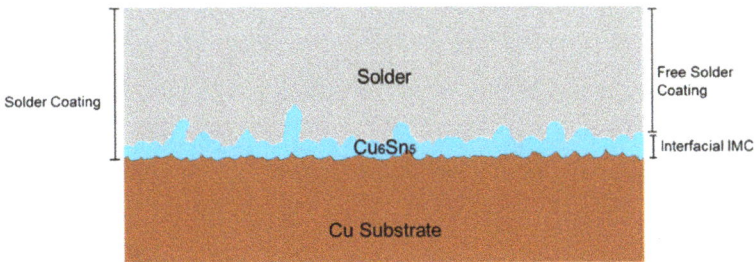

Figure 2. The thickness of measuring solder coating, free solder thickness, and Cu$_6$Sn$_5$ IMC interfacial layer.

The statistic evaluation of Sn whisker behavior was measured based on the whisker standards (JESD22-A121A) of the Joint Electron Device Engineering Council (JEDEC). The analyses of the length and density distribution of the Sn whisker were structured with ±5% accuracy and five preference average values of interpretations analyses using Image-J Software. In order to attain a statistically consistent assessment, the density of Sn Whiskers were analyzed using the binary thresholding image-segmentation technique. This technique separates the foreground pixels from the background pixels and produces binary segmentation images from Scanning electron microscopy (SEM) grayscale images as illustrated in Figure 3a,b. Respectively, Figure 3c,d indicate the validation of the elemental Sn whisker assessment using SEM-EDX analyses.

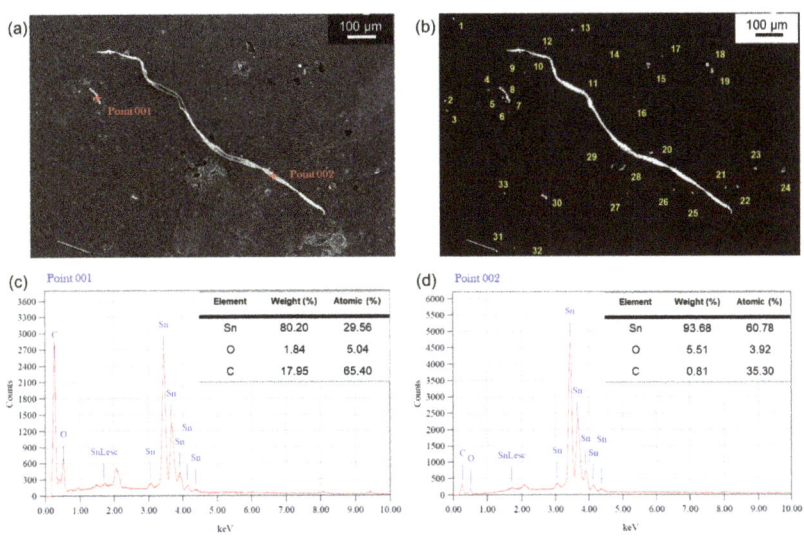

Figure 3. The method of image processing adapted to calculate density of Sn whisker growth and element analyses of Sn whisker (**a**) SEM-EDX grayscale image, (**b**) binary thresholding segmentation image, (**c**) EDX of point 001, (**d**) EDX of point 002.

For wettability analyses, the samples of the Sn0.7Cu and Sn0.7Cu0.05Ni solder were rolled and punched into a diameter of 0.2 mm, then reflowed onto the Cu substrates. The reflow time was conducted at 250 °C for 127 s. The samples were mounted, carried out using an optical microscope (OM) and Image-J software for wettability properties. This investigation was attained by measuring the wetting degree of the contact angle between the Sn-rich solder and the Cu substrate. The thermodynamic assessment of the phase diagram, phase equilibrium, and phase transformation was validated using the Calphad method and Thermo-Calc-2021a database TCSLD v3. 3. A binary crystal structure phase diagram of Sn-Cu and Sn-Ni preferred to justify the phase transformation of Sn0.7Cu and Sn0.7Cu0.05Ni evidently.

3. Results and Discussion

It should be noted that Sn whisker nucleation and growth is significantly correlated to total solder coating thickness. A thicker solder coating resulted in more residual stress and incubation time of Sn whisker growth on the solder surface [2,14–20]. Figure 4 shows a thickness observation of the total solder coating after hot-dip soldering with different pressures of hot air knives. The data clearly show the conditions of the equivalent solder coating thickness for Sn0.7Cu and Sn0.7Cu0.05Ni in the evaluation of Sn whisker growth. The total solder thickness decreased proportionally to the hot air knives' pressure, as well as dissimilar thicknesses with different type of solder. In order to obtain comparable solder thicknesses for both Sn0.7Cu and Sn0.7Cu0.05Ni, the hot air pressure was in the wide range of 0.1 to 10 MPa. At the lowest pressure of the hot air knives, 0.1 MPa, the total solder coating of Sn0.7Cu was 16.74 μm (A), and Sn0.7Cu0.05Ni was 13.82 (B) μm. Meanwhile, at the highest pressure of 10 MPa, the total solder coating of Sn0.7Cu was 13.20 μm and 8.68 μm for Sn0.7Cu0.05Ni. The similar total thickness coating of Sn0.7Cu was 13.34 μm (C) at 0.9 MPa, and Sn0.7Cu0.05Ni was 13.31 μm (D) at 0.2 MPa.

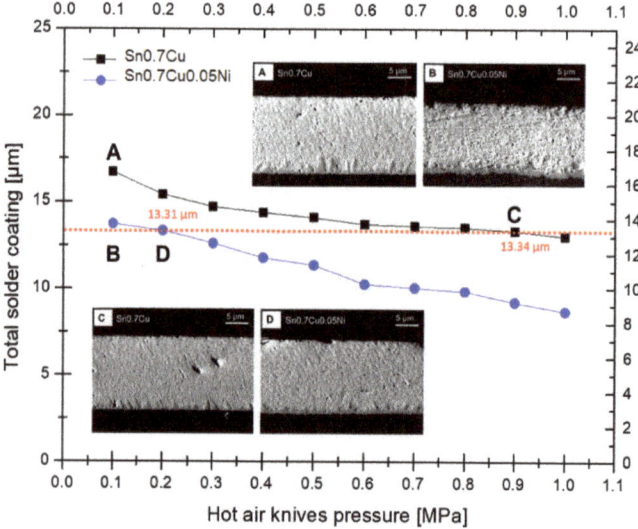

Figure 4. The solder coating thickness of Sn0.7Cu and Sn0.7Cu0.05Ni with hot air knives' pressure.

This consequence noticeably shows the purpose of hot air knives in controlling the thickness of the total solder coating. Additionally, the variation in total solder thickness significantly indicates the solder wettability properties of Sn0.7Cu and Sn0.7Cu0.05Ni. The wettability of the solder was evaluated through the wetting angle of the solder joint as presented in Figure 5. A larger wetting contact angle allows for a higher-thickness solder coating to be completed. The Sn0.7Cu0.05Ni solder joint showed better wettability with a smaller contact angle, $\theta = 16.8°$, related to the Sn0.7Cu solder joint with contact angle $\theta = 27.2°$. The superior fluidity properties of the Sn0.7Cu0.05Ni solder alloy are capable of producing reliable in the uniformity of the thickness of the Cu substrate. As agreed by Gain et al., Ni alloying enhances wettability, which may influence the mechanical reliability of the interconnection [16]. The optimum wettability that relates to lower surface-interfacial energy could be achieved by minimizing the contact angle value [17].

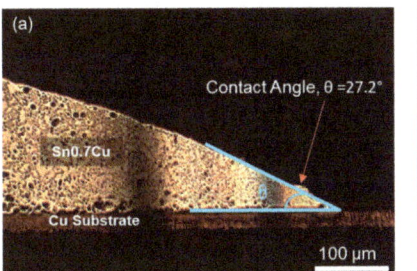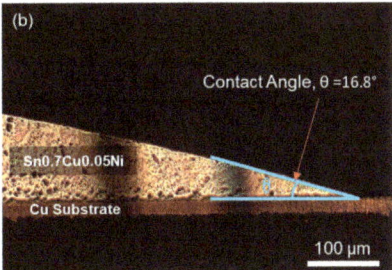

Figure 5. The wetting contact angle of solder joint (**a**) Sn0.7Cu, (**b**) Sn0.7Cu0.05Ni.

Figure 6 shows the average Sn whisker densities of the Sn0.7Cu and Sn0.7Cu0.05Ni solder joints with an average total solder thicknesses ±13.3 µm, ageing up to 600 h in room temperature, and annealed under 50 °C and 105 °C conditions. From the observation, most of the Sn whiskers were grown and noticed after 100 h of ageing under all conditions. After ageing up to 600 h at room temperature, the reference sample showed that the density of the Sn whisker reached 128 pcs/mm^2 on the Sn0.7Cu solder joint and 84 pcs/mm^2 on the Sn0.7Cu0.05Ni solder joint. Additionally, for the Sn0.7Cu and Sn0.7Cu0.05Ni solder joints, the observation also showed an increased growth in the Sn whiskers with annealed ageing time. Additionally, the density of Sn whisker growth indicated a reduction trend in the increase in isothermal annealing temperature. It is interesting to point out that the intensity of Sn whisker formation is higher at Sn0.7Cu related to the Sn0.7Cu0.05Ni solder joint.

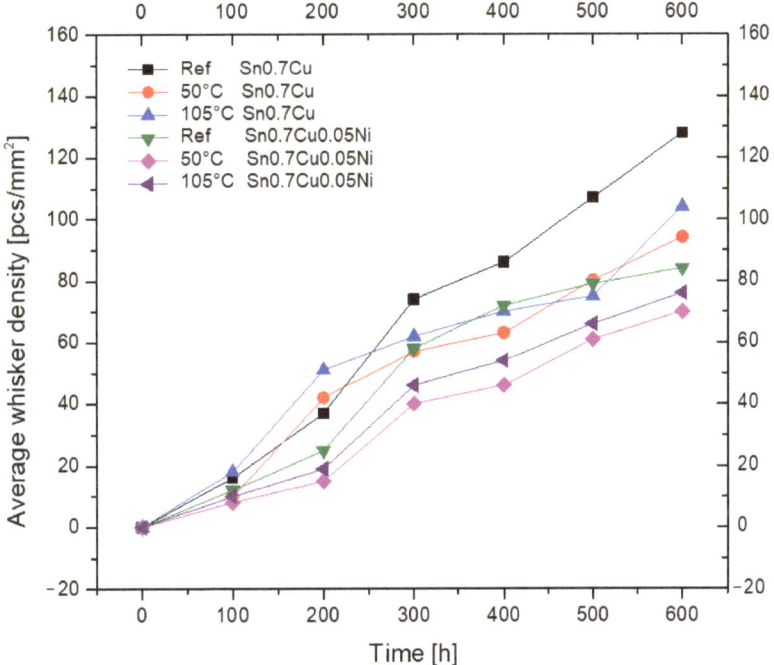

Figure 6. Average Sn whiskers density with isothermal ageing.

Figure 7 shows the surface morphology of Sn whisker nucleation and growth on Sn0.7Cu0.05Ni at ageing times 100 h, 300 h, and 600 h with room temperature ageing, 50 °C isothermal ageing, and 105 °C isothermal ageing. The number of Sn whiskers on the Sn0.7Cu0.05Ni solder joint with isothermal ageing increased scattering with ageing duration. Furthermore, the intensity of Sn whisker growth increased with ageing at room temperature of the reference sample, compared to the annealed samples under 50 °C and 105 °C conditions. After 100 h ageing time, the scattered nucleation of Sn whiskers, particularly under room temperature conditions, can be observed. It can also be observed that Sn whiskers significantly intensify to initiate after 300 h of ageing.

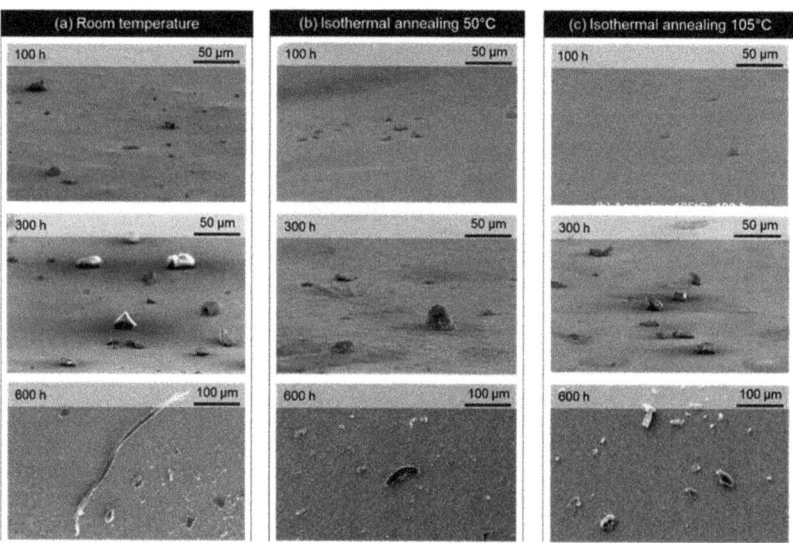

Figure 7. Surface morphology of Sn whisker growth on Sn0.7Cu0.05Ni aged at 100 h, 300 h, and 600 h with isothermal ageing at (**a**) room temperature, (**b**) 50 °C, and (**c**) 105 °C.

The length of the Sn whiskers after 600 h of ageing time at room temperature was obviously longer and had a greater density of Sn whisker growth per unit area. However, in the case of the annealed samples, the 105 °C condition increased the density of Sn whisker growth compared to the annealed samples 50 °C. It can be determined that isothermal annealing is functional for Sn whisker mitigation. The significant outcome was detected through the observation of the suppressing effect of Sn0.7Cu0.05Ni on Sn whisker growth in terms of density and length reduction.

The average Sn whisker lengths measured indicate that isothermal annealing is effective at reducing whisker lengths as observed in Figure 8. The length of the Sn whiskers increased with increasing ageing time and annealing temperature. The condition where the annealing temperature was elevated from 50 °C to 105 °C had a small length-reducing effect. This decrease effect was similar for both the annealed Sn0.7Cu and Sn0.7Cu0.05Ni solder joint samples. It was reported by Kim et al. that the behavior of Sn whiskers after 50 °C isothermal annealing for 6-month ageing is similar to Sn whisker growth in ambient storage for a 1-year-ageing duration in terms of size and of shape [21]. It is also noteworthy that Sn0.7Cu0.05Ni was seen to reduce the length of the Sn whiskers compared to Sn0.7Cu, either in room temperature or annealing temperature. This further confirms that Sn0.7Cu0.05Ni has a suppression effect in retarding the length of Sn whiskers.

Figure 9 identifies the quantitative assessment of nucleation and growth from a small nodule to the longest-length 763 µm Sn whiskers assessed from the surface of the Sn0.7Cu solder coating with 50 °C isothermal annealing and up to 600 h ageing time. The assessment of the kinetic growth of Sn whiskers are relative to the stress gradients induced in the solder coating layer [5]. Moreover, it can be clearly observed in Figure 9d that the conductive Sn whiskers able to grow longer may lead to reliability apprehension in miniaturization trends in the electronics industry [22]. This finding did not certainly occur toward the end of the observation as long Sn whiskers have a tendency to break off impulsively [13]. Additionally, Tu et al. validated that it is possible to short circuit as the Sn whiskers could break and fall between two neighboring conductors [8].

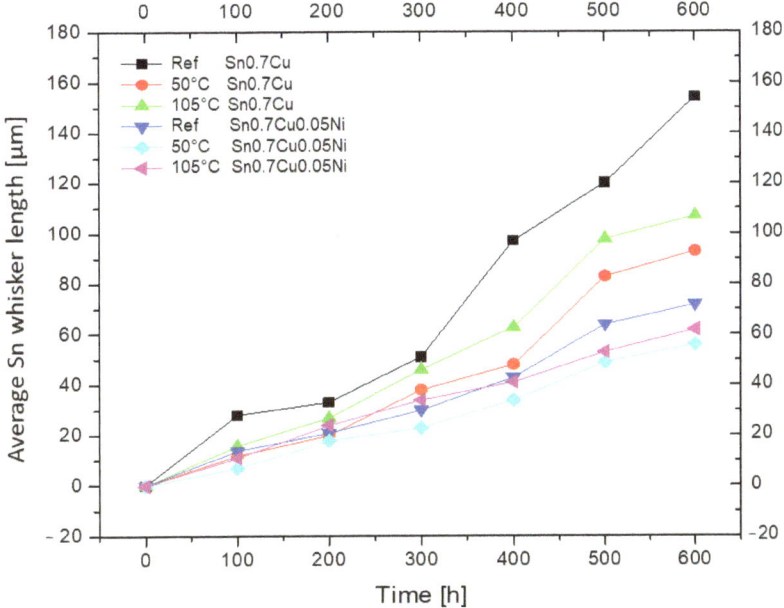

Figure 8. Average Sn whisker length with isothermal ageing.

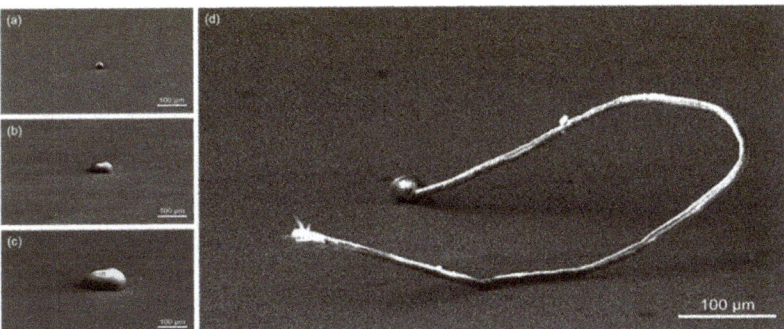

Figure 9. A quantitative evaluation of Sn whisker growth on the surface of the Sn0.7Cu solder joint with isothermal annealing at 50 °C for (**a**)100 h; (**b**) 200 h; (**c**) 400; and (**d**) 600 h.

Figures 10 and 11 show the comparison between Sn whisker growth behavior on the Sn0.7Cu and Sn0.7Cu0.05Ni solder joints with isothermal annealing at 105 °C for 400 h. Sn whiskers of the long filament with a very high aspect ratio were commonly observed on the surface of the Sn0.7Cu solder joint, whereas the short Sn whiskers or hillocks were spotted on the surface of Sn0.7Cu0.05Ni solder joint. Figure 10 presents kinked Sn whiskers growing from the nodule with fine striation marks along the length of the outer surface. The

Sn whisker displays fine striations along the whisker axis grooves along their length that are characteristic of typical Sn whiskers seen in the literature, as Sn whiskers are extruded and tend to propagate from the weakest points of grain boundaries [21] through surface cracks and imperfection sites.

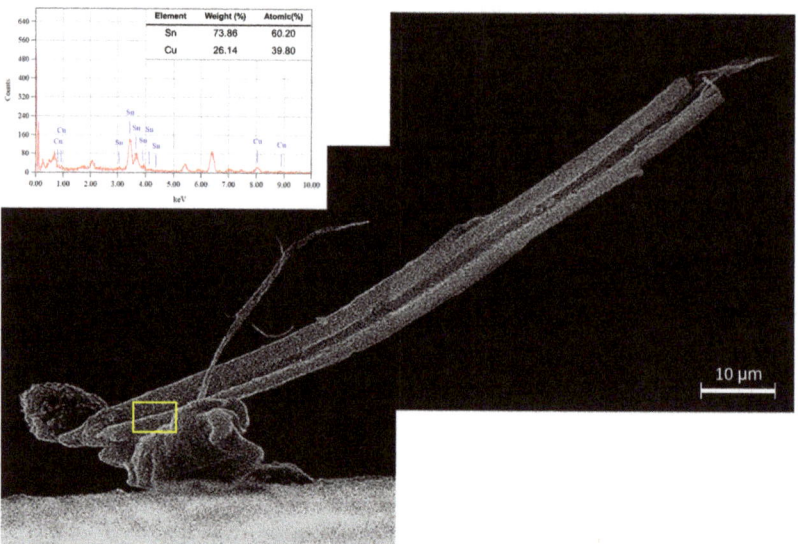

Figure 10. SEM-EDX analyses of kinked Sn whisker growing from nodule on Sn0.7Cu solder joint with isothermal annealing at 105 °C for 400 h.

Figure 11 shows the evaluation results of Sn whisker growth behavior on the Sn0.7Cu0.05Ni solder joint. However, a small composition of 0.05% Ni added to the Sn0.7Cu solder prompted a significant suppression effect on Sn whisker aspect-ratio growth. It was perceived that the Sn whiskers became denser and shorter than Sn whiskers on the Sn0.7Cu solder joint. The rate of Sn whisker growth on the Sn0.7Cu0.05Ni solder joint was much slower than that of the Sn0.7Cu solder joint with a divergent aspect ratio and morphology.

The formation of the Cu_6Sn_5 IMC interfacial layer between the Cu substrate and the Sn-rich solder was closely related to the Sn whisker growth behavior. In view of the formation of the Cu_6Sn_5 IMC interfacial layer as a main stimulus of the compressive stress in the solder coating layer, the effect of isothermal annealing had to be considered a substantial reliability aspect of the Sn whisker acceleration factor. An isothermal-annealing accelerated test provided the thermal kinetic mechanism of a lifetime at the device operation temperature.

Figure 12 presents the growth revolution of the thickness morphology of the $(Cu,Ni)_6Sn_5$ IMC interfacial layer on the Sn0.7Cu0.05Ni solder joint after isothermal ageing for 300 h at room temperature, annealed at 50 °C and 105 °C. It was revealed that the thickness of the $(Cu,Ni)_6Sn_5$ IMC interfacial layer was upsurged and uniform relative to the isothermal ageing temperature and duration. The average thickness of the $(Cu,Ni)_6Sn_5$ IMC interfacial layer aged for 300 h at room temperature was 1.34 µm and increased to 3.52 µm after annealing at 50 °C, before increasing to 5.48 µm after annealing at 105 °C. It proposes that Cu atoms diffuse more intensely through higher activation energy in the Sn solder layer at the isothermal annealing temperature. The activation energy and growth rate of the $(Cu,Ni)_6Sn_5$ IMC interfacial layer on the Sn0.7Cu0.05Ni solder joint were stimulated by the isothermal annealing temperature. At 105 °C, the growth of the $(Cu,Ni)_6Sn_5$ IMC interfacial layer with a higher Cu diffusion rate was faster compared to being annealed at 50 °C. Additionally, the uniform and stable IMC interfacial layers could be protective layers for Sn whisker nucleation from the solder joint [15]. In addition, isothermal annealing possibly significantly decelerates the irregular growth of Cu_6Sn_5 IMC interfacial layer against bulk diffusion and reduce lattices imperfections thus diminish residual stress in solder coating [15].

Figure 11. SEM-EDX analyses of faceted Sn whisker on Sn0.7Cu0.05Ni solder joint with isothermal annealing at 105 °C for 400 h.

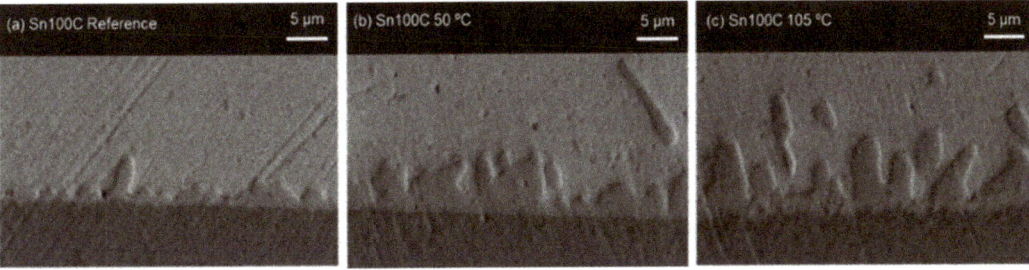

Figure 12. An assessment of the $(Cu,Ni)_6Sn_5$ IMC interfacial layer on the Sn0.7Cu0.05Ni solder joint after ageing for 300 h at (**a**) room temperature, (**b**) isothermal annealing 50 °C, and (**c**) isothermal annealing 105 °C.

In correlation with the stress generation due to the formation of the IMC interfacial layer and the growth rate of the Sn whiskers, it is supposed that the stress is relieved in sequence to the fast atomic diffusion of the IMC interfacial layer [8], and therefore reduces the stress gradient of Sn whisker growth. In this study, the formation of the Cu_3Sn IMC interfacial layer was not obviously apparent, possibly because Sn0.7Cu0.05Ni suppressed the formation of the Cu_3Sn IMC interfacial layer [23,24]. It was observed that the formation of the Cu_6Sn_5 and Cu_3Sn IMC interfacial layer during isothermal annealing was able to be function as a continuous diffusion barrier and was thus less prone to whisker formation [8,15].

It is marked that the grain structure of the IMC interfacial layer also intensely regulates the behavior of Sn whisker growth. Figure 13 show the microstructure and particle size analyses of the Cu_6Sn_5 and $(Cu,Ni)_6Sn_5$ IMC interfacial layer after isothermal ageing for 300 h at 50 °C. It is remarkable that the average grain size of $(Cu,Ni)_6Sn_5$ was 1.143 µm, which is considerably smaller than average grain size of Cu_6Sn_5 1.953 µm. It was established by Horvath et al. that a larger grain size induced more susceptibility to Sn whisker growth. The smaller grain sizes result in smaller grain boundary areas, which is a consequence for Sn whisker grains as it takes less stress to cause grain boundary sliding. Thus, the Sn0.7Cu solder joint is more vulnerable to whiskering than Sn0.7Cu0.05Ni [13].

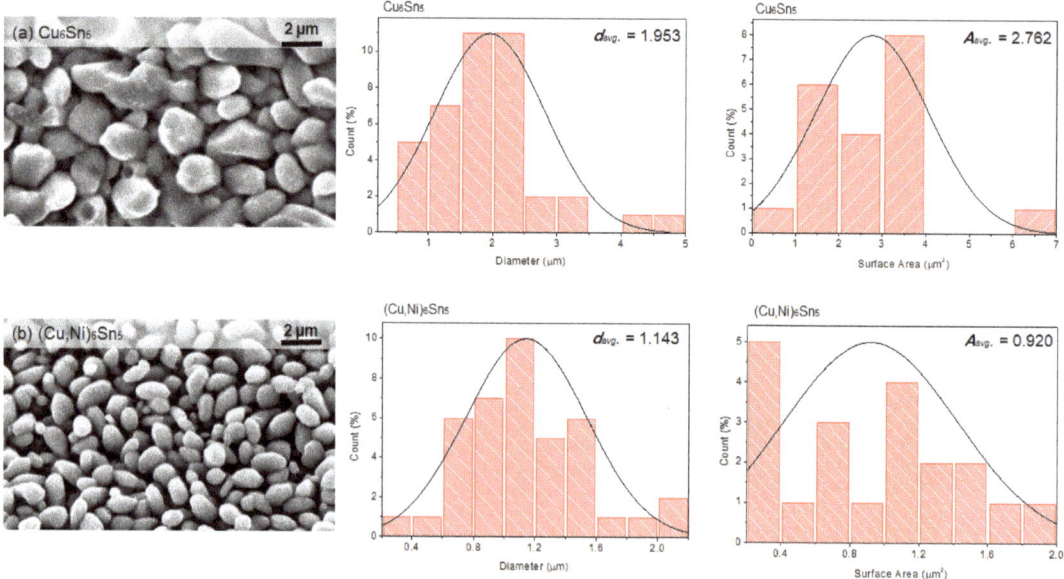

Figure 13. Morphology and particle size analyses of IMC interfacial layer after aging for 300 h at 50 °C: (**a**) Cu_6Sn_5 and (**b**) $(Cu,Ni)_6Sn_5$.

The accompanying aspect that inclines the invulnerable growth of the Sn whisker in the Sn0.7Cu0.05Ni solder joint is the phase stabilization of the IMC interfacial layer. Figure 14 shows a summary of the binary phase equilibrium for the Sn-Cu and Sn-Ni system using Thermo-Calc software. It presents the allotropic transformation of Cu_6Sn_5 IMC, occurring at a temperature that falls approximately below 186 °C, which causes a structural change from hexagonal η-Cu_6Sn_5 to monoclinic η'-Cu_6Sn_5 [24]. Meanwhile, the $(Cu,Ni)_6Sn_5$ IMC maintains stability in the hexagonal η-Cu_6Sn_5 phase. The stabilizing effect of the crystal structure the during cooling process in the hexagonal η-Cu_6Sn_5 phase inhibits volume modifications that may possibly come up with the residual stress diminished in the IMC interfacial layer [14,25,26]. The thermal reduction difference in the Sn solder and the substrate for the duration of the cooling process from the deposition soldering temperature correspondingly contributes to Sn whisker formation. In addition, the transformation of volume expansion in the intermetallic layer is significant to the properties of lower density of the IMC interfacial layer related to the Cu substrate. The volume expansion also produces compressive stresses of the Cu and Sn interface in the vertical direction against Sn solder coating [2]. Therefore, it is suggested that the stability characteristic of the $(Cu,Ni)_6Sn_5$ IMC is able to suppress Sn whisker growth on the Sn0.7Cu0.05Ni solder joint [14].

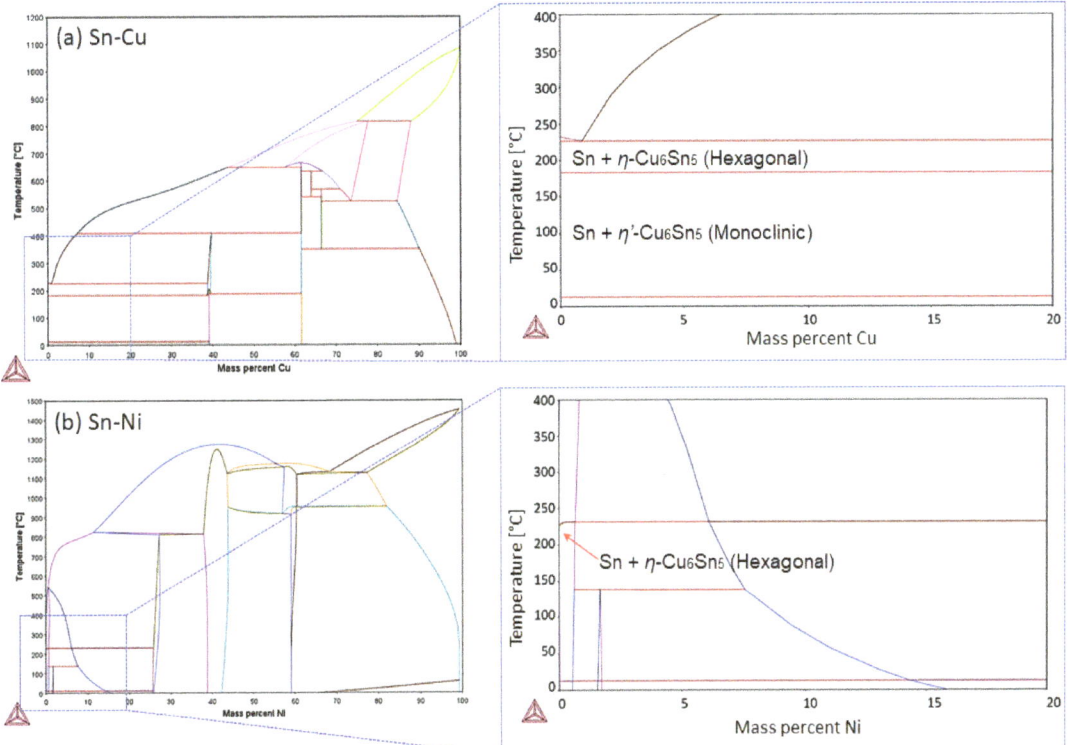

Figure 14. The binary phase diagram of (**a**) Sn-Cu system and (**b**) Sn-Ni system, simulated using Calphad method, Thermo-Calc-2021a database TCSLD v3.3.

4. Conclusions

The assessment of Sn whisker growth behavior on the surface of the Sn0.7Cu0.05Ni solder joint using the hot-dip soldering technique was examined after ageing to 600 h at room temperature, annealed under 50 °C and 105 °C conditions. The obtained conclusions are as follows:

(i) The Sn whisker growth on the Sn0.7Cu0.05Ni solder joint was significantly mitigated with isothermal annealing.

(ii) Isothermal annealing stimulates the activation energy and faster atomic diffusion of the IMC interfacial layer and hence reduces the stress gradient for Sn whisker growth on the Sn0.7Cu0.05Ni solder joint.

(iii) The stability characteristic of hexagonal η-Cu_6Sn_5 significantly contributes to the residual stress diminished in the IMC interfacial layer of $(Cu,Ni)_6Sn_5$ IMC and is able to suppress the formation and growth of Sn whiskers on the Sn0.7Cu0.05Ni solder joint.

The findings of this study provide environmental acceptance with the aim of suppressing Sn whisker growth and upsurging the reliability of the Sn0.7Cu0.05Ni solder joint at electronic-device-operation temperature.

Author Contributions: Conceptualization, methodology, and writing, A.N.H.; supervision and resources and editing, M.A.A.M.S.; methodology, formal analysis, and investigation, M.M.R.; methodology, data software, and formal analysis, M.M.A.B.A., P.V., I.G.S. and A.V.S. All authors have read and agreed to the published version of the manuscript.

Funding: This research was funded by Fundamental Research Grant Scheme (FRGS) Malaysia, grant number FRGS/1/2020/TK0/UNIMAP/03/18. This publication was also supported by TUIASI from the University Scientific Research Fund (FCSU).

Institutional Review Board Statement: Not applicable.

Informed Consent Statement: Not applicable.

Data Availability Statement: Not applicable.

Acknowledgments: The authors gratefully acknowledge the Centre of Excellent Geopolymer and Green Technology (CeGeoGTech), UniMAP, SIG Metal Processing & Metallurgy, Department of Materials Faculty of Chemical Engineering & Technology, Universiti Malaysia Perlis.

Conflicts of Interest: The authors declare no conflict of interest.

References

1. Boettinger, W.J.; Johnson, C.E.; Bendersky, L.A.; Moon, K.W.; Williams, M.E.; Stafford, G.R. Whisker and Hillock formation on Sn, Sn–Cu and Sn–Pb electrodeposits. *Acta Mater.* **2005**, *53*, 5033–5050. [CrossRef]
2. Illés, B.; Skwarek, A.; Bátorfi, R.; Ratajczak, J.; Czerwinski, A.; Krammer, O.; Medgyes, B.; Horváth, B.; Hurtony, T. Whisker growth from vacuum evaporated submicron Sn thin films. *Surf. Coat. Technol.* **2017**, *311*, 216–222. [CrossRef]
3. Lim, H.P.; Ourdjini, A.; Bakar, T.A.A.; Tesfamichael, T. The effects of humidity on tin whisker growth by immersion tin plating and tin solder dipping surface finishes. *Procedia Manuf.* **2015**, *2*, 275–279. [CrossRef]
4. Suganuma, K.; Baated, A.; Kim, K.-S.; Hamasaki, K.; Nemoto, N.; Nakagawa, T.; Yamada, T. Sn whisker growth during thermal cycling. *Acta Mater.* **2011**, *59*, 7255–7267. [CrossRef]
5. Jagtap, P.; Jain, N.; Chason, E. Whisker growth under a controlled driving force: Pressure induced whisker nucleation and growth. *Scr. Mater.* **2020**, *182*, 43–47. [CrossRef]
6. Pei, F.; Jadhav, N.; Chason, E. Correlation between surface morphology evolution and grain structure: Whisker/hillock formation in Sn-Cu. *JOM* **2012**, *64*, 1176–1183. [CrossRef]
7. Fukuda, Y.; Osterman, M.; Pecht, M. The impact of electrical current, mechanical bending, and thermal annealing on tin whisker growth. *Microelectron. Reliab.* **2007**, *47*, 88–92. [CrossRef]
8. Tu, K.; Chen, C.; Wu, A.T. Stress analysis of spontaneous Sn whisker growth. *J. Mater. Sci. Mater. Electron.* **2007**, *18*, 269–281. [CrossRef]
9. Chen, Y.-J.; Chen, C.-M. Mitigative tin whisker growth under mechanically applied tensile stress. *J. Electron. Mater.* **2009**, *38*, 415–419. [CrossRef]
10. Mathew, S.; Wang, W.; Osterman, M.; Pecht, M. Assessment of Solder-Dipping as a Tin Whisker Mitigation Strategy. *IEEE Trans. Compon. Packag. Manuf. Technol.* **2011**, *1*, 957–963. [CrossRef]
11. Baated, A.; Hamasaki, K.; Kim, S.S.; Kim, K.-S.; Suganuma, K. Whisker Growth Behavior of Sn and Sn Alloy Lead-Free Finishes. *J. Electron. Mater.* **2011**, *40*, 2278. [CrossRef]
12. Horvath, B.; Illes, B.; Harsanyi, G. Investigation of tin whisker growth: The effects of Ni and Ag underplates. In Proceedings of the 2009 32nd International Spring Seminar on Electronics Technology (ISSE 2009), Brno, Czech Republic, 13–17 May 2009; pp. 1–5.
13. Horváth, B.; Illés, B.; Shinohara, T.; Harsányi, G. Whisker growth on annealed and recrystallized tin platings. *Thin Solid Films* **2012**, *520*, 5733–5740. [CrossRef]
14. Hashim, A.N.; Salleh, M.A.A.M.; Sandu, A.V.; Ramli, M.M.; Yee, K.C.; Mokhtar, N.Z.M.; Chaiprapa, J. Effect of Ni on the Suppression of Sn Whisker Formation in Sn-0.7 Cu Solder Joint. *Materials* **2021**, *14*, 738. [CrossRef] [PubMed]
15. Lee, H.-X.; Chan, K.-Y.; Shukor, M.H.A. Effects of annealing on Sn whisker formation under temperature cycling and isothermal storage conditions. *IEEE Trans. Compon. Packag. Manuf. Technol.* **2011**, *1*, 1110–1115. [CrossRef]
16. Gain, A.K.; Zhang, L. Effects of Ni nanoparticles addition on the microstructure, electrical and mechanical properties of Sn-Ag-Cu alloy. *Materialia* **2019**, *5*, 100234. [CrossRef]
17. Ramli, M.; Salleh, M.; Abdullah, M.; Zaimi, N.; Sandu, A.; Vizureanu, P.; Rylski, A.; Amli, S. Formation and Growth of Intermetallic Compounds in Lead-Free Solder Joints: A Review. *Materials* **2022**, *15*, 1451. [CrossRef]

18. Ramli, M.; Salleh, M.; Sandu, A.; Amli, S.; Said, R.; Saud, N.; Abdullah, M.; Vizureanu, P.; Rylski, A.; Chaiprapa, J.; et al. Influence of 1.5 wt.% Bi on the Microstructure, Hardness, and Shear Strength of Sn-0.7Cu Solder Joints after Isothermal Annealing. *Materials* **2021**, *14*, 5134. [CrossRef]
19. Salleh, M.; Al Bakri, A.; Somidin, F.; Sandu, A.; Saud, N.; Kamaruddin, H.; McDonald, S.; Nogita, K. Comparative Study of Solder Properties of Sn-0.7Cu Lead-free Solder Fabricated Via the Powder Metallurgy and Casting Methods. *Rev. De Chim.* **2013**, *64*, 725–728.
20. Saleh, N.; Ramli, M.; Salleh, M.M. Effect of Zinc Additions on Sn-0.7 Cu-0.05 Ni Lead-Free Solder Alloy. In *IOP Conference Series: Materials Science and Engineering*; IOP Publishing: Bristol, UK, 2017; p. 012012.
21. Kim, K.-S.; Yu, C.-H.; Yang, J.-M. Behavior of tin whisker formation and growth on lead-free solder finish. *Thin Solid Films* **2006**, *504*, 350–354. [CrossRef]
22. Diyatmika, W.; Chu, J.; Yen, Y. Effects of annealing on Sn whisker formation: Role of Cu alloy seed layer. In Proceedings of the 2013 8th International Microsystems, Packaging, Assembly and Circuits Technology Conference (IMPACT), Taipei, Taiwan, 22–25 October 2013; pp. 319–322.
23. Salleh, M.M.; McDonald, S.; Nogita, K. Effects of Ni and TiO_2 additions in as-reflowed and annealed Sn0.7Cu solders on Cu substrates. *J. Mater. Process. Technol.* **2017**, *242*, 235–245. [CrossRef]
24. Nogita, K.; Gourlay, C.; Nishimura, T. Cracking and phase stability in reaction layers between Sn-Cu-Ni solders and Cu substrates. *JOM* **2009**, *61*, 45–51. [CrossRef]
25. Nogita, K.; Mu, D.; McDonald, S.; Read, J.; Wu, Y. Effect of Ni on phase stability and thermal expansion of $Cu_{6-x}Ni_xSn_5$ (X = 0, 0.5, 1, 1.5 and 2). *Intermetallics* **2012**, *26*, 78–85. [CrossRef]
26. Nogita, K. Stabilisation of Cu6Sn5 by Ni in Sn-0.7 Cu-0.05 Ni lead-free solder alloys. *Intermetallics* **2010**, *18*, 145–149. [CrossRef]

Disclaimer/Publisher's Note: The statements, opinions and data contained in all publications are solely those of the individual author(s) and contributor(s) and not of MDPI and/or the editor(s). MDPI and/or the editor(s) disclaim responsibility for any injury to people or property resulting from any ideas, methods, instructions or products referred to in the content.

Article

Sustainable Packaging Design for Molded Expanded Polystyrene Cushion

Normah Kassim [1,2,*], Shayfull Zamree Abd Rahim [1,2], Wan Abd Rahman Assyahid Wan Ibrahim [1,2], Norshah Afizi Shuaib [1,2], Irfan Abd Rahim [1,2], Norizah Abd Karim [1,2], Andrei Victor Sandu [3], Maria Pop [4], Aurel Mihail Titu [5,*], Katarzyna Błoch [6] and Marcin Nabiałek [6]

[1] Faculty of Mechanical Engineering and Technology, University Malaysia Perlis, Arau 02600, Perlis, Malaysia
[2] Center of Excellence Geopolymer and Green Technology (CEGeoGTech), Universiti Malaysia Perlis, Kangar 01000, Perlis, Malaysia
[3] Faculty of Material Science and Engineering, Gheorghe Asachi Technical University of Iasi, 41 D. Mangeron St., 700050 Iasi, Romania
[4] Faculty of Civil Engineering, Technical University of Cluj Napoca, Constantin Daicoviciu No 15, 40020 Cluj Napoca, Romania
[5] Industrial Engineering and Management Department, Faculty of Engineering, "Lucian Blaga" University of Sibiu, 10 Victoriei Street, 550024 Sibiu, Romania
[6] Faculty of Mechanical Engineering and Computer Science, Częstochowa University of Technology, 42-201 Częstochowa, Poland
* Correspondence: normah@unimap.edu.my (N.K.); mihail.titu@ulbsibiu.ro (A.M.T.)

Abstract: A molded expanded polystyrene (EPS) cushion is a flexible, closed-cell foam that can be molded to fit any packing application and is effective at absorbing shock. However, the packaging waste of EPS cushions causes pollution to landfills and the environment. Despite being known to cause pollution, this sustainable packaging actually has the potential to reduce this environmental pollution because of its reusability. Therefore, the objective of this study is to identify the accurate design parameter that can be emphasized in producing a sustainable design of EPS cushion packaging. An experimental method of drop testing and design simulation analysis was conducted. The effectiveness of the design parameters was also verified. Based on the results, there are four main elements that necessitate careful consideration: rib positioning, EPS cushion thickness, package layout, and packing size. These parameter findings make a significant contribution to sustainable design, where these elements were integrated directly to reduce and reuse packaging material. Thus, it has been concluded that 48 percent of the development cost of the cushion was decreased, 25 percent of mold modification time was significantly saved, and 27 percent of carbon dioxide (CO_2) reduction was identified. The findings also aided in the development of productive packaging design, in which these design elements were beneficial to reduce environmental impact. These findings had a significant impact on the manufacturing industry in terms of the economics and time of the molded expanded polystyrene packaging development.

Keywords: closed-cell foam; drop test; finite element analysis; design parameter; protective packaging

1. Introduction

Packaging design is an important element in the packaging value chain along the product distribution process because it will determine the effectiveness of the protected product [1,2]. The packaging design also functions as a crucial component in the packaging value chain because it determines the materials, manufacturing process, and environmentally friendly options for disposal [3,4]. The molded expanded polystyrene (EPS) cushion serves as protective packaging to prevent a product from shock and vibration impacts [4–6]. This packaging application has been extensively utilized across many industries, not only in food packaging, but also for all kinds of fast-moving consumer goods and household and electronic appliances [7–11]. Presently, the most popular polymer-based cushioning for home appliances is polystyrene (PS) foam [12]. Polystyrene foam is classified into two types: expanded rigid open-cell, which produces loose-fill foam, and expanded flexible closed-cell, which can be molded into a variety of shapes with a normal density range from 11 to 32 kg/m^3 [13–16]. Due to its high impact resistance, its ultralightweight, durable, low thermal conductivity, and its ability to be molded into any shape or size, this rigid foam has been used as a shock absorber (inserts) in a variety of packaging appliances [7,12,17–21].

Today's manufacturing industry is rapidly evolving to provide competitive products [6,22]. This product development is also improving in parallel with new technologies nowadays [23]. However, global production will continue wreaking havoc on society and the environment if these tendencies are not promptly addressed [24]. Hence, in the long run, packaging designers should consider designs that are sustainable and environmentally friendly to reduce pollution, prioritize environmental protection in design, and provide a design concept that adheres to ecological ethics [23]. Concurrently, the development of packaging designs involves a variety of disciplines such as history, science, engineering, economics, and social responsibility, all of which are heavily reliant on the engineer's knowledge and experience [25,26].

Packaging design has given a comprehensive surge, considering multiple factors appropriate to the complexity of sustainability challenges [27]. Thereby, packaging design for sustainability exemplifies the fact that it is made up through multiple elements: the natural environment, society, and economic performance [28]. Continuous improvement is necessary for packaging sustainability, and even minor modifications may lead to substantial gains for the environment (material), economy (cost), and society (manufacture) [29,30].

Hence, in the development phase, designers' decisions play a key role in ensuring that environmental repercussions are reduced. Approximately 80% of a product's environmental impact is defined at the design stages of the product development process [31]. Designers are in charge of specifying the material selection, how raw materials are processed or manufactured, and how products are packaged, distributed, used, and eventually disposed of [11]. Hence, this paper presents the investigation of design parameters that must be prioritized in producing a sustainable design of EPS cushion packaging. Currently, many literature reviews have emphasized the importance of EPS cushions as protection to ensure product safety [1,15,32–34]. Product packaging influences recycling behavior; thus, recyclability of the packaging should be considered a precious value of the packaging, allowing multidisciplinary research due to the complexity of the recycling behavior, between user waste management and the technical part of the system [35]. So, this study investigates the relationship between packaging design and environmental sustainability, discovering a positive impact on the industrial community and economy, particularly in the home appliance-manufacturing sector. The concept of sustainable design development entails using natural resources by protecting environmental values, reducing the use of resources, and elevating the quality of life [36]. Therefore, sustainable design is defined as environmentally responsible product design and development that integrates a product life-cycle perspective and incorporates work, culture, and organizational skill approaches. In terms of applying sustainable packaging design, the evaluation criteria applied to products are reducing material usage and diversity, reducing energy consumption, reusing the product, and reducing the weight and volume of the product [37]. Monteiro et. al. found that 58 percent of respondents believe that sustainable design is extremely important for the packaging industry [38].

Thus, destructive testing and a preliminary finite element analysis simulation are carried out to determine the main parameter of sustainable design of EPS cushion packaging. Cushion reliability impact is used to manage risks and understand how they affect the quality of the product, allowing design considerations to be made [39], whereby the acceleration response and strain histories of simulation results are correlated with experimental measurements [40,41]. The explicit method is preferred over other methods to simulate the drop tests [42]. Drop testing and the finite element analysis method are employed in the development process to evaluate design flaws and to see potential product damage caused by packaging design weakness (potential to be reused and reduced). The detailed results of both methods are then compared. The results will be used to produce a new molded expanded polystyrene packaging.

2. Methods

This study used two particular quantitative methods (see Figure 1), which are drop-test verification and finite element analysis validation. The drop test is performed to detect serious damage and to determine the dependability of the packaging design in order to protect products of varying weights or sizes [2]. Time and cost are major constraints, where the empirical approach is applied repeatedly [43]. Provisionally, the drop test was carried out in accordance with the standard procedure ASTMD5276-19 and met the requirements of ISO Standards 2206:1987 and 2248:1985. Furthermore, the finite element analysis (FEA) approach was carried out to compare the findings, and the effectiveness of using explicit analysis was demonstrated. The analysis was simulated using ANSYS software version 2019. Finally, the design parameters for sustainable cushion packaging were identified.

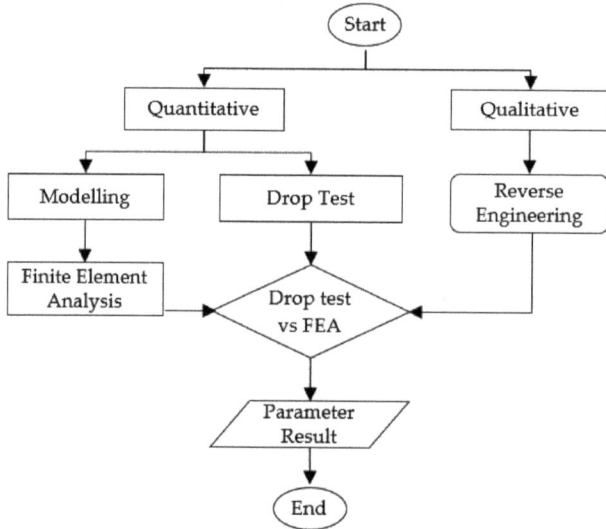

Figure 1. Flowchart of design and development method.

2.1. Drop Test

The design verification of cushion packaging was conducted using a drop-test apparatus (illustrated in Figure 2). The results were compared to FEA simulation analysis to identify the area of defects and possible impacts on products such as cracks, dents, or breaks. There are five designs of EPS cushion packaging that have been tested from different heights, depending on gross weight and packaging surface, as performed using the test apparatus mentioned in JIS Z0212 standard [41] and also as referenced in ASTM D5276 [16].

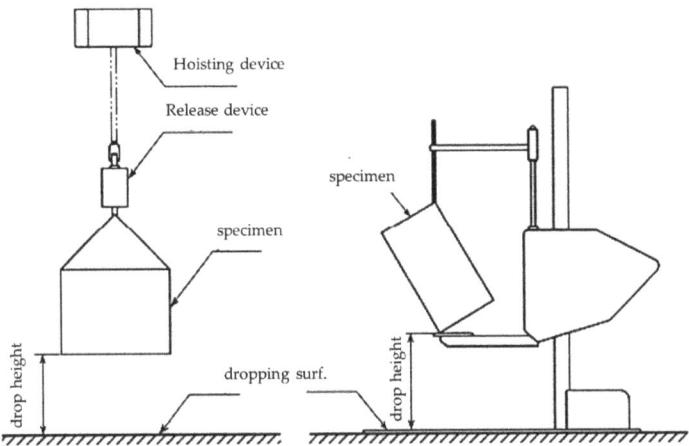

Figure 2. Drop-test apparatus [41].

The different sizes of packaging with similar appearance designs (see Table 1) were selected to study the drop impact against cushion packaging, as shown in Figure 3. This is the primary factor used to determine the most important design parameters to incorporate. These five types of packaging design were also chosen due to the variety of container loading efficiency (CLE) quantities, which contributes to comprehensive comparison and validates the main parameter of sustainable packaging design. These primary parameters will be followed in the design and manufacture of sustainable molded EPS packaging.

Table 1. Packaging size.

Model	Package Size (L × W × H)	Types of Rib	Rib Position
A	1684 × 506 × 1080	Inner and Outer	Symmetric
B	1332 × 170 × 827	Inner and Outer	Symmetric and Asymmetric
C	1292 × 177 × 764	Inner	Symmetric and Asymmetric
D	1187 × 158 × 732	Inner and Outer	Symmetric
E	1016 × 152 × 625	Inner	Symmetric and Asymmetric

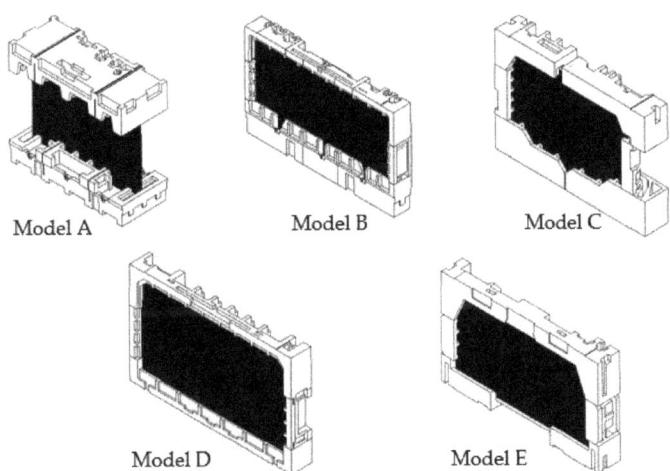

Figure 3. Molded EPS packaging design models.

Drop test was performed on pilot package sample. The complete assembly of molded cushion packaging and the product itself are used to assess the possibility of damage during transportation or handling processes. The drop height is specified in Table 2. The surfaces for drop test are shown in Figure 4. The drop-test packaging must then be successful in order to protect the product from deformation defects in the ±2.0 mm range and avoid defects on the product, such as cracks, bezel damage, scratches, light leakage from the screen, and component damages.

Table 2. Drop-test height.

Model	Gross Weight, W (kg)	Height, h (cm) of Surface Dropped			
		Bottom	Front Rear	Right Left	Corner Edges
A	57	30	25	25	25
B	23	40	36	36	36
C	19	50	40	40	36
D	14	55	45	45	36
E	12	55	45	45	36

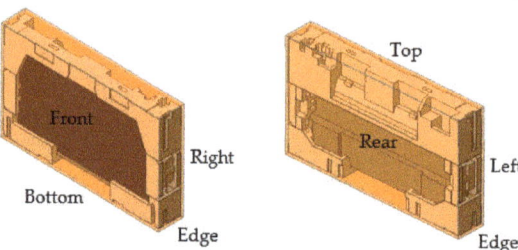

Figure 4. View of package surfaces.

The allowed height range should be ±2% or ±10 mm, whichever is greater.

Table 3 shows the test sequences that can be partially eliminated, depending on the type of packaged freight. Then, the test sequence can be modified, but prior agreement with the test requester is required. Furthermore, the chosen corners and edges to be tested will be the wake ones because corner drop testing helps determine the ability of the contents inside the packaging system to withstand rough handling [44–46].

Table 3. Drop sequence and number of times [41].

Sequence	Portion to Be Impacted	Test Times
1	Bottom adjacent corners Ex. Corner 2-3-5	1
2	Side adjacent edges Ex. Edge 3-5	1
3	Bottom-side face edge Ex. Edge 2-3	1
4	Front-side face edge Ex. Edge 2-5	1
5~10	All 6 faces	6
	Totals	10

The package must be placed in the position intended for transportation. However, if it is known, it must be placed vertically on the observer's right. When the package is positioned with one side facing the observer, the upper surface of the package is identified as No. 1, the side on the observer's right as No. 2, the bottom as No. 3, the surface on the observer's left as No. 4, the nearest side (front) as No. 5, and the side farthest away or rear as No. 6 (see Figure 5) [45].

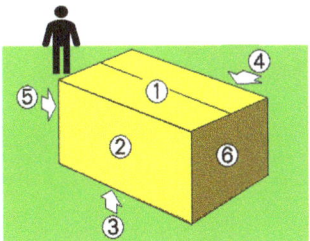

Figure 5. Drop-test placement if face direction are known [41].

The results of the drop test are as follows: Three EPS cushion packaging models encountered a serious design failure on the screen panel, as shown in Figure 6, while two other models faced a minor design failure on models B and E, as shown in Figure 7. The findings of the summary found that the drop-test results of models A, C, and D show that the screen panel is cracked in several different places. Model A demonstrated that almost the entire panel was cracked, beginning at the bottom of the product. For model D, the insufficient cushion thickness puts pressure on the bezel, which directly causes screen damage. When it was decided to use two pieces of packaging, the starting point of the screen panel crack of model C was at the top of the product. This area was not protected by cushion packaging, as shown in Figure 3. Next, a drop test for model B found a scratch defect on the screen surface of the product. Moreover, a light leakage defect occurred on the bottom packaging of model E. This illumination leak was caused by the design rib of cushion at the bottom, and this flaw also led to an unclear panel quality of the product while in use.

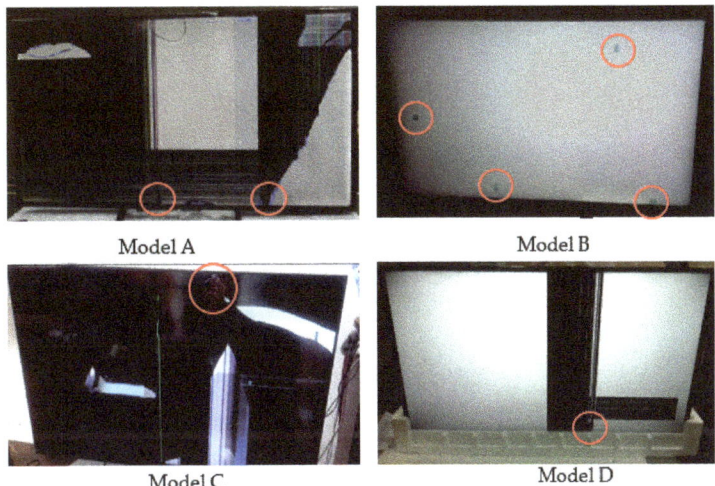

Figure 6. Drop test result.

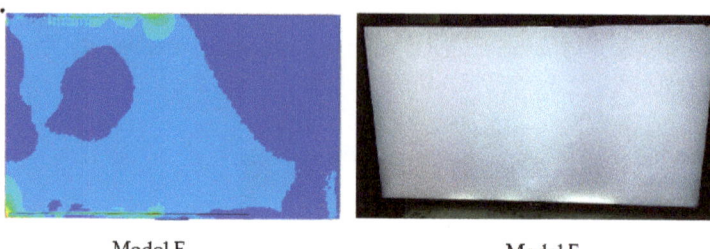

Figure 7. Drop-test results.

The root cause of failure was discovered to be caused by factors such as rib positioning, thickness, and cushion shape itself due to the disparate levels of expertise among packaging-design engineers. Therefore, finite element analysis is performed to compare the design parameters in detail, and the most critical design parameters are identified from the critical area in research findings (see Figures 8–13).

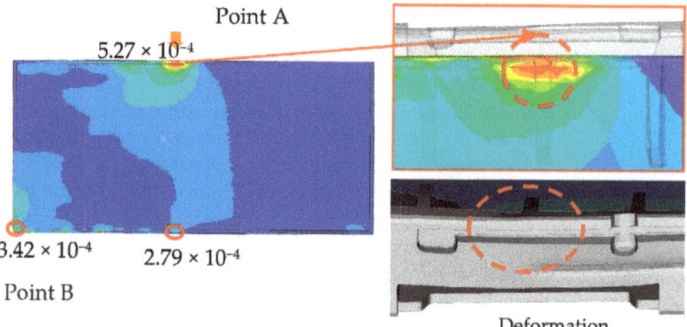

Figure 8. Impact of EPS cushion design on product performance.

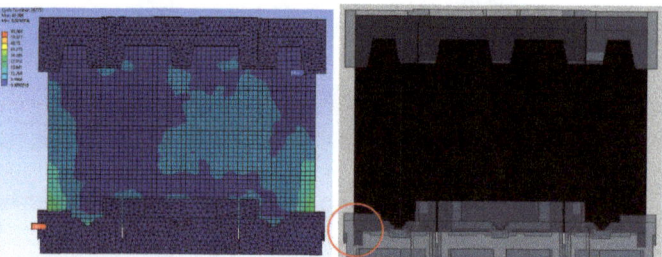

Figure 9. Analysis result of model A.

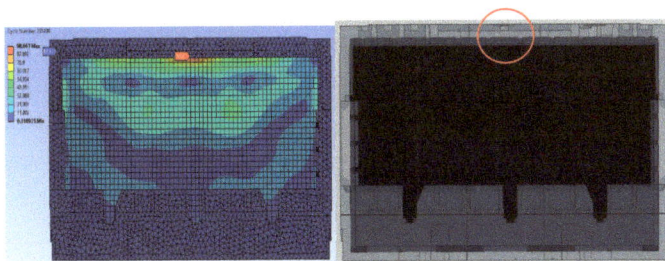

Figure 10. Analysis result of model B.

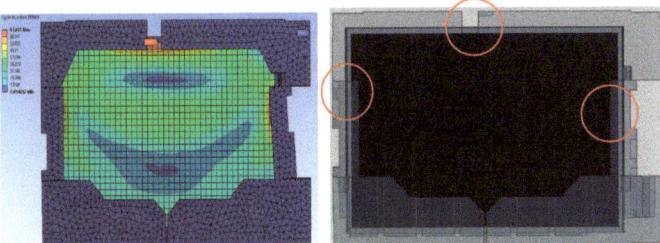

Figure 11. Analysis result of model C.

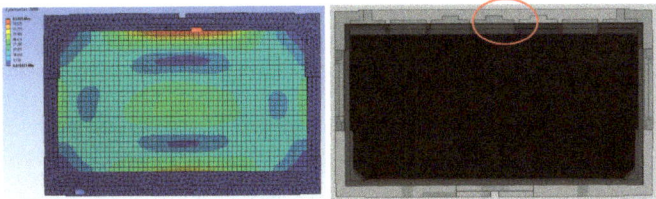

Figure 12. Analysis result of model D.

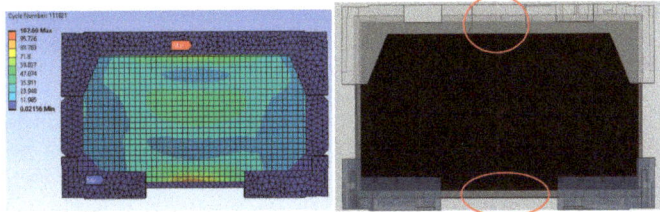

Figure 13. Analysis result of model E.

2.2. Finite Element Analysis

Expanded polystyrene foam is the best rigid substance to use as protective packaging for household appliances, but its sustainability is in doubt. Therefore, its reuse and reducing waste were the biggest contributions from packaging-design engineers [47]. Additionally, this study also agreed that the analysis is one of the effective testing methods that should be carried out during the product development process [5,10,12,42,48]. However, it is important to follow the cushion design parameters in order of importance. At the same time, analysis of structural appliances was also considered, such as positioning of electronic board, screen panel, speaker bracket, etc. [2,5,10]. So, the material properties of products and packaging were obtained, as shown in Table 4, for further analysis. The packaging engineer's expertise will also be at an equivalent standard. Packaging design weaknesses are overcome, and cushion waste is reduced.

Table 4. Material properties for analysis.

Components	Cushion	Box	Television	Screen
Material	EPS	Corrugated board	PPE + PS	Glass
Density (Kg/m^3)	18–20	610	1090	1170
Poisson's ratio	0.4	0.34	0.37	0.23

Explicit dynamic analysis was used in this finite element analysis. It is set that the product's packaging is dropped with a gravity parameter of 9806.6 mm/s and the floor is in a fixed rigid setting. Product damage is predicted. Improvements in sustainable design were implemented at an early stage of the product development cycle. The maximum stress and deformation direction are used as a reference in the analysis result to predict the critical area that would have design failure. The FEA analysis was compared to the results of the drop tests.

Figure 8 shows the impact of EPS cushion design on product performance, where the critical areas are located at point A and point B. The current design (design A) was modified before the molding tool development stage. To reduce the impact on the product's front surface, the thickness of the EPS cushion was reduced. High strain is thought to occur as a result of the pushing effect of the set locally, where the cushion ribs are located.

2.3. Analysis Result of Drop Test

The comparison of packaging design model analysis results in numerical maximum stress values. The maximum stress area was also compared, as shown in Figures 9 and 13, implying that design failure is possible. The maximum stress predicted during the drop test is highly mesh-size-dependent; there has to be a balance in choosing the mesh size, such that it yields accurate results and is computationally efficient. Contrary to this study, the optimum mesh size for five models of packaging design is chosen to be 20 mm, with skewness settings used to show that the mesh structure was close to its ideal form. The analysis results revealed that the maximum impact result on the cushion indicated that each of the packages required design improvement before molding development process.

The numerical value of the maximum stress was discovered to be marginally different for each model of packaging that has been studied. The highest value for model A was found on the bottom side of the packaging, while models B and D were found on the rear side of the cushion packaging, as stated in Table 5. Lastly, the maximum stress of model E and C cushion packaging was found on the front and left sides, respectively. This maximum stress was used to predict the possible areas of defects that exist, without going through the mold modification design repeatedly. As a result of the analysis, all of these cushion packages required design improvements. Firstly, the improvement design for model A focused on cushion thickness in order to reduce cushion stiffness and to withstand the hazards' impacts. Models B and D were both modified in terms of rib dimension, and uniform tolerance between the rib, screen panel, and rear cover was ensured. The product's accessory positioning must then be considered in order to determine the optimal rib dimensions (see Figure 14). Finally, models E and C demonstrated a proclivity for screen and clip components to be damaged. As shown in Figure 15, modifying the position of the cushion layout is required due to a lack of support that caused panel screen damage. Although the cushion packaging layout can be determined, packaging size must also be evaluated and calculated, especially when measuring container loading efficiency.

Table 5. Maximum stress result.

Model	Maximum Result of Equivalent (Von-Mises) Stress, MPa						
	Surface (Refer Figure 3)						
	Bottom	Front	Rear	Right	Left	Edge (Right)	Edge (Left)
A	13.464	6.045	7.079	20.47	5.371	2.682	8.259
B	13.45	6.141	11.619	4.81	6.237	4.336	4.873
C	4.5547	5.3831	4.2758	4.402	6.549	4.526	5.17
D	12.58	17.08	22.65	5.558	17.88	3.197	7.526
E	2.99	12.02	8.421	6.511	1.769	1.381	4.352

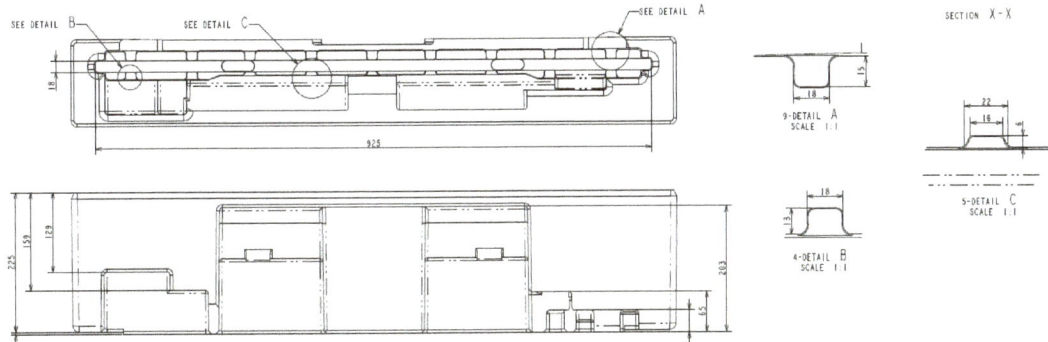

Figure 14. Dimension of rib.

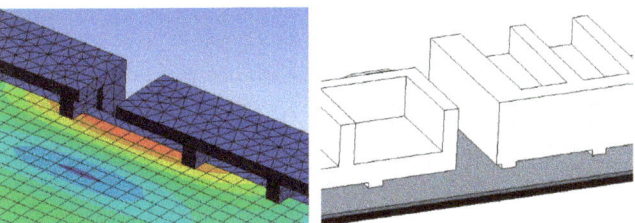

Figure 15. FEA result.

Differences in analysis results are caused by the cushion packaging's productive design. The high strain was close to the most critical component: the screen. This outcome demonstrates that it is likely to damage the screen and clip internal product components. Due to the lack of support on the front and back cushions, as shown in Figure 15, it is necessary to alter the position of the cushion layout on the upper cushion.

In general, the results of all analysis models are influenced by the EPS cushion design parameters as outstanding protective packaging. However, the primary important factors need to be ensured to produce sustainable cushion packaging. The analysis results show that the packaging design of models C and E are seen to be similar, but the cushion layout and length of bottom inner cushion are different. This difference in cushion layout has produced the same analysis results, where the possibility of damage is shown on the critical component, which is the screen panel, as illustrated in Figures 9 and 13.

On the other hand, the design considerations of molded packaging for models D and B are the same in terms of cushion layout design, dimensions, and thickness, even though the product size is different. Both models show the same analysis result, which is the emphasis on the position and dimension of ribs, predominantly on the top of the cushion packaging. However, because the installation of pre-existing accessory components must be taken into account, the cushion packaging model A that has been made differs greatly from all four models. Cushion thickness and rib positioning, moreover, have been identified as the main design factors to emphasize. The comparison results of destructive testing and finite element analysis are discussed below.

A damaged or cracked screen panel is a major problem that has been identified. This indirectly causes product deformation defects, as shown in Figure 16. It shows that the defect happened due to high stress impact at the edges of the cushion packaging, which initially intended to hold the product. Unfortunately, it has produced a cell pop-out defect on the screen panel. The number of ribs and their position have thus been identified as the primary design parameter. At the same time, it is vital to ensure that the mask gap between screen and bezel must be within the specs of the product standard.

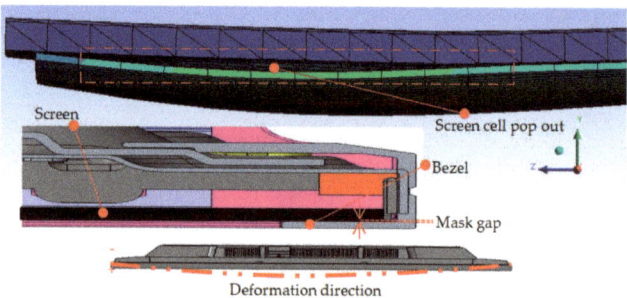

Figure 16. Design failure.

The analysis of design failure concludes that there is a lack of cushion packaging toughness to resist shock loads or impact. The design reliability is lower than expected due to several factors. First, the position of rib was placed on a sensitive area, such as on the middle chassis or on the clip of the panel board. This clip is placed in the front of the screen panel, which is used as a holder for each layer of the screen projector. This clip must avoid being exposed to any pressure or impact. Secondly, there is insufficient rib strength in order to absorb the impact, and the wall thickness of cushion is less than 10 mm (minimum thickness). In fact, the strength of cushion is influenced by the position, height, and thickness of ribs. In addition, the packaging weight also affects cushion stiffness. Therefore, the cushion weight must be reduced, and the design needs to be improvised if necessary. Minor consideration, such as tolerance between packages and goods, must be checked by considering the material types of protection bags. Commonly, the thickness of protection bags ranges from 0.02 mm to 0.4 mm.

In shear modulus (G), the shear stress (τ) is directly proportional to shear strain (γ) as expressed in Equation (1) [49]:

$$G = \tau/\gamma \tag{1}$$

The ratio of normal stress to normal strain within the elastic limit is known as young's modulus of elasticity (E) [13,21,25]. It is a material property and remains constant for a given material. When a material is subjected to shearing load, the ratio of shear stress induced to the corresponding shear strain is a constant within the elastic limit, and this constant is known as the rigidity modulus (G). Poisson's ratio (ν) is another material property which is the ratio of the lateral strain to longitudinal strain when loaded within the elastic limit [50]. The relationship between the two elastic constants is shown as follows.

$$E = 2G(1 + \nu) \tag{2}$$

where E is the Young's modulus, G is the rigidity modulus, and ν is the Poisson's ratio.

Practically, the rigidity modulus (G) value can also be defined through the experimental drop test. The machine used in the drop test was connected to a PC. All parameters were controlled and monitored by the PC. The value was discovered from sensor detector that are attached together inside appliances and packaging, as shown in Figure 17. The highest value among the channels would be selected as the shear modulus of rigidity (G) result. Product fragility is standardly determined by an actual drop test as provided in ASTM D3332. Shock testing can help determine a product or packaging system's level of fragility by measuring the amount of input acceleration required to damage the product's function or cosmetics. It is often measured in terms of fragility [25,43,46].

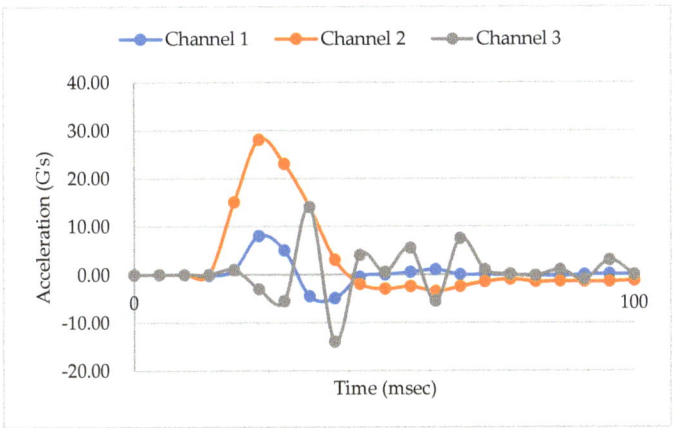

Figure 17. G's value result from drop test.

However, the comparison result shows that the maximum modulus of rigidity that is obtained through the destructive test is in the range of 8 G's to 30 G's per surface of each following model, as illustrated in Figure 18. The results show that the peak acceleration is dominated by the cushion packaging model E, while the lowest value is obtained from the cushion packaging design model A. Therefore, it has been proven that the cushion design will affect the fragility of EPS cushion packaging. The design of packaging model A is more complex than other cushion packaging designs.

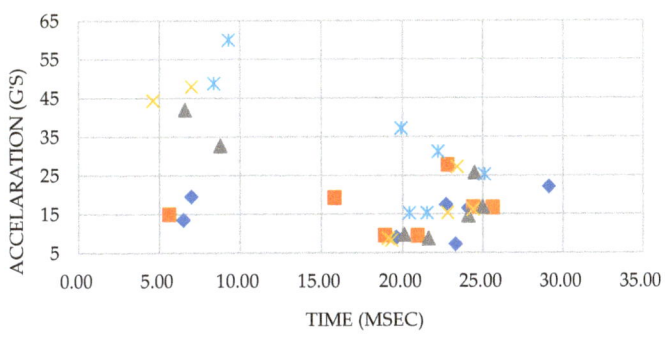

Figure 18. Acceleration (G'S) vs. Time (msec).

3. Results and Discussions

The string of this study has assumed that sustainable EPS cushion design should consider the weight and shape of the cushion (design optimization). Logically, the effectiveness of cushion packaging reliability is based on the complexity of the design or the tendency to reduce overall packaging material usage, which is in line with previous studies [5,10,16,17]. This study has proven that there are four main design parameters that need to be emphasized in designing cushion packaging, starting from the conceptual design phase. At once, the design for sustainability will be able to be implemented in product development, especially in the manufacturing industry. The main parameters to be considered in producing the sustainable design of EPS cushion packaging are graphically shown Figure 19.

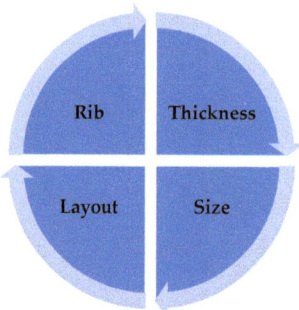

Figure 19. Parameter of cushion design.

Generally, waste from EPS cushions can be recycled in many ways once it comes to the end of its life. The choice of recycling methods is based on technology, environmental, or economic factors. The authors' views are similar to that of the study by Muralikrishna and Manickam (2017), where the interpreted life-cycle assessment is a cradle-to-grave analysis technique to assess environmental impacts associated with all the stages of a product's life: from raw material extraction through to material processing, manufacture, distribution, and use [51]. Accordingly, this study has discovered the main design parameters in the development process of sustainable EPS cushion packaging. Furthermore, the impact of CO_2 reduction that will be employed in the manufacturing sector is disclosed [52], and the waste-management options are evaluated from an economic perspective [53].

Knowledge of specific sustainable product design principles, such as designing for repair/reuse/remanufacture/recycling, is widely recognized as a vital skill and is supported by an earlier study [54,55]. The importance of understanding sustainable product design techniques, as well as how to select and implement the most appropriate ones based on the design challenge, was emphasized [24]. In this current study, an EPS waste hierarchy is introduced as a waste-management priority order: (1) preparing for reducing and reuse, and (2) design optimization, besides evaluating the waste-management options from an economic perspective [53]. Reusable packaging is the first option that designers should try to implement, if possible, because it does not mandate costs for recycling processing and remanufacturing. It is clear that this approach fits the circular economy concept [56].

A combination of the life-cycle assessment (LCA) and life-cycle cost (LCC) analysis was significant [57], allowing for cost savings, a reduction in manufacturing time, and a reduction in carbon dioxide emissions due to the sustainable design. Previous research has indicated that, in order to achieve sustainability, certain markers must be used, and the variables that affect the state of economic, social, and environmental problems must be controlled [58,59].

The ISO 14,040 rules state that an LCA study consists of four stages: defining goal and scope; inventory analysis; impact assessment; and interpretation. This guideline framework defines how much of a product's life cycle is engaged in sustainable design [60]. On the other hand, LCC allows for the identification of potential cost drivers and cost savings for a product or service throughout its entire life cycle (a project or a product from acquisition, installation, operation, and maintenance to the final disposal of the raw material) [61]. The majority of the contributions are focused on direct applications to product development and improvement. Due to this reason, this design parameter has been influencing the reduction in cushion packaging cost and time, as shown in Equations (3) and (4).

$$\text{Cost saving } (\%) = \frac{\text{Total development cost} - \text{total modification cost}}{\text{Total development cost}} \times 100\% \quad (3)$$

$$\text{Time saving } (\%) = \frac{\text{Modification time of mold}}{\text{Total development time}} \times 100\% \quad (4)$$

Finally, this paper was found to be beneficial in saving 48% of the development cost; had a 25% modification time reduction; and estimated a 27% carbon dioxide (CO_2) reduction. The summary of the findings has been summed up into the below accomplishments:

i. Costs were saved by 48% of the total cost of the cushion packaging development (see Figure 20). These average cost savings were identified from the five cushion packaging designs. This percentage will increase drastically if improvements in the cushion packaging designs can be made. At the same time, the implementation of sustainable design needs to be applied in the early stages of the product development life cycle. Four main parameters are emphasized without going through the trial-and-error process. So, it is able to reduce the EPS materials use in the product development cycle. Furthermore, it is proven that the relationship between cost development and effective design will contribute to sustainable EPS cushion packaging, as well as reduce EPS waste by avoiding the trial-and-error methods in the design validation process. Significant cost differences are found in the development of sustainable EPS cushion packaging. Modification costs are higher when improving the design if it is based on conventional practices. However, by incorporating these discoveries into a sustainable design (using the design parameters and reducing the use of EPS cushion), the massive amount of modification costs can be well organized.

ii. In the meantime, the average mold modification time was also saved by 43.5%, as shown in Figure 21, because repeated modification and retest verification can be avoided. This is because the possibility of design defects can be identified using the finite element analysis method as shown in Figure 22. Improvements will be made before the development of the molding tools for EPS cushion packing. This study was successful in validating the use of the analysis method in creating a new cushion design with sustainability elements. The results of the analysis, as shown in Figure 7, previously made it clear that cushion rib and thickness are the main parameters to be considered in order to reduce EPS waste disposal.

iii. The most significant finding is the reduction of carbon dioxide (CO_2) released from sustainable design method. It is estimated that improvements in packaging design, material, and multiple functional uses have cut CO_2 emissions by as much as 27%. The value of CO_2 emission per one set model (kg) was calculated. The comparison of the reduction is shown in Figure 23.

iv. Additionally, it was discovered that the grafting method, which entails using a dovetail technique in rib cushion design, is used in optimization design to input a new parameter (refer Figure 24). This design technique can also be interchanged to fit multiple sizes of products (see Figure 25) and together manage the cushion layout or align a product orientation position.

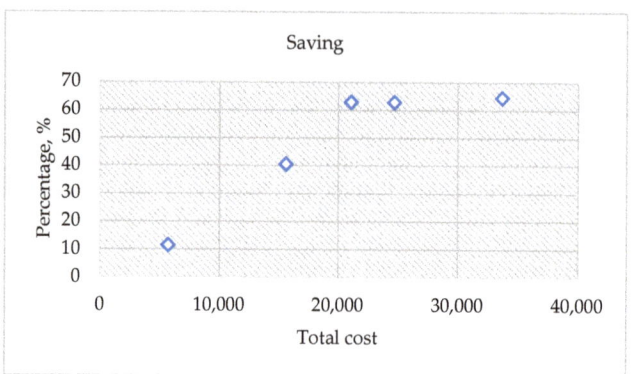

Figure 20. Percentage cost to be saved.

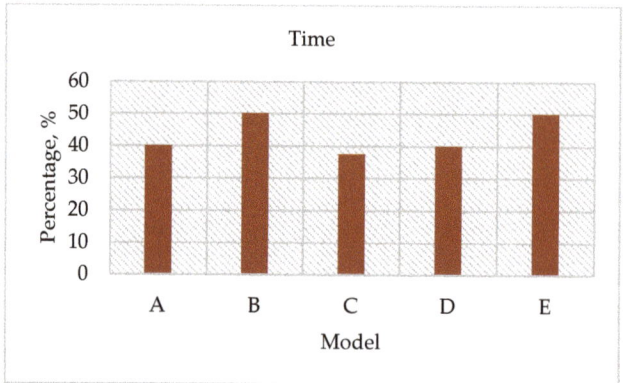

Figure 21. Modification time reduction.

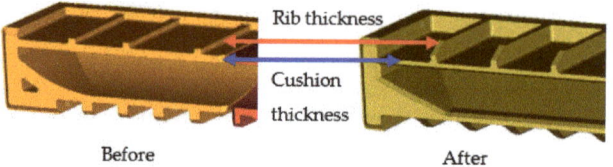

It is thought that high strain occurs due to the effect of pushing the set locally where the ribs are located.

Figure 22. Improvement finding using simulation analysis.

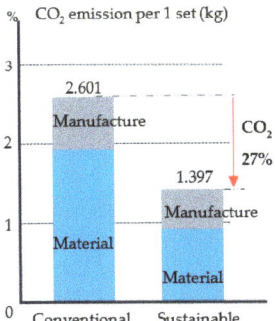

Figure 23. Carbon dioxide (CO_2) reduction.

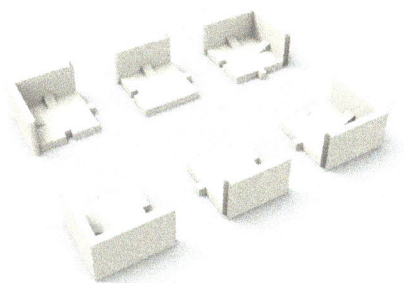

Figure 24. Design of dovetail technique.

Figure 25. Multi-size application for EPS cushion packaging.

The rib dimensions determined by the dovetailed technique are 40 mm × 20 mm × 20 mm, with a 25-degree slant. However, the detailed rib design consideration for new molded cushion packaging is depicted in Figure 26. If a cushion thickness of 18 mm is chosen, the cushion bead is expanded to normal size and the cushion's shape, or strength becomes stable.

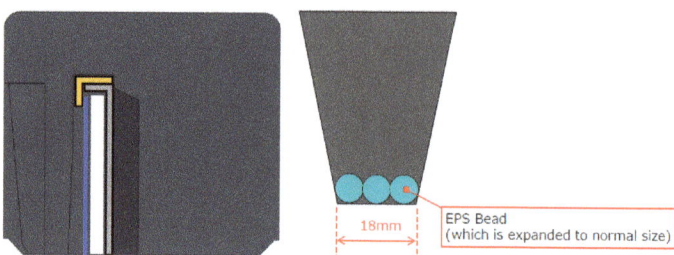

Figure 26. Cushion rib.

4. Conclusions

The intention of this study is to emphasize the design parameters that should be applied in cushion-packaging testing using destructive testing, finite element analysis, or risk analysis. The findings have a great potential to minimize the pollution impact on the environment and save development costs and materials, as well as realize an EPS cushion design sustainably. It would also be beneficial to the manufacturing industry in managing the EPS cushion waste. Therefore, the tendency for simulation analysis usage in industry is accepted to enhance designing process efficiency and improve skills in bringing forth sustainable protective packaging for electrical and electronic appliances. Hence, it can be concluded that:

i. The significant parameters of EPS cushion design increase the packaging reliability.
ii. The sustainable design can be implemented, and packaging design can be optimized through the manufacturing process and cost.
iii. Finite element analysis of the cushion design was a great idea for analyzing possible defects caused by design failure before the molding tool development begins.
iv. Optimization design through the reuse and reduction of EPS cushion usage contributed to increased design sustainability.
v. The challenge in this finding is that every packaging engineer must change their current design practice and analyze the packaging design prior to the development phase.

More research is needed to identify the biodegradable raw material or mixed composite based on these discovery parameters, especially for the molded expanded polystyrene cushion packaging of electrical and electronic appliances.

Author Contributions: Conceptualization, N.K., S.Z.A.R., W.A.R.A.W.I., M.N. and N.A.S.; data curation, N.K., I.A.R., M.P. and A.V.S.; formal analysis, N.K., I.A.R., K.B. and A.V.S.; investigation, N.K., I.A.R., N.A.K. and A.M.T.; methodology, N.A.K., W.A.R.A.W.I. and N.A.S.; project administration, N.A.K. and S.Z.A.R.; validation, N.K., W.A.R.A.W.I., I.A.R., N.A.S., M.P., M.N., K.B. and A.M.T.; writing—review and editing, N.K., S.Z.A.R., A.V.S. and A.M.T. All authors have read and agreed to the published version of the manuscript.

Funding: This study was supported by the Center of Excellence Geopolymer and Green Technology (CEGeoGTECH) UniMAP, and the Faculty of Mechanical Engineering &Technology, UniMAP. The author would like to acknowledge the support from Research Management Centre (RMC), UniMAP, and the Ministry of Education, Malaysia.

Institutional Review Board Statement: Not applicable.

Informed Consent Statement: Not applicable.

Data Availability Statement: Not applicable.

Acknowledgments: We would like to acknowledge the reviewer(s) for the helpful advice and comments provided.

Conflicts of Interest: The authors declare no conflict of interest.

References

1. Meents, S.; Verhagen, T. Reducing consumer risk in electronic marketplaces: The signaling role of product and seller information. *Comput. Human Behav.* **2018**, *86*, 205–217. [CrossRef]
2. Han, I.; Lee, Y.; Park, G. TV packaging optimization of the frontal drop impact using equivalent static loads. In Proceedings of the Eleventh World Congress on Structural and Multidisciplinary Optimisation, Sydney, Australia, 7–12 June 2015; pp. 2–6.
3. Zhu, Z.; Liu, W.; Ye, S.; Batista, L. Packaging design for the circular economy: A systematic review. *Sustain. Prod. Consum.* **2022**, *32*, 817–832. [CrossRef]
4. Emblem, A. Packaging and society. *Packag. Technol.* **2012**, 3–9. [CrossRef]
5. Lye, S.W.; Lee, S.G.; Chew, B.H. Virtual design and testing of protective packaging buffers. *Comput. Ind.* **2004**, *54*, 209–221. [CrossRef]
6. Hughes, K.; Vignjevic, R.; Corcoran, F.; Gulavani, O.; De Vuyst, T.; Campbell, J.; Djordjevic, N. Transferring momentum: Novel drop protection concept for mobile devices. *Int. J. Impact Eng.* **2018**, *117*, 85–101. [CrossRef]
7. P. Packaging, Cushion Curve Properties of Expanded Polystyrene Packaging. no. February, 2005. Available online: https://dlaegpxd16dn0.cloudfront.net/documents/EPS_Cushion_Curve_Properties-EPS_Industry_Alliance.pdf (accessed on 19 November 2017).
8. Low, K.H. Drop-impact cushioning effect of electronics products formed by plates. *Adv. Eng. Softw.* **2003**, *34*, 31–50. [CrossRef]
9. Lim, C.T.; Ang, C.W.; Tan, L.B.; Seah, S.K.W.; Wong, E.H. Drop impact survey of portable electronic products. *Proc. Electron. Compon. Technol. Conf.* **2003**, *1*, 113–120. [CrossRef]
10. Kim, W.J.; Kum, D.H.; Park, S.H. Effective design of cushioning package to improve shockproof characteristics of large-sized home appliances. *Mech. Based Des. Struct. Mach.* **2009**, *37*, 1–14. [CrossRef]
11. Chaukura, N.; Gwenzi, W.; Bunhu, T.; Ruziwa, D.T.; Pumure, I. Potential uses and value-added products derived from waste polystyrene in developing countries: A review. *Resour. Conserv. Recycl.* **2016**, *107*, 157–165. [CrossRef]
12. Kun, G.; Xi, W. Design and Analysis of Cushioning Packaging for Home Appliances. *Procedia Eng.* **2017**, *174*, 904–909. [CrossRef]
13. Mills, N.J. Chapter 11—Micromechanics of closed-cell foams BT—Polymer Foams Handbook. *Polym. Foam. Handb.* **2007**, *1*, 251–279. [CrossRef]
14. Patidar, H.; Singi, M.; Bhawsar, A.; Student, M.T.; Indore, M.P. Effect of Expanded polystyrene (EPS) on Strength Parameters of Concrete as a Partial Replacement of Coarse Aggregates. *Int. Res. J. Eng. Technol. (IRJET)* **2019**, *6*, 3779–3783.
15. Yi, J.W.; Park, G.J. Development of a design system for EPS cushioning package of a monitor using axiomatic design. *Adv. Eng. Softw.* **2005**, *36*, 273–284. [CrossRef]
16. Goodwin, D.; Young, D. *Protective Packaging for Distribution*; DEStech Publications Inc.: Lancaster, PA, USA, 2011; Volume 3.
17. Zhang, G.; Du, Y.; Li, X.; Che, X. Parametric design and multi-objective optimization of lcd packaging cushion foams. *Appl. Mech. Mater.* **2012**, *200*, 32–36. [CrossRef]
18. Tan, R.B.H.; Khoo, H.H. Life cycle assessment of EPS and CPB inserts: Design considerations and end of life scenarios. *J. Environ. Manag.* **2005**, *74*, 195–205. [CrossRef]
19. Liu, N.; Chen, B. Experimental study of the influence of EPS particle size on the mechanical properties of EPS lightweight concrete. *Constr. Build. Mater.* **2014**, *68*, 227–232. [CrossRef]
20. Kaya, A.; Kar, F. Properties of concrete containing waste expanded polystyrene and natural resin. *Constr. Build. Mater.* **2016**, *105*, 572–578. [CrossRef]
21. Chen, W.; Hao, H.; Hughes, D.; Shi, Y.; Cui, J.; Li, Z.X. Static and dynamic mechanical properties of expanded polystyrene. *Mater. Des.* **2015**, *69*, 170–180. [CrossRef]
22. Farris, A.; Pan, J.; Liddicoat, A.; Krist, M.; Vickers, N.; Toleno, B.J.; Maslyk, D.; Shangguan, D.; Bath, J.; Willie, D.; et al. Drop impact reliability of edge-bonded lead-free chip scale packages. *Microelectron. Reliab.* **2009**, *49*, 761–770. [CrossRef]
23. Zhang, S. Research on energy-saving packaging design based on artificial intelligence. *Energy Rep.* **2022**, *8*, 480–489. [CrossRef]
24. Watkins, M.; Casamayor, J.L.; Ramirez, M.; Moreno, M.; Faludi, J.; Pigosso, D.C.A. Sustainable Product Design Education: Current Practice. *She Ji* **2021**, *7*, 611–637. [CrossRef]
25. Bravington, C. *Packaging Technology*; Woodhead Publishing: Cambridge, UK, 2009; Volume 67. [CrossRef]
26. Fadiji, T.; Berry, T.M.; Coetzee, C.J.; Opara, U.L. Mechanical design and performance testing of corrugated paperboard packaging for the postharvest handling of horticultural produce. *Biosyst. Eng.* **2018**, *171*, 220–244. [CrossRef]
27. Watz, M.; Hallstedt, S.I. Towards sustainable product development—Insights from testing and evaluating a profile model for management of sustainability integration into design requirements. *J. Clean. Prod.* **2022**, *346*, 131000. [CrossRef]
28. Morana, J. *Sustainable Supply Chain Management*; John Wiley & Sons, Inc.: New York, NY, USA, 2013. [CrossRef]
29. Azzi, A.; Battini, D.; Persona, A.; Sgarbossa, F. Packaging Design: General Framework and Research Agenda By. *Packag. Technol. Sci.* **2012**, *29*, 399–412. [CrossRef]
30. Bengtsson, M.; Alfredsson, E.; Cohen, M.; Lorek, S.; Schroeder, P. Transforming systems of consumption and production for achieving the sustainable development goals: Moving beyond efficiency. *Sustain. Sci.* **2018**, *13*, 1533–1547. [CrossRef]
31. Bakırlıoğlu, Y.; McMahon, M. Co-learning for sustainable design: The case of a circular design collaborative project in Ireland. *J. Clean. Prod.* **2021**, *279*, 123474. [CrossRef]
32. Castiglioni, A.; Castellani, L.; Cuder, G.; Comba, S. Relevant materials parameters in cushioning for EPS foams. *Colloids Surf. A Physicochem. Eng. Asp.* **2017**, *534*, 71–77. [CrossRef]

33. Peache, R.J.; Sullivan, D.O. Current Trends in Protective Packaging of Computers and Electronic Components. *Curr. Trends Prot. Packag. Comput. Electron. Compon.* **1988**. [CrossRef]
34. Riley, A. *Plastics Manufacturing Processes for Packaging Materials*; Woodhead Publishing Limited: Cambridge, UK, 2012. [CrossRef]
35. Nemat, B.; Razzaghi, M.; Bolton, K.; Rousta, K. The role of food packaging design in consumer recycling behavior-a literature review. *Sustainability* **2019**, *11*, 4350. [CrossRef]
36. Charter, M.; Tischner, U. (Eds.) *Sustainable Solutions Developing Products and Services for the Future*, 1st ed.; Greenleaf Publishing: Austin, TX, USA, 2019.
37. Özgen, C. Sustainable design approaches on packaging design. *Lect. Notes Civ. Eng.* **2018**, *7*, 205–219. [CrossRef]
38. Monteiro, J.; Silva, F.J.G.; Ramos, S.F.; Campilho, R.D.S.G.; Fonseca, A.M. Eco-design and sustainability in packaging: A survey. *Procedia Manuf.* **2019**, *38*, 1741–1749. [CrossRef]
39. Emblem, A.; Emblem, H. *Packaging Technology: Fundamentals, Materials and Processes*; Elsevier: Amsterdam, The Netherlands, 2012. [CrossRef]
40. Liu, F.; Meng, G.; Zhao, M.; Zhao, J.F. Experimental and numerical analysis of BGA lead-free solder joint reliability under board-level drop impact. *Microelectron. Reliab.* **2009**, *49*, 79–85. [CrossRef]
41. Guideline, R.; Tests, P.; Level, C.; Level, R.; During, D.C. *Chapter 6 Establishment of the 'Reference Guideline for Packaging Tests'*; Japan International Cooperation Agency: Tokyo, Japan, 2007.
42. Balakrishnan, K.; Sharma, A.; Ali, R. Comparison of Explicit and Implicit Finite Element Methods and its Effectiveness for Drop Test of Electronic Control Unit. *Procedia Eng.* **2017**, *173*, 424–431. [CrossRef]
43. Zhou, C.Y.; Yu, T.X.; Lee, R.S.W. Drop/impact tests and analysis of typical portable electronic devices. *Int. J. Mech. Sci.* **2008**, *50*, 905–917. [CrossRef]
44. United Nations. *Requirements for the Construction and Testing of Packagings, Intermediate Bulk Containers (IBCS), Large Packagings, Tanks and Bulk Containers*; United Nations: New York, NY, USA, 2021; pp. 347–350. [CrossRef]
45. *International Standard ISO 3676*; ISO: Geneva, Switzerland, 1995.
46. Mills, N.J. Chapter 12 Product Packaging Case Study. Available online: https://www.researchgate.net/publication/302423817_Chapter_12_Product_packaging_case_study (accessed on 4 July 2022).
47. *ISO 14040*; Environmental Management-Life Cycle Assessment—Principles and Framework. ISO: Geneva, Switzerland, 2006.
48. Chandana, Y.V.N.; Kumar, N.V. Drop test analysis of ball grid array package using finite element methods. *Mater. Today Proc.* **2022**, *64*, 675–679. [CrossRef]
49. Al Hayek, M.A. *Rubber in Shear (Modulus of Rigidity)*; Islamic University of Gaza: Gaza, Palestine, 2016.
50. Greaves, G.N.; Greer, A.L.; Lakes, R.S.; Rouxel, T. Poisson's ratio and modern materials. *Nat. Mater.* **2011**, *10*, 823–837. [CrossRef]
51. Muralikrishna, I.V.; Manickam, V. Life Cycle Assessment. *Environ. Manag.* **2017**, 57–75. [CrossRef]
52. Cai, W.; Wang, L.; Li, L.; Xie, J.; Jia, S.; Zhang, X.; Jiang, Z.; Lai, K.H. A review on methods of energy performance improvement towards sustainable manufacturing from perspectives of energy monitoring, evaluation, optimization and benchmarking. *Renew. Sustain. Energy Rev.* **2022**, *159*, 112227. [CrossRef]
53. Schleier, J.; Simons, M.; Greiff, K.; Walther, G. End-of-life treatment of EPS-based building insulation material—An estimation of future waste and review of treatment options. *Resour. Conserv. Recycl.* **2022**, *187*, 106603. [CrossRef]
54. Faludi, J.; Gilbert, C. Best practices for teaching green invention: Interviews on design, engineering, and business education. *J. Clean. Prod.* **2019**, *234*, 1246–1261. [CrossRef]
55. Bhamra, T.; Lilley, D.; Tang, T. Design for Sustainable Behaviour: Using Products to Change Consumer Behaviour. *Des. J.* **2015**, *14*, 427–445. [CrossRef]
56. Lofthouse, V.; Trimingham, R.; Bhamra, T. Reinventing refills: Guidelines for design. *Packag. Technol. Sci.* **2017**, *30*, 809–818. [CrossRef]
57. Petrillo, A.; De Felice, F.; Jannelli, E.; Autorino, C.; Minutillo, M.; Lavadera, A.L. Life cycle assessment (LCA) and life cycle cost (LCC) analysis model for a stand-alone hybrid renewable energy system. *Renew. Energy* **2016**, *95*, 337–355. [CrossRef]
58. Wan, C.K.; Lin, S.Y. Negotiating social value, time perspective, and development space in sustainable product design: A dialectics perspective. *Des. Stud.* **2022**, *81*, 101121. [CrossRef]
59. Suppipat, S.; Hu, A.H. Achieving sustainable industrial ecosystems by design: A study of the ICT and electronics industry in Taiwan. *J. Clean. Prod.* **2022**, *369*, 133393. [CrossRef]
60. Šenitková, I.; Bednárová, P. Life cycle assessment. *JP J. Heat Mass Transf.* **2015**, *11*, 29–42. [CrossRef]
61. Albuquerque, T.L.M.; Mattos, C.A.; Scur, G.; Kissimoto, K. Life cycle costing and externalities to analyze circular economy strategy: Comparison between aluminum packaging and tinplate. *J. Clean. Prod.* **2019**, *234*, 477–486. [CrossRef]

Disclaimer/Publisher's Note: The statements, opinions and data contained in all publications are solely those of the individual author(s) and contributor(s) and not of MDPI and/or the editor(s). MDPI and/or the editor(s) disclaim responsibility for any injury to people or property resulting from any ideas, methods, instructions or products referred to in the content.

Review

Solidification/Stabilization Technology for Radioactive Wastes Using Cement: An Appraisal

Ismail Luhar [1], Salmabanu Luhar [2,*], Mohd Mustafa Al Bakri Abdullah [3,*], Andrei Victor Sandu [4,5,6,*], Petrica Vizureanu [4,7], Rafiza Abdul Razak [2], Dumitru Doru Burduhos-Nergis [4] and Thanongsak Imjai [2,8]

1. Department of Civil Engineering, Shri Jagdishprasad Jhabarmal Tibrewala University, Rajasthan 333001, India
2. Center of Excellence Geopolymer and Green Technology (CEGeoGTech), Universiti Malaysia Perlis (UniMAP), Perlis 01000, Malaysia
3. Faculty of Chemical Engineering Technology, Universiti Malaysia Perlis (UniMAP), Perlis 01000, Malaysia
4. Faculty of Material Science and Engineering, Gheorghe Asachi Technical University of Iasi, 41 D. Mangeron St., 700050 Iasi, Romania
5. Romanian Inventors Forum, Str. Sf. P. Movila 3, 700089 Iasi, Romania
6. National Institute for Research and Development for Environmental Protection INCDPM, 294 Splaiul Independentei, 060031 Bucharest, Romania
7. Technical Sciences Academy of Romania, Dacia Blvd 26, 030167 Bucharest, Romania
8. School of Engineering and Technology, Walailak University, Nakhon Si Thammarat 80160, Thailand
* Correspondence: ersalmabanu.mnit@gmail.com (S.L.); mustafa_albakri@unimap.edu.my (M.M.A.B.A.); sav@tuiasi.ro (A.V.S.)

Abstract: Across the world, any activity associated with the nuclear fuel cycle such as nuclear facility operation and decommissioning that produces radioactive materials generates ultramodern civilian radioactive waste, which is quite hazardous to human health and the ecosystem. Therefore, the development of effectual and commanding management is the need of the hour to make certain the sustainability of the nuclear industries. During the management process of waste, its immobilization is one of the key activities conducted with a view to producing a durable waste form which can perform with sustainability for longer time frames. The cementation of radioactive waste is a widespread move towards its encapsulation, solidification, and finally disposal. Conventionally, Portland cement (PC) is expansively employed as an encapsulant material for storage, transportation and, more significantly, as a radiation safeguard to vigorous several radioactive waste streams. Cement solidification/stabilization (S/S) is the most widely employed treatment technique for radioactive wastes due to its superb structural strength and shielding effects. On the other hand, the eye-catching pros of cement such as the higher mechanical strength of the resulting solidified waste form, trouble-free operation and cost-effectiveness have attracted researchers to employ it most commonly for the immobilization of radionuclides. In the interest to boost the solidified waste performances, such as their mechanical properties, durability, and reduction in the leaching of radionuclides, vast attempts have been made in the past to enhance the cementation technology. Additionally, special types of cement were developed based on Portland cement to solidify these perilous radioactive wastes. The present paper reviews not only the solidification/stabilization technology of radioactive wastes using cement but also addresses the challenges that stand in the path of the design of durable cementitious waste forms for these problematical functioning wastes. In addition, the manuscript presents a review of modern cement technologies for the S/S of radioactive waste, taking into consideration the engineering attributes and chemistry of pure cement, cement incorporated with SCM, calcium sulpho–aluminate-based cement, magnesium-based cement, along with their applications in the S/S of hazardous radioactive wastes.

Keywords: radioactive wastes; cement; solidification/stabilization (S/S); biochar; waste form; supplementary cementitious materials (SCM); magnesia-based cement; calcium sulphoaluminate cement; calcium aluminate cement

Citation: Luhar, I.; Luhar, S.; Abdullah, M.M.A.B.; Sandu, A.V.; Vizureanu, P.; Razak, R.A.; Burduhos-Nergis, D.D.; Imjai, T. Solidification/Stabilization Technology for Radioactive Wastes Using Cement: An Appraisal. *Materials* 2023, *16*, 954. https://doi.org/10.3390/ma16030954

Academic Editor: Francisco Agrela

Received: 18 December 2022
Revised: 10 January 2023
Accepted: 16 January 2023
Published: 19 January 2023

Copyright: © 2023 by the authors. Licensee MDPI, Basel, Switzerland. This article is an open access article distributed under the terms and conditions of the Creative Commons Attribution (CC BY) license (https://creativecommons.org/licenses/by/4.0/).

1. Introduction

Arthur Schopenhauer, a great German philosopher, once stated that "each and every single truth passes all the way through three phases prior to being documented, at first, it is ridiculed; secondly, it is aggressively opposed, and in third and ultimate stage, it is recognized as being self-evident". For the last more than three decades, this is exactly a fitting statement in the context of the history of stabilization/solidification (S/S) technology. The S/S of perilous and death-defying radioactive wastes provides a grand necessitate in the field of civil engineering with a view to consolidating the research, and general practices in high-tech work and especially in technology. Simply speaking, solidification/stabilization (S/S) technology can lend a hand to make the radioactive wastes physically more stable in order to manage them under atmospheric conditions.

1.1. Radioactive Wastes

Nuclear energy, i.e., atomic energy, is that energy which lies in the nucleus of an atom, which is not only a promising non-fossil-based energy source but also exhibits minimal emissions of carbon dioxide (CO_2) [1,2] proving it more beneficial [3,4]. In developing nations such as China, there is a foreseeable development of nuclear energy in the coming decades owing to the exigency for energy boosts and a net zero carbon footprint policy [3,4]. Therefore, nuclear energy is heading towards being established as one of the alternatives that may lend a hand to mankind to address the global energy crisis in the upcoming time. In 2018, the production of nuclear power was reported as more or less 10.2% of the total electricity generation in the world and it is likely to reach 13% in 2030 [5]. Radioactive wastes are generated from several sources, viz., nuclear power plants, nuclear armament, or nuclear fuel treatment plants, however, the majority of waste originates from the nuclear weapons reprocessing and nuclear fuel cycle. The nuclear industries are generating radioactive wastes during different processes of production and that is the reason for the accumulation of roughly 200,000 m^3 of nuclear waste with low- and intermediate-level global generation each year, in harmony with the statistics of the World Nuclear Energy Association [6]. Consequently, the development of safer and competent low- as well as intermediate-level radioactive waste treatment technology has turned out to be a great dilemma for nuclear power plants. The low-level radioactive wastes, i.e., the technical wastes, generated during the processes of maintenance include incompressible and compressible components, viz., plastic, absorbent paper, gloves, rags, scrap work clothes and gas jackets, which account for about 90% of the total volume of radioactive nuclear waste. The bulk quantity of the low-level radioactive wastes in nuclear power plants can be classified as combustible wastes; as a result, the thermal treatment technologies could lend a hand to accomplish higher volume reduction, inorganic alteration and lower residue radioactivity levels. Amongst the thermal treatment technologies, counting incineration, melting, solidification and molten salt-oxidation technology, dry oxidation as well as thermal plasma [7], dry oxidation technologies such as pyrolysis and gasification could attain a higher volume reduction of low-level radioactive wastes with the safer treatment of radionuclide [8].

Several nuclear reactors with graphite as a moderator and reflector are facing being decommissioned sooner or later, and the waste of radioactive graphite is a huge part of the concerned wastes. By 2021, more than 2.5 million tons of irradiated graphite, i.e., i-graphite, generated from nuclear reactors, is for the time being stored in provisional storage facilities and reactor stores [9]. According to International Atomic Energy Agency (IAEA) [10], the i-graphite inventory is chiefly concerted in USA, Russia and UK. The process of production of nuclear power generates radioactive wastes of low-level radioactive waste (LLW), intermediate-level radioactive waste (ILW) and high-level radioactive waste (HLW) depending on its radioactivity attributes. The radionuclide present in the radioactive wastes is for the most part made up of ^{235}U and its fission yields, namely, ^{137}Cs, ^{90}Sr, ^{60}Co, ^{140}Ba, ^{129}I, etc. Amongst them, the half-life periods of ^{137}Cs and ^{90}Sr are the greatest figures at 33 and 29.9 years, in that order. The radiation occurring all throughout their decay will gravely contaminate the ecosystem and threaten the life safety measures of lives on the planet. The conventional management move towards radioactive wastes is to solidify them first of all and after that dispose of them below the subsurface or in deep geological dumping sites. The quantity of the low and intermediate radioactive waste liquids is much bigger in comparison with the high radioactive waste liquid, which accounts for beyond 90% of the total radioactive wastes. The technique for cement solidification is far and wide employed in the solidification treatment of low-level and intermediate-level radioactive liquids because of the saving of raw material and its uncomplicated progression [11,12]. Two huge Chinese nuclear plants are exercising cement solidification technology to cope up with low- and intermediate-level radioactive waste liquids for lots of years successfully and lucratively. During decommissioning of nuclear power plants and their operation, the generation of low- and intermediate-level radioactive wastes takes place. A large quantity of this waste is dumped—the nonconforming part that necessitates solidification and packaging, which is hard to decontaminate and mainly encloses dispersive particulate wastes such as the debris of spent concretes, slurry or sludge, and fine-grained polluted soil. The published report of IAEA has focused on such dispersive/particulate radioactive wastes and is directed to treat and package them correctly for permanent clearance [13]. Quite a lot of stabilization/solidification techniques were developed to treat the referred radioactive wastes with cement, polymer and asphalt. The said agents of solidification alter the dispersive/particulate wastes into non-dispersive solidified wastes, formed all the way through their definite process [14]. In the meantime, the key setback is a noteworthy boost in their volume when the solidification of the dry particles (powder) is performed by grouping with cement, asphalt, or polymer because each particulate matter becomes encapsulated and goes through the solidification agent [15]. This augment in the volume increases the cost of waste disposal and the capability of disposal facilities is also filled up swiftly. With a view to addressing the referred issues, the investigations on minimizing the volume of the powdered wastes and the solidification technology are essential.

Primarily, Liquid Radioactive Waste (LRW) is found generated in the nuclear power industry and also in unexpected nuclear accidents. Its immobilization is one of the most efficient measures for managing radioactive waste. In order to generate liquid radioactive waste, nuclear waste is dissolved in boiling nitric acid (HNO_3), and both uranium (U) and plutonium (PUREX) are recovered. The leftover liquid is considered liquid radioactive waste, which contains radioactive substances, namely, Cs^{+1}, Sr^{+2}, and Co^{+2}. Customarily, the liquid radioactive waste is managed in three stages [16,17]:

- Cooling the leftover liquid;
- Drying and concentrating;
- Mixing it with silicate or borate.

The waste materials are acquired in a form of glass at towering temperatures. In the context of 90Sr in liquid radioactive waste, there are some moves to separate it from LRW, including ion exchange, precipitation and solvent extraction [18]. Alternatively, a solid matrix can immobilize 90Sr. Every single technique should spotlight the application of low-priced energy. Globally, the liquid radioactive waste immobilization in solid products has been well-studied [19–21]. The chemical immobilization of Sr^{+2} in calcium–silicate–hydrate (C–S–H) does not occur, and this is the disadvantageous root cause of the application of traditional Portland cement to immobilize Sr^{+2} ions in this medium [22]. The phosphate has an ideal effect in the solidification of Sr^{+2}, assuggested in preceding investigations. For this reason, phosphate-based materials are regarded as ideal matrices for the aqueous immobilization of Sr^{+2}. The radioactive waste management chart is depicted in Figure 1.

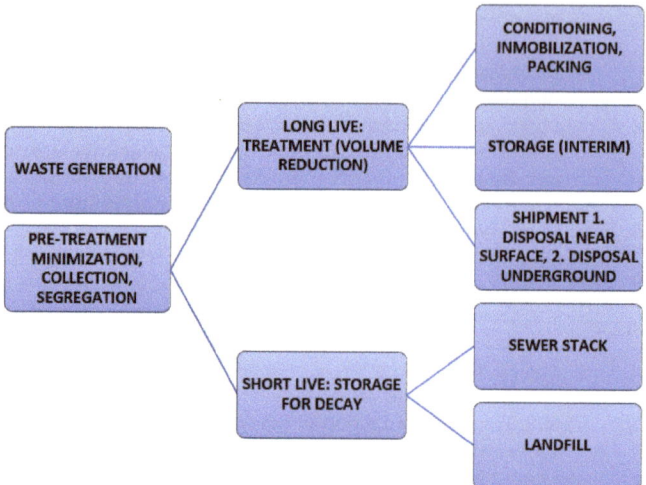

Figure 1. Radioactive waste management chart.

1.2. Stabilization/Solidification (S/S) Technology

"Stabilization" refers to the chemical methods, which can mitigate the perilous potential of a particular waste by converting the pollutants into less soluble, poisonous or movable forms. However, the stabilization does not essentially alter the handling attributes and physical nature of the definite type of waste. On the other hand, "Solidification" refers to the methods of encapsulation of the wastes, which form a solid material. However, the solidification does not essentially engage a chemical interaction amongst the pollutants and the hardening additives. The S/S can be accomplished through chemical reaction kinetics among hardening reagents and the waste, or by means of courses of action of a mechanical kind. The waste form or the solidification produced might be a clayey or argillaceous material, a granular particulate, or in the form of a monolithic block, as well as a few other physical forms, which are normally regarded as "solid." The migration of pollutants is frequently confined via coating the wastes with the help of materials having inferior permeability, or by slimming down the surface area exposed to leaching. Frequently, both terms are referred to as S/S and can be utilized interchangeably. Inorganic binders such as cement are brilliantly efficient for immobilizing heavy metals using mechanisms of physical and chemical containment. Significantly, loads of substances present in the wastes influence the setting and solidifying attributes of the binders, in particular, the cement-based systems for cementing. There are varied processes and equipment, which are developed to serve the purposes. The ex situ or in situ processes can be useful to carry out the grouping of binders and wastes, of which, the in situ techniques are getting much positive response for the remediation of polluted sites, and they can be further categorized as backhoe-based methods, shallow area methods and drilling, augering, jetting, or trenching methods. While, in the case of ex situ processes, the mixing, mortar mixers, pug mills, or concrete mixers are mostly utilized. The depth of the pollution and the attributes of the tainted media are regarded as the bases for the choice of the kind of mixing method.

As we know, the inappropriate management of unsafe radioactive wastes creates a solemn threat to human health and other breathing organisms and their surroundings. For illustration, the toxic leachates, which include perilous wastes from unacceptably maintained unsafe waste landfill, could rigorously pollute the subsurface and surface waters; leakage or accidents of nuclear power plant blasts can gravely kill or injure the neighboring public. Therefore, these waste management projects must be regulated with adequately knowledgeable, competent, and reliably dependable regulations and legislation. The fundamentals of the management of risky, radioactive, and mixed wastes are essential for enough definition, classification, designation, and characterization to offer bounds to the crisis, which vary from nation to nation. Since 1980, the U.S. Environmental Protection Agency (EPA) has developed an all-inclusive program for perilous waste to make certain that risky waste can be managed unharmed. A "cradle-to-grave" strategy for such wastes from the point of generation to final removal is established for their identification, recycling, storage, and dumping. In 1976, RCRA, i.e., The Resource Conservation and Recovery Act, was motivated by apprehension over the indecent discarding of hazardous wastes. RCRA Subtitle C provides a wide-ranging program concerning the identification, generation, transportation, treatment, storage, and safe removal of risky wastes. The Atomic Energy Act (AEA) of 1954 is the fundamental law governing the production, utilization, possession, accountability, and disposal of radioactive materials in the U.S.A. Additionally, several laws state radioactive waste management methods and authorities such as the Low-Level Radioactive Waste Policy Act (LLWPA) and the Nuclear Waste Policy Act (NWPA). The Nuclear Regulatory Commission (NRC) or the U.S. Department of Energy (DOE) under the AEA regulates hazardous radioactive wastes. The characterization and classification of radioactive wastes are done by and large specified by related regulations and laws.

Characteristically, stabilization/solidification (S/S) is a process that entails the mixing of the waste with a binder in order to slim down the contaminant leachability by both chemical and physical means and to transform the perilous waste into an eco-acceptably waste form for landfill dumping or construction utilizations. The S/S is extensively employed to dispose of low-level radioactive wastes (LLRW), hazardous, and mixed wastes, as well as remediation of tainted sites. The S/S is regarded as the Best Demonstrated Available Technology (BDAT) for 57 harmful wastes in accordance with The United States Environmental Protection Agency (USEPA) [23]. The report of the USEPA for 1996 revealed that more or less 30% of the Superfund remediation sites applied S/S technologies [24]. Cementitious materials are the most commonly exercised for S/S among all the binders. The cement solidification method is extensively employed in the solidification treatment of the LLW and ILW liquids due to the economy of raw material and its easy process [11,12]. Two large nuclear plants utilizing cement solidification technology are located in China, to deal with lower and medium radioactive waste liquids for several years and, they are proven to be flourishing and money-spinning. Considering the many benefits, hydraulic cement is extensively employed for S/S of low-level radioactive wastes, perilous wastes, mixed wastes, and in the remediation of polluted sites. The cement-based S/S technology exhibits the following benefits in comparison with the rest of the other technologies [25]: (1) excellent impact and compressive strength, (2) good-quality and long-term physical and chemical stability, (3) relatively cost-effective, (4) documented application and compatibility with various wastes over decades, (5) familiar material and technology, (6) extensive accessibility of the chemical ingredients, (6) non-toxicity of the chemical ingredients, (7) effortlessness of application in processing because usually carried out at ambient temperature and pressure and no need of unique or very special equipment at all, (8) higher loadings of waste are feasible, (9) inert to ultraviolet (UV) radiation, (10) elevated resistance to bio-degradation, (11) lower water solubility and leachability of some pollutants, (12) comparatively lower water permeability, (13) capacity of most aqueous wastes to bind chemically with matrix, (14) good mechanical and structural attributes, (15) good self-guarding for radioactive wastes, (15) fast, controllable setting, with no segregation or settling during curing, (16) absence of free water provided correctly formulated, and (17) a longer shelf-life of cement powder.

The S/S of pollutants using cements includes the following three features:

(1) The chemical fixation of pollutants—chemical interactions among the cement hydration yields and the pollutants;
(2) The physical adsorption of the pollutants on the surface of cement hydration yields;
(3) The physical encapsulation of polluted waste or soil.

Of these, Items1 and 2 rely upon the nature of the yields of hydration and pollutants while Item 3 relates to both the nature of the hydration yields as well as the paste density and its physical structure. The cement-based waste forms may be fitting for controlled construction exercises provided the leaching and other performance of the cement-solidified waste forms meet suitable eco-criteria.

For the S/S of waste, the choice of cementing materials must be made considering the following criteria on the basis of the waste properties:

(1) The cement and the waste compatibility;
(2) The pollutant's chemical fixation;
(3) The physical encapsulation of polluted waste and soil;
(4) The durability of ultimate waste forms;
(5) The waste form leachability;
(6) The gainfulness of S/S in terms of cost.

Practically, numerous additives are frequently employed with cementing materials to resolve all of the above aspects.

Quite recently, the application of cementitious materials for the solidification of dangerous matters has proven to be significantly promising. The benefits of solidification/stabilization include the following [26]:

- Safe transport and easy burial;
- Enhanced physical attributes of the wastes for effortless handling;
- Lesser eco-pollution by leaching and evaporation of risky constituents;
- Potential for recycling wastes into construction material;
- Detoxification of substances for safe-guarding workers.

The Portland cement, correct additives and in some incidents fly ash are grouped together with the waste with a view to produce a solidified mass for dumping during solidification. A gel is initially developed, then fibrils formed as silicate compounds hydrate when the waste is grouped together with cement. The referred inter-locking fibrils bind various hydration yields and the cement into a hardened mass. The process of solidification using the Portland cement is most appropriate to inorganic wastes such as incinerator residue, heavy metal enclosing wastes and road wastes. The keeping of metals in the form of insoluble hydroxide of carbonate salts is assigned to the higher pH of the cement. Whereas plastic, metal filings, and asbestos-like materials boost the strength of the matrix, other organic and inorganic compounds can retard setting, decline the final strength and cause swelling. The additives enclosing clay, sodium silicate and vermiculite may be integrated with the mixes to counterbalance the influences of the said materials. Additionally, the low-level radioactive wastes and organic wastes can be solidified using Portland cement.

The S/S is a potentially promising technology that utlizes the supplement of a binding agent to encapsulate and trim down the mobility of the dangerous waste elements [27]. The S/S can act as a significant potential process for making wastes acceptable for land dumping, since the constraints on filling the lands turn out to be stronger and wastes particularly hazardous ones are banned from land dumping. Inferior permeability and lesser pollutant leaching rates can make the forbidden wastes acceptable for their disposal in landfills subsequent to the S/S process [28]. In the S/S process, the method of using Portland cement as a binding agent mixed with water and the heat has been developed. The mixture turns out to be strongly alkaline. The chemical reactions become sluggish after a few minutes. After that, an induction period or dormant period usually lasts for several hours. The anhydrous clinker grains develop as coating proceeds, with an early amorphous precipitate throughout the initial reaction period, which plays a role of semi-protective film and slows reaction during the induction period. The breakdown of the film marks the onset of swift hydration towards the finish of the induction period. Also, the breakdown of the film initiates the growth of a constant but initially lower strength gel network of linking particles, ensuing in physical hardening of the cement matrix. As the gel carries on the process of solidification and densification, the cement achieves maximal strength. The archetypal contemporary Portland cement attains around two thirds of hydration in

28 days [29]. The chief chemical that is regarded in hydrated cement is colloidal calcium–silicate–hydrate (C-S-H) gel. This gel is developed at the surfaces of particles of cement [30]. The C-S-H gel has significant implications for the mechanisms of fixation during the process of solidification and it is largely accountable for strength development [31,32].

Most frequently, HE employs cement as a binder for an assortment of wastes for solidification. The utilization of cement for solidification is beneficial for the simplicity of the process and it taking place under normal temperature, however, the quite high cost of this binder and also the eco-aspect of anthropogenic CO_2 footprints associated with cement production are the prominent setbacks [33–35]. The said process is apposite for inorganic materials, viz., ashes and dehydrated sludge from industrial wastewater treatment plants, and also for the solidification of waste going to landfills. At present a broad range of combinations of diverse kinds of binders is being exercised. The cement itself is an energy-intensive product; as a result, it is essential to ensure the optimum composition of the solidification mix. The drawback lies with the sensitivity to the presence of definite substances that influence the hydration reaction kinetics and the solid structure development. There are other disadvantages, such as a boost in the volume of solidified waste, which is unsuitable for depositing in landfills, as well as a low resistance to corrosion agents. More often than not, the wastes are mixed with Portland cement and additives, which affect, in a positive way, the characteristics of the cement, and with enough water content, this begins the hydration reaction kinetics. Subsequently, the S/S process initiates and waste is added into the cement structure. The waste reacts with water and cement to develop hydroxides of metals or carbonates. Normally, the same are less soluble than the original metal compounds in the waste. The cementation technology can predominantly be executed on the accessible equipment—solidification technology lines (mobile/stable). The cement can be employed as an activator for other potentially binding materials, e.g., low-priced fly ashes or glassy slags. Ultimately, the referred to secondary binders have turned out to be an integral part of the cement matrix, which uses one kind of waste to immobilize other sorts of more hazardous wastes. The development of less soluble hydroxides of metals or carbonates all through the hydration reactions results in meeting the requisite limits for leachability examinations. The benefit of solidification technology also lies in the likelihood of processing amorphous metals. Also, solidified/stabilized waste can be managed with no trouble and the danger of dust creation is very little. The discharge of heavy metals from the product is also moderately lower. The output solidification product can often be exercised as construction material or backfill in transport construction engineering or mining operations [36].

Naturally, radioactive elements are found in the crust of the earth. The large unstable atoms turn out to be more stable ones by emitting radiation to eliminate surplus atomic energy, i.e., radioactivity. The said radiation can be emitted in the form of positively charged (+ve) alpha and negatively charged (−ve) beta particles, as well as gamma or X-rays. The radiation from radioactive materials of alpha and gamma rays affect the body very badly. The waste materials that either enclose or get contaminated with radionuclide at concentrations or activities beyond nationwide regulatory authority-established clearance levels for which no application is at this time predicted are included under the roof of "radioactive wastes". In other words, radioactive waste comprises of any material which is either inherently radioactive, or gets contaminated with a dose of radioactivity, and that is deemed to have no more utilization. Characteristically, an amount of radioactive waste consists of numerous radionuclides that are unstable isotopes of elements, which experience decay and thereby emit ionizing radiation. The said radiation is very much injurious to humans and the eco-system. The diverse isotopes emit dissimilar kinds and levels of radiation that persist for unlike periods of time. However, the radioactive nature of all radioactive waste grows weaker with time. Every radionuclide present in the waste possesses a half-life, i.e., the time it uses for half of the atoms to decay into another nuclide. Ultimately, all radioactive waste decays into non-radioactive elements, meaning "stable nuclides". Thus, the radioactive wastes decay naturally over the time similar to all radioactive material. Therefore, once the radioactive material decays adequately, the waste is regarded as non-hazardous; however, the time ranges widely for the purpose right from a few hours to thousands of years, as found in the case of plutonium (Pu), which is highly radioactive. These wastes are being produced by industries, namely, in the fields of nuclear power generation, mining, mineral exploration, medicine, agriculture, manufacturing, defense, definite kinds of scientific researches, non-destructive testing, and reprocessing of nuclear weapons. It is a known fact that the nuclear power is pigeonholed by the very huge quantity of energy extended from a very little quantity of fuel, and the amount of waste generated throughout this progression is also comparatively tiny. Nevertheless, the bulk of the waste generated is radioactive and consequently it must be cautiously and methodically managed as a harmful material. Obviously, the nature of these kinds of wastes is hazardous to human health and exposure to high doses of radiation causes vomiting, nausea, hair loss, diarrhea, hemorrhage, cell and DNA damage, destruction of the intestinal lining and central nervous system damage, increases in the possibility of cancer and even death, as well as contamination of the ecosystem, since it emits radioactive particles, along with causing soil infertility and genetic mutations, seeds to not sprout, slow growth, losses in fertility that can alter attributes of the plant, etc. The radioactive wastes cannot be destroyed, and for this reason, they stay for a prolonged time in the eco-system, escorting a high risk to lives if not correctly managed. Not only this, but the contemporary process of mining uranium (U) discharges elevated quantities of carbon dioxide (CO_2) into the atmosphere. Additionally, CO_2 is being emitted into the open air when newer nuclear power plants are constructed and the transport of radioactive waste is being performed. The direct consequences of exposure to ionizing radiation in air, water and food are reported as being very much dodgy. Radioactive wastes may be found in all three forms, i.e., as gas, or liquid or solid, and surprisingly their level of radioactivity also varies. Incredibly, these sorts of wastes may remain at constant levels of radiation for a few hours to several months or even hundreds of thousands of years!

2. Research Methodology

A comprehensive literature review was conducted to identify and appraise allied available information on record, which comprises pedagogic ideas and referenced examples of the fusion work. Recently, one of the rapidly expanding study disciplines in recent years has become a crucial sub-discipline of solidification/stabilization technology for radioactive wastes using cement. To comprehend in-depth the most recent and emerging drift of radioactive waste as edifice material, the keywords "radioactive waste", "solidification/stabilization technology", and "solidification/stabilization of radioactive waste using cement", have been methodically recovered, using bibliographic databases of "Springer", "Elsevier", "Taylor and Francis", "Wiley" and "Hindawi". Furthermore, comprehensive data analysis and categorization were carried out based on a thorough understanding of titles, graphical abstracts, highlights, abstracts, keywords, entire texts, conclusions, and impressions. The cited literature data represent a comprehensive description of the progress, portrayal, and application of cement in stabilization technology for radioactive waste.

3. Classification of Radioactive Wastes

Broadly speaking, the radioactive wastes are classified as low level, e.g., paper, tools, clothing, rags, etc., which enclose petite quantities of chiefly short-lived radioactivity; intermediate-level wastes with elevated amounts of radioactivity, therefore, necessitate some shielding. The LLW and ILW are the wastes generated from general operations such as the cleaning of reactor cooling systems as well as fuel storage ponds, and the decontamination of tools, filters, and metal components get polluted and become out to be radioactive on account of their utilization in or near the reactor. Lastly, the third group comprises high level wastes which are extremely radioactive and hot, owing to decay heat, and hence they need both cooling and shielding. The storage time period of radioactive wastes relies upon the type of waste and radioactive isotopes. The long-term storage of high level wastes requires burial in deep geological formations; however, short-term storage can be implemented on or near to the Earth's surface. Most low-level and short-lived intermediate level wastes, in general, experience land-based disposal at once after packaging. Mostly, at present, the near-surface disposal facilities are in operation, viz., a few low-level liquid wastes from reprocessing plants are disposed of in the sea, including radionuclides; however, this is regulated and controlled, and the uppermost radiation dose anyone gets from them is a small fraction of natural background radiation only. The long-lived ILW and HLW include spent fuel when regarded as a waste, which stays radioactive and is subjected to deep geological disposal. The safe techniques for the ultimate dumping of high-level radioactive waste are verified technically and the global consensus is "the geological disposal is the most excellent feasible systematic solution". Significantly, with a view of ultimate dumping, the "multiple barriers" geological disposal is planned to make sure that no noteworthy environmental releases take place for more than tens of thousands of years. This is a valuable method to immobilize the radioactive elements in HLW and long-lived ILW, and to isolate them from the bio-sphere. Notably, nuclear power is the only huge-scale energy-producing technology that has the potential to take complete responsibility for all its radioactive waste and fully encapsulate it into the product. The monetary provisos are planned for the management of every single civilian's radioactive waste. The cost of management and disposal of nuclear power plant wastes is normally more or less 5% of the total cost of the generated electricity. Figure 2 displays the Nuclear waste inventory for LLW, ILW, HLW, VLLW.

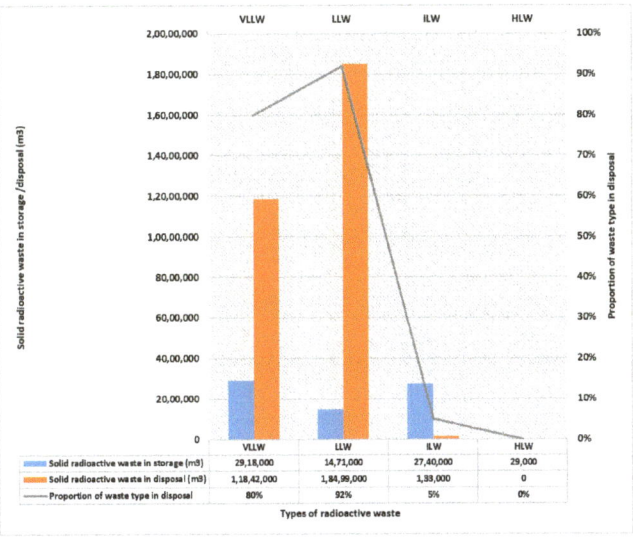

Figure 2. Nuclear waste inventory.

There are three general categories of radioactive wastes:

3.1. Low (Including Very Low)—Level Radioactive Wastes

The exempt waste and very-low-level waste comprises radioactive materials at a harmless level for lives or the ecosystem. For the most part it consists of demolished material such as concrete, bricks, metal, piping, plaster, valves, etc., produced during rehabilitation or dismantling in nuclear industries. As a result of the concentration of natural radioactivity present in definite minerals exercised in their manufacturing, some other industries, namely, food processing, chemical, steel, hospitals, industry, as well as the nuclear fuel cycle, etc., also produce this sort of radioactive wastes. The low-level waste is radioactively contaminated industrial or research waste, i.e., short-lived radioactivity, which includes general items such as paper, protective clothing, plastic bags, cardboard, tools, clothing, filters, and packaging material, etc., when they get in touch with radioactive materials in any industry using radioactive material such as government, manufacturing unit, medical, utility, research facilities, etc. The near-surface disposal of low-level wastes is commonly done since they possess a radioactive content not within the limit of 4 gigabecquerels (GBq) per tonne (GBq/t) of alpha activity or 12 GBq/t beta-gamma activity. For this reason, no shielding is essential during handling and transport. It is fit for dumping in near-surface facilities. It contains roughly 90% of the volume but possesses merely 1% of the radioactivity of all radioactive wastes. The total LILW generated in different countries are presented in Figure 3.

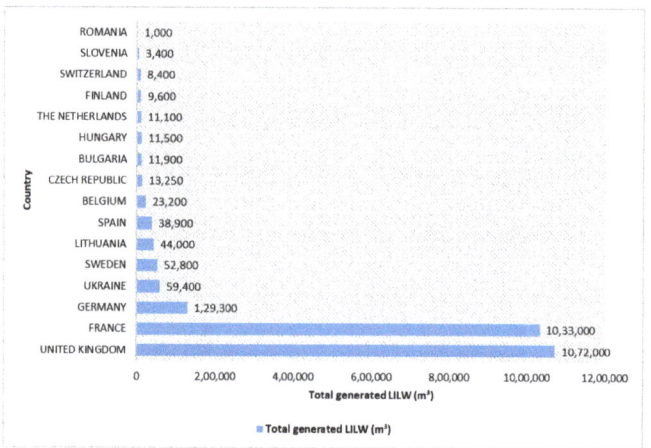

Figure 3. Total generated LILW in different countries.

3.2. Intermediate-Level Waste

Intermediate-level radioactive waste (ILW) is reported to be more radioactive than low-level radioactive waste (LLW), however, the heat generation is <2 kW/m^3, which is insufficient to be considered for the design or selection of storage and dumping facilities. The ILW necessitates some protection because of its high levels of radioactivity. Characteristically, the ILW comprises resins, chemical sledges, metal fuel cladding, and the polluted materials from decommissioning of the reactor. The small items and any non-solids may be solidified in bitumen or concrete for dumping. ILW constitutes about 7% of the total volume and possesses 4% of the radioactivity of all radioactive waste.

3.3. High-Level Waste

The high-level waste (HLW) is full of extremely radioactive and for the most part comparatively short-lived fission products, creating a concern. If the waste is stored, possibly in deep geological storage, after several years the fission products decay, lessening the radioactivity of the waste. Significantly, the HLW is adequately radioactive for its decay heat of >2 kW/m^3 to elevate its temperature, as well as the temperature of its surroundings. For these reasons, the HLW needs both cooling and shielding. The radioactive wastes are being produced at each stage of the production of electricity from nuclear materials, i.e., the nuclear fuel cycle, which involves the mining and milling of uranium(U) ore, its processing and fabrication into nuclear fuel, its utilization in the reactor, its reprocessing, the treatment of the utilized fuel coming from the reactor, and the waste dumping. Thus, they arise from the "burning" of uranium (U) fuel in a nuclear reactor and encloses the fission products and trans-uranic elements generated in the core of the reactor when electricity is produced. Statistically, the HLW accounts just for 3% of the total volume; however, it provides 95% of the total radioactivity of produced waste. Highly radioactive fission products and transuranic elements come from uranium (U) and plutonium (Pu) during the operations of reactors, and are enclosed inside the spent fuel. When used fuel is not reprocessed, it is regarded as a waste of the HLW type.

There are two different types of HLW, as follows:

A. Utilized fuel, which is designated as the waste.
B. Separated waste from the reprocessing of utilized fuel.

The HLW has both types of components, i.e., long-lived and short-lived ones, relying upon the length of time period needed to decrease the radioactivity of definite radionuclides to levels that are regarded as safe for the public and the neighboring atmosphere. The difference turns out to be vital for the management and dumping of HLW, if normally short-lived fission products can be separated from long-lived actinides. The HLW is the center of noteworthy attention concerning nuclear power, and is managed in view of that. The HLW includes utilized nuclear fuel from nuclear reactors along with the waste generated from the reprocessing of used up nuclear fuel. The majority of used up nuclear fuel comes from nuclear power plant reactors of the commercial type. At this time, mostly high-level waste is stored at the site itself where the waste is generated.

4. Nuclear Power and Defense Operations—The Sources of Radioactive Wastes

Commonly, the radioactive wastes are further divided into numerous precise categories relying upon their activity on the whole as briefed through the Figure 4, i.e., very-low-level waste (VLLW), low-level waste (LLW), intermediate-level waste (ILW) and high-level waste (HLW).

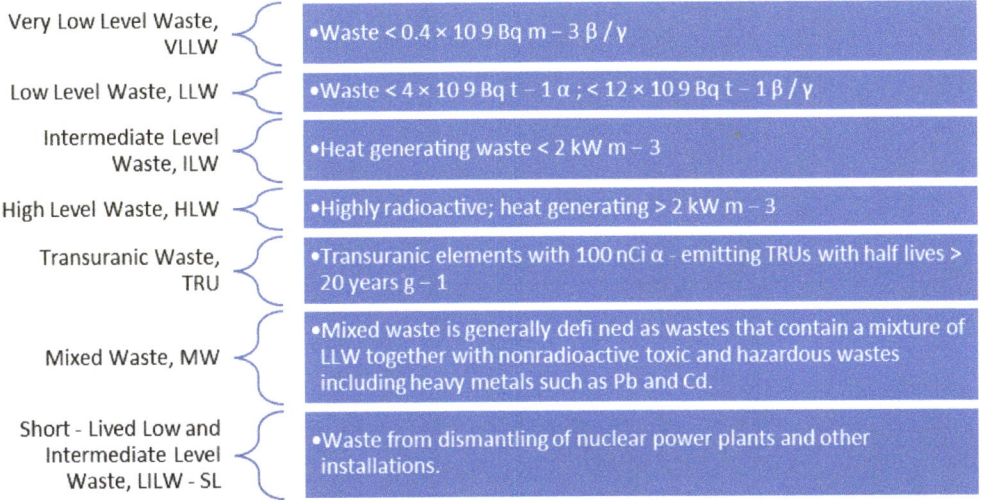

Figure 4. Different Categories of radioactive waste.

The high-level waste generates heat and hence, the temperature of this waste may rise drastically owing to the result of radioactive decay courses at least in the shorter time-scales. The sources of HLW include high-level liquid waste (HLLW) produced in the duration of the reprocessing of used up nuclear fuel that encloses a lot of short-lived fission products along with actinides and longer-lived fission products. One more source of HLW comes from the production of plutonium (Pu) metal and tritium (H-3 or 3H, or T)—the only radioactive isotope1 of hydrogen, used in weapon applications.

Additionally, the ILW may be heat-generating, however to a lesser degree than HLW, and chiefly comprises items such as the components within nuclear reactors, including graphite from reactor cores, fuel element debris, and fuel cladding along with radioactive sources employed in experimental instruments or medical equipment, chemical sludges and filters, which is defined as waste with an activity of >4 × 10 9 Bq t—1 α—radiation and >12 × 10 9 Bq t—1 β—and γ—radiation. By and large, this sort of waste is encased in concrete inside steel containers and put into storage in anticipation of final disposal. Some other categories of treatments of ILW that are not very heat-generating include pyrochemical, electro-refining, and associated wastes from the reprocessing of plutonium (Pu) metal for weapon applications, and these also necessitate particular types of treatment.

The activity of LLWs, including mainly worn shields and a few pieces of equipment or materials employed in the radioactive facilities, along with polluted soil and construction materials and varied organic and inorganic materials, though, comparatively low, cannot be dumped off as ordinary waste. The activity of LLW is set at <4 × 10 9 Bq t—1 α—radiation and <12 × 10 9 Bq t—1 β—and γ—radiation. At present, it is compacted into steel drums which are positioned within the bulky boxes and packed with concrete. The activity is very low in the case of VLLW and hence, it can normally be dumped off as ordinary waste, either in domestic landfills or by undergoing incineration treatment. It is, in general, defined as waste enclosing <0.4 × 10 9 Bq m—3 β—and γ—activity.

In U.S.A., an additional category of radioactive waste is defined for particular wastes with a lower activity than HLW that comprises transuranic elements, called "transuranic waste (TRU)" and encloses 100 nCi of α—emitting transuranic elements with half-lives > 20 years per gram of waste [37].

Several supplementary materials such as surplus uranium (U) and plutonium (Pu) from both civilian and military nuclear programs were at one stage throughout the 1990s well thought-out as wastes and may at some future date be affirmed as wastes again, although this is nowadays becoming more and more unlikely. At this time, used up nuclear fuel is being stored with no reprocessing, which is also considered as nuclear waste, although, the newly altering situation, brought about by the rising universal exigency for energy, and especially for energy sources that do not release considerable amounts of greenhouse gases (GHG), meticulously, CO_2—carbon dioxide—is now leading to an international drive to construct new nuclear power stations, and it will need fuel recycling. In 2008, the UK Government announcement supported principally the construction of newfangled plants. It is reported that the total quantity of nuclear power generated globally at present is of the order of 370 GW [38], which is estimated to augment between 447 GW and 670 GW by 2030. China and no-one else is setting up to construct 30 brand-new reactors by 2020, whilst India is now structuring seven novel nuclear plants. In Finland, one is under construction at Olkiluoto and in the U.S.A., many states pointed to the awareness of the construction of newer nuclear power stations. It is approximated that three to four novel plants would have to be constructed annually, opening in 2015, just to uphold the existing 20% nuclear power supply share of U.S.A. [39]. The exigency for electricity in U.S.A. is predicted to rise to 30% by 2030. Thirty-five, brand-new reactors are planned to be built. Although the proposed costing of a new-fangled facility is amplified to US$12–18 billion, the people are currently primarily in favour of nuclear power. Consequently, it is very obvious that nuclear power is certainly came back on the agenda, with lots of articles also appearing in the well-liked press stressing this modification in political trends. This is posing a very solemn challenge for scores of nations that are short of a big enough pool of skilled personnel and fresh graduates in the nuclear sciences, counting waste management specialists too.

Additionally, there also exist a few remarkable types of radioactive wastes. They include the following:

4.1. Transuranic Waste (TRUW)

In accordance with the U.S. regulations, Transuranic waste (TRUW) is defined as, with no regard to form or origin, waste that is polluted with alpha-emitting transuranic radionuclides with half-lives beyond 20 years and concentrations more than 100 nCi/g

(3.7 MBq/kg), exclusive of HLW. Those elements, which possess an atomic number bigger than uranium, are called transuranic, meaning "beyond uranium". On account of their long half-lives, TRUW is disposed of with more care than either LLW or ILW. That simply means that the "transuranic wastes" are synthetic radioactive elements possessing an atomic number of 92 (uranium) or higher. In the U.S.A., mostly, transuranic waste is found generated from nuclear weapons productions, which encloses common items, viz., tools, and lab equipment polluted for the duration of the initial age of nuclear weapons research and development. Presently, this kind of waste is stored at federal facilities and finally disposed of.

4.2. Uranium (U) Or Thorium (Th) Mill Tailings

The mill tailings of radioactive wastes stay around following the mining and milling of uranium (U) or thorium (Th) ores. They are stored at the sites of their generation in specifically designed ponds known as "impoundments".

4.3. Technologically Enhanced Naturally-Occurring Radioactive Material (TENORM)

A few radiological materials can subsist naturally in the eco-system. Other sources of radioactive wastes include medical and industrial wastes, as well as naturally occurring radioactive materials (NORM). At times, these naturally-occurring radiological materials (NORM) can turn out to be concentrated in the course of human activities such as mining or extraction of natural resource, the processing or consumption of coal, oil, and gas, and some minerals, bringing coal to the surface or burning it to produce concentrated ash, etc. This concentrated or relocated NORM is called "Technologically Enhanced NORM, or TENORM", which is generated from plenty of industries and processes such as oil and gas drilling, mining, and production, as well as water treatment. The said waste must be disposed of or managed systematically as per the rules and regulations of authorities. Globally, the highest Tenorm waste stream is found to comprise coal ash, with by and large 280 million tons annually, which includes uranium-238 and all its non-gaseous decay products in addition to thorium-232 and its progeny. More often than not, the referred to ash is just buried, or may be employed as a constituent for construction materials. TENORM is not regulated as limitedly as nuclear reactor waste, although there are no major differences in the radiological perils of these materials.

4.4. Legacy Waste

Apart from the routine radioactive waste from current nuclear power generation, there is one more kind of radioactive waste, known as, "legacy waste" that exists in pioneered nuclear power and particularly where power programs were developed out of military programs. At times it is found in volume, creating difficulty for its management. Owing to momentous activities typically associated with uranium mining, military programs, and the radium industry, copious sites enclose or are polluted with radioactivity. Single-handedly in the United States, the Department of Energy reports that there are "millions of gallons of radioactive waste" and "thousands of tons of spent nuclear fuel and material" along with the "huge amounts of polluted soil and water".

5. Impacts of Exposure to Radioactive Wastes

Exposure to radiation from radioactive wastes may cause adverse health impacts due to ionizing radiation, such as:

A dose of 1 sievert carries a 5.5% risk to mankind of developing cancer.
Ionizing radiation can cause deletions in chromosomes of human beings.
A possible birth defect in children.
May influence DNA, mRNA and protein repair.
The thyroid gland may be injured.
A propensity to have long biological half-lives and a high relative biological effectiveness makes it far more damaging to tissues, etc.

6. Illustrations of Accidents While Dealing with Radioactive Wastes

Unfortunately, a few incidents have taken place in the world's history where radioactive material was disposed of inappropriately, such as in the case of defective shielding during transport, or due to it simply being abandoned or even stolen from a waste store. The few remarkable accidents are:

- The Goiânia accident, which involved radioactive scrap originating from a hospital.
- In Japan, the nuclear substances were found in the waste of Japanese nuclear facilities.
- The waste stored in Lake Karachay in the old USSR was blown over the region during a dust storm following the lake partially drying out. In a low-level radioactive waste facility located at Maxey Flat, in Kentucky, containment trenches were covered up with dirt, instead of cement or steel, and fell down under the action of heavy rainfall into the trenches and ultimately filled with water, which invaded the trenches and turned out to be radioactive.
- Quite a lot of Italian deposits of radioactive waste have run into river water; therefore, domestically useful water has become polluted. Several accidents occurred in France during the summer of 2008. They are:
    (i) At the Areva plant of Tricastin, the liquid enclosing untreated uranium (U) overflowed in a faulty tank during a drain exercise and more or less 75 kg of the radioactive material percolated into the ground and finally into two nearby rivers;
    (ii) More than 100 staff members became contaminated with lower doses of radiation after the deterioration of the nuclear waste site on the Enewetak Atoll in the Marshall Islands and a prospective radioactive spill.

7. Radiation Concerns

As illustrated in Figure 4, there exist scores of various public apprehensions in the context of radioactivity and the diverse radioactive waste materials that exist, since the radioactive substances and radiation are being recognized as exceedingly hazardous. To add to this, these concerns also exist with regard to the wastes generated from nuclear power stations that they may get discharged into the eco-system. Of course, radiation is a natural occurrence and can be handled safely by using appropriate protections. Chiefly, uranium (U), thorium (Th) and potassium-40 (^{40}K) radioactive isotopes present in the crust of the earth are making basic grounds for natural radiation jointly with their decay products, such as radon gas (Rn), significantly [40]. There exist copious and diverse sources of background ionizing radiation to which each one is generally exposed. The maximum levels take place in geological regions whereby granitoid rocks or mineralized sands are found predominantly on account of traces of natural radioactive materials such as minerals enclosing U and Th as well as Rn (radon gas)—a decay product. Additionally, living at higher altitudes escalates the cosmic radiation level received. The applications of the X-rays in medical and dental can also boost exposure. The background radiation consists of roughly 87% of total natural radiation in association with 13% from synthetic resources. In order of severity, the radiation comes from medicinal applications, an assortment of varied sources, fallout from early atmospheric nuclear weapons examinations, job-related exposure, and radiation caused owing to nuclear discharges. At this juncture, a note should be made that there do not exist differences in the effects on materials among natural and non-natural radiation, including bio-systems. The effects of radiation on human beings are gauged in an international (SI) unit known as the Sievert, symbolized as Sv, which relates to the absorbed dose in human tissue to the effective bio-damage of the radiation. Sievert (Sv) is a unit of the effective dose received (1 Sv = 100 rem, i.e., Roentgen Equivalent Man (rem); or we can say, 1 rem = 0.01 Sv). It is well-known that the smallest dose received by each person breathing on the planet earth consists of a typical natural background level of approximately 2 mSv per year. There does not exist any scientific evidence of risk such as the development of cancers, etc., at doses less than 50 mSv per year in the short-term or 100 mSv per annum in the long-term. Nevertheless, it is also identified that a short-term dose of more than 1000 mSv is regarded as a great enough dose to cause instant radiation illness, whereas a dose of 10,000 mSv by and large results in death within a few weeks only.

8. Interventions with Hydration of Cement

The vast varieties of industrial courses of actions can generate wastes, which may get treated by stabilization/solidification (S/S). Of these, the inorganic wastes have a propensity to be more well-suited with cementitious binders. For the most part, the wastes treated by S/S with cement enclose metallic pollutants in an inorganic matrix made up largely of calcium (Ca), aluminum (Al), and silicon (Si), viz., dusts from air-contamination control systems, sludge, and soils. The USEPA declared that organic compounds interfere with cement-based S/S, in particular when the organic concentration goes beyond 1% of total organic carbon (C) by mass. Nevertheless, there exist several illustrations whereby the organic wastes have hardened with cement [41]. Numerous inorganic wastes enclose a few organic stains. In point of fact however that a few wastes can productively be incorporated or treated with cement in more elevated quantities than others. Virtually, there are no limitations on their chemical and physical attributes, other than that they must be finely divided solids or liquids at ambient temperature. The wastes possibly will enclose small additions of chemicals such as $CaCl_2$ or gypsum, which can be used as supplements and admixtures in traditional cement and concrete utilizations, but the wastes may also hold lots of dissimilar compounds that would not otherwise be incorporated with cement, including destructive pollutants. Each and every waste components can be coined as "impurities". This means that they are potentially reactive materials, which would not customarily be present in industrial cement. In the case of S/S, the interaction of wastes with binders is of much importance from two key points of views:

1. The interferences of impurities with hydration of cement together with setting and strength development, as well as matrix durability;
2. Immobilization of pollutants.

The referred to features are correlated in the sense that the development of a well-built durable matrix of inferior permeability is more often than not significant for the immobilization of pollutants, and is consequently a target of S/S treatment. Both beneficial and detrimental interactions can take place among wastes and hydraulic binders, i.e., "cement".

Still, the comprehensive progressions of hydration of cement stay as a topic of rigorous study and are at variance for dissimilar kinds of cement. Generally, it advances by dissolution of the phases of anhydrous cement from the surface of the particles of cement in the mixing water, followed by precipitation of products of hydration to build a strong cohesive matrix. The impurities can modify the customary hydration of cement at diverse phases; consequently, the dissimilar compounds have unlike, and at times multiple, interaction mechanisms with the binder. The impurities can be modified through altering the chemical composition of the developing pore solution in context of the pH, ionic strength, and chemistry:

- Solubility of anhydrous stages;
- Dissolution kinetics of anhydrous stages;
- Rate of development of yields of hydration;
- Rate of nucleation of yields of hydration;
- Morphology of the hydration yields;
- Chemistry of the hydration yields.

Often, the referred to influences rely on concentration and can differ in accordance with the conditions of curing. While impurities with different interference mechanisms as pure compounds are united, as what probably happens in both waste applications and solidification utilizations, the overall influence can be tricky to forecast [42]. Both the setting and hardening are physical demonstrations of the chemical course of action of hydration of cements. The setting is defined by Taylor [43] as "hardening with no noteworthy improvement in compressive strength," and stiffening as "momentous growth of compressive strength." Accordingly, interfering with the hydration of cement might result in the mass effects as follows:

- The acceleration/activation of setting or solidifying, counting, or flash setting, whereby the matrix loses its plasticity instantaneously upon the mixing;
- The false setting, whereby the plasticity of matrix is lost swiftly upon the mixing; however, it can be recovered by supplementary mixing;
- An exigency of altered water;
- The setting or hardening retardation, counting, and absolute inhibition of hydration;
- The modified strength growth counting disruption of matrix;
- The changed chemistry of the pore solution.

A speeding up of setting and false setting can create challenges in the cement's handling, considering the usage point of view, and can promote equipment failure as well as pitiable workability. The demand for increased water can cause an increase in the speed of setting or false setting, or change rheological attributes of the mixture, which can result in a more permeable structure and reduced durability. Often, the modified strength growth is considered as taking the form of enhanced or reduced early strength, however, later strength can also be influenced, and even a yield that seems to set and achieve strength generally at first can deteriorate swiftly afterwards. Several illustrations of interactions of impurities with cement, in either cement and concrete yields or S/S, can be found in the literature review [44–51], which is abridged in the following sections, for all three types of compounds, i.e., organic compounds, inorganic compounds, and other impurities. The referred to outcomes were generated by experimental studies making the use of pure compounds. Additionally, there are a number of research investigations accessible in the literature conducted in order to treat real wastes. It is much more complicated to simplify real wastes because of their heterogeneity and complex structure and chemistry as well. In the literature studies, plenty of investigations are based on the physical demonstrations of setting and strength growth, or at times, on the extents of the heat evolved by hydration reactions. At the simplest level, the retardation reduces the amounts of the hydration yields at a given time, while the acceleration escalates them. Nevertheless, the impurities present in the pastes of cement may also modify the proportions of the hydration yields, alter the ratio of Ca/Si in the C-S-H, create solid solutions with the hydration yields, or craft completely brand-new hydration yields. Quite a few current investigations revealed the influences of wastes and pollutants on leachate pH and S/S product acid neutralization capability; these are suggestive of modifications to the hydration yields [48,49]. These kinds of alterations may bring about vital consequences for the immobilization of pollutants by all of the mechanisms; however, this is a characteristic of S/S with limited studies. Thus, it is obvious that often cements respond radically to small supplements of impurities; it is feasible to employ additives and admixtures to regulate the hydration properties of an S/S yield to trim down the unwanted influences of components present in the wastes.

8.1. Influences of Inorganic Impurities on Hydration of Cement

The influences monitored for in several compounds enclosing inorganic and heavy metal components during the hydration of cement are explained in the literature [50–52]. The terms of "acceleration" and "retardation" express influences on setting, hardening, or both, as they are applied in the literature in the same fashion. The previous research data reveal the preponderance of studies on Portland cement in the literature, although, at times, the same is also correct for other kind of cement, or incorporated cement systems, e.g., the metal cations of zinc (Zn^{+2}), lead (Pb^{+2}), cadmium (Cd^{+2}), and chromium (Cr^{+2}) also shrink the strength of geopolymer slag cement [53]. The little influences are talked about in-depth in the following notes that refer to the superscripts in the tables:

I.   The C3A and C3S are accelerators, and they also have a propensity to be the center of attention of action by other accelerators and retarders. These are the uppermost reactive phases in Portland cement, and also most imperative for setting and early strength growth. The other reactive calcium aluminates or Ca–aluminate cements are also accelerators for Portland cement, and vice versa. The referred to species play the role of accelerators or activators for pozzolans. Additionally, lime or cement kiln dust can be employed as activators for pozzolans.

II. The carbonates of alkali demonstrate the surprising nature of the few compounds that interfere with the cement setting. The small proportions of less than 0.1% of alkali carbonates were found to retard the Portland cement setting; an augmented quantity results in flash setting, and further boosted quantities can have no influence on setting, while flash setting takes place at very elevated proportions. So far, the interference mechanism is not well-comprehended, however, it is put forward that the influence of carbonates is owing in part to the production of thaumasite—a calcium silicate mineral, rather than ettringite—a hydrous calcium aluminum sulphate mineral.

III. Generally, the salts of potassium (K) and sodium (Na) are believed to increase the pH and cause precipitation of amorphous CH that interferes with C3A hydration, and, up until now, a lot of alkali salts play the role of accelerators of Portland cement. They can be supplemented as activators to cement enclosing pozzolans, whereby they boost the solubility of the anhydrated phases. A few salts, namely sodium chloride (NaCl), accelerate at lower concentrations, though retard at very elevated concentrations.

IV. Sodium silicate (Na_2SiO_4) is a well-liked additive for waste solidification. It is useful as an accelerator or activator for pozzolans, or to devour surplus water as it extends both silicate and a higher pH. In the case of the latter, or in the incident that excessive quantities are added on, silica gel development may promote a physically unstable matrix, with shrinkage and swelling due to the humidity modifications of the nearby environment.

V. Sulphates have quite a lot of probable influences on the hydration of Portland cement. Perhaps owing to acceleration or retardation by reaction with C3A and C4AF, they may also lead to false or flash setting, by formulating gypsum in place of ettringite, or by matrix destruction via late ettringite development and having a bulk volume on account of its waters of hydration. Additionally, thiosulphate is accounted as an accelerator. The salts of chloride can also outline enormous chloro-aluminates, which are harmful to the matrix if their formulation is late. Both chlorides and sulphates, as well as carbonates and other anions, can destruct the matrix, provided the solubility of one of their salts in the pore solution is high and crystallization takes place [54]

VI. An increase in MgO by a couple of percentage points can destroy the cement matrix through gradually hydrating to more bulky $Mg(OH)_2$; $MgSO_4$ also reacts to give rise to more voluminous products, gypsum and $Mg(OH)_2$, as well as degrading C-S-H.

VII. Evolution of gas can cause matrix destruction, for example, from the reaction of aluminum metal or ammonia at a higher pH.

VIII. While soluble chromium (Cr) salts speed up the hydration of Portland cement, chromium (Cr^{+3}) oxide has a modest influence on setting. A review of Mattus [44] suggested that chromium (Cr) replaces silicon (Si) in C-S-H, however, the ultimate strength of the matrix is slimmed down. Chromate, (CrO_4^{-2}) is believed to play a similar role to sulphate and develop chromo-aluminate crystals which cover C3A grains. Like sulphate, chromate can also play as an accelerator.

IX. The influence of $ZnSO_4$ exemplifies the significance of taking into consideration the collective effect of both the anion and cation. While, in general, most Zn-compounds are believed to be retarders of Portland cement. $ZnSO_4$ is found to be an accelerator at concentrations < 2.5% and a retarder at concentrations between 2.5% and 5.5%; it entirely slows down the hydration of cement at elevated concentrations.

X. Boric acid has a strong accelerating impact on the setting of Portland cement, however, it retards solidifying. The carbonates, hydroxides, aluminates and silicates accelerate setting, but retard solidifying or leave it unaltered; nitrates and halides accelerate both the setting and solidifying. With a view to simplify the outcomes of diverse examinations, we separated out the influences of cations and anions, and searched for interaction impacts; data from twelve literature studies of pure compounds added to Portland cement paste were gathered and utilized to build neural network models of Unconfined Compressive Strength (UCS) as a function of mixture chemistry [55]. The utilization of the most excellent neural network using other information proposed that Cs is a retarder and Cr^{+4} has no influence. The accessible information is distinguished for Hg, K, Mn, Na, and SO_4^{-2}. Impacts were monitored for in the literature for Ca–aluminate cements [56–62], which are normally lesser utilized in waste management because of the better-quality characteristics of cements based on calcium silicate.

XI. The literature suggests that Li-salts, with the only exception of Li_2BO_2, are the strongest accelerators for Ca-aluminate cements of all kinds and are competent enough to cause flash setting. Li-salts are believed to play a role by creating nucleation substrates for the hydration yields.

XII. Nevertheless, for the other anions and cations, the wide-ranging impacts on Ca–aluminate cement are accounted for. Quite a lot of researchers have attempted to rank the influences of a variety of cations and anions without harmony among themselves [60]. Some others have given general statements to explain this behavior, e.g., Parker [63] explained that alkalis are accelerators and acids are retarders of Ca–aluminate cement, however, sulfuric acid is well-known to be as an accelerator; Sharp et al. [64] proposed that those compounds which augment the C/A ratio play the role of accelerators, however, $CaCl_2$ is well-known retarder.

XIII. The probable variety of the influences is depicted by monitoring for Ca–sulphate [65] and they found gypsum with no impact on the setting time of Ca–aluminate cement, while hemihydrate is an accelerator and anhydrite is a retarder. They proposed that the relations among the dissolution kinetics of the referred to diverse forms of calcium sulphate control both the kinetics and the hydration yields.

XIV. The suggestion has been that CH plays a role as an accelerator for calcium aluminates by boosting the C/A ratio of the solution, and driving precipitation of the calcium-rich phases C2AH8 and C4AH13. The Portland cement is believed to be an accelerator since hydration of C2S and C3S generates CH. Yet again, the neural network analysis was employed to construct models of setting time as a function of mixture composition, utilizing available information for pure compound additions to Ca–aluminate cements. This eveealed that the reproducibility of setting time measurement is pitiable, and that this accounts for a few of the differences in the outcomes in the literature.

8.2. Influences of Organic Impurities on Hydration of Cement

So far, numerous organic compounds are supplemented as purposeful admixtures to cement and concrete. Their selection and effects are derived from Massazza and Testolin [47]. Some other organic compounds are found usually in waste, such as oil and grease, chelating agents such as ethylene diamine tetraacetate (EDTA), and chlorinated hydrocarbons—trichlorobenzene, phenols, alcohols, carbonyls and glycols. The impacts of organic compounds on the hydration of Portland cement are changeable and extremely concentration-reliant, e.g., generally sugars and triethanolamine are retarders for C3S hydration, but can play the role as both retarders and accelerators for C3A hydration, and will promote flash setting at higher concentrations [44]. The polar solvents hinder hydration to a much larger extent than non-polar solvents as accounted for by Nestle et al. [66]. The findings for definite pollutants were examined by them as well as others also. Additionally, scores of organic compounds result in the progressive worsening of cement yields over time. The majority of organic compounds are retarders of Ca–aluminate cements. Particularly, citrate salts are usually employed as retarders. Nevertheless, Bier et al. [67] presented a technique of utilizing Li and citrate jointly to optimize the setting attributes of Ca–aluminate cements, such that retardation essentially did not occur, while Baker and Banfill [68] utilized trilithium citrate as an accelerator. The trilithium citrate was found to have more or less no effect in experiments carried out by Damidot et al. [56]. Even though pozzolans are accelerators of C3S hydration, they themselves hydrate, and pozzolanic cements are inclined to achieve strength more slowly than Portland cement, unless an activator, viz., an alkali or CH, is employed. Pozzolans may reduce the impacts of other impurities [69] and are well-known to diminish the hydraulic conductivity and enhance the strength and durability of the ultimate yield in the long duration. While an addition is made to Ca–aluminate cements, pozzolans lend a hand to formulate C2ASH8 that resists conversion to lesser voluminous C3AH6 [70,71]. At last, the cement hydration is also influenced by the physical characteristics of the cement and any impurities, i.e., crystal defects, crystallinity, as well as the size of particle and surface area of the dissimilar mineral phases. The reactivity of binders is augmented with fineness, however, finer impurities such as colloidal matter or clay can retard the setting.

9. Solidification of Strontium (SR) through Diverse Kinds of Cements

Looking into earlier studies, the research investigations have thrown light chiefly on the leaching of radionuclides in PC-based systems with supplement of diverse minerals such as cement systems based on bentonite, zeolite, and tobermorite [72–74]. The radionuclides' stability in the matrices is found to be improved when the integration of minerals is made. The minerals are chosen on account of their higher adsorption competences. The higher concentration of salts in the LLW and ILW radioactive waste liquid has an immense impact upon the hydration process, hydration yields, rheology behavior, and setting time of cement paste. For these reasons, there are a few issues for the solidification of radionuclides from nuclear plants. Consequently, the leaching rate of radionuclides becomes modified with the product performance and the pore-structure of the matrices [11,75–87]. Sodium hydroxide's (NaOH) influence on the structure of toughened Portland cement matrices and the leachability of Sr^{+2} was studied by Zheng et al. [12]. The outcomes pointed towards the improvement in the adsorption competence of C–S–H gel by the use of NaOH and mitigated the porosity of matrices that lead to the decline in the leaching rate of Sr^{+2}. Conversely, sodium nitrate ($NaNO_3$) boosted the increase in the leaching rate of Sr^{+2} and the adsorption capability of C-S-H gel was also found to be enhanced since the porosity get improved.

Apart from PC, the magnesium phosphate ($MgSO_4$) cement (MPC) is considered as a chemically bonded ceramic with good solidification performance for both LLW and ILW liquids.. The application of MPC in the interest to solidify Pu-contaminated ash where the mass quantity of Pu is up to 5% in form of PuO2 was performed in an early research study by Wagh et al. [77]. The mechanism of immobilization comprises the physical encapsulation in the denser matrix together with the lower solubility of PuO2.

The sulpho–aluminate cement possesses the potential to immobilize Sr^{+2} and Cs^+ owing to its good retention capability and resistance to salts. Xu et al. [78] have concentrated on the mechanism to immobilize Sr^{+2} and Cs^+ in the form of radioactive ion exchange resin by the sulpho–aluminate cement. The results of experiments revealed that the competence of immobilization with regard to Sr^{+2} and Cs^+ by sulpho–aluminate cement is superior to PC. In context of the stages in cement, ettringite can proficiently absorb Sr^{+2} (88%) and Cs^+ (70%) when the ratios of retention of the aluminum hydroxide gel for the two ions were merely 10% and 40%, correspondingly. The Sr^{+2} immobilization by ettringite is assigned to the substitution of Ca^{+2} in crystal lattice by Sr^{+2}.

Type 1 Standard Portland cement (PC), is general-purpose cement fit for all applications where the particular characteristics of other kinds are not essential. Often, this kind is most useful for S/S systems.

Type II Moderately Resistant to Sulphate attack PC is employed where preventative measures against reasonable sulphate attack is vital.

The resistance to sulphate attack is achieved by forming a cement with a lower enclosure of tricalcium aluminate, with the highest proportion reaching up to 8%. Generally, the Type II cement will achieve strength and produce heat at a rate inferior to Type I. In S/S utilizations this might be considered, because, here, there is an involvement of volatile organics. The low temperature of the S/S mixture might reduce the discharge of organic species with a volatile nature.

Type III High Initial Strength PC offers more initial strengths than Type I or Type II, although the eventual long-standing strengths are almost analogous. Both chemically and physically, it is alike to Type I cement, except that its particles are grounded finer. Though rich mixtures of Type I cement can be employed to obtain increased early strength, Type III cement might offer this more agreeably and more cost-effectively. The swifter hydration will normally discharge heat more rapidly and grounds for a somewhat higher rise in temperature than Type I.

Type IV Lower Hydration Heat PC is utilized where the quantity of heat produced should be minimized as found in the case of great structures of concrete, viz., huge foundations and dams. However, accessibility is tremendously restricted mostly since parallel properties can be attained from a Type I cement, more often than not comprising of a mix of Type I cement and fly ash (FA). Since it lacks in accessibility and fitting substitution, Type IV cement perhaps has low utilization in S/S applications.

Type V Higher Resistant to Sulphate attack PC is utilized where S/S systems either enclose or are subjected to rigorous sulphate action—chiefly whereby soils, waste, or subsurface water own higher quantities of sulphate. It obtains strength more leisurely than Type I or Type II cement. On account of its low content (highest up to 5%) of tricalcium aluminate, it resists more against sulphate attack than Type II cement. By and large, the amount of iron is greater in Type V cement, and this might be sought after if the species present in the waste form are insoluble iron complexes.

The magnesium–potassium phosphate cement (MKPC) is an efficient agent for S/S technology. Although MKPC demonstrated optimistic impacts on S/S, some studies were carried out on the mechanism of the S/S of heavy metals by MKPC. This was in accordance with the suggestion of Singh et al. [79], who projected that the cause for the immobilization of chemically bonded phosphate ceramics is chemical stabilization, the reaction kinetics among pollutant metal salts and the phosphate solution, followed by physical encapsulation inside the denser matrix of phosphate. The particles of fly ash were encapsulated in the binder matrix as found by Rao et al. [86]. In the context of the hydration yields of heavy metal, Jeong et al. [81] unearthed that PbO can structure the PbHPO4 in magnesium phosphate cements (MPCs).

The MKPCs solidify cement paste powders produced from the heating and grinding of radioactive concrete wastes. The benefits of the application of MKPCs are their higher early strength, higher bonding strength, little drying shrinkage, lower permeability, and higher resistance to sulphate attack [82–84]. Quite a lot of investigations on the applications of MKPCs for the solidification and stabilization of radioactive or heavy metal wastes have been conducted [85–87]. The compressive strengths of MKPC waste forms are reported to be just about two-fold greater than the ordinary Portland cement (OPC) grouts, and the porosities are mitigated by half [83]. Additionally, the MKPCs display a greater resistance against radioactivity in comparison with PC [88,89], authenticating the potential of MKPCs for radioactive waste solidification. To add to this, unlike vitrification or ceramic solidification procedures, MKPCs are cost-effective and energy-proficient and employ uncomplicated equipment; particularly, there is an absence of volatilization of nuclides.

On the other hand, MPCs are employed to stabilize and solidify an assortment of lower level radioactive mixed wastes, Pu-polluted ashes, and technetium enclosing waste solutions at the Argonne National Laboratory, USA [83,90]. The stabilization of a higher content of sodium waste streams with MPC is studied by Colorado et al. [91]. Vinokurov et al. [92]

investigated the immobilization of simulated liquid alkaline high-level waste (HLW) enclosing actinides and fission as well as corrosion yields with magnesium (Mg)–potassium(K)–phosphate (PO_4) matrices. Therefore, the viability of MPC for the S/S technology for risky and radioactive wastes is verified by prior studies.

Calcium Aluminate Cement (Cac) and Cac Modified With Phosphate (Cap) cements are some of the substitute cements in PC and are renowned materials, which are exercised for solidification and for the encapsulation of radioactive wastes [90]. Usually, it is incorporated with supplementary cementitious materials (SCMs) such as fly ash or blast furnace slag. The CAC is characterized by the internal environment possessing a considerably lower pH, ranging from 10.5 to 1,1 than that of PC and its integrated systems, displaying more than 13 [93]. The encapsulation of the reactive metallic radioactive wastes in CAC is examined through making the use of this property. The alteration of CAC with phosphate (CAP) can lower the pH in a range of 9.0 to 10.5 [93]. Additionally, the alteration of CAC by phosphate is identified to stop the traditional hydration of CAC, and results in solidification through an acid–base reaction amongst acidic phosphate solution and CAC powders playing the role as a base [94–100]. Recently, Caps have been examined for their use in S/S technology of dangerous as well as radioactive materials, because of their capabilities of solidification reactions and/or internal environment pH lowering [101–105].

10. Impact of Cement Solidification Technology

10.1. Compressive Strength

In a study by Laili et al. [106] on the solidification of radioactive waste resins, they utilized cement mixed with dissimilar percentages (0%, 5%, 8%, 11%, 14% and 18 %) of organic material, called "biochar". The outcomes demonstrated that the biochar content had an influence on the compressive strength of the solidified resins. The outcomes for the effect of biochar dosage displayed that the addition of 14 wt % of biochar augments the value of compressive strength from 6.2 MPa (5 wt %) to 10.1 MPa. Further than this percentage, the compressive strength was found to be somewhat diminished to 9.2 MPa (18 wt %). Biochar can absorb water and this absorbed water in biochar does not bind chemically with carbon. As a result, the absorbed water would be discharged all throughout the hydration process, which might support the hydration procedure during the initial age of cemented waste. The competence of water retention in biochar may cut the water loss through evaporation that offers good conditions for curing designed cemented waste.

Nishi and Natsuda [107,108] worked on an improvement in resins load and compressive strength of solidified waste and pointed out that resin loading for solidification into OPC must be limited to lesser than 20% to put a stop to the development of cracks at elevated loadings that will result in an unsteady waste yield. The cracks of the cement waste form are assigned to the produced swelling pressure of resin of 50 MPa (max)), which goes beyond the tensile strength of the cement waste form. Pan [109] investigated resin encapsulation using furnace slag and fly ash as admixtures and pointed out that a loading of 24 wt. % of resin can be encapsulated when the use of a combination of furnace slag (24 wt. %) and fly ash (24 wt. %), and OPC (8 wt. %) is made. The supplementation of zeolite into OPC in order to be encapsulated into the resins was focused on by Bagosi [110], who designated that 24 mL resin can be encapsulated into a mix of 55.9 g OPC, and 37.3 g zeolite. Supposedly, the weight of 24 mL resins is 40 g, then the proportion of resin loading is more or less 43%. With a view to boost the resin loading and stability of the waste products, fibers of stainless steel, glass or carbon can be supplemented. The upshots [111] unearthed that the addition of definite fibers enhanced the stability of the solidified waste product, although the fibers of glass and carbon are not found fitting for resin solidification into cement because fibers of glass are not alkali-resistant and get dissolved in the cement matrix. One challenge with the carbon fiber application is the formation of a homogenous waste form. At this time, Durafiber or polypropylene is exercised in the cement industry to minimize the cracks in the cement structures, and may be regarded for the encapsulation of spent radioactive resins provided this fiber is radiation-resistant.

Natsuda [108–112] conducted research with a view to enhance resin load and compressive strength of the solidification products. In a final conclusion, it has been reported that the resin load in the solidification products of OPC must be regularized at lower than 20% volume of wet resins to solidification product. While resins load is greater than 20%, the cracking is engendered often and the solidification products display inferior stability. Additionally, the resins' expansion in water was studied, and it was reported that the expansion of resins is one of the key causes which results in cracking. The expansion pressure range is reported to be between 0 to 50 MPa. The resins must be saturated in water previous to solidification with a view to avoid cracking. An expansion model was set up, and from this model, when capabilities for ions and water of resins achieve saturation, the expansion pressure must be zero. There must be no cracking and/or fractures, however, the experiments exhibited that when resins loads are greater, there would be the development of fractures in the solidification products even though the resins are saturated.

10.2. Behavior of Leaching Process

"Leaching" is a procedure through which a liquid dissolves and takes away the soluble constituents of a material. At times, "leaching" occurs by means of percolation or the flow of water, which can smash up cement-based structures rigorously, e.g., dams, pipes, conduits, etc. It can potentially degrade the cement-based engineered barriers employed for long-standing storage of nuclear wastes.

Simply speaking, the diffusion of radioactive ions into outer medium through conjoint pores is known as "leaching". The rate of leaching of pollutants is confirmed by the characteristics of the waste form, such as the pore structure, quantity of water, hydraulic conductivity, and homogeneity. Numerous research studies are conducted on the relations among ion leaching and the micro-structure of the solidification yields. The researchers have found that minimizing pore size in solidification products and enhancing pore structures are the efficient and valuable techniques in the interest to mechanically enclose wastes in solids. Hence, a lower ratio of water/cement (W/C) and compressive molding are exercised for minimizing the pores in solidification product, and surface painting is also utilized to clog pores in the products.

The release rates are estimated for quite a lot of radionuclides such as Cs-134, Mn-54, Co-60, Zn-65 and Eu-152. The release data designate that merely one of the formations exhibit the leaching of Cs-134 from the hardened spent resins. Additionally, there are no other radionuclides such as Mn-54, Co-60, Zn-65 and Eu-152 spotted in any leachates. The bulk of research carried out on leaching by world researchers has utilized a much higher total activity of radionuclides. For instance, Rudin et al. [113] have spiked 75 µL of a NIST-perceptible SeCl2, with Se-75 having a total activity of 19,443 Bq in the paste of cement. The samples taken for use for the leaching test by Papadokostaki and Savidou [114] possessed the total activity of roughly 370 KBq of Cs-137. Though the early activity of radionuclides in spent resins is regarded as low the waste form produced in the said research studies appears to be capable of immobilizing the radionuclides [115]. Accordingly, the findings put forward that further research of leaching tests utilizing spiked solutions with a higher total activity of radionuclides into fresh ion exchange resins must be carried out.

Interestingly, Li [116] monitored the rate of leaching of nuclides. The dumping of spent ion exchange resins was performed subsequent to the encapsulation into cement. The heavy metals were then precipitated in the matrix of cement whereas alkali metals, namely, cesium (Cs), stayed considerably soluble and leach out from the waste form under conditions of cement encapsulation. The findings designated that even though Cs-loaded resins are encapsulated, the matrix will have a leaching rate that is one or two orders of magnitude more in comparison with the leaching rate for Cs from the resins themselves. The recent research concentrates on the diminution of Cs leaching in terms of the total Cs adsorbed on the resin through amalgamation of natural and chemically treated zeolites into the cement as an admixture. The study results indicate that the supplement of natural zeolites slimmed down the Cs discharge to around 70% to 75% of the amount initially bound in the resin over a leaching period of three years. Kaolin clay has an influence on the leaching attributes and strength of the cemented waste form. The results of leaching examinations uncovered that the addition of kaolin into cement mitigates the rates of leaching of radionuclides considerably. Nevertheless, the outcomes unveiled that supplement of clay in a surplus of 15 wt. % trims down the hydrolytic stability of the cemented waste form. The lowest rate of leaching and highest strength were recorded when the addition of 5 % kaolin content to the cement matrix was made.

Favorably, the exhausted ion exchange resins are cemented for dumping. The heavy metals precipitate readily in the higher pH environment of cements, however, the alkali metals such as cesium (Cs) stay significantly soluble. The cementation of Cs-loaded resins has setbacks because when they cemented, they display rates of leaching that are one or two orders of magnitude more elevated in the cement matrix than in the resins themselves. The investigations by Bagosi [117] have thrown lights on the diminution of noteworthy Cs leaching in terms of the total Cs adsorbed on the resin through adding natural untreated and chemically treated zeolites to the cement. Finally, they concluded that the supplement of natural zeolites declined Cs discharge by up to 70% to 75% of the amount first bound in the resin during the course of a leaching period of 3 years. Osmanlioglu [118] supplemented kaolin clay into cement with a view to bring down the rates of leaching. The influences of kaolin clay on the leaching characteristics of the cemented waste forms were evaluated, and the impact of the kaolin addition on the strength of the cemented waste form was also tested. Notably, the long-lasting examinations of leaching put on show that kaolin addition into the cement mitigates the rates of leaching of the radionuclides. Nevertheless, the supplements of clay in surplus of 15 wt.% were found to cause an imperative dwindle in the hydrolytic stability of cemented waste forms. Furthermore, it is reported that paramount waste isolation, sans causing a loss in the mechanical strength, was achieved when the kaolin content in cement is 5%. El-Kamash [119] worked on the leaching of 137Cs and 60Co radionuclides fixed in cement and cement-based materials and found that the leaching examinations of 137Cs and 60Co radionuclides symbolize the behavior of leaching and a few of the archetypal radionuclides detected in lower level solid waste forms. The addition of 0 to 15% silica fumes and ilmenite to cement resulted in a reduction in the rate of leaching of each nuclide at diverse studied temperatures. The examinations were carried out for a research study on the leaching of heavy metal ions from cementitious waste. A variety of mathematical models were then employed to gauge the behavior of embedded radioactive wastes. An iterative model was suggested by Krishnamoorthy et al. [120] in order to simulate the rates of discharge of radionuclides from cylindrical-shaped blocks of cement. Two expressions of the leach rate for the diffusive discharge of radioactive components from both cylindrical- and rectangular-shaped waste forms are derived by Pescatore [117]. As an ensemble view, our collaborators managed to use alternative materials for radiation shielding and as well as utilize wastes towards obtaining sustainable materials [121–127].

11. Traditional Treatment of Organic Liquid Radioactive Waste

Effluents from nuclear power plants and some medical research institutes contain radioactive heavy metals as well as complex combinations of dangerous organic chemicals and irradiated surfactants. In comparison to other types of radioactive waste, the volume of organic liquid radioactive waste created is minimal. Figure 5 lists typical organic waste kinds, sources, and characteristics. Figure 5 also depicts the characteristics and limits of several technologies used in the treatment of organic liquid wastes. Nuclear power plant liquid radioactive waste typically comprises soluble and insoluble radioactive components (fission and corrosion products) as well as nonradioactive chemicals. The overarching goal of waste treatment procedures is to disinfect liquid waste to the point that the decontaminated bulk volume of aqueous waste may be discharged into the environment or recycled. The waste concentrate is further processed, stored, and disposed of. Because nuclear power facilities create practically every type of liquid waste, nearly every procedure is used to treat radioactive effluents. Liquid waste streams are frequently decontaminated using standard procedures. Each procedure has a distinct impact on the liquid's radioactive concentration. The extent to which they are employed together is determined by the amount and source of contamination. Evaporation, chemical precipitation/flocculation, solid-phase separation, and ion exchange are the four basic technological procedures for treating liquid waste. These therapy methods are well-known and frequently used. Nonetheless, several nations are making attempts to increase safety and economy via the use of modern technology. Evaporation achieves the best volume reduction impact when compared to the other procedures. Decontamination factors ranging from 10^4 to 10^6 are achieved depending on the content of the liquid effluents and the kind of evaporators.

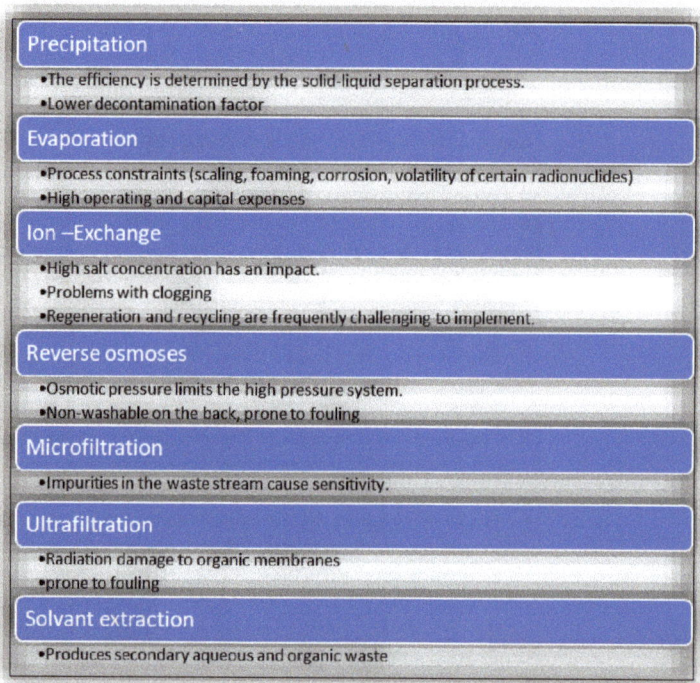

Figure 5. The limitations of various aqueous liquid treatment solutions.

12. Conclusions

The safer and systematic disposal of radioactive waste is of the highest precedence and, hence, the development of secure engineered barriers requires greater focus and attention. The cementitious binders for the use of radioactive waste immobilization offer a solution, which is not only stable but lucrative. The key researchers of resins solidification are concentrated on escalating the loading of spent resins, mitigating the leaching of nuclides, and enhancing the compressive strength of the matrix. The chemistry of Ca–sulpho–aluminate cement is a promising option for the cementation of radioactive spent resins. This kind of cement production is characterized by a lower leaching rate of nuclides, higher spent resin loadings, and stability through wet and dry curing, together with higher compressive strength, and negligible hurdles in manufacturing. The biochar content and spent resin loading affect the compressive strength of the waste form. The review suggests that the magnesium potassium phosphate cements can be utilized to immobilize radioactive concrete wastes generated during the decommissioning of nuclear power plants. The magnesium phosphate cement is competent enough to obtain the rapid solidification of higher content and high-level liquid wastes and radioactive substances in nuclear emergency incidents. The radioactive ions of higher content and high-level can be solidified chemically and encapsulated by the magnesium phosphate cement matrix physically. The potential research in the context of solidification desirably would be concentrated on the control of costs,, solidification competence of perilous components, the optimization of mechanical attributes, examinations of the action mechanisms of diverse components in solid waste, and long-standing stability explorations.

Author Contributions: Conceptualization and methodology, I.L. and S.L.; validation, formal analysis and visualization M.M.A.B.A., A.V.S., P.V. and D.D.B.-N.; data curation, R.A.R. and T.I. writing—original draft preparation, S.L. and I.L.; writing—review and editing and funding acquisition, A.V.S. All authors have read and agreed to the published version of the manuscript.

Funding: This research publication was supported by the Gheorghe Asachi Technical University of Iasi (TUIASI) from the University Scientific Research Fund (FCSU).

Institutional Review Board Statement: Not applicable.

Informed Consent Statement: Not applicable.

Data Availability Statement: Not applicable.

Conflicts of Interest: The authors declare no conflict of interest.

Abbreviations

SCM	Supplementary cementitious material
PC	Portland cement
S/S	Solidification/stabilization
CO_2	Carbon Dioxide
IAEA	International Atomic Energy Agency
LLW	low-level radioactive waste (LLW)
ILW	intermediate-level radioactive waste (ILW)
HLW	high radioactive waste (HLW)

References

1. Baek, J. Do nuclear and renewable energy improve the environment? Empirical evidence from the United States. *Ecol. Indic.* **2016**, *66*, 352–356. [CrossRef]
2. Goh, T.; Ang, B. Quantifying CO_2 emission reductions from renewables and nuclear energy—Some paradoxes. *Energy Policy* **2018**, *113*, 651–662. [CrossRef]
3. Li, X.-D.; Qu, Y.-H.; Zhang, L.-H.; Gong, Y.; Dong, Y.-M.; Li, G.-H. Forecast of China's future nuclear energy development and nuclear safety management talents development. *IOP Conf. Ser. Earth Environ. Sci.* **2021**, *691*, 012022.
4. Nathaniel, S.P.; Alam, S.; Murshed, M.; Mahmood, H.; Ahmad, P. The roles of nuclear energy, renewable energy, and economic growth in the abatement of carbon dioxide emissions in the G7 countries. *Environ. Sci. Pollut. Res.* **2021**, *28*, 47957–47972. [CrossRef] [PubMed]
5. IEA. *Key World Energy Statistics 2020, International Energy Agency Statistics Report*; IEA: Paris, France, 2020.
6. Palmer, R.A. Radioactive Waste Management. In *Waste: A Handbook for Management*, 2nd ed.; Letcher, T.M., Vallero, D., Eds.; Academic Press: Cambridge, MA, USA, 2019; pp. 225–234.
7. Li, J.; Liu, K.; Yan, S.; Li, Y.; Han, D. Application of thermal plasma technology for the treatment of solid wastes in China: An overview. *Waste Manag.* **2016**, *58*, 260–269. [CrossRef]
8. Duan, Z.; Fiquet, O.; Ablitzer, C.; Cassayre, L.; Vergnes, H.; Floquet, P.; Joulia, X. Application of pyrolysis to remove hydrogen from an organic nuclear waste. *J. Hazard. Mater.* **2021**, *401*, 12336. [CrossRef]
9. Vulpius, D.; Baginski, K.; Fischer, C.; Thomauske, B. Location and chemical bond of radionuclides in neutron-irradiated nuclear graphite. *J. Nucl. Mater.* **2013**, *438*, 163–177. [CrossRef]
10. IAEA. *Processing of Irradiated Graphite to Meet Acceptance Criteria for Waste Disposal*; IAEA-TECDOC-1790; IAEA: Vienna, Austria, 2016.
11. Zheng, Z.; Li, Y.; Zhang, Z.; Ma, X. The impacts of sodium nitrate on hydration and microstructure of Portland cement and the leaching behavior of Sr^{2+}. *J. Hazard. Mater.* **2020**, *388*, 121805. [CrossRef]
12. Zheng, Z.; Li, Y.; Cui, M.; Yang, J.; Wang, H.; Ma, X.; Chen, Y. Insights into the effect of NaOH on the hydration products of solidified cement-$NaNO_3$ matrices and leaching behavior of Sr^{2+}. *Sci. Total Environ.* **2021**, *755*, 142581. [CrossRef]

13. International Atomic Energy Agency. *Selection of Technical Solutions for the Management of Radioactive Waste*; IAEA-TECDOC-1817; IAEA: Vienna, Austria, 2017.
14. International Atomic Energy Agency. *Strategy and Methodology for Radioactive Waste Characterization*; IAEA-TECDOC-1537; IAEA: Vienna, Austria, 2007.
15. International Atomic Energy Agency. *Treatment of Low-and Intermediate Waste Concentrates*; Technical Report Series No. 82 1985; IAEA: Vienna, Austria, 1968.
16. Modolo, G.; Wilden, A.; Geist, A.; Magnusson, D.; Malmbeck, R. A review of the demonstration of innovative solvent extraction processes for the recovery of trivalent minor actinides from PUREX raffinate. *Radiochim. Acta* **2012**, *100*, 715–725. [CrossRef]
17. Taylor, R.J.; Gregson, C.R.; Carrott, M.J.; Mason, C.; Sarsfield, M.J. Progress towards the full recovery of neptunium in an advanced PUREX process. *Solvent Extr. Ion Exch.* **2013**, *31*, 442–462. [CrossRef]
18. Li, X.; Liu, B.; Yuan, J.; Zhong, W.; Mu, W.; He, J.; Ma, Z.; Liu, G.; Luo, S. Ion-exchange characteristics of a layered metal sulfide for removal of Sr2þ from aqueous solutions. *Separ. Sci. Technol.* **2012**, *47*, 896–902. [CrossRef]
19. Gin, S.; Jollivet, P.; Tribet, M.; Peuget, S.; Schuller, S. Radionuclides containment in nuclear glasses: An overview. *Radiochim. Acta* **2017**, *105*, 927–959. [CrossRef]
20. Zhang, Y.; Zhang, Z.; Thorogood, G.; Vance, E. Pyrochlore based glass-ceramics for the immobilization of actinide-rich nuclear wastes: From concept to reality. *J. Nucl. Mater.* **2013**, *432*, 545–554. [CrossRef]
21. Wu, P.; Dai, Y.; Long, H.; Zhu, N.; Li, P.; Wu, J.; Dang, Z. Characterization of organomontmorillonites and comparison for Sr(II) removal: Equilibrium and kinetic studies. *Chem. Eng. J.* **2012**, *191*, 288–296. [CrossRef]
22. Guangren, Q.; Yuxiang, L.; Facheng, Y.; Rongming, S. Improvement of metakaolin on radioactive Sr and Cs immobilization of alkali-activated slag matrix. *J. Hazard. Mater.* **2002**, *92*, 289. [CrossRef]
23. USEPA. *Technology Resource Document—Solidification/Stabilization and Its Application to Waste Materials*; EPA/530/R-93/012; USEPA: Washington, DC, USA, 1993.
24. USEPA. *Innovative Treatment Technologies: Annual Status Report*, 8th ed.; EPA/542/R-96/010; USEPA: Washington, DC, USA, 1996.
25. Conner, J.R. *Chemical Fixation and Solidification of Hazardous Wastes*; Van Nostrand Reinhold: New York, NY, USA, 1990.
26. Dinchak, W.G. Solidification of hazardous wastes using portland cement. *Concr. Constr.* **1985**, *33*, 781–783.
27. Lange, L.C.; Hills, C.D.; Poole, A.B. Preliminary investigation into the effects of carbonation on cement-solidified hazardous wastes. *Environ. Sci. Technol.* **1996**, *30*, 25–30. [CrossRef]
28. Wiles, C.C. A review of solidification/stabilization technology. *J. Hazard. Mater.* **1987**, *14*, 5–21. [CrossRef]
29. Glasser, F.P. Fundamental aspects of cement solidification and stabilization. *J. Hazard. Mater.* **1997**, *52*, 151–170. [CrossRef]
30. Cocke, D.L. The binding chemistry and leaching mechanism of hazardous substances in cementitious solidification/stabilization systems. *J. Hazard. Mater.* **1990**, *24*, 231–253. [CrossRef]
31. Yousuf, M.; Mollah, A.; Vempati, R.K.; Lin, T.-C.; Cocke, D.L. The interfacial chemistry of solidification/stabilization of metals in cement and pozzolanic material systems. *Waste Manag.* **1995**, *15*, 137–148. [CrossRef]
32. Cartledge, F.K.; Butler, L.G.; Chalasani, D.; Eaton, H.C.; Frey, F.P.; Herrera, E.; Tittlebaum, M.E.; Yang, S. Immobilization mechanisms in solidification/stabilization of Cd and Pb salts using portland cement fixing agents. *Environ. Sci. Technol.* **1990**, *24*, 867–873. [CrossRef]
33. Malviya, R.; Chaudhary, R. Factors affecting hazardous waste solidification/stabilization: A review. *J. Hazard. Mater. B* **2006**, *137*, 267–276. [CrossRef] [PubMed]
34. EPA; Office of Solid Waste and Emergency Response. *Solidification/Stabilization Use at Superfund Sites*; n. 68-W-99-003; USEPA: Washington, DC, USA, 2000.
35. Hills, C.D. Introduction to the science behind stabilisation/solidification technology. In Proceedings of the Stabilisation/Solidification Symposium, Halifax, NS, Canada, 29 August 2007.
36. European Commission, INT. *Pollution Prevention and Control, Reference document on Best Available Techniques for the Waste Treatments Industries Spain*; European Commission: Brussels, Belgium, 2006.
37. Ewing, R.C. Nuclear waste forms for actinides. *Proc. Natl. Acad. Sci. USA* **1999**, *96*, 3432–3439. [CrossRef]
38. Banks, M. IAEA sees bright future for nuclear power. *Phys. World* **2007**, *20*, 10. [CrossRef]
39. Sheppard, L.M. Is nuclear power the answer to U.S. energy needs? *Am. Ceram. Soc. Bull.* **2008**, *87*, 20–24.
40. Hart, M.M. Radiation in the environment. *Radiat. Prot. Manag.* **2005**, *22*, 13–19.
41. Trussell, S.; Spence, R.D. A review of solidification/stabilization interferences. *Waste Manag.* **1994**, *14*, 507–519. [CrossRef]
42. Mattus, C.H.; Mattus, A.J. Literature review of the interaction of select inorganic species on the set and properties of cement and methods of abatement through waste pretreatment. In *Stabilization and Solidification of Hazardous, Radioactive, and Mixed Wastes*; ASTM: West Conshohocken, PA, USA, 1996; Volume 3.
43. Taylor, H.F. *Cement Chemistry*; Thomas Telford: London, UK, 1997.
44. Jones, L. Interference mechanisms in waste stabilization/solidification processes. *J. Hazard. Mater.* **1990**, *24*, 83–88. [CrossRef]
45. Massazza, F.; Testolin, M. Latest developments in the use of admixtures for cement and concrete. *Il Cem.* **1980**, *2*, 73–146.
46. Mattus, C.H.; Gilliam, T.M. *A Literature Review of Mixed Waste Components: Sensitivities and Effects Upon Solidification/Stabilization in Cement-Based Matrices*; U.S. Department of Energy: Washington, DC, USA, 1994.
47. Hills, C.D.; Pollard, S.J. The influence of interference effects on the mechanical, microstructural and fixation characteristics of cement-solidified hazardous waste forms. *J. Hazard. Mater.* **1997**, *52*, 171–191. [CrossRef]

48. Polettini, A.; Pomi, R.; Sirini, P. Fractional factorial design to investigate the influence of heavy metals and anions on acid neutralization behavior of cement based products. *Environ. Sci. Technol.* **2002**, *36*, 1584–1591. [CrossRef] [PubMed]
49. Stegemann, J.A.; Buenfeld, N.R. Prediction of leachate pH for cement paste containing pure metal compounds. *J. Hazard. Mater. B* **2002**, *90*, 169–188. [CrossRef] [PubMed]
50. Stegemann, J.A.; Perera, A.S.; Cheeseman, C.; Buenfeld, N.R. 1/8 fractional factorial design investigation of metal effects on cement acid neutralisation capacity. *J. Environ. Eng.* **2000**, *126*, 10. [CrossRef]
51. Dumitru, G.; Vazquez, T.; Puertas, F.; Blanco-Varela, M.T. Influence of BaCO3 on hydration of Portland cement. *Mater. Constr.* **2000**, *49*, 43–48. [CrossRef]
52. Fernandez Olmo, I.; Chacon, E.; Irabien, A. Influence of lead, zinc, iron (III) and chromium (III) oxides on the setting time and strength development of Portland cement. *Cem. Concr. Res.* **2001**, *31*, 1213–1219. [CrossRef]
53. Malolepszy, J.; Deja, J. Effect of Heavy Metals Immobilization on Properties of Alkali-Activated Slag Mortars. In Proceedings of the 5th International Conference on the Use of Fly Ash, Silica Fume, Slag & Natural Pozzolans in Concrete, Milwaukee, WI, USA, 4–9 June 1995; pp. 1087–1102.
54. Malone, P.G.; Poole, T.S.; Wakeley, L.D.; Burkes, J.P. Salt-related expansion reactions in Portland-cement-based wasteforms. *J. Hazard. Mater.* **1997**, *52*, 237–246. [CrossRef]
55. Stegemann, J.A.; Buenfeld, N.R. Prediction of unconfined compressive strength of cement paste with pure metal compound additions. *Cem. Concr. Res.* **2002**, *32*, 903–913. [CrossRef]
56. Damidot, D.; Rettel, A.; Capmas, A. Action of admixtures on fondu cement: Part 1. Lithium and sodium salts compared. *Adv. Cem. Res.* **1996**, *8*, 111–119. [CrossRef]
57. Matusinovic, T.; Vrbos, N. Alkali metal salts as set accelerators for high alumina cement. *Cem. Concr. Res.* **1993**, *23*, 177–186. [CrossRef]
58. Nilforoushan, M.R.; Sharp, J.H. The effect of additions of alkaline earth metal chlorides on the setting behavior of a refractory calcium aluminate cement. *Cem. Concr. Res.* **1995**, *25*, 1523–1534. [CrossRef]
59. Banfill, P.F. The effect of sulphate on the hydration of high alumina cement. *Cem. Concr. Res.* **1986**, *16*, 602–604. [CrossRef]
60. Currell, B.R.; Grezeskowlak, R.; Midgley, H.G.; Parsonage, J.R. The acceleration and retardation of set high alumina cement by additives. *Cem. Concr. Res.* **1987**, *7*, 420–432. [CrossRef]
61. Matusinovic, T.; Curlin, D. Lithium salts as set accelerators for high alumina cement. *Cem. Concr. Res.* **1993**, *23*, 885–895. [CrossRef]
62. Rodger, S.A.; Double, D.D. The chemistry of hydration of high alumina cement in the presence of accelerating and retarding admixtures. *Cem. Concr. Res.* **1984**, *14*, 73–82. [CrossRef]
63. Parker, T.W. The constitution of aluminous cement. In Proceedings of the 3rd International Symposium on the Chemistry of Cement, London, UK, 15–20 September 1952; Cement and Concrete Association: London, UK, 1954.
64. Sharp, J.H.; Bushnell-Watson, S.M.; Payne, D.R.; Ward, P.A. The effect of admixtures on the hydration of refractory calcium aluminate cements. In *Calcium Aluminate Cements*; Mangabhai, R.J., Ed.; Chapman & Hall: London, UK, 1990; pp. 127–141.
65. Bayoux, J.P.; Bonin, A.; Marcdargent, S.; Verschaeve, M. Study of the hydration properties of aluminous cement and calcium sulphate mixes. In *Calcium Aluminate Cements*; Mangabhai, R.J., Ed.; Chapman & Hall: London, UK, 1990; pp. 320–334.
66. Nestle, N.; Zimmermann, C.; Dakkouri, M.; Niessner, R. Action and distribution of organic solvent contaminations in hydrating cement: Time-resolved insights into solidification of organic waste. *Environ. Sci. Technol.* **2001**, *35*, 4953–4956. [CrossRef] [PubMed]
67. Bier, T.A.; Mathieu, A.; Espinosa, B.; Marcelon, C. Admixtures and their interactions with high range calcium aluminate cement. In Proceedings of the UNITECR Congress, Kyoto, Japan, 19–22 November 1995.
68. Baker, N.C.; Banfill, P.F. Properties of fresh mortars made with high alumina cement and admixtures for the marine environment. In *Calcium Aluminate Cements*; Mangabhai, R.J., Ed.; Chapman & Hall: London, UK, 1990; pp. 142–151.
69. Gervais, C.; Ouki, S.K. Performance study of cementitious systems containing zeolite and silica fume: Effects of four metal nitrates on the setting time, strength and leaching characteristics. *J. Hazard. Mater.* **2002**, *93*, 187–200. [CrossRef]
70. Majumdar, A.J.; Singh, B. Properties of some blended high-alumina cements. *Cem. Concr. Res.* **1992**, *22*, 1101–1114. [CrossRef]
71. Majumdar, A.J.; Singh, B.; Edmonds, R.N. Hydration of mixtures of cement fondu aluminous cements and granulated blast furnace slag. *Cem. Concr. Res.* **1990**, *20*, 197–208. [CrossRef]
72. Rahman, R.A.; El Abidin, D.Z.; Abou-Shady, H. Assessment of strontium immobilization in cement–bentonite matrices. *Chem. Eng. J.* **2013**, *228*, 772–780. [CrossRef]
73. Shrivastava, O.P.; Shrivastava, R. Cation exchange applications of synthetic tobermorite for the immobilization and solidification of cesium and strontium in cement matrix. *Bull. Mater. Sci.* **2000**, *23*, 515–520. [CrossRef]
74. El-Kamash, A.; El-Naggar, M.; El-Dessouky, M. Immobilization of cesium and strontium radionuclides in zeolite-cement blends. *J. Hazard. Mater.* **2006**, *136*, 310–316. [CrossRef] [PubMed]
75. Nicoleau, L.; Schreiner, E.; Nonat, A. Ion-specific effects influencing the dissolution of tricalcium silicate. *Cem. Concr. Res.* **2014**, *59*, 118–138. [CrossRef]
76. Mota, B.; Matschei, T.; Scrivener, K. The influence of sodium salts and gypsum on alite hydration. *Cem. Concr. Res.* **2015**, *75*, 53–65. [CrossRef]

77. Wagh, A.; Strain, R.; Jeong, S.; Reed, D.; Krause, T.; Singh, D. Stabilization of Rocky Flats Pu-contaminated ash within chemically bonded phosphate ceramics. *J. Nucl. Mater.* **1999**, *265*, 295–307. [CrossRef]
78. Xu, X.; Bi, H.; Yu, Y.; Fu, X.; Wang, S.; Liu, Y.; Hou, P.; Cheng, X. Low leaching characteristics and encapsulation mechanism of Cs^+ and Sr^{2+} from SAC matrix with radioactive IER. *J. Nucl. Mater.* **2021**, *544*, 152701. [CrossRef]
79. Singh, D.; Wagh, A.S.; Cunnane, J.C.; Mayberry, J.L. Chemically bonded phosphate ceramics for low-level mixed-waste stabilization. *J. Environ. Sci. Health. Part A Environ. Sci. Eng. Toxicol.* **1997**, *32*, 527–541. [CrossRef]
80. Rao, A.J.; Pagilla, K.; Wagh, A.S. Stabilization and solidification of metal-laden wastes by compaction and magnesium phosphate-based binder. *J. Air Waste Manag. Assoc.* **2000**, *50*, 1623–1631. [CrossRef]
81. Jeong, S.Y.; Wagh, A.; Singh, D. Stabilization of lead-rich low-level mixed waste in chemically bonded phosphate ceramic. *Ceram. Trans.* **1999**, *107*, 189–197.
82. Yang, Q.; Zhu, B.; Wu, X. Characteristics and durability test of magnesium phosphate cement-based material for rapid repair of concrete. *Mater. Struct.* **2000**, *33*, 229–234. [CrossRef]
83. Soudee, E.; Pera, J. Influence of magnesia surface on the setting time of magnesia-phosphate cement. *Cem. Concr. Res.* **2002**, *32*, 153–157. [CrossRef]
84. Abdelrazig, B.; Sharp, J.; El-Jazairi, B. The microstructure and mechanical properties of mortars made from magnesia-phosphate cement. *Cem. Concr. Res.* **1989**, *19*, 247–258. [CrossRef]
85. Buj, I.; Torras, J.; Rovira, M.; de Pablo, J. Leaching behaviour of magnesium phosphate cements containing high quantities of heavy metals. *J. Hazard. Mater.* **2010**, *175*, 789–794. [CrossRef]
86. Singh, D.; Mandalika, V.; Parulekar, S.; Wagh, A. Magnesium potassium phosphate ceramic for 99Tc immobilization. *J. Nucl. Mater.* **2006**, *348*, 272–282. [CrossRef]
87. Qiao, F.; Chau, C.; Li, Z. Property evaluation of magnesium phosphate cement mortar as patch repair material. *Construct. Build. Mater.* **2010**, *24*, 695–700. [CrossRef]
88. Vinokurov, S.E.; Kulikova, S.A.; Krupskaya, V.V.; Myasoedov, B.F. Magnesium Potassium Phosphate Compound for Radioactive Waste Immobilization: Phase Composition, Structure, and Physicochemical and Hydrolytic Durability. *Radiochemistry* **2018**, *60*, 70–78. [CrossRef]
89. Chartier, D.; Sanchez-Canet, J.; Antonucci, P.; Esnouf, S.; Renault, J.P.; Farcy, O.; Lambertin, D.; Parraud, S.; Lamotte, H.; Coumes, C.C.D. Behaviour of magnesium phosphate cement-based materials under gamma and alpha irradiation. *J. Nucl. Mater.* **2020**, *541*, 152411. [CrossRef]
90. Glasser, F. Progress in the immobilization of radioactive wastes in cement. *Cem. Concr. Res.* **1992**, *22*, 201–206. [CrossRef]
91. Colorado, H.; Singh, D. High-sodium waste streams stabilized with inorganic acidbase phosphate ceramics fabricated at room temperature. *Ceram. Int.* **2014**, *40*, 10621–10631. [CrossRef]
92. Vinokurov, S.; Kulyako, Y.; Slyuntchev, O.; Rovny, S.; Myasoedov, B. Low temperature immobilization of actinides and other components of high-level waste in magnesium potassium phosphate matrices. *J. Nucl. Mater.* **2009**, *385*, 189–192. [CrossRef]
93. Kinoshita, H.; Swift, P.; Utton, C.; Carro-Mate, R.; Marchand, G.; Collier, N.; Milestone, N. Corrosion of aluminium metal in OPC- and CAC-based cement matrices. *Cem. Concr. Res.* **2013**, *50*, 11–18. [CrossRef]
94. Sugama, T.; Carciello, N. Strength development in phosphate-bonded calcium aluminate cements. *J. Am. Ceram. Soc.* **1991**, *74*, 1023–1030. [CrossRef]
95. Irisawa, K.; Garcia-Lodeiro, I.; Kinoshita, H. Influence of mixing solution on characteristics of calcium aluminate cement modified with sodium polyphosphate. *Cem. Concr. Res.* **2020**, *128*, 10591. [CrossRef]
96. Sugama, T.; Carciello, N. Sodium phosphate-derived calcium phosphate cements. *Cem. Concr. Res.* **1995**, *25*, 91–101. [CrossRef]
97. Chavda, M.; Kinoshita, H.; Provis, J. Phosphate modification of calcium aluminate cement to enhance stability for immobilisation of metallic wastes. *Adv. Appl. Ceram.* **2014**, *113*, 453–459. [CrossRef]
98. Ma, W.; Brown, P. Hydration of sodium phosphate-modified high alumina cement. *J. Mater. Res.* **1994**, *9*, 1291–1297. [CrossRef]
99. Ma, W.; Brown, P. Mechanical behaviour and microstructural development in phosphate modified high alumina cement. *Cem. Conc. Res.* **1992**, *22*, 1192–1200. [CrossRef]
100. Chavda, M.; Bernal, S.; Apperley, D.; Kinoshita, H.; Provis, J. Identification of the hydrate gel phases present in phosphate-modified calcium aluminate binders. *Cem. Concr. Res.* **2015**, *70*, 21–28. [CrossRef]
101. Swift, P.; Kinoshita, H.; Collier, N.; Utton, C. Phosphate-modified calcium aluminate cement for radioactive waste encapsulation. *Adv. Appl. Ceram.* **2013**, *112*, 1–8. [CrossRef]
102. Kamaluddin, S.; Garcia-Lodeiro, I.; Kinoshita, H. Strontium in phosphate-modified calcium aluminate cement. *Key Eng. Mater.* **2019**, *803*, 341–345. [CrossRef]
103. Navarro-Blasco, I.; Duran, A.; Perez-Nicolas, M.; Fernandez, J.M.; Sirera, R.; Alvarez, J.I. A safe disposal phosphate coating sludge by formation of an amorphous calcium phosphate matrix. *J. Environ. Manag.* **2015**, *159*, 288–300. [CrossRef] [PubMed]
104. Fernandez, J.; Navarro-Blasco, I.; Duran, A.; Sirera, R.; Alvarez, J. Treatment of toxic metal aqueous solutions: Encapsulation in a phosphate-calcium aluminate matrix. *J. Environ. Manag.* **2014**, *140*, 1–13. [CrossRef] [PubMed]
105. Garcia-Lodeiro, I.; Irisawa, K.; Jin, F.; Meguro, Y.; Kinoshita, H. Reduction of water content in calcium aluminate cement with/out phosphate modification for alternative cementation technique. *Cem. Concr. Res.* **2018**, *109*, 243–253. [CrossRef]
106. Laili, Z.; Yasir, M.S.; Wahab, M.A. Solidification of radioactive waste resins using cement mixed with organic material. *AIP Conf. Proc.* **2015**, *1659*, 050006.

107. NishI, T. Advanced solidification system using high performance cement. In Proceedings of the Fifth International Conference on Radioactive Waste Management and Environmental Remediation, Berlin, Germany, 3–9 September 1995; pp. 1095–1098.
108. Natsuda, M.; Nishi, T. Solidification of ion exchange resins using new cementitious material (1) swelling pressure of ion exchange resin. *J. Nucl. Sci. Technol.* **1992**, *29*, 883–889. [CrossRef]
109. Pan, L.K.; Chang, B.D.; Chou, D.S. Optimization for solidification of low-level-radioactive resin using Taguchi analysis. *Waste Manag.* **2001**, *21*, 767–772. [CrossRef]
110. Bagosi, S.; Csetenyi, L.J. Immobilization of caesium-loaded ion exchange resins in zeolite-cement blends. *Cem. Concr. Res.* **1999**, *29*, 479–485. [CrossRef]
111. Huang, W.H. Properties of cement-fly ash grout admixed with bentonite silica fume or organic fiber. *Cem. Concr. Res.* **1997**, *27*, 395–406. [CrossRef]
112. Natsuda, M.; Nishi, T. Solidification of ion exchange resins using new cementitious material. (2) Improvement of resin content fiber reinforced cement. *J. Nucl. Sci. Technol.* **1992**, *29*, 1093–1099. [CrossRef]
113. Rudin, M.J. Leaching of selenium from cement-based matrices. *Waste Manag.* **1996**, *16*, 305–311. [CrossRef]
114. Papadokostaki, K.G.; Savidou, A. Study of leaching mechanisms of caesium ions incorporated in Ordinary Portland Cement. *J. Hazard. Mater.* **2009**, *171*, 1024–1031. [CrossRef] [PubMed]
115. Zhou, Y.Z. Study on Cementation Technology of Radioactive Spent Resins and Mechanism Investigation. Ph.D. Thesis, Tsinghua University, Beijing, China, November 2002.
116. Li, J. *Cementation of Radioactive Waste from a PWR with Calcium Sulfoaluminate Cement*; International Atomic Energy Agency: Vienna, Austria, 2013.
117. Pescatore, C. Leach rate expressions for performance assessment of solidified low-level radioactive waste. *Waste Manag.* **1991**, *11*, 223–229. [CrossRef]
118. Osmanlioglu, A.E. Immobilization of radioactive waste by cementation with purified kaolin clay. *Waste Manag.* **2002**, *22*, 481–483. [CrossRef]
119. El-Kamash, A.M.; El-Dakroury, A.; Aly, H. Leaching kinetics of 137Cs and 60Co radionuclides fixed in cement and cement-based materials. *Cem. Concr. Res.* **2002**, *32*, 1797–1803. [CrossRef]
120. Krishnamoorthy, T.M.; Joshi, S.N.; Doshi, G.R.; Nair, R.N. Desorption kinetics of radionuclides fixed in cement-matrix. *Nucl. Technol.* **1993**, *104*, 351–357. [CrossRef]
121. Azeez, A.B.; Mohammed, K.S.; Abdullah, M.M.A.B.; Zulkepli, N.N.; Sandu, A.V.; Hussin, K.; Rahmat, A. Design of Flexible Green Anti Radiation Shielding Material against Gamma-ray. *Mater. Plasstice* **2014**, *51*, 300–308.
122. Luhar, I.; Luhar, S.; Abdullah, M.M.A.B.; Razak, R.A.; Vizureanu, P.; Sandu, A.V.; Matasaru, P.-D. A State-of-the-Art Review on Innovative Geopolymer Composites Designed for Water and Wastewater Treatment. *Materials* **2021**, *14*, 7456. [CrossRef]
123. Azeez, A.B.; Mohammed, K.S.; Abdullah, M.M.A.B.; Hussin, K.; Sandu, A.V.; Razak, R.A. The Effect of Various Waste Materials' Contents on the Attenuation Level of Anti-Radiation Shielding Concrete. *Materials* **2013**, *6*, 4836–4846. [CrossRef]
124. Burduhos Nergis, D.D.; Vizureanu, P.; Corbu, O. Synthesis and Characteristics of Local Fly Ash Based Geopolymers Mixed with Natural Aggregates. *Rev. Chim.* **2019**, *70*, 1262–1267. [CrossRef]
125. Azeez, A.B.; Mohammed, K.S.; Sandu, A.V.; Al Bakri, A.M.M.; Kamarudin, H.; Sandu, I.G. Evaluation of Radiation Shielding Properties for Concrete with Different Aggregate Granule Sizes. *Rev. Chim.* **2013**, *64*, 899–903.
126. Shahedan, N.; Abdullah, M.; Mahmed, N.; Kusbiantoro, A.; Tammas-Williams, S.; Li, L.-Y.; Aziz, I.; Vizureanu, P.; Wysłocki, J.; Błoch, K.; et al. Properties of a New Insulation Material Glass Bubble in Geopolymer Concrete. *Materials* **2021**, *14*, 809. [CrossRef] [PubMed]
127. National Academies of Sciences, Engineering, and Medicine. Low-Level Radioactive Waste Management and Disposition: Background Information. In *Low-Level Radioactive Waste Management and Disposition*; National Academies Press: Washington, DC, USA, 2017.

Disclaimer/Publisher's Note: The statements, opinions and data contained in all publications are solely those of the individual author(s) and contributor(s) and not of MDPI and/or the editor(s). MDPI and/or the editor(s) disclaim responsibility for any injury to people or property resulting from any ideas, methods, instructions or products referred to in the content.

Article

Computer Simulations of End-Tapering Anchorages of EBR FRP-Strengthened Prestressed Concrete Slabs at Service Conditions

Chirawat Wattanapanich [1], Thanongsak Imjai [1,*], Reyes Garcia [2], Nur Liza Rahim [3], Mohd Mustafa Al Bakri Abdullah [3], Andrei Victor Sandu [4,5,6,*], Petrica Vizureanu [4,7], Petre Daniel Matasaru [8] and Blessen Skariah Thomas [9]

Citation: Wattanapanich, C.; Imjai, T.; Garcia, R.; Rahim, N.L.; Abdullah, M.M.A.B.; Sandu, A.V.; Vizureanu, P.; Matasaru, P.D.; Thomas, B.S. Computer Simulations of End-Tapering Anchorages of EBR FRP-Strengthened Prestressed Concrete Slabs at Service Conditions. *Materials* 2023, *16*, 851. https://doi.org/10.3390/ma16020851

Academic Editors: Angelo Marcello Tarantino and Baoguo Han

Received: 18 December 2022
Revised: 8 January 2023
Accepted: 13 January 2023
Published: 15 January 2023

Copyright: © 2023 by the authors. Licensee MDPI, Basel, Switzerland. This article is an open access article distributed under the terms and conditions of the Creative Commons Attribution (CC BY) license (https://creativecommons.org/licenses/by/4.0/).

[1] School of Engineering and Technology, Walailak University, Nakhon Si Thammarat 80161, Thailand
[2] Civil Engineering Stream, School of Engineering, The University of Warwick, Coventry CV4 7AL, UK
[3] Faculty of Chemical Engineering Technology, Universiti Malaysia Perlis, Arau 02600, Malaysia
[4] Faculty of Material Science and Engineering, Gheorghe Asachi Technical University of Iasi, 41 D. Mangeron St., 700050 Iasi, Romania
[5] Romanian Inventors Forum, Str. Sf. P. Movila 3, 700089 Iasi, Romania
[6] National Institute for Research and Development for Environmental Protection INCDPM, 294 Splaiul Independentei, 060031 Bucharest, Romania
[7] Technical Sciences Academy of Romania, Dacia Blvd 26, 030167 Bucharest, Romania
[8] Faculty of Electronics, Telecommunications and Information Technology, Gheorghe Asachi Technical University of Iasi, Carol I 11A, 700506 Iasi, Romania
[9] National Institute of Technology Calicut, Calicut 673601, India
* Correspondence: thanongsak.im@wu.ac.th (T.I.); sav@tuiasi.ro (A.V.S.)

Abstract: This article examines numerically the behavior of prestressed reinforced concrete slabs strengthened with externally bonded reinforcement (EBR) consisting of fiber-reinforced polymer (FRP) sheets. The non-linear finite element (FE) program Abaqus® is used to model EBR FRP-strengthened prestressed concrete slabs tested previously in four-point bending. After the calibration of the computational models, a parametric study is then conducted to assess the influence of the FRP axial stiffness (thickness and modulus of elasticity) on the interfacial normal and shear stresses. The numerical analysis results show that increasing the thickness or the elastic modulus of the FRP strengthening affects the efficiency of the FRP bonding and makes it susceptible to earlier debonding failures. A tapering technique is proposed in wet lay-up applications since multiple FRP layers are often required. It is shown that by gradually decreasing the thickness of the FRP strengthening, the concentration of stress along the plate end can be reduced, and thus, the overall strengthening performance is maximized. The tapering is successful in reducing the bond stress concentrations by up to 15%, which can be sufficient to prevent concrete rip-off and peel-off debonding failure modes. This article contributes towards a better understanding of the debonding phenomena in FRP-strengthened elements in flexure and towards the development of more efficient computational tools to analyze such structures.

Keywords: FEA; end effect; EBR; plate bonding; tapering technique; prestressed concrete slab

1. Introduction

The need to increase service loads, changes or updates in design codes, design errors or aging make the strengthening of reinforced concrete elements necessary. Due to several advantages over more traditional techniques, externally bonded reinforcement (EBR) in the form of fiber-reinforced polymers (FRP) has gained a special interest in strengthening applications of reinforced concrete (RC) structures [1–4]. The most common technique to strengthen reinforced concrete elements in flexure consists of the simple bonding of FRP plates or fabrics on the elements' soffit. The two basic techniques for FRP strengthening normally used in practical applications are (1) prefabricated systems by means of cold cured adhesive bonding, or (2) wet lay-up systems. The application depends on strengthening needs and the type of structure. In the case of FRP strips and laminates (prefab type), the adhesive ensures bonding, and therefore, a high viscosity thixotropic adhesive is applied. In the case of FRP sheets or fabrics (wet lay-up application), a resin ensures both bonding and impregnation of the sheets, and therefore, a low viscosity resin is often required. Since the tensile strength of FRP materials is several times higher than that of more traditional strengthening solutions (e.g., steel plates), the tensile strength of EBR FRP systems is much higher than its bonding strength to the concrete face. As a result, FRP debonding between concrete and EBR can become dominant in flexural elements strengthened with EBR FRP systems.

Previous research indicates that EBR FRP strengthening can increase the flexural capacity of elements in both service and ultimate conditions [5–8]. However, reinforced concrete elements strengthened with EBR FRP can also suffer from premature and brittle debonding of the FRP plate or sheets [6–10]. Typical EBR FRP debonding failure modes of RC slabs can include plate end interfacial debonding (Figure 1a), concrete cover separation (Figure 1b), intermediate crack-induced interfacial debonding (Figure 1c) or shear failure due to diagonal cracking (Figure 1d). An FRP debonding failure can prevent the strengthened element from reaching its theoretical ultimate capacity, while also reducing its ductility. Therefore, bond failures such as those shown in Figure 1a–d must be prevented. This is necessary because the design procedure for FRP-strengthened flexural elements assumes that the design moment after strengthening results from a full composite action between the EBR FRP system and the concrete element. Indeed, in a real design, designers only have to verify that bond failures (due to shear cracks, end anchorage or along the FRP) do not occur.

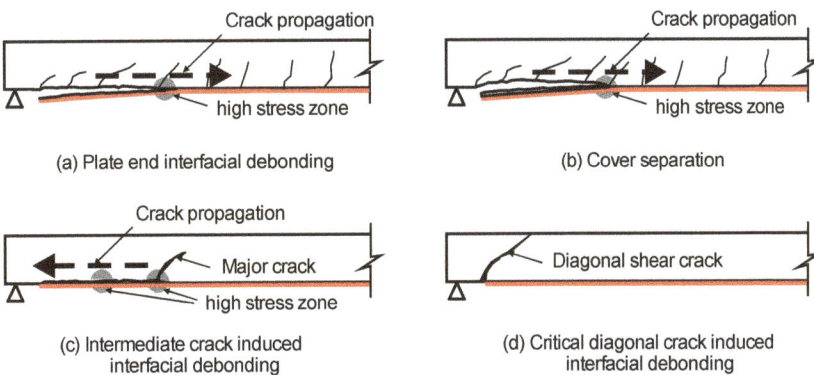

Figure 1. Typical EBR FRP debonding failure modes of RC slabs (adapted from [7]). (**a**) Plate end interfacial debonding; (**b**) Cover separation; (**c**) Intermediate crack induced interfacial debonding; (**d**) Critical diagonal crack induced interfacial debonding.

In order to prevent debonding failures at the end anchorage of a plate or sheet of an EBR FRP-strengthened slab (Figure 2a), several solutions are used in practical applications [9]. For instance, steel plates and bolts (Figure 2b) or bolted steel angle systems (Figure 2c) were proposed. However, this implies that drilling (or fiber unknitting) is necessary to accommodate the bolts and anchor them into the slab or walls. Corrosion issues may also affect the plates or bolts, thus compromising the effectiveness of the FRP strengthening system. U-shaped anchorage systems were also used (Figure 2d). In this solution, the FRP sheets are extended beyond the edge of the slab into the support walls. The FRP sheets are then folded and anchored with a transverse bar or rod glued with epoxy resin. Another typical solution is the bonding of transverse FRP sheets to the slab soffit, placed normal to the (main) longitudinal direction of the flexural EBR FRP plate or sheets.

The latter solution is by far the simplest to prevent end anchorage debonding failures, but it is also prone to experience premature debonding.

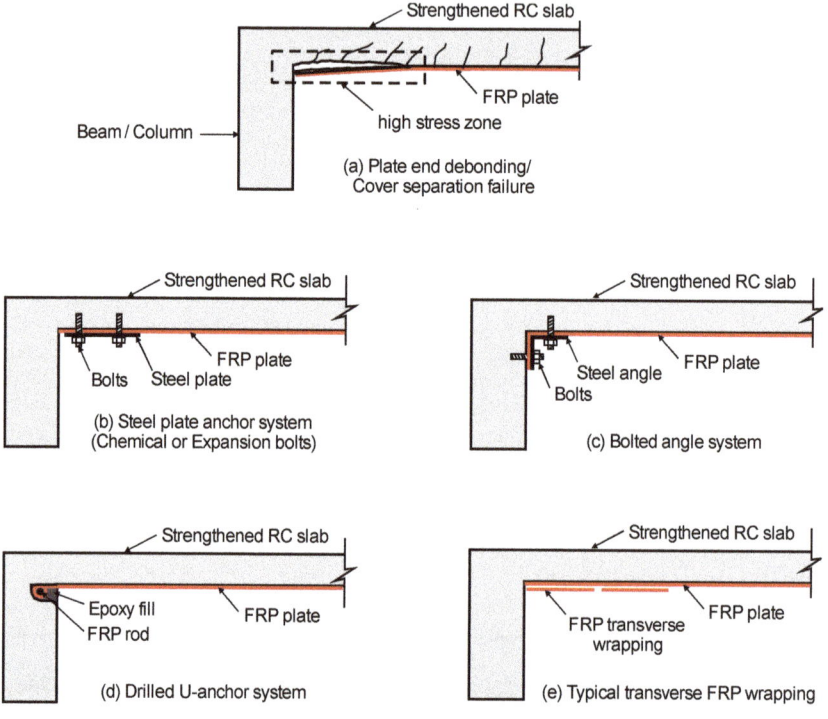

Figure 2. Schematic anchorage systems of EBR of RC slabs using FRP plate. (**a**) Plate end debonding/Cover separation failure; (**b**) Steel plate anchor system (Chemical or Expansion bolts); (**c**) Bolted angle system; (**d**) Drilled U-anchor system; (**e**) Typical transverse FRP wrapping.

Several models/guidelines exist to predict the bond behavior of FRP-concrete systems, but they are still unable to give accurate predictions [6,7]. This inaccuracy can be attributed to a lack of understanding of the development of bond stress along the FRP–concrete interface, especially at the end plate section where high interfacial stress mobilizes and weakens the EBR FRP system [6]. Recent research [10] has found that, while numerous studies have examined the debonding of EBR FRP in flexural elements [11–18], much less research has focused on modeling numerically the behavior of such elements [19–25]. Moreover, parametric studies based on finite element analysis are particularly scarce and very much needed [10], as performing computer simulations is much more cost-effective than performing actual experiments.

Past studies have adopted different approaches to simulate debonding phenomena. For example, smeared and discrete approaches can be used to simulate cracking and debonding in plain and FRP-strengthened concrete structures [26,27]. Other numerical techniques to model the interaction between concrete and FRP systems assume a full composite action at the FRP–concrete interface so that no relative movement occurs between them [8]. Using this type of contact modeling, a full composite action develops and debonding is prevented, which has proven appropriate to model EBR FRP in reinforced concrete elements up to the service load level [5–8]. Despite the above advancements in computational approaches, the investigation of the modeling capabilities built in current finite element (FE) software was recently identified as a research need in EBR FRP-strengthened structures [11].

This article examines numerically the end-plate behavior of prestressed concrete slabs strengthened in flexure with EBR FRP sheets. This is achieved by using previous data from four-point bending tests on EBR FRP-strengthened prestressed concrete slabs. Non-linear FE analyses was first performed to calibrate computational models of the strengthened slabs. Subsequently, a parametric study was conducted to assess the influence of the FRP axial stiffness (i.e., thickness and modulus of elasticity) on the normal and shear interfacial stresses. The article also provides some recommendations to enhance the capacity of FRP bonded to prestressed concrete slabs, as well as to improve the effectiveness of FRP plate bonding techniques. This article contributes towards a better understanding of the debonding phenomena in FRP-strengthened elements in flexure and towards the development of more efficient computational tools to analyze such structures.

2. Experimental Investigation

Five prestressed concrete slabs tested previously by the authors [8] were taken as case studies for calibrating FE models. Prestressed slabs are commonly used as floor systems in houses and small buildings in Southeast Asia, and, in many cases, such slabs need to be strengthened due to the change in use of the structure. The slabs had an effective span length of 3300 mm (Figure 3a) and a cross-section of 50 × 350 mm² (Figure 3b). Four f4 mm prestressed tendons (f_{sy} = 1860 MPa, E_s = 205 GPa) were used as flexural bottom reinforcement. Such tendons complied with the ASTM A421 [28] specifications. The slabs were subjected to four-point bending (Figure 3c), with a shear span to effective depth ratio equal to 4.5.

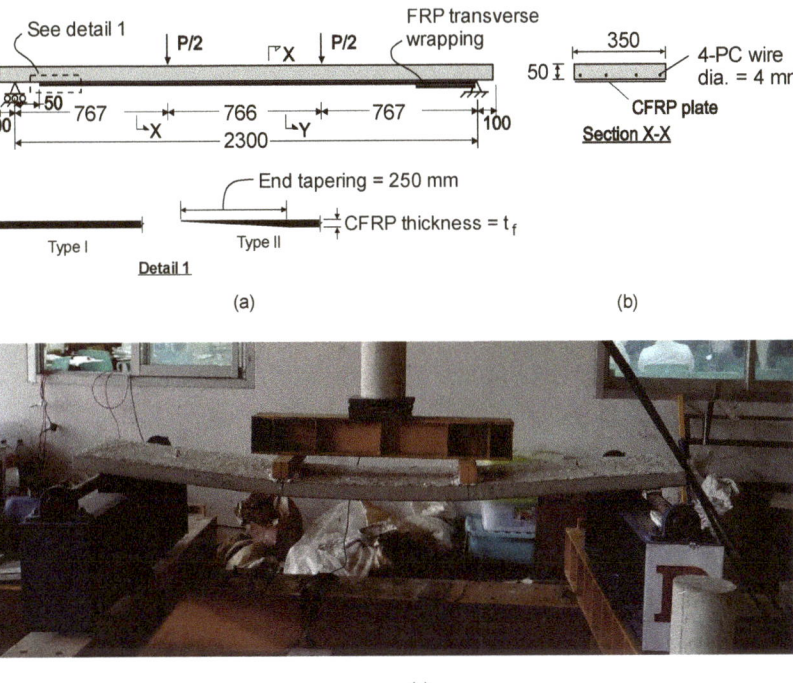

Figure 3. (**a**) Elevation and (**b**) cross section of slabs tested in four-point bending and (**c**) view of slab during test.

One layer of carbon FRP (CFRP) strip with single or double layouts, having a thickness t_f, of 1.4 mm, was used to strengthen the slabs' soffit. The FRP sheets were bonded with the slabs in an upside-down position, as shown in Figure 4, that shows the following specimens:

- PS-EBR-1-250-TP: EBR CFRP-strengthened slab, strip width 250 mm, tapered at one end anchorage (tapering length = 250 mm).
- PS-EBR-2-100-TP: EBR CFRP-strengthened slab, strip width 100 mm each, tapered at one end anchorage (tapering length = 250 mm).
- PS-EBR-1-250 and PS-EBR-2-100: same as above but without tapering.
- PS-C: unstrengthened control slab.

Figure 4. Application of EBR CFRP sheets at the slab's soffit (PS-EBR-2-100).

The properties of the CFRP were modulus of elasticity E_f = 200 GPa, Poisson ratio = 0.29, ultimate strength f_{fu} = 2590 MPa and rupture strain ε_{fu} = 0.015, as given by the manufacturer. The FRP was bonded to the slabs' soffit using a two-part epoxy adhesive bonding agent. The properties of the epoxy resin reported by the producer were modulus of elasticity E_m = 5 GPa, tensile strength f_m = 20 MPa and Poisson ratio = 0.35. The slabs were cast using C30 concrete with a mean compressive strength of f_c = 30.8 MPa, tensile splitting strength f_{cr} = 3.6 MPa and modulus of elasticity E_c = 22 GPa, which were calculated according to *fib* Model Code [29].

Table 1 summarizes the material parameters used in the FE modeling. As shown in Figure 4, the FRP sheets were anchored at one end with transverse CFRP sheets so as to promote FRP debonding on the opposite end anchorage of the slabs.

Table 1. Properties of material used in FE modeling of tested slabs.

Parameters	Tensile Steel	Compressive Steel	CFRP Sheets	Resin
Modulus of elasticity [GPa]	200	206	200	5
Poisson's ratio	0.29	0.29	0.29	0.35
Yield stress [MPa]	470	360	-	-
Ultimate strength [MPa]	560	400	2590	20
Ultimate plastic strain	0.023	0.0082	-	-

The five slabs were subjected to monotonic load until failure. The mid-span deflection behavior of the slabs was monitored using vertical displacement transducers located at the soffit. As expected, the ultimate failure of the slabs was controlled by debonding of the CFRP sheets at the free end anchorage for all of the strengthened specimens. Figure 5 shows the intermediate crack-induced interfacial debonding failure of specimens with tapering-end CFRP sheets (PS-EBR-2-100-TP and PS-EBR-1-250-TP). For the specimen without tapering (PS-EBR-2-100), end-debonding failure of CFRP was observed at the left side of the specimen, as shown in Figure 6.

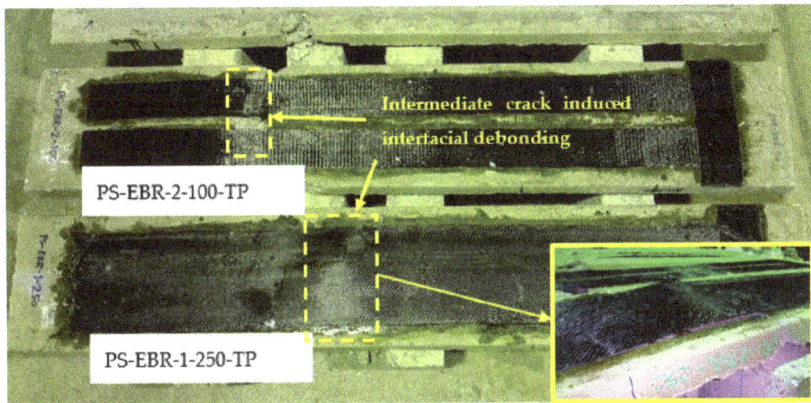

Figure 5. Intermediate crack-induced interfacial debonding failure of specimens with end-tapering.

Figure 6. End-debonding failure of specimen without tapering (PS-EBR-2-100).

3. Numerical Investigations

3.1. Modeling Assumptions

The prestressed concrete slabs were modeled in Abaqus® FE software [30]. The concrete, adhesive layer (two-part epoxy resin) and CFRP sheets were modeled using two-dimensional solid biquadratic elements (CPS8) with eight-nodes (two degrees of freedom per node), as shown in Figure 7a. It should be noted that CPS4 or CPS4R elements consume less computation cost, but they were deemed unsuitable as they can cause shear stress deformation instead of bending deformation [30]. Figure 7b,c show the constitutive models of concrete and steel tendons used in Abaqus®. The internal reinforcement was modeled using 1D elements and is embedded in the concrete matrix, (E_s = 206 GPa and Poisson's ratio = 0.3 (see Figure 7c). The prestressed tendons were ignored in the modeling. Instead, an equivalent prestressing force was applied as surface loads at both sides of the concrete elements, as reported in the previous studies by Okumus et al. [31]. Bond slip and dowel action were not explicitly considered in this study, primarily because the numerical simulations focused on the behavior of the slabs at the serviceability level where no major cracks or local slip were not evident in the experiments [4]. The concrete and CFRP sheets were assembled by a tie type of constraint so that no relative movement between them occurred. Using this type of contact modeling, a full composite action developed. As a result, the modeling of links and transverse CFRP sheets could be omitted. The analysis was performed by incremental loading, with integration in each increment. Since considerable nonlinearity was expected (including the possibility of instability as the concrete cracks or failure occurred at the FRP–concrete bond interface), the load magnitudes were covered by a single scalar parameter. The modified Riks algorithm with automatic increments was used. This method uses the "arc length" along the static equilibrium path in load-displacement space. This method in general worked well and provided a conservative solution for similar problems [30]. The FE method has proven effective at predicting with reasonable accuracy experimental results in previous studies [23,31,32].

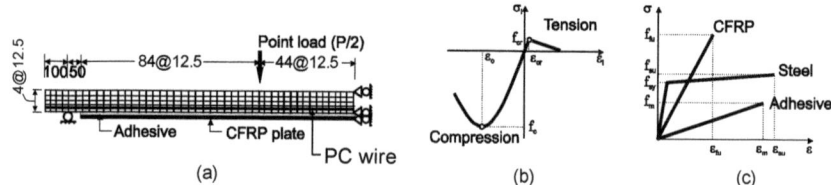

Figure 7. (a) FE mesh and geometry, (b) concrete constitutive model and (c) reinforcement (CFRP/steel/adhesive).

3.2. Concrete Constitutive Model

Concrete was assumed to behave nonlinearly in both compression and tension, as shown in Figure 7b. The constitutive curve for concrete in compression is used as input parameters to the FE code in a form of stress–strain ($\sigma_c - \varepsilon_c$) series of values, as summarized in Table 2. By using the *fib* Model Code 2010 [29] approach, the FE analysis adopted the following stress–strain ($\sigma_c - \varepsilon_c$) constitutive model for concrete in compression:

$$\sigma_c = -f'_c \frac{\left(\frac{E_{ci}}{E_{c1}}\right)\left(\frac{\varepsilon_{ci}}{\varepsilon_{c1}}\right) - \left(\frac{\varepsilon_{ci}}{\varepsilon_{c1}}\right)^2}{1 + \left(\frac{E_{ci}}{E_{c1}} - 2\right)\left(\frac{\varepsilon_{ci}}{\varepsilon_{c1}}\right)} \quad (1)$$

where $E_1 = \left|\frac{f'_c}{\varepsilon_1}\right|$, $E_{ci} = E_{cm}\left(\frac{f'_c}{f_{cm0}}\right)^{\frac{1}{3}}$; $\varepsilon_1 = -0.0022$ and $f_{cm0} = 10$ Mpa.

Table 2. Concrete parameters adopted in numerical analysis.

Initial elastic modulus	E_0	22,750	MPa
Poisson's ratio	ν	0.15	-
Compressive cylinder strength	f_{ck}	30.8	MPa
Tensile strength	σ_{cr}	0.037	MPa
Tension stiffening(pre-load)	ε_{max}	0.0007	-
Tension stiffening	ε_{max}	0.0031	-

In Equation (1), ε_c is the strain at extreme fiber that depends on the compressive strength of concrete f'_c. The values of E_{cm} adopted in this study correspond to the secant elastic modulus of concrete, which was derived according to *fib* Model Code 2010 [29].

4. Predictions of Slab Deflections

Figure 8a–e compares experimental load-deflection with the numerical predictions given by the FE models. For the control specimen (PS-C), the specimen failed by concrete crushing at the ultimate load of 3.98 kN and deflection of 32.8 mm. When considering the deflection limit of L/360 = 2300/360 = 6.38 mm, this occurs at a load level of 2.15 kN. The load level increases to 2.56 kN at L/240, as well as the deflection limit for a strengthened prestressed slab. The design live load of this slab is 1.5 kN for typical houses, and therefore, the test result evidenced that the design of the control PS-C was satisfactory. In the case of the strengthened slabs without tapering PS-EBR-1-250 and PS-EBR-2-100 (Figure 8b,c), it was found that the end-debonding failure (i.e., Figure 6) occurred. This occurred at the ultimate loads of 4.35 kN and 4.20 kN, respectively, as shown in Figure 8b,c. When considering the load level at deflection limit of L/360, it was found that the load level of PS-EBR-1-250 and PS-EBR-2-100 increased to 2.71 kN and 2.82 kN, respectively, which were 126% and 131% higher compared to the control PS-C.

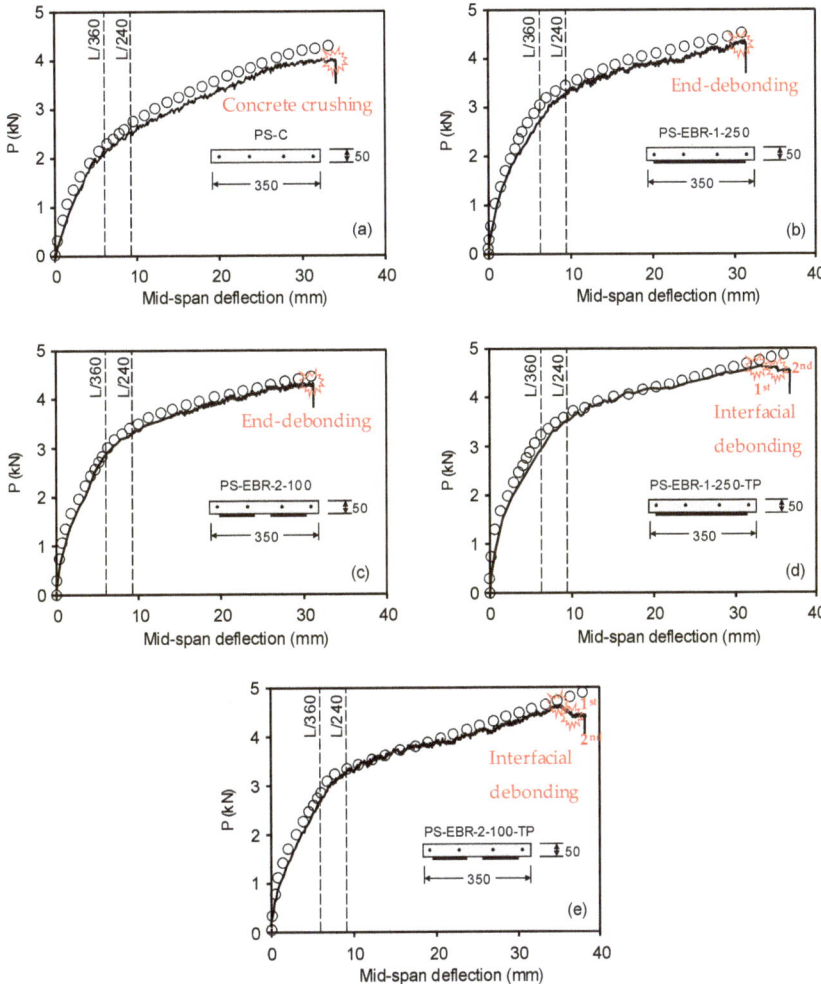

Figure 8. Comparison of experimental load-deflection and numerical predictions. (**a**) PS-C; (**b**) PS-EBR-1-250; (**c**) PS-EBR-2-100; (**d**) PS-EBR-1-250-TP; (**e**) PS-EBR-2-100-TP.

For CFRP-strengthened specimens with tapering CFRP sheets at the end anchorage (PS-EBR-1-250-TP and PS-EBR-2-100-TP), it was found that plate end bonding failure was well prevented, and instead, multiple intermediate cracks induced interfacial debonding failures (i.e., Figure 5). This is reflected in the load-mid-span deflection curves in Figure 8d,e as sudden drops in the load towards the end of the tests. Generally, the first interfacial debonding failures occurred at the ultimate load, i.e., at 4.50 kN and 4.52 kN. This was followed by the second interfacial debonding failures at 4.23 kN and 4.12 kN for specimens PS-EBR-1-250-TP and PS-EBR-2-100-TP, respectively. When considering the load level at deflection limit of L/360 for PS-EBR-1-250-TP and PS-EBR-2-100-TP, it was found that the load level increased to 3.91 kN and 3.78 kN, respectively, which were 155% and 150% more than the equivalent loads of PS-C. From the results in Figure 8, it is concluded that all EBR CFRP-strengthened slabs improved both service and ultimate loads (by up to 15%) and stiffness (by up to 7%) over the control PS-C slab. It is also evident that the end tapering technique with CFRP sheets improved the ultimate load by up to 5% compared to their identical specimens, thus, also preventing end debonding failure by allowing a more gradual interfacial debonding failure.

The results in Figure 8a–e also show that the higher the load, the stiffer the response is obtained from the FE results compared to the experimental results (e.g., defection at failure from FE model is 13.3 mm vs. 17.1 mm from PS-C). Nevertheless, the FE models predict reasonably well the load and corresponding deflection up to 2.25 kN, which is considered sufficient to represent the point of service condition (i.e., a deflection = L/360). This is also consistent with the findings reported in previous studies [6–8]. At higher load levels, the deflections obtained from FE models are lower than the experimental values due to several cracks being developed after the service load level is reached. For CFRP-strengthened slabs, the FE models predict better the results compared to unstrengthened specimens up to a load level of approximately 75% of the ultimate load level. Moreover, the FE models predicts 20–25% lower deflections at ultimate load. It is also noted that the numerical prediction in this study did not capture the failure due to end-debonding or interfacial debonding (i.e., Figure 1), since perfect bonding between CFRP and concrete surface was assumed in the analysis, and thus, no separation was allowed to occur in the FE analysis. It is also interesting that numerical predictions agreed well with experimental results after service load level. This can be attributed to the fact that major cracks are prevented by the CFRP sheets bonded at the slabs' soffit.

5. Parametric Studies

A parametric study is performed on slab PS-EBR-1-250-TP to investigate the stress in the flexural CFRP sheets and find solutions to make the strengthening more effective by varying the axial CFRP stiffness (thickness and modulus of elasticity). Figure 9a shows the load-deflection curves of strengthened slabs with different fiber thicknesses (t_f = 0.7, 1.4, 2.0 and 4.0 mm), whereas Figure 9b shows equivalent results for different moduli of elasticity (E_f = 150, 200 and 250 GPa). The results in these figures indicate that, as E_f increases, the overall stiffness of the slab also increases, and less deflection is recorded (Figure 9b). When the stress in the CFRP is compared, a thicker CFRP develops less stress than the thinner plate. Therefore, it can be concluded that increasing E_f gives the same results as increasing t_f. However, the results show that the load at which diagonal crack develops changes marginally if E_f changes. The results of the parametric study show that increasing the thickness or the elastic modulus of the FRP plate can reduce the efficiency of the FRP plate bonding technique and make it susceptible to earlier debonding failure or to allow multiple interfacial debonding failure. Therefore, it is proposed to consider end tapering to investigate the increase in effectiveness of the CFRP sheets. The CFRP strengthening is designed according to the bending moment diagram in order to achieve the effective use of the material by tapering the thickness of both sides (see tapering detail in Figure 3a) at the loading point of both ends of the longitudinal CFRP plates.

Table 3 shows the normal stress of CFRP sheets at the concrete–plate interface and tensile failure load for different CFRP thickness. It can be seen that the normal stress in the case of 0.7 to 1.4 mm thickness plate increases by 31% from the case of a 1.4 mm plate. For the case of 2.0 to 4.0 mm thickness, a 37% increase is obtained when compared with the case of a 4.0 mm CFRP. It reveals that end-tapering technique has an advantage in making the FRP plate more effective by reducing the thickness of the FRP plate at the plate end. Another advantage of tapering is that in strengthened slabs without tapering, the end plate or sheets are found to have a high concentration of principal stress and shear stress. Indeed, the termination of a plate or sheet with constant thickness creates an abrupt change in the element stiffness, which creates both shear stress and normal stress concentrations near the end plate or sheet.

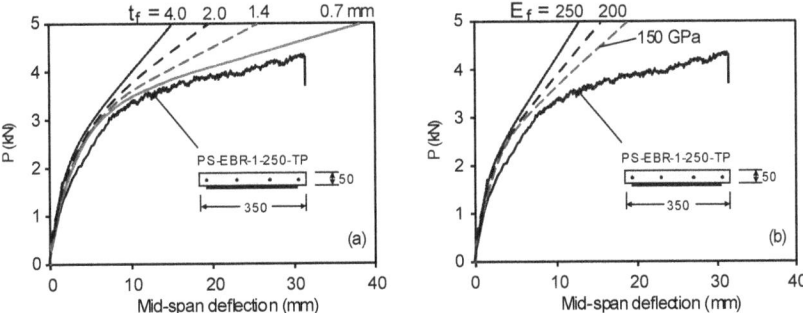

Figure 9. Effects of (a) thickness of CFRP, and (b) and modulus of elasticity of CFRP.

Table 3. Normal stress of CFRP sheets at concrete and tensile failure load for different CFRP thickness.

FRP Thickness t_f (mm)	Normal Stress in FRP S_{11} (MPa)	Load (kN)
0.7	413	4.72
1.4	281	4.35
2.0	254	4.30
4.0	160	4.25

6. Performance of End Effect for EBR FRP-Strengthened Slabs

Figure 10a,b show, correspondingly, the contour plots of principal tensile stress (S_{max}) and shear stress (S_{11}) of the concrete–adhesive interface of the CFRP-strengthened slabs without end-tapering (PS-EBR-250) and with end-tapering (PS-EBR-250-TP). In these plots, the thickness of the CFRP varied from 1.0 to 1.4 mm in PS-EBR-250-TP to compare with the 1.4 mm of CFRP in PS-EBR-250. The development of the principal stresses in the FE model can be used to identify the concentration of stresses in the concrete, CFRP sheets and adhesive resin. As a failure ratio of 0.037 was used (i.e., the tensile stress of concrete input in Abaqus® was 1.14 MPa), a diagonal crack in the concrete would occur when the tensile stress in the concrete (at the end of the plate) reaches its ultimate strength. However, the full composite action is assumed in this modeling exercise, which makes the analysis run completely until a compressive failure of concrete occurs. At this point, the maximum tensile stress in the CFRP sheets can be used to examine the effectiveness of the FRP strengthening before the beam starts to develop diagonal cracking, loses its composite action and eventually fails due to FRP debonding (i.e., Figure 10a). The results confirm that the use of end-tapering at the end anchorage of the CFRP sheets reduces the stresses compared to the slab without end tapering (see Figure 10b).

Figure 11a compares the CFRP–concrete interfacial principal stresses of slabs without (PS-EBR-250) and with tapering (PS-EBR-250-TP) as a function of the length of the CFRP plate sheets. Likewise, Figure 11a compares analogous results but for shear stresses of such slabs. The results in these figures show that the end-tapering technique has an advantage in making the FRP plate by reducing both von Mises and shear stresses and make it susceptible to early debonding. In these plots, two FRP plates with different thicknesses (constant

t_f = 1.4 mm and tapering t_f = 1–1.4 mm) are used to investigate the principal and shear stresses of the slabs and the FRP plates. Figure 11a,b indicate that the principal stress and shear stress concentrate at the end of the CFRP sheets if a constant thickness (t_f = 1.4 mm) is considered. The concentration of these stresses is found to be the cause of debonding due to rip-off or peel-off. Conversely, the tapering reduces the principal stresses (by up to 15%) and shear stresses (by up to 10%) along the anchorage length, thus, re-distributing stresses towards the beam mid-span. Therefore, tapering is extremely successful in reducing the stresses by up to 15%, which can be sufficient to prevent concrete rip-off and peel-off FRP debonding failures.

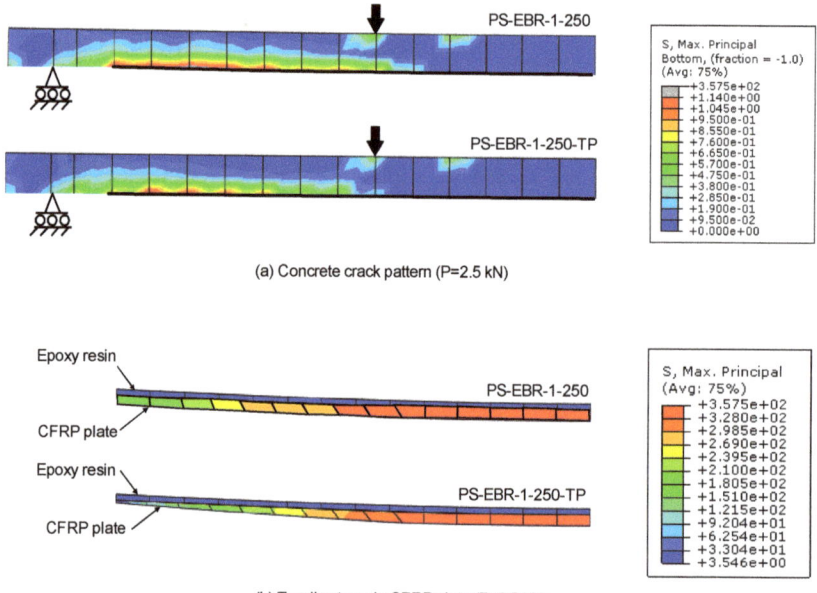

Figure 10. Comparison of principal tensile stress at load P = 2.5 kN of slabs without (PS-EBR-250) and with tapering (PS-EBR-250-TP), (**a**) stress along CFRP sheets and (**b**) detail of tensile stresses at end anchorage of CFRP sheets.

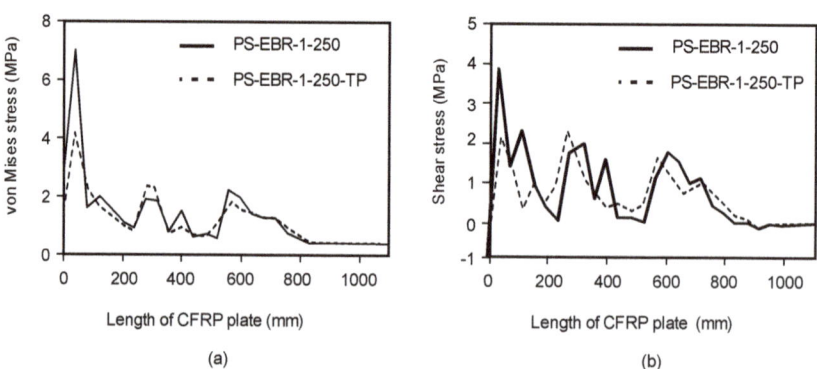

Figure 11. CFRP-concrete interfacial stress (**a**) principal stress and (**b**) shear stress.

7. Summary and Conclusions

This article numerically examines the behavior of prestressed reinforced concrete slabs strengthened with EBR CFRP sheets in flexure. Computer simulations based on finite element (FE) analysis were conducted on five specimens tested previously by the authors to investigate stress concentrations in the slabs, as well as at the CFRP–concrete bonded interface. The simulation is then used to carry out a parametric study that shows that increasing the thickness or the elastic modulus of the FRP plate affects the efficiency of the FRP plate bonding. Further analysis investigated the efficiency of tapering the EBR FRP on reducing the stress concentrations at the end anchorage to the strengthening system. Based on the results of this study, the following conclusions can be drawn:

- In all CFRP-strengthened prestressed slabs, the ultimate failure was controlled by debonding of the CFRP sheets at the end anchorage. End debonding failure and intermediate crack-induced interfacial debonding failure were observed in the slabs without and with tapered CFRP sheets, respectively.
- The results from the tests showed that an end tapering technique improved the ultimate load by up to 5% compared to slabs without tapering. The tapering also prevented end debonding failure by allowing multiple interfacial debonding failures.
- For CFRP-strengthened slabs, the FE models predicted better the results compared to unstrengthened specimens up to a load level of approximately 75% of the ultimate load. However, the numerical predictions showed approximately 20–25% fewer deflections at ultimate loads. Further research is necessary to achieve better predictions at ultimate loads.
- Results from a parametric study showed that increasing the thickness or the elastic modulus of the CFRP strengthening sheets can affect the efficiency of the FRP plate bonding technique, thus, making it susceptible to potential premature debonding failures.
- The FE results confirmed that a tapering technique at the end anchorage of the CFRP sheets can increase the capacity of a CFRP strengthening system. For the CFRP-strengthened slabs analyzed in this study, the end tapering reduced the stresses by up to 15%, which can be sufficient to prevent concrete rip-off and peel-off debonding failures. Nonetheless, further numerical analyses are necessary to extend the validity of these observations to other structural elements strengthened with EBR CFRP sheets.

Author Contributions: Conceptualization, C.W., T.I. and B.S.T.; validation, N.L.R., M.M.A.B.A., A.V.S., P.V., P.D.M. and B.S.T.; formal analysis, N.L.R., M.M.A.B.A., A.V.S., P.V. and P.D.M.; data curation, B.S.T.; writing—review and editing, R.G. and A.V.S.; visualization, N.L.R., M.M.A.B.A., A.V.S., P.V. and P.D.M.; funding acquisition, R.G. and A.V.S. All authors have read and agreed to the published version of the manuscript.

Funding: This research was funded by National Research Council of Thailand (NRCT5-RSA63019-04). This research publication was also supported by the Gheorghe Asachi Technical University of Iasi (TUIASI) from the University Scientific Research Fund (FCSU).

Institutional Review Board Statement: Not applicable.

Informed Consent Statement: Not applicable.

Data Availability Statement: Not applicable.

Conflicts of Interest: The authors declare no conflict of interest.

References

1. Garcia, R.; Helal, Y.; Pilakoutas, K.; Guadagnini, M. Bond behaviour of substandard splices in RC beams externally confined with CFRP. *Constr. Build. Mater.* **2014**, *50*, 340–351. [CrossRef]
2. Garcia, R.; Helal, Y.; Pilakoutas, K.; Guadagnini, M. Bond strength of short lap splices in RC beams confined with steel stirrups or external CFRP. *Mater. Struct.* **2015**, *48*, 277–293. [CrossRef]

3. Imjai, T.; Setkit, M.; Garcia, R.; Figueiredo, F.P. Strengthening of damaged low strength concrete beams using PTMS or NSM techniques. *Case Stud. Constr. Mater.* **2020**, *13*, e00403. [CrossRef]
4. Imjai, T.; Setkit, M.; Figueiredo, F.P.; Garcia, R.; Sae-Long, W.; Limkatanyu, S. Experimental and numerical investigation on low-strength RC beams strengthened with side or bottom near surface mounted FRP rods. *Struct. Infrastruct. Eng.* **2022**, 1–16. [CrossRef]
5. Smith, S.T.; Teng, J.G. FRP-strengthened RC beams. I: Review of debonding strength models. *Eng. Struct.* **2002**, *24*, 385–395. [CrossRef]
6. Täljsten, B. Strengthening of beams by plate bonding. *J. Mater. Civ. Eng.* **1997**, *9*, 206–212. [CrossRef]
7. Teng, J.G.; Smith, S.T.; Yao, J.; Chen, J.F. Intermediate crack-induced debonding in RC beams and slabs. *Constr. Build. Mater.* **2003**, *17*, 447–462. [CrossRef]
8. Imjai, T.; Garcia, R. Performance of Damaged RC Beams Repaired and/or Strengthened with FRP Sheets: An Experimental Investigation. In *Mechanics of Structures and Materials XXIV*, 1st ed.; Hao, H., Zhang, C., Eds.; CRC Press: London, UK, 2016; p. 1963. [CrossRef]
9. Grelle, S.V.; Sneed, L.H. Review of anchorage systems for externally bonded FRP laminates. *Int. J. Concr. Struct. Mater.* **2013**, *7*, 17–33. [CrossRef]
10. Godat, A.; Chaallal, O.; Obaidat, Y. Non-linear finite-element investigation of the parameters affecting externally bonded FRP flexural-strengthened RC beams. *Results Eng.* **2020**, *8*, 100168. [CrossRef]
11. Askar, M.K.; Hassan, A.F.; Al-Kamaki, Y.S. Flexural and Shear Strengthening of Reinforced Concrete Beams Using FRP Composites: A State of The Art. *Case Stud. Constr. Mater.* **2022**, *17*, e01189. [CrossRef]
12. Ashour, A.F.; El-Refaie, S.A.; Garrity, S.W. Flexural strengthening of RC continuous beams using CFRP laminates. *Cement. Concr. Compos.* **2004**, *26*, 765–775. [CrossRef]
13. Pham, H.; Al-Mahaidi, R. Experimental investigation into flexural retrofitting of reinforced concrete bridge beams using FRPComposites. *Compos. Struct.* **2004**, *66*, 617–625. [CrossRef]
14. Lundquist, J.; Nordin, H.; Täljsten, B.; Olafsson, T. Numerical analysis of concrete beams strengthened with CFRP: A study of anchorage lengths. In Proceedings of the BBFS: International Symposium on Bond Behaviour of FRP in Structures, Hong Kong, China, 7–9 December 2005.
15. Zhang, A.H.; Jin, W.L.; Li, G.B. Behavior of preloaded RC beams strengthened with CFRP laminates. *J. Zhejiang Univ. Sci. A* **2006**, *7*, 436–444. [CrossRef]
16. Coronado, C.A.; Lopez, M.M. Sensitivity analysis of reinforced concrete beams strengthened with FRP laminates. *Cement. Concr. Compos.* **2006**, *28*, 102–114. [CrossRef]
17. Arduini, M.; Nanni, A. Behavior of precracked RC beams strengthened with carbon FRP sheets. *J. Compos. Constr.* **1997**, *1*, 63–70. [CrossRef]
18. Hu, H.T.; Lin, F.M.; Jan, Y.Y. Nonlinear finite element analysis of reinforced concrete Beams strengthened by fibre reinforced plastics. *Compos. Struct.* **2001**, *63*, 271–281. [CrossRef]
19. Teng, J.G.; Zhang, J.W.; Smith, S.T. Interfacial stresses in reinforced concrete beams bonded with a soffit plate: A finite element study. *Constr. Build. Mater.* **2002**, *16*, 1–14. [CrossRef]
20. Pesic, N.; Pilakoutas, K. Concrete beams with externally bonded flexural FRP reinforcement: Analytical investigation of debonding failure. *Compos. B Eng.* **2003**, *34*, 327–338. [CrossRef]
21. Yang, Z.; Chen, J.; Proverbs, D. Finite element modelling of concrete cover separation failure in FRP plated RC beams Construct. *Build. Mater.* **2003**, *17*, 3–13. [CrossRef]
22. Li, L.J.; Guo, Y.C.; Liu, F.; Bungey, J. An experimental and numerical study of the effect of thickness and length of CFRP on performance of repaired reinforced concrete beams. *Constr. Build. Mater.* **2006**, *20*, 901–909. [CrossRef]
23. Camata, G.; Spacone, E.; Zarnic, R. Experimental and nonlinear finite element studies of RC beams strengthened with FRP plates. *Compos. B Eng.* **2007**, *38*, 277–288. [CrossRef]
24. Kotynia, R.; Abdel Baky, H.M.; Neale, K.W.; Ebead, U.A. Flexural strengthening of RC beams with externally bonded CFRP systems: Test results and 3-D nonlinear finite element analysis. *J. Compos. Constr.* **2008**, *12*, 190–201. [CrossRef]
25. De Maio, U.; Greco, F.; Leonetti, L.; Blasi, P.N.; Pranno, A. A cohesive fracture model for predicting crack spacing and crack width in reinforced concrete structures. *Eng. Fail. Anal.* **2022**, *139*, 106452. [CrossRef]
26. Rimkus, A.; Cervenka, V.; Gribniak, V.; Cervenka, J. Uncertainty of the smeared crack model applied to RC beams. *Eng. Fract. Mech.* **2020**, *233*, 107088. [CrossRef]
27. De Maio, U.; Greco, F.; Leonetti, L.; Blasi, P.N.; Pranno, A. An investigation about debonding mechanisms in FRP-strengthened RC structural elements by using a cohesive/volumetric modeling technique. *Theor. Appl. Fract. Mech.* **2022**, *117*, 103199. [CrossRef]
28. *ASTM A421/A421M-21*; Standard Specification for Stress-Relieved Steel Wire for Prestressed Concrete. ASTM International: West Conshohocken, PA, USA, 2021.
29. CEB-FIB. *Model Code. Comité Euro-International du Béton*; Thomas Telford Services Ltd.: London, UK, 2010.
30. Hibbitt, Karlsson & Sorensen, Inc. *ABAQUS User's Manual, Version 6.13*; Hibbitt, Karlsson & Sorensen, Inc.: Providence, RI, USA, 2013.

31. Okumus, P.; Oliva, M.; Becker, S. Nonlinear finite element modeling of cracking at ends of pretensioned bridge girders. *Eng. Struct.* **2012**, *40*, 267–275. [CrossRef]
32. Imjai, T.; Guadagnini, M.; Pilakoutas, K. Bend Strength of FRP Bars: Experimental investigation and Bond Modelling. *J. Mater. Civ. Eng.* **2017**, *29*, 04017024. [CrossRef]

Disclaimer/Publisher's Note: The statements, opinions and data contained in all publications are solely those of the individual author(s) and contributor(s) and not of MDPI and/or the editor(s). MDPI and/or the editor(s) disclaim responsibility for any injury to people or property resulting from any ideas, methods, instructions or products referred to in the content.

Article

Mixing of Excitons in Nanostructures Based on a Perylene Dye with CdTe Quantum Dots

Yuri P. Piryatinski [1], Markiian B. Malynovskyi [1], Maryna M. Sevryukova [1], Anatoli B. Verbitsky [1], Olga A. Kapush [2], Aleksey G. Rozhin [3] and Petro M. Lutsyk [3,*]

[1] Institute of Physics, National Academy of Sciences of Ukraine, 46 Prospekt Nauky, 03680 Kyiv, Ukraine
[2] V. Lashkaryov Institute of Semiconductors Physics, National Academy of Sciences of Ukraine, 41 Prospekt Nauky, 03680 Kyiv, Ukraine
[3] Aston Institute of Photonic Technologies, College of Engineering and Physical Sciences, Aston University, Aston Triangle, Birmingham B4 7ET, UK
* Correspondence: p.lutsyk@aston.ac.uk

Abstract: Semiconductor quantum dots of the A_2B_6 group and organic semiconductors have been widely studied and applied in optoelectronics. This study aims to combine CdTe quantum dots and perylene-based dye molecules into advanced nanostructure system targeting to improve their functional properties. In such systems, new electronic states, a mixture of Wannier–Mott excitons with charge-transfer excitons, have appeared at the interface of CdTe quantum dots and the perylene dye. The nature of such new states has been analyzed by absorption and photoluminescence spectroscopy with picosecond time resolution. Furthermore, aggregation of perylene dye on the CdTe has been elucidated, and contribution of Förster resonant energy transfer has been observed between aggregated forms of the dye and CdTe quantum dots in the hybrid CdTe-perylene nanostructures. The studied nanostructures have strongly quenched emission of quantum dots enabling potential application of such systems in dissociative sensing.

Keywords: perylene dye; nanoparticles; quantum dots; cadmium telluride; photoluminescence; time-resolved spectroscopy; exciton

1. Introduction

Self-assembly of molecules and the formation of one-dimensional molecular structures with an atomically close distance between the molecules in one direction attracts considerable attention in recent decades [1–3]. For example, planar aromatic molecules organize into one-dimensional face-to-face stacks with a strong intermolecular overlap of π-orbitals. Such structures are interesting for applications in photoelectronic devices such as solar cells [4], light-emitting diodes and transistors [5], etc. Perylene derivatives, such as N,N-dimethylperylene-3,4,9,10-dicarboximide (MePTCDI), or 3,4,9,10-perylenetetracarboxylic dianhydride (PTCDA), are well-known examples of materials forming one-dimensional crystals with an extremely small distance between molecular planes in one-dimensional stacks (3.37 Å for PTCDA, 3.40 Å for MePTCDI) [6,7]. The properties of perylene derivatives crystals are similar to features of both conventional inorganic semiconductors and organic molecular crystals. Therefore, perylene-based molecular structures are an ideal model material to study fundamental excitonic processes linking inorganic and organic semiconductor classes of materials that may ultimately prove useful for applications in optoelectronic devices.

Inorganic semiconductor nanoparticles or quantum dots (QDs) of the A_2B_6 group, (CdS, CdTe) have wide practical applications in optoelectronics, for example as labels for biological research [8–10]. Combining two objects, such as CdTe QDs and perylene dyes, into one system has the potential to significantly improve their functional properties, therefore a comprehensive understanding of their fundamental properties is needed. In such systems, new (mixed) electronic and excitonic states may appear at the interface of organic and inorganic materials. Various models of excitons are used to classify excitons: small-radius Frenkel exciton (FE) model [11,12], charge transfer exciton (CTE) model [13–15], and the large-radius Wanier-Mott exciton (WME) model [16]. FE and CTE models are used to classify excitons in organic semiconductor materials, while the WME model is typical for inorganic semiconductors. FE is a neutral excited state in which an electron and a hole are located on the same molecule. The intermolecular interaction leads to a finite transition integral for the transfer of electronic excitation from one molecule to another, and as a result, FEs propagate through a crystal as coherent waves. CTE consists of a pair of charge carriers located on different neighboring molecules. Such arrangements are ensured in organic crystals because (in contrast to inorganic semiconductors) the binding energy of lower CTEs is greater than the width of the valence and the conduction bands. WME model considers the Coulomb interaction between an electron and a hole and is based on an approximation of the effective mass for them in a periodic lattice potential. The main characteristic of WME is hydrogen absorption and emission behavior in crystals with a large dielectric constant. The average distance between the electron and the hole for this type of exciton is much larger than the lattice constant. Mixing the different excitons is an appealing area of fundamental research allowing one to pursue enhancement of resonant optical nonlinearity, fluorescence efficiency and relaxation processes [17].

The studied perylene derivative, perylene-3,4:9,10-bis(dicarboximide)-N,N-bis(1-methyl-3pyridinium) bis-n-toluenesulfonate (2416SL) is similar in its structural and electronic properties to the well-investigated PTCDA. The difference is that 2416SL molecules have ionic groups on the periphery (Figure 1), which makes them soluble in water [18], therefore semiconductor crystalline structures with 2416SL can be prepared from solutions. In water, 2416SL molecules associate into aggregates, which at high concentrations assemble into structures with long-range orientational order. Such structures can form lyotropic chromonic liquid crystals (LCLC), the elementary building blocks of the metaphase of which are elongated disk-like molecular aggregates [19]. 2416SL can form oriented nanostructured films preserving the LCLC orientational order in solid crystalline film [18,20,21]. The main motivation for researching 2416SL in different aggregated states stems from promising excitonic and related optical properties, which appear due to the regular planar organization of 2416SL molecules in thread-like H-aggregates having π-π stacking with an unusually small intermolecular distance of 0.34 nm [6,7]. The excitonic properties of neat 2416SL have not been studied comprehensively, and this aspect will be investigated here as well in more detail. Thus, 2416SL is a relevant model object that allows us to study a wide range of fundamental phenomena involved in the operation of various electronic devices made of organic and inorganic materials. Functionalizing CdTe QDs by perylene dye, such as 2416SL, will allow one to create nanostructures with novel features, and fundamental spectral characteristics of such new nanostructures have to be analyzed. The study of the nature of exciton states in the bulk and at the interface of organic/inorganic materials is one of the fundamental objectives for the application of such hybrid functional systems in the future.

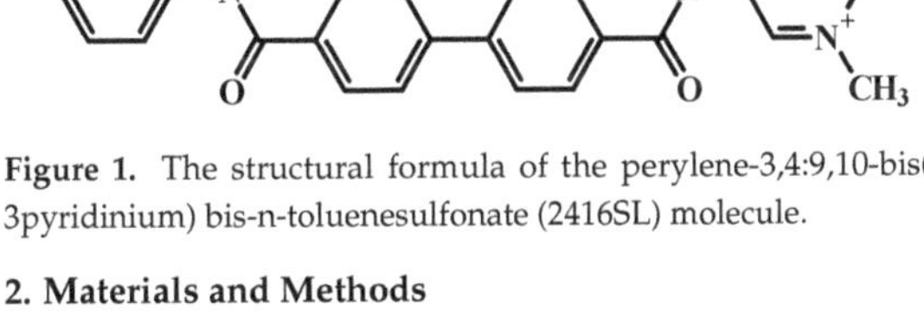

Figure 1. The structural formula of the perylene-3,4:9,10-bis(dicarboximide)-N,N-bis(1-methyl-3pyridinium) bis-n-toluenesulfonate (2416SL) molecule.

2. Materials and Methods

The structural formula of the studied 2416SL molecule is shown in Figure 1. 2416SL was synthesized at the Institute of Organic Chemistry, the National Academy of Sciences of

Ukraine using the methodology described before [20]. Solutions of 2416SL were prepared in dimethyl sulfoxide (DMSO) and water. 2416SL dissolves well in DMSO. In water, even at low concentrations, 2416SL molecules aggregate. The aggregation of 2416SL molecules in aqueous solvents can be associated with the high hydrophobicity of their perylene core-chromophore [22]. The aqueous solutions in water were heated to 90 °C and then cooled down before measurement. The concentrations of the solutions were in the range of $(10^{-3}$–$5 \cdot 10^{-6})$ M.

2416SL films were obtained by drop casting of high concentration (0.1 M) aqueous solution on a quartz substrate and drying at room temperature. The films were annealed at 470 K to improve crystalline structure.

A dispersion of CdTe quantum dots (QDs) in deionized water was obtained in the presence of thioglycolic acid (TGA) [10]. All reagents and solvents obtained from commercial suppliers were of reagent grade quality. Milli-Q water, CdI_2, NaOH, and thioglycolic acid (TGA $\geq$ 90%) were purchased from Himlaborreactive (Ukraine). In the synthesis of the QDs, each chemical element was introduced into the reactor in the form of a precursor: a molecule or complex containing at least one constituent element. In our case, the Cd_2^+ source was the CdI_2 salt, and the Te_2^- source was H_2Te gas prepared electrochemically in a galvanostatic cell. The low-temperature colloidal synthesis has been performed in the reactor of complete mixing in the presence of TGA as a stabilizer. CdI_2 was dissolved in water, and TGA was added under stirring, followed by adjusting the pH to 10 by dropwise addition of NaOH solution. H_2Te gas was passed through the solution using argon as a carrier gas. The size of CdTe QD increased with the duration of the synthesis. The size of CdTe QDs was determined by dynamic light scattering and the ratio of the particle diameter, d, and the absorption wavelength of the first exciton maximum. For studied CdTe QDs, the average diameter d was 2.5 and 3.5 nm, having a relatively narrow distribution of QDs sizes characterized by dynamic light scattering (for d = 2.5 nm, the distribution range is 1.5–6.0 nm, and 3.5 nm QDs sizes spread slightly wider over 1.8–9.0 nm). The concentration of CdTe QDs in the initial dispersions was approx. 10^{-5} M.

The formation of hybrid nanostructures of CdTe-2416SL took place by admixing initial dispersions of CdTe in portions of $V_n = 0.1 \cdot n$ mL, where n was taken from 1 to 15, to 1 mL of an aqueous solution of 2416SL (with a concentration of $5 \cdot 10^{-5}$ M). The mixture of CdTe-2416SL at n = 15 has been studied in two forms due to abundant aggregation and precipitate formation. One form was a CdTe-2416SL supernatant where all precipitate was allowed to go down for 24 h and only the top half of the mixture was studied. Another form was a freshly mixed CdTe-2416SL with all the micro and nano-aggregations present in the dispersion.

The structure of electron-vibrational and excitonic transitions in the studied samples (2416SL solutions, CdTe QDs dispersions, 2416SL films, and hybrid systems based on CdTe-2416SL) was studied by analyzing electronic absorption spectra, steady-state, and time-resolved photoluminescence (PL) spectra. The complex use of spectral techniques allowed us to identify molecular and exciton signatures in the studied systems.

Absorption spectra were measured using a Lambda 1050UV/VIS/NIR spectrophotometer (PerkinElmer, US). Steady-state PL spectra were measured using a USB2000+UV-VIS-ES spectrometer through an optical fiber with a diameter of 600 μm. LLS-385 LED (Ocean Optics, US) and EPL-405 laser (Edinburgh Instruments Ltd., Livingston, UK) were used to excite steady-state PL with the corresponding λ_e wavelength.

Time-resolved PL emission spectra (TRES) were measured using a LifeSpec II spectrofluorimeter (Edinburgh Instruments Ltd., Livingston, UK). An EPL-405 laser with a wavelength of λ_e = 405 nm and a pulse duration of 40 ps was used to excite time-resolved PL in the visible range. The frequency of excitation pulses can be adjusted in the range of 10 kHz–20 MHz. To determine the lifetimes τ of excited states of molecules, the time-correlated photon counting with picosecond time resolution was used allowing us to measure PL decay kinetics of weakly emitting samples with characteristic lifetimes of $(10^{-6}$–$10^{-11})$ s. To excite PL in this method, a sequence of short excitation pulses of radiation from the lasers with a strictly fixed follow-up period is used. The probability of PL detection is kept below one photon when the object is excited by a single pulse, and the repetition frequency of the exciting pulse is set as high as possible. On the other hand, the sequence of pulses is maintained in such a way that the time interval between pulses is at least 5–10 times longer than the decay time of PL being recorded. The obtained time dependence of PL kinetics, $I(t)$, was approximated by the expression:

$$I(t) = IRF * \sum_{i=1}^{n} A_i \exp\left(-\frac{t}{\tau_i}\right)$$

where IRF is the instrument response function of the detector, i is a serial number, τ is the lifetime of the excited state, and A_i is the weighting factor. To measure IRF, the certified colloidal LUDOX solution was used.

To establish the true PL attenuation curve according to experimental data, it is necessary to solve the integral equation $I(t) = \int_0^t F(t-t')G(t')dt'$, where $I(t)$ is the experimental dependence of PL intensity on time, $F(t)$ is the true dependence of PL attenuation on time, $G(t)$ is the IRF. Using a sequence of PL kinetic curves for different emission wavelengths, TRES maps were constructed representing the spectral dependence of PL on the delay time, t. TRES map is a functional dependence of the PL intensity on two variables—the radiation wavelength (λ_{EM}; Y-axis) and the delay time (t; X-axis). In the TRES map, as instantaneous PL spectra are measured, the corresponding t are recorded with the reference to the maximum of the laser pulse in the IRF, when t = 0 ns. To calculate τ and plot the TRES maps, the F900 software package (version 7.2, Edinburgh Instruments Ltd., Livingston, UK) was used.

3. Results and Discussion
3.1. Solutions and Films of 2416SL

To determine the structure of electron-vibrational and exciton transitions in 2416SL, the absorption and PL spectra of aqueous and DMSO solutions at different concentrations were studied. Furthermore, the spectra of 2416SL films deposited on quartz provided additional insight. Absorption (Figure 2, curves 1,3) and PL (Figure 2, curves 2,4) spectra for low concentration ($5 \cdot 10^{-6}$ M) solutions of 2416SL in water and DMSO at 296 K have variations, which can be associated with their different solubility. The positions of the maxima of the electronic and electronic-vibrational bands in the absorption and PL spectra (Figure 2) of solutions of 2416S in water and DMSO are summarized in Tables 1 and 2.

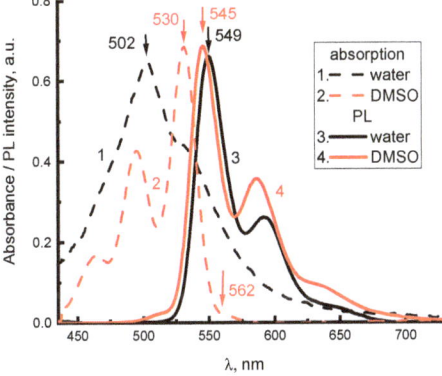

Figure 2. Spectra of absorption (1,2) and steady-state PL (3,4) of molecular solutions of 2416SL in DMSO (2,4) and water (1,3) at the concentration of $5 \cdot 10^{-6}$ M, λ_e = 385 nm, T = 296 K.

Table 1. Positions of electronic absorption band maxima for solutions of 2416SL.

Solution	Transitions (nm/cm^{-1})		
	$S_0(0) \to S_1(0)$	$S_0(0) \to S_1(1)$	$S_0(0) \to S_1(2)$
DMSO	530/18,870	495/20,200	464/21,550
water	532/18,800 *	502/19,920	471/21,220

* The transition is estimated from the comparison of the absorption spectra of aqueous and DMSO solutions, the values of their vibrational repetitions, and Stokes shifts.

Table 2. Positions of PL band maxima of 2416SL solutions.

Solution	Transitions (nm/cm^{-1})		
	$S_0(0) \leftarrow S_1(0)$	$S_0(0) \leftarrow S_1(1)$	$S_0(0) \leftarrow S_1(2)$
DMSO	545/18,340	586/17,070	633/15,800
water	549/18,190	591/16,920	639/15,650

The spectra in DMSO (Figure 2) are mirror symmetric, have a Stokes shift of 530 cm^{-1}, and oscillating repetitions with a frequency close to 1330 cm^{-1} forming solid evidence that the spectra are of molecular origin. Absorption spectra for aqueous solutions of 2416SL (Figures 2 and 3a) are lacking mirror symmetry between the absorption and PL spectra. There is a significantly reduced light absorption in the region of the purely electronic $S_0(0) \to S_1(0)$ optical transition, and a high intensity 502 nm band, approximately in the region of $S_0(0) \to S_1(1)$ transition. In the present work, only aqueous solutions of 2416SL are studied in detail, because aqueous solutions of CdTe and their mixtures with 2416SL are the focus of this study. The absence of mirror symmetry between the absorption and PL spectra of aqueous solutions, additional absorption in the short wavelength region, and reduced contribution of molecular spectral signatures evidence that, even at low concentrations, H-aggregates are formed. Therefore, the absorption and PL spectra are formed not only by molecular but also by collective excitations in the aggregates, and disk-like molecules of 2416SL aggregate into thread-like columnar structures with a diameter equal to the size of the molecule. Furthermore, the perylene derivatives are well-known for forming H-aggregates, where molecules are positioned almost parallel to each other [23]. Such one-dimensional molecular aggregates have features characteristic of collective excitations (excimers, FE [11,12], CTE [24–26]).

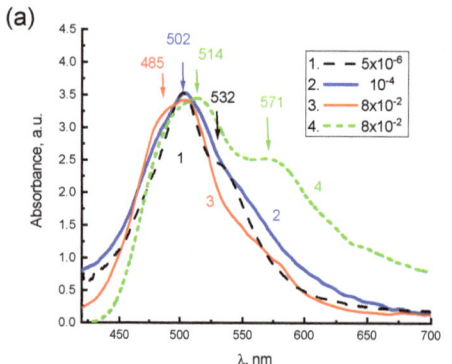
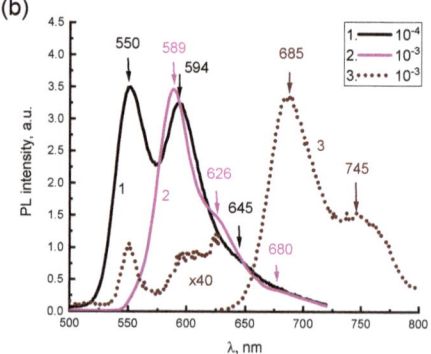

Figure 3. (a) Absorption spectra of aqueous solutions for 2416SL at the concentrations of $5 \cdot 10^{-6}$ M (1), 10^{-4} M (2), and $8 \cdot 10^{-2}$ M (3,4); λ_e = 405 nm; T = 296 K (1–3) and 77 K (4). (b) Steady-state PL spectra of aqueous solutions for 2416SL at the concentrations of 10^{-4} M (1) and 10^{-3} M (2,3); λ_e = 405 nm; T = 296 K (1,2) and 4.2 K (3).

Due to the stacking parallel orientation of molecules and their dipoles in molecular H-aggregates in solutions and films, their PL of FE is significantly quenched in comparison to an isolated molecule. This is happening because optical transitions for FE in H-aggregate absorption and PL between $S_0(0) \leftrightarrow S_1(0)$ electronic states are forbidden [16]. However, the strict prohibition of $S_0(0) \leftrightarrow S_1(0)$ transitions is valid, only for excitation delocalized along an infinite crystal [16]. In real crystals, the effect of dipole ordering depends on the exciton coherence length. In molecular aggregates and thin polycrystalline films, the coherence length decreases due to thermal and structural disorder [26–28], and optical $S_1(0) \rightarrow S_0(1)$ transitions to higher electronic vibrational states of the $S_0(1)$ ground state are allowed, albeit with a smaller intensity. This can be related to the characteristic features of the absorption spectra for aqueous solutions of 2416SL at different concentrations (Figure 3), there is a lack of mirror symmetry between the electronic absorption and PL spectra at room temperature, as well as very low absorption in the region of the purely electronic $S_0(0) \rightarrow S_1(0)$ transition and a significant intensity of the 502 nm band, approximately in the region of the allowed $S_0(0) \rightarrow S_1(1)$ transition. The above features of absorption and PL spectra for aqueous solutions of 2416SL can be explained by the manifestation of FE in H-aggregates, which are formed in these solutions.

For aqueous solutions of 2416SL of different concentrations, starting from $5 \cdot 10^{-6}$ M and more, changes in the PL spectra at room temperature are much more significant than in the absorption spectra (Figure 3). Two bands with maxima at 550 and 594 nm can be distinguished in the steady-state PL spectrum of aqueous solutions for 2416SL at concentrations of 10^{-4} M (Figure 3b, curve 1). The position of the maximum of the first PL band corresponds to the $S_1(0) \rightarrow S_0(0)$ transition of the molecular solution, and the second maximum reflects the $S_1(0) \rightarrow S_0(1)$ transition of the H-aggregate, as this band is getting dominant in the high concentration solutions. An increase in the concentration from $5 \cdot 10^{-6}$ M to 10^{-4} M leads to a significant drop in the intensity of the purely electronic $S_1(0) \rightarrow S_0(0)$ transition, and the intensity of the 594 nm band of the $S_1(0) \rightarrow S_0(1)$ transition relatively increases (Figure 3b, curve 1). At the concentration increased to 10^{-3} M, only a band with a maximum of 589 nm and a weak shoulder at 626 nm is observed in the PL spectrum (Figure 3b, curve 2).

The absorption spectrum of the aqueous solution of 2416SL ($8 \cdot 10^{-2}$ M) changes significantly with a temperature drop from 297 K (Figure 3a, curve 3) to 77 K (Figure 3a, curve 4). In the absorption spectrum at 77 K, the intensity of absorption at a longer wavelength increases dramatically and a new band of 571 nm appears. The presence of this band at low temperatures can be associated with structural changes in H-aggregates, a manifestation of their excitonic properties and the formation of low-temperature CTEs.

For the 10^{-3} M solution at 4.2 K, the intensity of the PL bands at 589 and 626 nm decrease to practically zero, and only bands with maxima at 685 and 745 nm are observed in the PL spectrum (Figure 3b, curve 3). The dramatic reduction of the 589 and 626 nm bands is characteristic of excimer emission [12]. In solutions, 2416SL molecules form elongated disc-like aggregates and, in such one-dimensional molecular aggregates, the formation of excimers has already been established [12]. Excimers should not be confused either with CTE states, which involve significant charge transfer between molecules, or with FEs, which characterize the coherent excitation of a crystal. An excimer is an optically excited dimer stabilized by resonance interaction. A necessary condition for the formation of excimers is a small distance between molecules, which is usually achieved due to effective π-stacking [16] and the convergence of molecules in an excited state. The ground state of excimers is antibonding; therefore, excimers have no direct absorption in the ground state and must be excited by energy transfer. The excimer radiation is characterized by a broad structureless band of PL. In 2416SL nanoaggregates, as in pyrene and α-perylene, several types of excimers can be realized [12]. In our case, the 2416SL excimers are featured by PL bands of 589 and 626 nm.

Changes in the PL spectra of the aqueous solutions at different concentrations can be associated with a manifestation of collective excitations of FEs, excimers, and CTEs in H-aggregates [7,12]. For molecular aggregates of perylene derivatives, CTEs play a significant role in PL [6,7,12,24–26,29]. Comparative analysis of absorption and PL spectra for 2416SL in solutions and condensed state allows us to determine the nature of these collective excitations. Figure 4 shows the spectra of absorption (Figure 4, curves 2,3) and steady-state PL (Figure 4, curves 4,5) of 2416SL films, before (Figure 4, curves 2,5) and after thermal annealing (Figure 4, curves 3,4). The absorption spectra of the films after thermal annealing (Figure 4) are more structured than the absorption spectra of nanoaggregates in solutions and are more like the absorption spectra of PTCDA films [6,7]. Such spectral changes can be associated with structural changes in the aggregates after thermal annealing. In the spectra of thermally annealed films, clearly expressed CTE absorption maxima at 556 (CTE_1) and 589 nm (CTE_2) appear (well evidenced by differential spectrum in Figure 4, curve 6). These maxima correspond to Franck-Condon's non-relaxed optical CTE states [14,15]. The PL bands, which correspond to the emission of relaxed exciton states, have maxima at 636, 680, and 750 nm. The nature of the absorption and PL spectra in Figure 4 will be analyzed below.

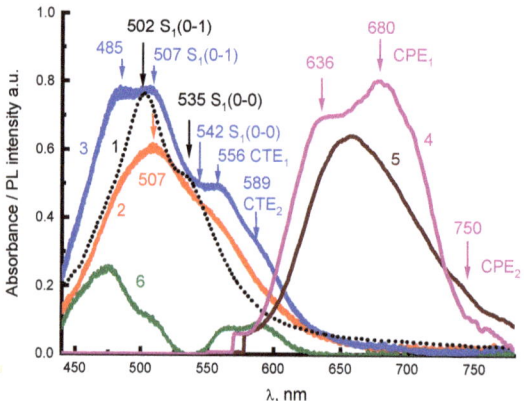

Figure 4. Absorption (1–3) and steady-state PL (4,5) of aqueous solution (1) of 2416SL at the concentration of $5 \cdot 10^{-6}$ M (1) and 2416SL films before (2,5) and after thermal annealing at 470 K (3,4); differential spectrum, Δ, (6) obtained by subtraction of the spectrum (2) from the spectrum (3). λ_e = 405 nm; T = 296 K.

CTE occupying an intermediate place in the classification of excitons based on their internal structure have the charge or its part transferred to a neighboring molecule due to photoinduced electron transfer [13–15]. With incomplete charge transfer, the wave function of the resulting state can be delocalized within two or more molecules and have both excitonic and ionic features. If the excitonic character prevails, CTE can coherently move along the crystal. Unrelaxed CTEs are formed directly upon optical excitation and appear in the absorption spectra. Due to a large static dipole moment (up to 25 Debye on the nearest molecules), CTEs can be a cause of a large nonlinear second-order polarizability and a strong electroabsorption. CTEs have a strong tendency to self-localize. They polarize the surrounding molecules, which leads to the relaxation of the crystal lattice into a new equilibrium. If the time of CTE excitation at the lattice nodes is longer than the lattice relaxation time, the excitation is accompanied by local deformation of the lattice and the formation of the excitonic polaron—CPE [14]. Relaxed molecular-polaron excitons (CPE) are a characteristic feature of molecular crystals. CPEs appear as intermediate states in the processes of photogeneration and radiative recombination of CTE states.

In addition to excimer radiation in the long-wavelength region of the PL spectrum for concentrated solutions of 2416SL (Figure 3), weakly intense bands of 680 and 750 nm can be distinguished. These bands are also observed in the spectra of 2416SL films at room temperature (Figure 4). When the 2416SL solution is cooled to 4.2 K, the PL intensity in this spectral region increases significantly and the bands at 685 and 745 nm appear in the spectra (Figure 3b, curve 3). The PL emission of the bands at 680 and 750 nm can be associated with CPE_1 and CPE_2, respectively.

Electronic states in quasi-one-dimensional molecular crystals of the PTCDA type with a strong overlap of molecular orbitals were comprehensively analyzed [3,6,7,12,24–26,29]. In such quasi-one-dimensional crystals, due to the small, less than 0.35 nm, intermolecular distance, there is a strong overlap of the π-orbitals of neighboring molecules. In such crystals, the difference between FE and CTE energies becomes small, and their strong mixing determines the nature of the lowest exciton states [7,13,30,31]. As soon as the energy difference between CTE and FE becomes small, both types of excitons can interact, and new mixed excitonic states are formed. At FE and CTE being close in energy, the FE band admixes some CTE states and shifts down, and CTE also shifts up, acquiring some energy of the FE state, and becomes optically allowed. These mixed FE-CTE states exhibit the properties of two types of excitons: FE provides a high oscillator strength, and CTEs lead to high sensitivity in external electric fields. Such exciton mixing can also result in a noticeable transition dipole for CTE [7]. This can explain the spectral dependence of the PTCDA films on their thickness in the range of 0.3–10 nm when the effects of quantum confinement become important [32]. In finite chains, simultaneously with excitonic "bulk" states, "surface" states can also arise [33]. The "surface" states are localized at the end of the chain and can be shifted to the blue or red region of the spectrum compared to the bulk states. Thus, PL in 2416SL nanoaggregates can be caused by direct excitation due to the borrowing of some transition oscillator strength from intense transitions with subsequent radiative recombination or thermally activated decay into free charge carriers. In 2416SL having a strong tendency for mixing of FEs and CTEs, the CTEs play a significant role in PL emission.

Registration of the dependence of the PL intensity on the decay time t was carried out for radiation wavelengths λ_{EM} = 550 and 590 nm corresponding to molecular and H-aggregate emissions, respectively. The PL decays for λ_{EM} = 550 nm at the concentration of 10^{-4} M are well described by a single-exponential function and feature molecular emission. The kinetics of PL decay for λ_{EM} = 590 nm at the concentration of 10^{-3} M have a good fit by a biexponential function with fast and slow components of lifetimes τ. This can be attributed to the emission of molecular and aggregated forms of 2416SL. Table 3 shows the values of τ calculated for the PL kinetics of 2416SL.

Table 3. PL lifetimes of aqueous solutions of 2416SL for various concentrations and λ_{EM}; λ_e = 405 nm; T = 296 K.

2416SL Concentrations	λ_{EM}, nm	τ_1, ps	%	τ_2, ps	%	χ^2
10^{-4} M	550	-		4050		1.162
10^{-3} M	590	1710	3	4760	97	1.068

The presence of two emission components in the PL spectra of concentrated 2416SL solutions is demonstrated in TRES maps (Figure 5a,b) and instantaneous PL spectra (Figure 5c,d). In TRES maps, the dependence of the PL intensities vs. t and λ_{EM} for aqueous solutions of 2416SL at the concentrations of 10^{-4} (Figure 5a) and 10^{-3} M (Figure 5b) is shown. PL emission wavelengths (520–700 nm) and delay times (0–20 ns) are plotted on the vertical and horizontal axes, respectively, whereas PL intensity is a function of color in relative units. For 2416SL concentration of 10^{-4} M (Figure 5a), two PL bands with maxima at 551 and 594 nm can be distinguished in the TRES map. When the concentration of 2416SL increases to 10^{-3} M (Figure 5b), the intensity of the band at 551 nm decreases significantly, and only the band at 589 nm remains.

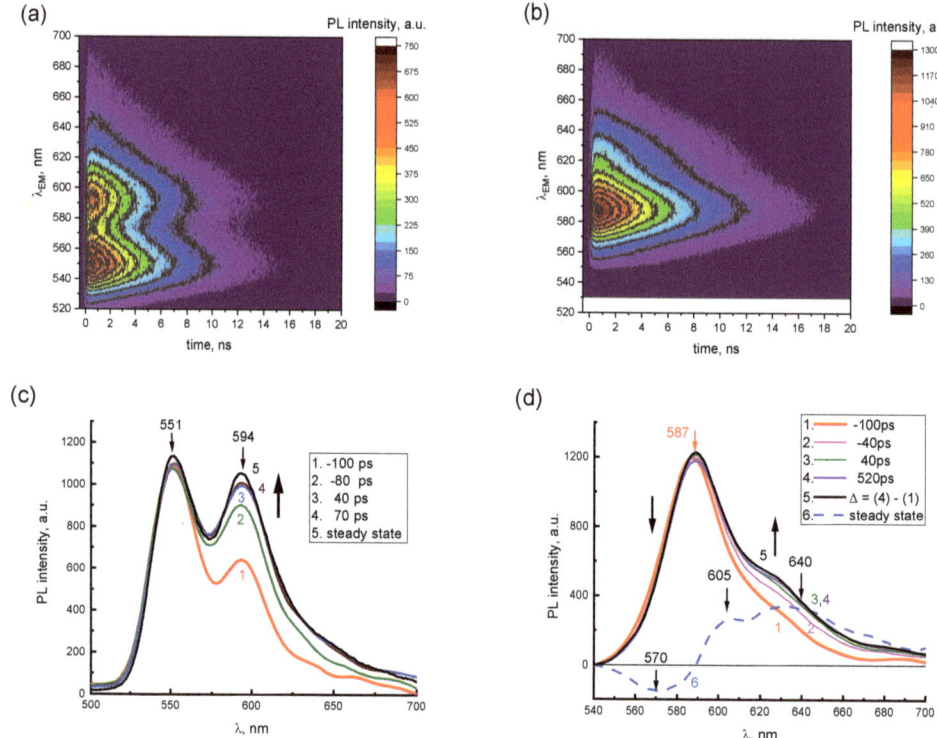

Figure 5. TRES maps (**a**,**b**) and instantaneous PL spectra at various delay times (**c**,**d**) of aqueous solutions of 2416SL for two concentrations 10^{-4} M (**a**,**c**) and 10^{-3} M (**b**,**d**). The following delay times are presented: (**c**) −100 (1), −80 (2), 40 (3), 70 ps (4); (**d**) −100 (1), −40 (2), 40 (3), 520 ps (4). The steady-state spectra (5) are shown as a reference. (**d**) Differential spectrum, Δ, (6) is obtained by subtraction of spectrum (1) from spectrum (4). λ_e = 405 nm; T = 296 K.

In Figure 5c,d, instantaneous PL spectra are presented for aqueous solutions of 2416SL at various delay times t and concentrations, using the data of TRES maps in Figure 5a,b. The instantaneous PL spectra are normalized by the most intense PL bands in the steady-state spectra. These are bands with emission maxima at 551 nm and 589 nm for the concentration of 10^{-4} (c) and 10^{-3} M (d), respectively. Negative values of delay times mean that instantaneous PL spectra were recorded at the leading edge of the laser pulse. It can be seen from Figure 5c that at short delay times of −100 ps, the instantaneous PL spectra are similar to the molecular spectra of Figure 3, and already at delay times of 40 ps, they coincide with the steady-state PL spectra (Figure 5c, curves 3,5), which is also manifesting contribution of the aggregated form of 2416SL.

Figure 5d shows the instantaneous PL spectra for aqueous solutions of 2416SL at the concentration of 10^{-3} M, which reflect the time dependence of the instantaneous spectra of the aggregated form of 2416SL. In the time delay of around 40 ps, the instantaneous PL spectra reach an equilibrium value and become similar to the steady-state PL spectra (Figure 5d, curves 3,5). A relative decrease in PL on the long-wavelength side of the 589 nm band is also observed. The presented difference spectrum (Figure 5d, curve 6), obtained by subtracting the instantaneous spectra at the time delay of 520 and −100 ps (Figure 5d, curves 4,1), has bands with maxima at 570, 605, and 640 nm. Analysis of the instantaneous PL spectra (Figure 5) and the lifetimes (Table 3) allows ones to conclude that the band maxima at 551, 589, 626, 685 and 745 nm are of different natures. PL in these bands has different lifetimes and differs in temperature dependence.

3.2. Dispersions of CdTe QDs

In Figure 6a, absorption and PL spectra are shown for initial aqueous dispersions of CdTe QDs with a diameter of d = 2.5 nm (curves 1,3) and 3.5 nm (curves 2,4). The spectra evidence the positions of exciton transitions for CdTe QDs. In the absorption spectra of CdTe QDs, bands with maxima at 525 and 583 nm can be distinguished, which correspond to the excitonic absorption of the QDs with d = 2.5 and 3.5 nm, respectively [34]. Exciton emission bands with maxima at 557 and 619 nm can be distinguished in the PL spectra (Figure 6, curves 3,4) of these QDs. As the size of QDs increases, the spectra of exciton absorption and emission shift to the long-wavelength side (Figure 6a), having a clear manifestation of the quantum-size effect [34].

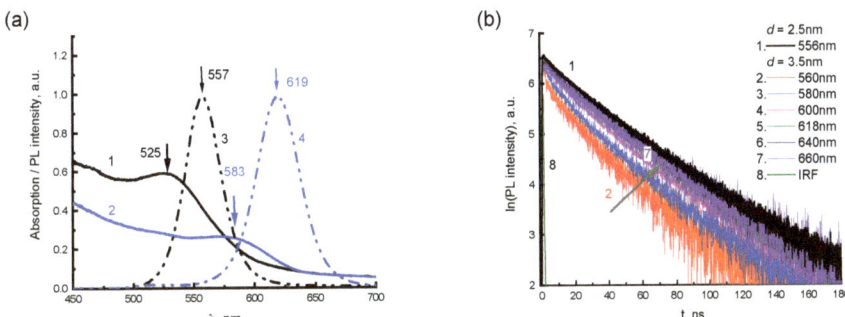

Figure 6. (**a**) Absorption (1,2) and PL (3,4) spectra of aqueous dispersions of CdTe QDs with a diameter of d = 2.5 (1,3) and 3.5 (2,4) nm. (**b**) PL kinetics at various λ_{EM} for aqueous dispersions of CdTe QDs with a diameter of 2.5 (1) and 3.5 (2–7) nm. λ_e = 405 nm. T = 296 K.

Measurements of time-resolved PL spectra for the dispersions of CdTe showed that, for QDs with a diameter of 2.5 nm, the band at 557 nm does not change with delay time. This allows us to confirm that the studied dispersions of CdTe QDs at d = 2.5 nm have a very narrow distribution of their sizes. For CdTe QDs with a diameter of 3.5 nm, the time-resolved PL (TRES) depends on the time delay. At small delay times, in addition to the dominant emission band of 619 nm, an additional shoulder was observed on the short-wavelength side of the PL spectra evidencing broader distribution of sizes for CdTe QDs in the range of 3.5 nm diameter.

The dependence of PL lifetimes for CdTe QDs on their diameter (d) and λ_{EM} is shown in Figure 6b and Table 4. Kinetics of PL decays for the dispersions of CdTe QDs at different λ_{EM} are well described by a three-exponential function: $I_{PL}(t) = A_1 e^{-t/\tau_1} + A_2 e^{-t/\tau_2} + A_3 e^{-t/\tau_3}$. As can be seen from Table 4, the shorter λ_{EM} is, the faster the PL lifetime is for CdTe QDs at d = 3.5 nm. The PL spectra for the 3.5 nm CdTe QDs are dominated by components with lifetimes from 31 to 62 ns.

3.3. Mixing of Excitons in Hybrid Systems of CdTe-2416SL

QDs of CdTe stabilized by thioglycolic acid are negatively charged and can adsorb on the surface positively charged molecules by Coulombic attraction. 2416SL molecules have a positive charge (Figure 1) and can be attached to the surface of CdTe due to electrostatic interaction.

Table 4. PL lifetimes for aqueous dispersions of CdTe QDs depending on their diameter d and λ_{EM}. λ_e = 405 nm. T = 296 K.

CdTe Diameter	λ_{EM}, nm	τ_1, ns	%	τ_2, ns	%	τ_3, ns	%	χ^2
2.5 nm	556	2.3	1.3	20	38.7	42	60	1.355
3.5 nm	560	0.160	2.3	7.7	11	31	66.7	0.936
	580	0.150	1.7	13.7	24.3	40.7	74	1.077
	600	0.180	1.1	17	29	46	69.9	1.230
	618	0.160	1	20	32	48	67	1.065
	640	0.137	0.8	22.6	34.2	49	65	1.072
	660	0.134	0.6	14	9.4	62	90	1.058

After the mixture of CdTe-2416SL at n = 15, a broad absorption band (Figure 7a, curve 3) with a maximum at 504 nm was observed. In Figure 7a, the absorption spectra of the aqueous solutions of neat 2416SL (Figure 7a, curve 1) and initial dispersions of CdTe (Figure 7a, curve 2) are shown as a reference. A part of the formed CdTe-2416SL nanoaggregates gradually precipitate. To analyze the spectral features of the nanoaggregates, absorption spectra of CdTe-2416SL supernatant without precipitate (Figure 7a, curve 3) and freshly mixed CdTe-2416SL with all the nano-aggregations present (Figure 7a, curve 4) were compared. The spectrum of freshly mixed CdTe-2416SL represent a superposition of the absorption of the supernatant and precipitated aggregate. The differential spectrum, Δ, (Figure 7a, curve 5) obtained by subtraction of the supernatant spectrum from freshly mixed CdTe-2416SL spectrum demonstrate spectral features of the CdTe-2416SL nanoaggregates. The differential absorption spectrum, which is characterized by a new absorption band with a maximum at 569 nm, can be associated with light absorption by CdTe-2416SL nanoaggregates. The new absorption band that appears in the hybrid structures of CdTe-2416SL (Figure 7a) can be associated with the hybridization of CTE and WME at the interface of 2416SL and CdTe.

The addition of the CdTe dispersions to the solution of 2416SL resulted in significant quenching of 2416SL emission with the most intense relative quenching of PL in the 550 nm band (Figure 7b). The band with a maximum at 595 nm featuring H-aggregated 2416SL remains in the PL spectrum of CdTe-2416SL mixtures and becomes more prominent upon excitation with λ_e = 510 nm (Figure 7b, curve 5) evidencing efficient absorption responsible for this emission. This behavior of the absorption and PL spectra can be associated with the formation of CdTe-2416SL nanoaggregate structures, as the observed emission is different to the features of H-aggregate and molecular PL of neat 2416SL. Importantly, the PL of CdTe QDs, which have a high quantum yield in the dispersions, was not observed. Strong quenching of QDs emission can be associated with the hybridization of WME in the CdTe QDs with the CTE states at the interface of 2416SL and CdTe and the formation of mixed exciton states—CTE-WME.

PL spectra of CdTe-2416SL nanoaggregates based on CdTe QDs with a diameter of 3.5 and 2.5 nm (Figure 7c,d) are similar featuring intense PL bands with maxima at 550, 594 and 640 nm. The bands are associated with the formation of CdTe-2416SL nanoaggregates. The difference between these spectra can be attributed to the change of spectral positions of excitonic transitions for QDs of various diameters.

Figure 8 shows the PL kinetics of CdTe-2416SL mixtures for two λ_{EM} (550 nm in Figure 8a and 620 nm in Figure 8b) depending on the concentration of CdTe QDs. The measured PL lifetimes for the above wavelengths are shown in Table 5. According to the PL kinetics, the PL lifetime in the 620 nm band does not depend on the CdTe concentration. The PL intensity and lifetime in the 550 nm band decrease with increasing CdTe concentration, which can be associated with the processes of CdTe-2416SL formation and energy transfer in them. Förster resonant energy transfer (FRET) mechanism could be involved in studied CdTe-2416SL nanostructures.

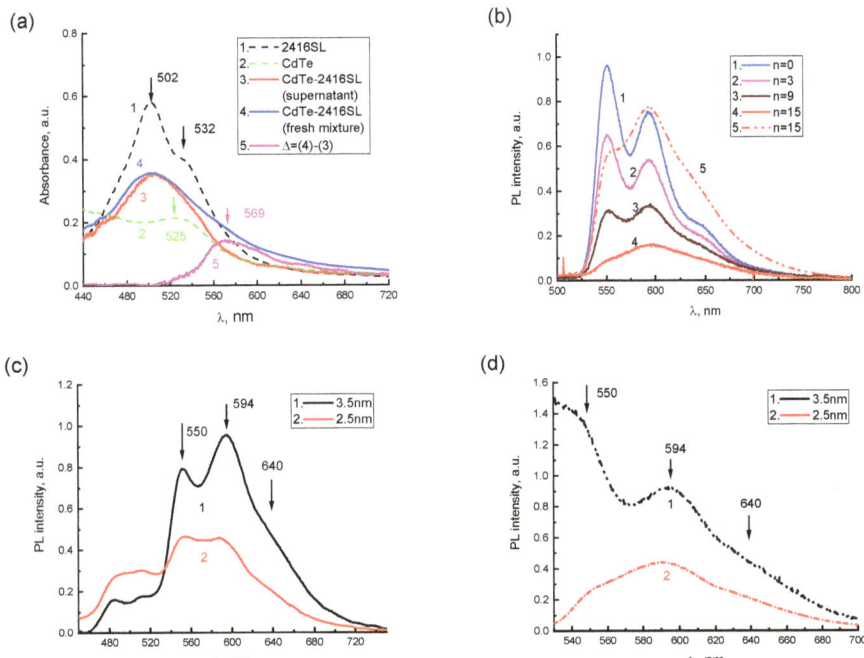

Figure 7. (**a**) Absorption spectra of the aqueous solution of 2416SL ($5 \cdot 10^{-5}$ M) (1), initial dispersion of CdTe QDs ($d = 2.5$ nm) (2), the mixture of CdTe-2416SL at $n = 15$ (3,4) in the form of supernatant (nanoaggregates precipitated) (3) and freshly mixed having nanoaggregates in the dispersion (4). Differential spectrum, Δ, (5) obtained by subtraction of the spectrum (3) from the spectrum (4). (**b**) PL quenching dynamics for the aqueous solution of 2416SL ($5 \cdot 10^{-5}$ M) with fresh admixing of CdTe QDs ($d = 2.5$ nm) dispersions at $n = 0$ (1), $n = 3$ (2), $n = 9$ (3), $n = 15$ (4,5); $\lambda_e = 385$ (1–4) and 510 (5) nm. (**c,d**) PL spectra for (supernatant) mixtures of CdTe-2416SL at $n = 8$ for CdTe QDs with $d = 3.5$ nm (1) and 2.5 (2) nm; $\lambda_e = 385$ (**c**) and 510 nm (**d**). T = 296 K.

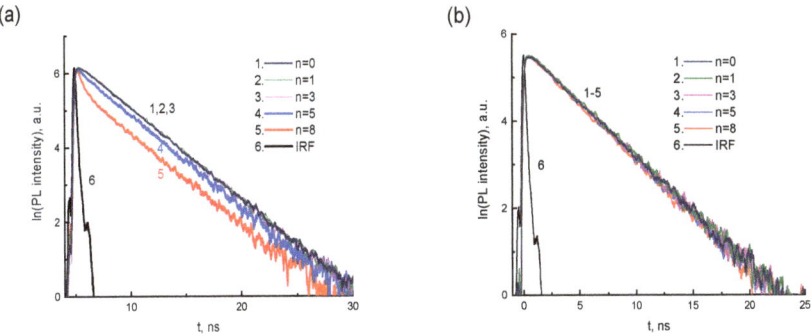

Figure 8. PL kinetics for CdTe-2416SL mixtures at $\lambda_{EM} = 550$ nm (**a**) and 620 nm in (**b**) depending on the concentration of CdTe ($d = 3.5$ nm) QDs (V_n): $n = 0$ (1), $n = 1$ (2), $n = 3$ (3), $n = 5$ (4), $n = 8$ (5); $\lambda_e = 405$ nm. T = 296 K.

Table 5. PL lifetimes for CdTe-2416SL mixtures depending on λ_{EM} and the concentration of CdTe (d = 3.5 nm) QDs (n); λ_e = 405 nm. T = 296 K.

λ_{EM}, nm	n	τ_1, ps	%	τ_2, ps	%	χ^2
550	1, 3			4040		0.967
550	8	350	11	3900	89	1.037
620	1, 3			4450		1.105
620	8			3770		1.037

2416SL-CdTe nanocomposites arise due to the electrostatic interaction between the columnar nanoaggregates of 2416SL and CdTe QDs. A new band with a maximum of 569 nm appears in the absorption spectra of the mixtures (Figure 7a, curve 5), which can be associated with the mixed exciton states—CTE-WME. The band with a maximum at 594 nm in the PL spectra of mixtures can also be attributed to the emission of mixed CTE-WME states. Such states arise due to the close position of CTE of 2416SL and WME in CdTe QDs. Electronic states and resonant energy transfer in hybrid nanostructures containing organic and inorganic semiconductor materials were studied before [35–39]. The high efficiency of non-radiative energy transfer from semiconductor nanostructures (quantum wells/QDs) to organic material with overlapping electronic excitation spectra has been demonstrated [36–39]. The time of energy transfer for WME to organic matter is less than the exciton lifetime in the absence of an organic coating [35–39]. In our case, significant changes of the PL spectra in the emission ranges of 2416SL and CdTe QDs are observed for the mixtures. The intensity and lifetime of PL in the emission band of CdTe-2416SL nanoaggregates (550 nm band) decrease with increasing CdTe concentration, which can be associated with FRET and formation of CdTe-2416SL nanostructures.

For FRET, the rate of energy transfer depends on the degree of overlapping of the PL spectrum of the donor and the absorption spectrum of the acceptor, the mutual orientation of the transition dipole moments, and on the distance R between the interacting molecules [40,41]. As a result of FRET, the fluorescence quantum yield of the donor φ_d and the lifetime of the excited state of the donor τ_d decrease compared to the intrinsic radiation time τ_D, since an additional channel for reducing the population of the excited state of the donor with the migration constant k_m appears. If the donor and acceptor molecules are at a distance $R \neq R_0$ from each other, then the ratio between the characteristic migration time $\tau_m = k_m^{-1}$ and the intrinsic radiation lifetime of the excited state of the donor:

$$\tau_m = \tau_D \cdot (R/R_0)^6$$

where R_0 (Förster radius) is the characteristic distance, at which the probability of FRET is equal to the probability of spontaneous fluorescence of the donor molecule and is determined by the condition $k_m \cdot \tau_D = 1$. In the first case of potential FRET, the 2416SL with absorption of 502 and 532 nm and PL emission peaking at 549 and 591 might be a donor, and the QDs would act as an acceptor. Absorption levels of QDs with d = 2.5 nm are not having much of overlap for the above FRET conditions to be met properly. However, for 2416SL adsorbed on the QDs with a diameter of 3.5 nm, the above conditions for energy transfer are reasonably satisfied. The emission band of 2416SL overlaps with the absorption band of 3.5 nm CdTe QDs having a maximum of about 583 nm. Furthermore, it needs to be admitted that there is practically no PL emission from the QD levels for the mixtures due to formation of CTE-WME states with low quantum yield. Another case of potential FRET process might involve CTE states of 2416SL (absorption maxima at 556 and 589 nm) that could act as acceptors of energy from QDs levels. The photon energy absorbed by CTEs could quickly relax to CPE levels of emission, but due to CTE-WME hybridization the absorbed energy more likely to relax on the CTE-WME levels as these levels have much longer lifetime, and the emission would be observed from these mixed levels. Therefore, the FRET from the QDs to CTE is unlikely process. Overall, the hybrid CdTe-2416SL nanostructures with strong quenched emission of QDs might be applied in dissociative sensing. Such sensors would work by enabled interaction of the perylene dye with an analyte, leading to the dissociation of the nanostructures and an emergence of strong PL emission of the QDs. Current research provides a fundamental understanding for the emission of the hybrid CdTe-2416SL nanostructures and further studies toward sensing would be pursued to provide clear insight into the above applications.

The energy diagram is proposed in Figure 9 to elucidate various types of exciton transitions for 2416SL in the condensed phase and mixed excitons in 2416SL nanocomposites with CdTe QDs. The nature of excitonic transitions has been discussed above, and the diagram is aiming to group the transitions and visualize the complexity of studied transitions. The first group is associated with FE transitions in 2416SL H-aggregates that occur in the solutions and films. In addition to FE, at optical Frank-Condon transitions, it is possible to excite an unrelaxed electron-polaron pair: excitons with charge transfer (CTE_1 and CTE_2). Relaxed molecular-polaron pairs (CPE_1 and CPE_2) appear as intermediate states in the processes of photogeneration and radiative recombination of the CTEs. In the hybrid 2416SL-CdTe nanostructures, the unrelaxed CTE_1 state of the aggregated 2416SL and the WME of CdTe QDs form mixed exciton states. As can be seen from the diagram, the interaction of exciton states leads to the appearance of new levels above the bottom of the exciton zone for CdTe WME and the exciton zone of CTE_1. The mixed exciton transition is indicated on the diagram as CTE_1-WME_1 transition. The emission of hybrid 2416SL-CdTe nanostructures (e.g., 594 nm) strongly overlaps with the emission of 2416SL aggregates and is not included in the diagram. Table 6 summarizes the nature and values of exciton transitions for absorption and PL in the studied systems.

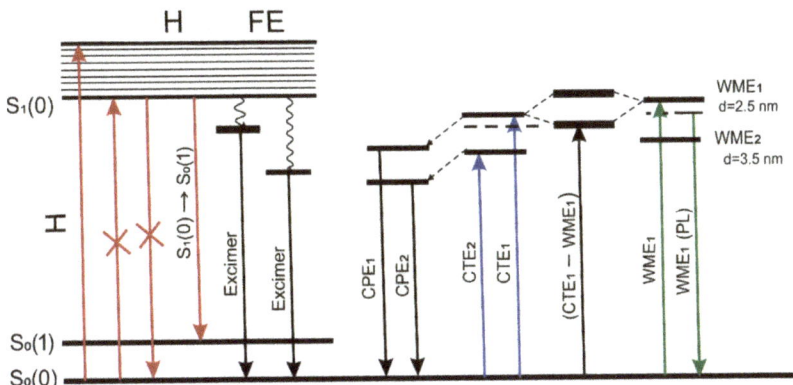

Figure 9. Energy diagram of exciton transitions for 2416SL in the condensed phase and CdTe-2416SL nanoaggregates. More details about each transition are summarized in Table 6.

Table 6. Exciton transitions for 2416SL in the condensed phase and CdTe-2416SL nanoaggregates.

#	Nature of the Transitions		λ, nm	E, eV	Comments
1	$S_0(0) \rightarrow S_1(1)$, FE	Absorption	507	2.44	
2	$S_0(0) \rightarrow S_1(0)$, FE	Absorption	542	2.29	forbidden
3	$S_1(0) \rightarrow S_0(0)$, FE	PL	550	2.26	forbidden
4	$S_1(0) \rightarrow S_0(1)$, FE	PL	594	2.09	
5	Excimer-1, FE	PL	589	2.11	
6	Excimer-2, FE	PL	636	1.95	
7	CTE_1	Absorption	556	2.23	
8	CTE_2	Absorption	589	2.11	
9	CPE_1	PL	680	1.82	

Table 6. *Cont.*

#	Nature of the Transitions		λ, nm	E, eV	Comments
10	CPE$_2$	PL	750	1.65	
11	WME$_1$	Absorption	525	2.36	d = 2.5 nm
12	WME$_2$	Absorption	583	2.13	d = 3.5 nm
13	WME$_1$	PL	557	2.23	d = 2.5 nm
14	WME$_2$	PL	619	2.00	d = 3.5 nm
15	CTE$_1$–WME$_1$	Absorption	569	2.18	

4. Conclusions

Comprehensive studies of the optical properties of water-soluble perylene derivative, 2416SL, were carried out using absorption spectroscopy and techniques of steady-state and picosecond time-resolved PL. Comparison of the absorption and PL spectra of 2416SL in solutions and films revealed the nature of the molecular and aggregated excitonic states. Spectral and lifetime analysis allowed us to identify optical bands of FE, CTE and mixed FE-CTE states. Furthermore, the pathways of non-radiative and radiative relaxation of the indicated collective electronic excitations were determined in steady-state and picosecond time-resolved PL spectra. The emission of excimer and localized polaron states with charge transfer has been identified.

In the aqueous mixtures of 2416SL and CdTe QDs, the aggregation of 2416SL molecules on the surface of CdTe results in the hybridization of CTE and WME and the formation of mixed CTE-WME states. The new absorption and PL bands that appear in the mixtures of CdTe-2416SL as well as strong quenching of QDs emission are associated with such hybridization. FRET from the dye to the CdTe QDs in CdTe-2416SL nanostructures has been analyzed proving its feasibility for hybrid nanostructures made of CdTe QDs of 3.5 nm in diameter. The energy diagram of possible exciton transitions leading to mixed excitons in CdTe-2416SL nanocomposites is proposed to help in understanding the nature of excitonic transitions. Learning more about the fundamental nature of mixed excitons at the interface of organic and inorganic nanostructures makes us a step closer to the application of excitonic elements in molecular electronics and optoelectronics.

Author Contributions: Conceptualization, Y.P.P., A.B.V., A.G.R. and P.M.L.; Methodology, Y.P.P., O.A.K. and P.M.L.; Validation, Y.P.P., M.M.S., A.B.V. and A.G.R.; Formal analysis, Y.P.P., M.B.M., A.B.V., A.G.R. and P.M.L.; Investigation, Y.P.P., M.B.M., M.M.S. and O.A.K.; Resources, O.A.K.; Data curation, Y.P.P., M.B.M. and M.M.S.; Writing—original draft, Y.P.P., A.B.V., O.A.K. and P.M.L.; Writing—review & editing, Y.P.P. and P.M.L.; Visualization, M.B.M., M.M.S. and P.M.L.; Supervision, A.B.V. and A.G.R. All authors have read and agreed to the published version of the manuscript.

Funding: This research received no external funding.

Institutional Review Board Statement: Not applicable.

Informed Consent Statement: Not applicable.

Data Availability Statement: The data presented in this study are available on request from the corresponding author.

Conflicts of Interest: The authors declare no conflict of interest.

References

1. Ma, S.; Du, S.; Pan, G.; Dai, S.; Xu, B.; Tian, W. Organic molecular aggregates: From aggregation structure to emission property. *Aggregate* **2021**, *2*, e96. [CrossRef]
2. De Sio, A.; Sommer, E.; Nguyen, X.T.; Groß, L.; Popović, D.; Nebgen, B.T.; Fernandez-Alberti, S.; Pittalis, S.; Rozzi, C.A.; Molinari, E.; et al. Intermolecular conical intersections in molecular aggregates. *Nat. Nanotechnol.* **2021**, *16*, 63–68. [CrossRef] [PubMed]
3. Hoffmann, M.; Hasche, T.; Schmidt, K.; Canzler, T.W.; Agranovich, V.M.; Leo, K. Excitons in quasi-one-dimensional crystalline perylene derivatives: Band structure and relaxation dynamics. *Int. J. Mod. Phys. B* **2001**, *15*, 3597–3600. [CrossRef]

4. Cao, J.; Yang, S. Progress in perylene diimides for organic solar cell applications. *RSC Adv.* **2022**, *12*, 6966–6973. [CrossRef] [PubMed]
5. Chaudhry, M.U.; Muhieddine, K.; Wawrzinek, R.; Sobus, J.; Tandy, K.; Lo, S.-C.; Namdas, E.B. Organic light-emitting transistors: Advances and perspectives. *Adv. Funct. Mater.* **2020**, *30*, 1905282. [CrossRef]
6. Bulovic, V.; Burrows., P.E.; Forrest, S.R.; Cronin, J.A.; Thompson, M.E. Study of localized and extended excitons in 3,4,9,10-perylenetetracarboxylic dianhydride (PTCDA) I. Spectroscopic properties of thin films and solutions. *Chem. Phys.* **1996**, *210*, 1–12. [CrossRef]
7. Hoffmann, M.; Schmidt, K.; Fritz, T.; Hasche, T.; Agranovich, V.M.; Leo, K. The lowest energy Frenkel and charge-transfer excitons in quasi-one-dimensional structures: Application to MePTCDI and PTCDA crystals. *Chem. Phys.* **2000**, *258*, 73–96. [CrossRef]
8. Lee, T.; Enomoto, K.; Ohshiro, K.; Inoue, D.; Kikitsu, T.; Hyeon-Deuk, K.; Pu, Y.-J. Controlling the dimension of the quantum resonance in CdTe quantum dot superlattices fabricated via layer-by-layer assembly. *Nat. Commun.* **2020**, *11*, 5471. [CrossRef]
9. Korbutyak, D.V.; Kalytchuk, S.M.; Geru, I.I. Colloidal CdTe and CdSe quantum dots: Technology of preparing and optical properties. *J. Nanoelectron. Optoelectron.* **2009**, *4*, 174–179. [CrossRef]
10. Kapush, O.A.; Trishchuk, L.I.; Tomashik, V.N.; Tomashik, Z.F. Effect of thioglycolic acid on the stability and photoluminescence properties of colloidal solutions of CdTe nanocrystals. *Inorg. Mater.* **2014**, *50*, 13–18. [CrossRef]
11. Davydov, A.S. *Theory of Molecular Excitons*; Springer: Berlin/Heidelberg, Germany, 2013; p. 313.
12. Barashkov, N.N.; Sakhno, T.V.; Nurmukhametov, R.N.; Khakhel, O.A. Excimers of organic molecules. *Russ. Chem. Rev.* **1993**, *62*, 579–593. [CrossRef]
13. Agranovich, V.M.; Hochstrasser, R.M. *Spectroscopy and Excitation Dynamics of Condensed Molecular Systems (Modern Problems in Condensed Matter Sciences)*; Elsevier Science: Amsterdam, The Netherlands, 1983.
14. Silinsh, E.A.; Capek, V. *Organic Molecular Crystals. Interaction, Localization, and Transport Phenomena*, 1st ed.; American Institute of Physics: New York, NY, USA, 1994.
15. Silinsh, A.; Kurik, M.V.; Capek, V. *Electronic Processes in Organic Molecular Crystals. Localization and Polarization Phenomena*; Zinatne: Riga, Latvia, 1988. (In Russian)
16. Pope, M.; Swenberg, C.E. *Electronic Processes in Organic Crystals and Polymers*, 2nd ed.; Oxford University Press: Oxford, UK, 1999.
17. Agranovich, V.M.; La Rocca, G.C.; Bassani, F.; Benisty, H.; Weisbuch, C. Hybrid Frenkel-Wannier-Mott excitons at interfaces and in microcavities. *Opt. Mater.* **1998**, *9*, 430–436. [CrossRef]
18. Boiko, O.; Komarov, O.; Vasyuta, R.; Nazarenko, V.; Slominskiy, Y.; Schneider, T. Nano-architecture of self-assembled monolayer and multilayer stacks of lyotropic chromonic liquid crystalline dyes. *Mol. Cryst. Liq. Cryst.* **2005**, *434*, 305/[633]–314/[642]. [CrossRef]
19. Lydon, J. Chromonic mesophases. *Curr. Opin. Colloid Interface Sci.* **2004**, *8*, 480–490. [CrossRef]
20. Camorani, P.; Furier, M.; Kachkovskii, O.; Piryatinskiy, Y.; Slominskii, Y.; Nazarenko, V. Absorption spectra and chromonic phase in aqueous solution of perylenetetracarboxylic bisimides derivatives. *Semicond. Phys. Quantum Electron.* **2001**, *4*, 229–238. [CrossRef]
21. Nazarenko, V.G.; Boiko, O.P.; Anisimov, M.I.; Kadashchuk, A.K.; Nastishin, Y.A.; Golovin, A.B.; Lavrentovich, O.D. Lyotropic chromonic liquid crystal semiconductors for water-solution processable organic electronics. *Appl. Phys. Lett.* **2010**, *97*, 263305. [CrossRef]
22. Winnik, F.M. Fluorescence studies of aqueous solutions of poly(N-isopropylacrylamide) below and above their LCST. *Macromolecules* **1990**, *23*, 233–242. [CrossRef]
23. Czikkely, V.; Försterling, H.D.; Kuhn, H. Light absorption and structure of aggregates of dye molecules. *Chem. Phys. Lett.* **1970**, *6*, 11–14. [CrossRef]
24. Spano, F.C. The fundamental photophysics of conjugated oligomer herringbone aggregates. *J. Chem. Phys.* **2003**, *118*, 981–994. [CrossRef]
25. Spano, F.C. Analysis of the UV/Vis and CD spectral line shapes of carotenoid assemblies: Spectral signatures of chiral H-aggregates. *J. Am. Chem. Soc.* **2009**, *131*, 4267–4278. [CrossRef]
26. Spano, F.C. The spectral signatures of Frenkel polarons in H- and J-aggregates. *Acc. Chem. Res.* **2010**, *43*, 429–439. [CrossRef]
27. Varghese, S.; Das, S. Role of molecular packing in determining solid-state optical properties of π-conjugated materials. *J. Phys. Chem. Lett.* **2011**, *2*, 863–873. [CrossRef] [PubMed]
28. Zhao, Z.; Spano, F.C. Vibronic fine structure in the absorption spectrum of oligothiophene thin films. *J. Chem. Phys.* **2005**, *122*, 114701. [CrossRef]
29. Vertsimakha, Y.; Lutsyk, P.; Palewska, K.; Sworakowski, J.; Lytvyn, O. Optical and photovoltaic properties of thin films of N,N′-dimethyl-3,4,9,10-perylenetetracarboxylic acid diimide. *Thin Solid Films* **2007**, *515*, 7950–7957. [CrossRef]
30. Piryatinski, Y.P. Vlijanie smeshyvaniya eksitonnykh I ionizirovannykh sostoyanij na spektry pogloshcheniya I fotoprovodimosti kristalov pentacena. *Fizika Tverdogo Tela* **1989**, *31*, 208–219. (In Russian)
31. Petelenz, P. Mixing of frenkel excitons and ionic excited states of a linear molecular crystal with two molecules in the unit cell. II. Physical consequences. *Phys. Stat. Sol. B* **1977**, *79*, 61–70. [CrossRef]
32. Agranovich, V.M.; Kamchatnov, A.M. Quantum confinement and superradiance of one-dimensional self-trapped Frenkel excitons. *Chem. Phys.* **1999**, *245*, 175–184. [CrossRef]

33. Agranovich, V.M.; Schmidt, K.; Leo, K. Surface states in molecular chains with strong mixing of Frenkel and charge-transfer excitons. *Chem. Phys. Lett.* **2000**, *325*, 308–316. [CrossRef]
34. Yu, W.W.; Qu, L.; Guo, W.; Peng, X. Experimental determination of the extinction coefficient of CdTe, CdSe, and CdS nanocrystals. *Chem. Mater.* **2003**, *15*, 2854–2860. [CrossRef]
35. Agranovich, V.M.; Basko, D.M.; La Rocca, G.C.; Bassani, F. Excitons and optical nonlinearities in hybrid organic-inorganic nanostructures. *J. Phys. Condens. Matter* **1998**, *10*, 9369–9400. [CrossRef]
36. Agranovich, V.M.; La Rocca, G.C.; Bassani, F. Efficient electronic energy transfer from a semiconductor quantum well to an organic material. *JETP Lett.* **1997**, *66*, 748–751. [CrossRef]
37. Basko, D.; La Rocca, G.; Bassani, F.; Agranovich, V.M. Förster energy transfer from a semiconductor quantum well to an organic material overlayer. *Eur. Phys. J. B* **1999**, *8*, 353–362. [CrossRef]
38. Agranovich, V.M.; Basko, D.M. Resonance energy transfer from a semiconductor quantum dot to an organic matrix. *JETP Lett.* **1999**, *69*, 250–254. [CrossRef]
39. Basko, D.M.; Bassani, F.; La Rocca, G.C.; Agranovich, V.M. Electronic energy transfer in a microcavity. *Phys. Rev. B* **2000**, *62*, 15962. [CrossRef]
40. Förster, T. Transfer mechanisms of electronic excitation energy. *Radiat. Res. Suppl.* **1960**, *2*, 326–339. [CrossRef]
41. Lakowicz, J.R. *Principles of Fluorescence Spectroscopy*, 3rd ed.; Springer: New York, NY, USA, 2006. [CrossRef]

Disclaimer/Publisher's Note: The statements, opinions and data contained in all publications are solely those of the individual author(s) and contributor(s) and not of MDPI and/or the editor(s). MDPI and/or the editor(s) disclaim responsibility for any injury to people or property resulting from any ideas, methods, instructions or products referred to in the content.

Article

Near-Infrared (NIR) Silver Sulfide (Ag₂S) Semiconductor Photocatalyst Film for Degradation of Methylene Blue Solution

Zahrah Ramadlan Mubarokah [1], Norsuria Mahmed [1,2,*], Mohd Natashah Norizan [2,3], Ili Salwani Mohamad [2,3], Mohd Mustafa Al Bakri Abdullah [1,2], Katarzyna Błoch [4], Marcin Nabiałek [4], Madalina Simona Baltatu [5,*], Andrei Victor Sandu [5,6,7] and Petrica Vizureanu [5,8]

1. Faculty of Chemical Engineering & Technology, Universiti Malaysia Perlis (UniMAP), Arau 01000, Malaysia
2. Centre of Excellence Geopolymer and Green Technology (CEGeoGTech), Universiti Malaysia Perlis (UniMAP), Arau 01000, Malaysia
3. Faculty of Electronic Engineering & Technology, Universiti Malaysia Perlis (UniMAP), Arau 02600, Malaysia
4. Faculty of Mechanical Engineering and Computer Science, Częstochowa University of Technology, 42-201 Częstochowa, Poland
5. Department of Technologies and Equipments for Materials Processing, Faculty of Materials Science and Engineering, Gheorghe Asachi Technical University of Iași, Blvd. Mangeron, No. 51, 700050 Iasi, Romania
6. National Institute for Research and Development in Environmental Protection INCDPM, Splaiul Independentei 294, 060031 Bucharest, Romania
7. Romanian Inventors Forum, Str. Sf. P. Movila 3, 700089 Iasi, Romania
8. Technical Sciences Academy of Romania, Dacia Blvd 26, 030167 Bucharest, Romania
* Correspondence: norsuria@unimap.edu.my (N.M.); cercel.msimona@yahoo.com (M.S.B.)

Abstract: A silver sulfide (Ag₂S) semiconductor photocatalyst film has been successfully synthesized using a solution casting method. To produce the photocatalyst films, two types of Ag₂S powder were used: a commercialized and synthesized powder. For the commercialized powder (CF/comAg₂S), the Ag₂S underwent a rarefaction process to reduce its crystallite size from 52 nm to 10 nm, followed by incorporation into microcrystalline cellulose using a solution casting method under the presence of an alkaline/urea solution. A similar process was applied to the synthesized Ag₂S powder (CF/syntAg₂S), resulting from the co-precipitation process of silver nitrate (AgNO₃) and thiourea. The prepared photocatalyst films and their photocatalytic efficiency were characterized by Fourier transform infrared spectroscopy (FTIR), X-ray diffraction (XRD), and UV-visible spectroscopy (UV-Vis). The results showed that the incorporation of the Ag₂S powder into the cellulose films could reduce the peak intensity of the oxygen-containing functional group, which indicated the formation of a composite film. The study of the crystal structure confirmed that all of the as-prepared samples featured a monoclinic acanthite Ag₂S structure with space group P_{21}/C. It was found that the degradation rate of the methylene blue dye reached 100% within 2 h under sunlight exposure when using CF/comAg₂S and 98.6% for the CF/syntAg₂S photocatalyst film, and only 48.1% for the bare Ag₂S powder. For the non-exposure sunlight samples, the degradation rate of only 33–35% indicated the importance of the semiconductor near-infrared (NIR) Ag₂S photocatalyst used.

Keywords: near-infrared irradiation; silver sulfide; cellulose film; photocatalysis; methylene blue

1. Introduction

Since the prehistoric age, humans have used dyes from natural resources for coloring purposes. The lengthy process, poor colorfastness, excessive cost of producing natural dyes, and the increasing demand for textiles led to the discovery of synthetic dyes from petroleum compounds that outperformed the properties of natural dyes [1–3]. Statistically, more than 7×10^5 tons of synthetic dyes are produced worldwide, and about 1×10^4 of them are used in industry [4]. However, these synthetic dyes have a negative impact on the environment, such as water pollution. Textile industries discharge around 35% of dye wastewater into water bodies as effluent annually [5]. These dyes can contaminate the aquatic habitat, which may enter the food chain [6,7]. Methylene blue (MB), or $C_{16}H_{18}N_3SCl$, is one of the synthetic dye materials used in textile industries. When dissolved in water, it results in the formation of a blue-colored solution. The discharging of MB dye into the environment is considered as a significant threat for aesthetical and toxicological reasons. In terms of aquatic life, its existence in the water could reduce sunlight transmittance and decrease oxygen solubility due to the high molar absorption coefficient (~8.4×104 L·mol^{-1}·cm^{-1} at 664 nm) of MB dyes [8–10]. At a certain concentration, it can cause serious threats to human health, including respiratory distress, abdominal disorders, blindness, and digestive and mental disorders [11].

In recent years, tremendous attempts have been made to improve industrial wastewater treatment methods. These methods can be divided into physical [12–16], biological [17–21], and chemical methods. For the chemical methods, there are several conventional chemical dye-removal processes that have been used, e.g., the coagulant flocculation, electrochemical destruction, advanced oxidation process (AOP), etc. [22–25]. Amongst these, the advanced oxidation processes (AOPs) have received the most attention due to their significant competence in acting toward a wide range of organic or inorganic dye pollutants in the aqueous phase, by converting those pollutants into stable inorganic compounds such as water, carbon dioxide, and salt without any footprint and sludge production [26,27]. The AOPs also rely on the in situ production of greatly reactive hydroxyl radicals (OH•) [28].

In the midst of the AOPs, heterogeneous photocatalysis has been considered a leading method and desired breakthrough to cope with organic contaminants. Photocatalysis is a reaction that involves light (photoreaction) and accelerates the reaction due to the presence of a catalyst that absorbs light energy to form the reducing and oxidizing (electron–hole pairs) ions on the surface of the catalyst. The electron's handover occurs during the oxidation–reduction process [29]. The reduction of an acceptor occurs when the electrons (e^-) combine with oxygen in the water to generate an anion (O_2^-), which oxidizes the hydroxyl radical (OH•), while the hole (h^+) will oxidize the dissolved hydroxyl and convert it into a radical with great energy [30]. These processes work simultaneously. The photocatalytic properties of various metal oxides (e.g., titania, TiO_2, and zinc oxide, ZnO) have been extensively studied. The results of these studies show that the metal oxides become active potential photocatalysts in wastewater treatment [31–34]. However, due to a wide bandgap, they can only absorb the ultraviolet (UV) light and show poor performance in the visible–near-infrared (NIR) range of the solar spectrum [29,35–37]. In addition, the solar spectrum itself is dominated by the visible (46%) and NIR spectral (49%) rather than the UV spectral (5%) [38]. In addition, the manufacturing cost is relatively high for many of the metal oxide materials [39,40]. Therefore, metal sulfides have become an option.

Among all of the metal sulfide photocatalyst materials, silver sulfide (Ag_2S) shows more efficient charge separation, attributed to the remarkable synergistic effects of strong NIR light absorption and excellent surface properties [41,42]. Furthermore, compared to the other metal sulfide compounds such as cadmium selenide, CdS, and lead selenide, PbS, Ag_2S is found to be acceptable for wastewater treatment due to its non-toxic properties [43–45]. Silver sulfide has an eminent performance in the degradation of pollutants, solar energy conversion, and production of hydrogen with various approaches that have been well established for synthesizing a variety of visible–NIR-light driven Ag_2S photocatalysts. For example, Ag_2S synthesized using the facile ion-exchange method at room temperature can be used as an effective photocatalyst for the decomposition of methylene orange (MO). The results of a study confirmed the excellent photo-oxidation performance of Ag_2S since MO can be completely photodegraded with Ag_2S in just 30 min, and 70 min under visible light and NIR light irradiation. This performance is due to the narrow band gap of Ag_2S, 1.078 eV, and the lower recombination efficiency and photogenerated electron–hole pairs of Ag_2S during the photocatalytic process [46]. The Ag_2S photocatalyst has usually been used in the powder form [43,47–49]. This is because the photocatalyst powder can be well dispersed in suspensions [50].

However, this photocatalyst powder exhibits certain drawbacks especially for nano-sized powders/particles. During the degradation process, the photocatalyst may undergo coagulation due to the instability of the nano-sized particles, which will hamper the light incidence on the active centers, consequently reducing its catalytic activity [51]. Furthermore, for the slurry system, the main challenge is to recover the nano-sized photocatalyst particles from the treated water [52–54]. This condition leads to the requirements of high filtration costs of catalyst removal [55], hindering its industrial application and impractical. In addition, the Ag_2S nanoparticles are susceptible to photo-corrosion, which means that the sulfide ions have the potential to be oxidized into sulfur by photogenerated holes when they are exposed to irradiation [56]. There have been a lot of efforts towards increasing the photocatalytic efficiency and preventing the photo-corrosion of Ag_2S [57–60]. A photocatalyst film is one of the plausible material technologies to diminish the impacts of photocatalyst powder on the environment [61]. A photocatalyst film has promising performance in terms of adsorption and efficient mass transport. The adsorption ability is maxed out by the larger surface area that allows more catalyst deposit in the film, resulting in higher photocatalytic activity. Moreover, the mass transport of reactants, intermediates, and products is enhanced by the compact substrate that allows better contact with the catalyst, and prevents the leakage of the powder of the photocatalyst material into clean water [62,63]. Table 1 shows the cost comparison for some types of photocatalyst, especially titania (TiO_2), which is a common photocatalyst used for industrial purposes. Those types of photocatalyst are used under a UV lamp-assisted reactor for the photocatalytic processes and consume a certain amount of energy (power). For an industrial batch process, it will consume a lot of energy with a higher cost of operation. Thus, our study introduces a composite film (cellulose/Ag_2S) that can effectively degrade methylene blue concentrations using only sunlight, which is cost-effective. Furthermore, the composite film can also be easily removed from the water after the treatment process without a costly separation method.

Table 1. Comparison of the cost-effectiveness assessment of several photocatalysts.

Type of Photocatalyst	Types of Wastewater	Energy Sources	Catalyst Dosage (g/L)	Maximum Photodegradation Efficiency	Time to Obtain Maximum Photodegradation (min.)	Total Power Consumed (kWh)	Operational Cost (USD/kg) [b]	Refs.
Rutile TiO_2	Acetaldehyde	UV lamps (56 W)	5.01	80	100	not available	127.13	[64]
GAC^a–TiO_2	Livestock wastewater	UV lamps (56 W)	6	100	6	0.524	0.68	[65]
V_2O_5–TiO_2	Methylene blue	UV lamps	4.75	92	120	1	8.7205	[66]
TiO_2/O_3	Tert-butyl alcohol	UVA lamps (15W)	not available	75	10	37	not available	[67]
Cellulose/Ag_2S	Methylene blue	Solar energy	7	100	120	0	0	Current study

[a] Granular activated carbon. [b] Energy costs (USD) per kg of wastewater removal.

To date, bio-based products such as cellulose film have become attractive compounds due to their excellent properties that are harmless to the environment. A cellulose film has distinctive characteristics such as transparency, robustness, low water content, etc. These properties lead to the wide usage of cellulose film as a renewable alternative to petroleum-based materials [68]. Furthermore, it is hypothesized that cellulose may become one of the solutions to improve the efficiency of electron donors and to establish the hole (h^+) transporter during the light irradiation process. To the best of our knowledge, the incorporation of Ag_2S powder into cellulose film has not been studied, although the integration between semiconductor catalyst and cellulose film is a promising innovation for photocatalytic applications. Therefore, the integration of Ag_2S in cellulose-derived film is hoped to enhance the photocatalytic efficiency. Thus, this study aimed to synthesize cellulose/Ag_2S film in the presence of an alkali/urea solution using a simple solution casting method to describe the transformation of the functional group and the structure of the cellulose thin film due to the deposition of Ag_2S. We also compared the photocatalytic efficiency of the commercial Ag_2S and synthesized Ag_2S doped into the cellulose film via the degradation of the MB solution.

2. Materials and Methods

The current study successfully synthesized films that were stated as CFs for cellulose film, cellulose films with commercial Ag_2S (CF/comAg_2S), and cellulose films with synthesized Ag_2S (CF/syntAg_2S). The materials and methods used for synthesizing the Ag_2S photocatalyst film are described in the section below.

2.1. Materials and Reagents

The materials and reagents used consist of microcrystalline cellulose, MCC ($\leq$100%), commercial silver sulfide, Ag_2S, silver nitrate, $AgNO_3$, thiourea, $(NH_2)_2CS$, and polyvinyl alcohol, PVA. All of these materials were bought from Sigma-Aldrich. The PVA was dissolved with distilled water at the temperature of 90 °C until the transparent PVA solution with a concentration of 5.0% (w/w) occurred. Chemicals including sodium hydroxide (NaOH), glycerol ($C_3H_8O_3$), and acetone (C_3H_6O) were supplied by HmbG chemicals. The acetone was diluted in distilled water with a ratio of 2:1 as the agent of regeneration for cellulose films, called an acetone bath. Urea ($CO(NH_2)_2$) was obtained from Bendosen Laboratory Chemicals.

2.2. Preparation of Sodium Hydroxide/Urea (NaOH/urea) Aqueous Solution

To obtain the NaOH/urea solution, the NaOH pellets and urea powder were weighed to obtain 7 g and 12 g of mass, respectively [47,69]. Subsequently, each sample of NaOH and urea was diluted in a separate beaker glass before being mixed in a 100 mL volumetric flask. The balanced chemical reaction worked in the mixture can be written as:

$$2NaOH(aq) + CO(NH_2)_2(aq) \rightarrow 2NH_3(g) + Na_2CO_3(aq) \tag{1}$$

The solution was agitated to obtain a homogenous solution. Once the homogeneity was obtained, the solution was stored for the next process.

2.3. Liquid Phase Rarefaction of Commercial Ag_2S by Magnetic Stirring Technique

Prior to the incorporation of the commercial Ag_2S into the cellulose film, it was essential to break down the bulk Ag_2S into smaller particle sizes. In this step, the commercial Ag_2S powder was weighed according to the mass variation that is displayed in Table 2.

Table 2. Sample code and composition of refracted commercial Ag_2S.

Sample Code	Ag_2S Mass (Gram)	Volume of Distilled Water (mL)
CF/comAg$_2$S1	0.075	6.0
CF/comAg$_2$S2	0.100	6.0
CF/comAg$_2$S3	0.300	6.0

After the desired amount of Ag_2S was obtained, the powder was placed into a beaker containing 6 mL of distilled water and vigorously stirred using a magnetic stirrer under ambient temperature for 24 h. The crystallite size of rarefaction Ag_2S was then investigated using X-ray diffraction (XRD) to confirm its size reduction.

2.4. Synthesis of the Cellulose Film

The cellulose films were prepared using a solution casting method. The cellulose with NaOH/urea aqueous solution was used as the main reagent to synthesize the cellulose films. Both commercial Ag_2S and synthesized Ag_2S were incorporated into the cellulose system to study the different behavior of the cellulose film with commercial and synthesized Ag_2S. The composition of NaOH, urea, and water was different for each process as shown in Table 3. Figure 1 shows the overall schematic procedure to synthesize the CFs, CF/comAg$_2$S, and CF/syntAg$_2$S samples.

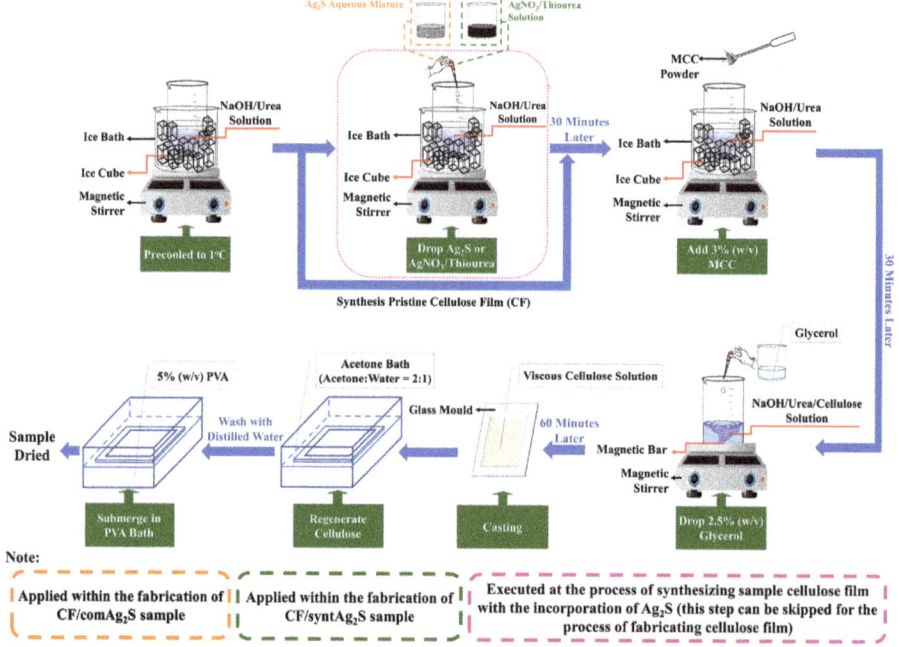

Figure 1. Schematic illustration of sequence process to synthesize the samples.

2.4.1. Synthesis Cellulose Film (CF) and Cellulose Film/Commercial Ag_2S (CF/comAg$_2$S)

At the beginning of the process, 15 mL of NaOH/urea aqueous solution was precooled to a temperature of ~1 °C in the ice bath. After that, 3 wt% of microcrystalline cellulose (MCC) was added into the precooled solvent, and the mixture was rapidly agitated for 20 min. Then, the cellulose solution was taken out from the ice bath and 2.5 wt% of glycerol [70] was wisely dropped into the solution. Afterwards, the solution was spun for another hour to achieve the homogeneity. Once the dope and viscous solution was obtained, it was cast onto an 8.5 cm × 6 cm × 0.2 cm glass mold. Subsequently, the casted sample was regenerated in an acetone coagulant bath [71] until opaque-sheet-like cellulose hydrogel was formed. The hydrogel was soaked in distilled water for neutralization. Before drying, the hydrogel was submerged in a 5 wt% polyvinyl alcohol (PVA) bath for 2 h [72]. Finally, a transparent cellulose film was obtained. The same steps were also conducted in order to synthesize the CF/comAg$_2$S specimens, except the 6mL of refracted Ag$_2$S produced from Section 2.3 was first poured into the NaOH/urea solution prior to the addition of the desired amount of MCC into the precooled solution.

2.4.2. Synthesis of Cellulose Film/Synthesized Ag$_2$S (CF/syntAg$_2$S)

The co-precipitation method was used to synthesize the Ag$_2$S particles. Silver nitrate and thiourea were the main precursors to produce the Ag$_2$S slurry. The presence of silver and sulfide ions in the solution was required for the chemical deposition process to occur in order to complete the formation of Ag$_2$S. The chemical reaction involved during the deposition process is shown in the following formulas [73,74]:

$$(NH_2)_2CS + H_2O \leftrightarrow (NH_2)_2CO + H_2S \qquad (2)$$

$$2AgNO_3 + H_2S \rightarrow Ag_2S + 2HNO_3 \qquad (3)$$

Table 3 shows the final composition that was utilized in this section for the whole synthesis process:

Table 3. Sample code and sample composition to synthesize samples CF/syntAg$_2$S.

Sample Code	Molarity of AgNO$_3$	AgNO$_3$ Solution		Thiourea Solution			
		AgNO$_3$	Water	NaOH	Thiourea	Urea	Water
CF/syntAg$_2$S1	0.1 M	0.084 g	5.0 mL	1.6 g	1.3 g	1.6 g	10.5 mL
CF/syntAg$_2$S2	0.3 M	0.254 g	5.0 mL	1.6 g	1.3 g	1.6 g	10.5 mL
CF/syntAg$_2$S3	0.5 M	0.422 g	5.0 mL	1.6 g	1.3 g	1.6 g	10.5 mL

According to Table 3, the total volume of solution that was used in the synthesis of CF/syntAg$_2$S samples was 20 mL (5 mL AgNO$_3$ solution and 15 mL thiourea solution). The solutions were prepared in two different beakers. At the beginning of the process, 15 mL of thiourea solution was poured into a 50 mL glass beaker while stirring on a magnetic stirrer. After that, the 5 mL of AgNO$_3$ solution was added dropwise into the stirred thiourea solution. At this stage, a black precipitate was obtained, which indicated the formation of Ag$_2$S particles. The solution was left to stir for another 15 min to ensure all of the substances completely reacted. Then, the precipitated solution was precooled in the ice bath for 20 min. Subsequently, 3% w/v of MCC powder was added into the black cold-precipitated solution and continuously stirred until the MCC was well dispersed in the solution. The mixed solution was then removed from the ice bath, dripped with glycerol, and spun for an entire hour. The cellulose/synthesized AgNO$_3$ solution was transferred into the glass mold and then regenerated by using an acetone bath and PVA bath, in sequence. Lastly, the cellulose film obtained after the regenerated film was open-air dried for 2 days. All of the procedures above were repeated for different molarities of synthesized AgNO$_3$.

2.5. Photocatalytic Activity Test

For photocatalytic testing, a concentration of 10 ppm of methylene blue (MB) solution was prepared. An amount of 0.1 g of MB powder was diluted with distilled water in a 100 mL volumetric flask. Afterwards, 0.7 g of each sample was submerged into a glass beaker consisting of 70 mL of 10 ppm MB and immediately exposed to sunlight for a total of 300 min. Every 30 min, about 40 mL of the irradiated solution was transferred into a glass vial with a dropper to investigate its degradation behavior using an ultraviolet-visible

(UV-Vis) spectrophotometer, (Lambda 25, Perkin Elmer, Waltham, MA, USA) with 650 nm of wavelength. The degradation ratio of MB was determined using Equation (4); where A_t is the degradation ratio, I_0 is base absorbance, and I_t is absorbency after time t [33,73,75].

$$A_t = \frac{I_0 - I_t}{I_0} \times 100 \qquad (4)$$

2.6. Characterizations

To identify chemical bonds and to locate the functional group of the cellulose films, the samples were characterized using a Fourier-transform infrared (FTIR) analysis. The analysis was carried out using an ATR-Perkin Elmer Spectrometer 2000 FT-IR, based on the attenuated total reflection (ATR) approach to ensure non-destructive analysis. The measurement was conducted under 650 cm^{-1}–4000 cm^{-1} and 4 cm^{-1} resolution. The phase determination and crystallite size measurement were conducted using a Rigaku RINT 2000 X-ray diffraction (XRD). The reflection mode with monochromator-filtered Cu Kα radiation (λ = 0.15418 nm) at 30 kV and 10 mA was applied. Samples were prepared in two different ways depending on the type of tested sample. For the powder sample, it was loaded onto the XRD sample holder with a small circle cavity in the middle of the holder. For film samples, the cellulose film was cut into a size of 25 mm, thickness $\leq$ 8 mm, and stuck on the holder. The prepared samples were subsequently scanned with 2theta in the range between 15° and 80°. The Debye-Scherrer Equation (5) was generated to compute the crystallite size and the degree of crystallinity was obtained from Equation (6) as follows:

$$D = \frac{K\lambda}{\beta \cos \theta} \qquad (5)$$

$$\text{Degree of Crystallinity} = \frac{\text{Area of all crystalline peaks}}{\text{Total area under XRD peaks}} \times 100 \qquad (6)$$

where, D = crystallite size (nm); K = Scherrer's constant (K = 0.94); λ = the wavelength of X-ray (1.54178 Å); β = full width half maximum (FWHM); θ = angle of diffraction (rad) [76].

3. Results and Discussion

3.1. The Formation of Cellulose-Photocatalyst Film

In nature, the structure of cellulose includes both axial C–H bonds and equatorial hydroxyl groups. Hence, it has amphiphilic properties with the inter-layer region being hydrophobic and the intra-layer region being hydrophilic. The intra- and intermolecular hydrogen bonds should be broken to a large extent when the cellulose is dissolved in the solvent [77]. Thus, when cellulose is dissolved in the NaOH/urea solution, it is expected that the NaOH interacts with the hydrophilic hydroxyl groups of cellulose, breaking the hydrogen bond between the chains, and the urea reacts with the hydrophobic portions, resulting in a solution that dissolves cellulose [78].

In this study, it was also discovered that the semi-crystalline properties of the cellulose can impact the end-product of the prepared films. During the drying process, the cellulose films will experience severe shrinkage [72,79,80]. Nevertheless, when 2.5% w/v glycerol and 5% PVA bath was applied, the shrinkage of the film samples became less. This is because the existence of PVA as a plasticizer in the film can improve the toughness of the cellulose glycerol composite films [73]. The commercial Ag$_2$S powder was micron-sized and might affect the distribution of particles when incorporated in the MCC film. Thus, the rarefaction process was conducted as shown in Figure 2.

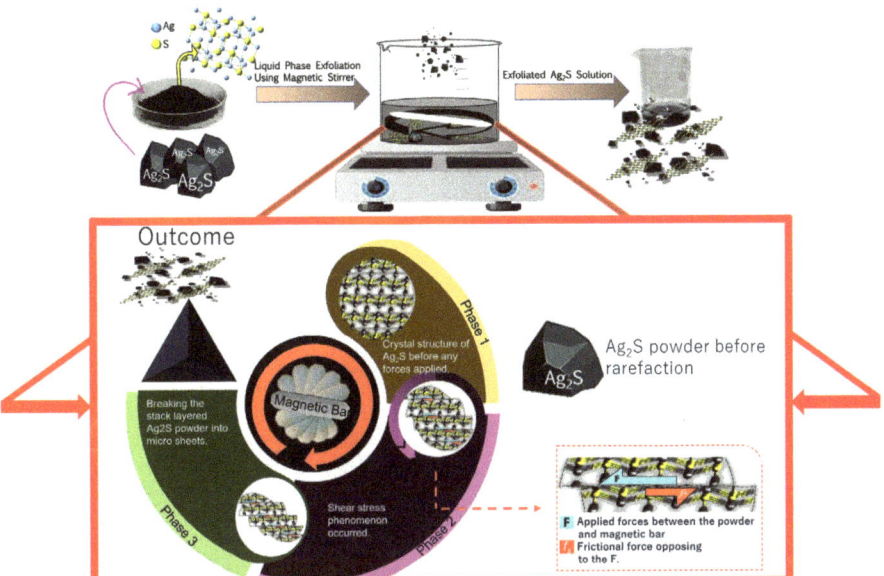

Figure 2. Illustrated scheme of the rarefaction process of commercial semiconductor Ag_2S.

The particle size transformations were involved during the process. The stirring action from the magnetic stirrer was deployed in a clockwise direction at a certain speed in the liquid while applying fluidic shear forces among the involved substances [81]. The existence of opposing forces between the applied forces and the friction forces within the system (Figure 2) caused the occurrence of shear stress in the Ag_2S powders, which can reduce the Ag_2S particle size. The longer the magnetic bar rotates, the more forces work, resulting in more energy provided to wear down the Ag_2S powder. Thus, smaller crystal Ag_2S were formed. According to the peak analysis and crystallite size calculation, it was confirmed that before stirring, the Ag_2S particle has a crystallite size of 52 nm and this size was reduced to 10 nm after the rarefaction process. Thus, the purpose of the rarefaction in this study was succeeded.

The distribution of both the rarefacted and synthesized Ag_2S powder in the cellulose film can be observed in Figure 3. According to Figure 3d–f, the particles from the synthesized Ag_2S particles evenly spread in the whole area of the cellulose matrix. Meanwhile, the samples that were loaded with the commercial Ag_2S had diverse particle distributions (Figure 3a–c). It was shown that part of the cellulose film was not affected by the Ag_2S. The degree of crystallinity and the crystallite size of the synthesized Ag_2S can be categorized as one of the factors that affected this particle distribution. Therefore, further discussion about the structure and chemical function of the samples will be explained in the following section.

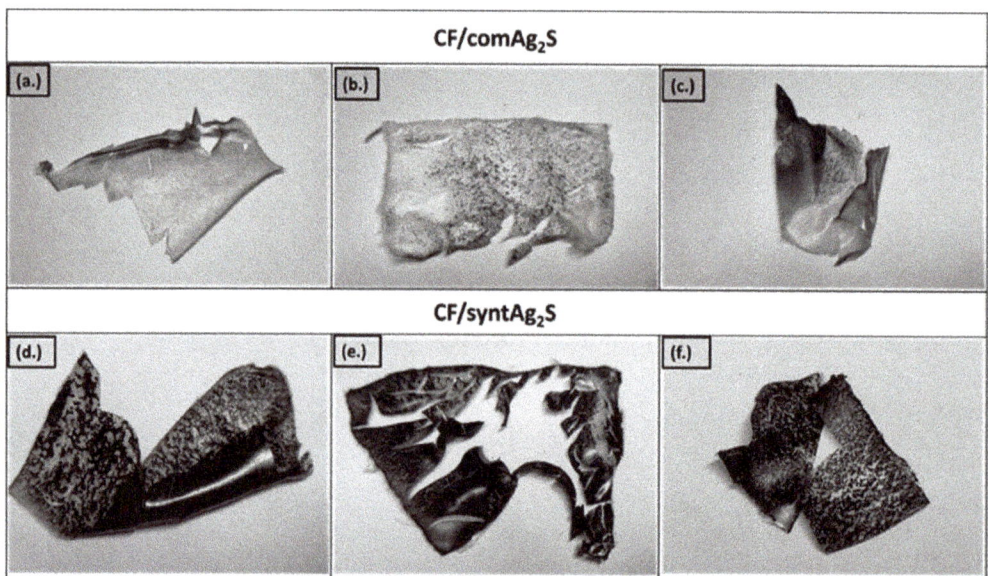

Figure 3. Prepared cellulose films: (**a**) CF/comAg$_2$S1, (**b**) CF/comAg$_2$S2, (**c**) CF/comAg$_2$S3, (**d**) CF/syntAg$_2$S1, (**e**) CF/syntAg$_2$S2, and (**f**) CF/syntAg$_2$S3.

3.2. Crystalline Phase Investigation of the Synthesized Ag$_2$S powder

Figure 4 shows the XRD spectra of the synthesized Ag$_2$S particles that were obtained from the reaction of 0.3 M AgNO$_3$ and thiourea. The diffraction peaks of Ag$_2$S that were observed at 2theta = 22.4°, 25.9°, 28.9°, 31.5°, 34.4°, 36.8°, 37.7°, 40.7°, 43.6°, and 46.2° showed similar patterns with the commercial Ag$_2$S and confirmed the formation of Ag$_2$S phase (ICDD No. 00-014-0072) with monoclinic structure.

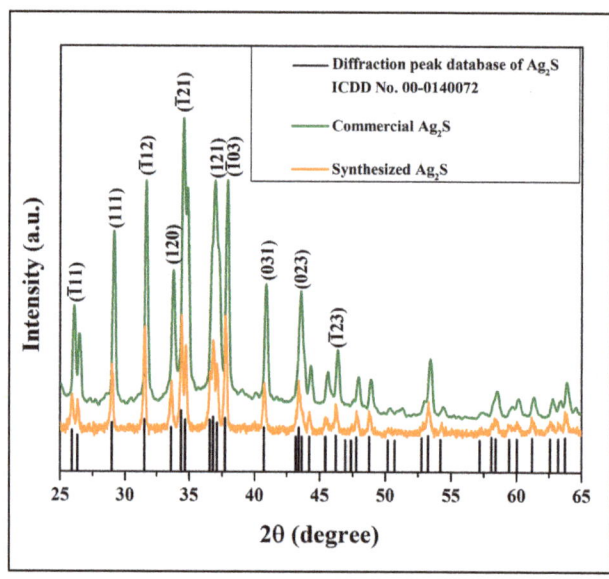

Figure 4. X-ray diffraction spectrum of synthesized Ag$_2$S.

3.3. Crystal Structure and Phase Identification of Cellulose Film/Synthesized Ag₂S (CF/syntAg₂S)

To ensure the formation of the synthesized silver sulfide in the cellulose film samples, it was necessary to delve into the characterization of their crystal structure and phase identification. Furthermore, the influence of the $AgNO_3$ concentration on the crystal structure of the sample was also studied. The crystallite is crucial because the crystallinity of the catalyst in the film will have a direct impact on the photocatalytic properties.

The study of the crystal structure of the CF/syntAg₂S sample confirmed that all as-prepared samples have diffraction peaks of a monoclinic Acanthine Ag_2S structure with space group $P2_1/C$. The specific lattice parameter of these samples can be indexed as a = 9.5200 Å, b = 6.9300 Å, and c = 8.2900 Å. Moreover, the peak of (002) due to the reflection of the cellulose plane also existed in all of the samples that had undergone different $AgNO_3$/thiourea loadings during the synthesis. The acquired peaks of the Ag_2S and cellulose were matched well with the references ICDD No. 00-009-0422 and ICDD No. 00-050-2241, respectively. The average grain size and degree of crystallinity of the samples are shown in Table 4.

Table 4. Phase identification and crystallinity of the CF/syntAg₂S sample.

Sample Code	Index Miller Cellulose (002)		Index Miller Ag₂S (311)		(220)		(022)		Crystallite Size (Å)	Degree of Crystallinity (%)
	2θ Stnd.	2θ Ob.	2θ Stnd.	2θ Ob.	2θ Stnd.	2θ Ob.	2θ Stnd.	2θ Ob.		
CF/syntAg₂S1	22.78	22.76	31.56	-	34.50	34.44	36.92	37.18	221	47.93
CF/syntAg₂S2	22.78	22.30	31.56	31.18	34.50	33.99	36.92	36.47	188	76.76
CF/syntAg₂S3	22.78	22.78	31.56	31.48	34.50	34.49	36.92	36.99	174	37.75

In terms of peak intensity, it was observed that the higher the concentration, the lower and broader the peak (see Figure 5), which is attributed to the low crystallite size of the sample [82]. Table 4 illustrates that the smallest yield of crystallite size was found from the highest concentration of the substituent (CF/syntAg₂S). Furthermore, Table 4 also depicts that the substituent was responsible for the shifted peak that occurred in the sample with concentrated loadings. The shifted peak towards the lower 2theta indicated that the lattice parameter of the pristine cellulose had been incorporated by the synthesized Ag_2S [83,84].

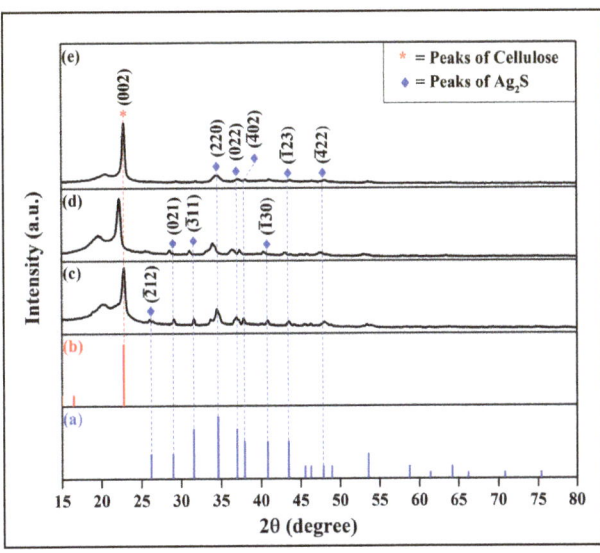

Figure 5. X-ray diffraction spectra of: (**a**) database of Ag₂S (ICDD No. 00-009-0422), (**b**) database of cellulose (ICDD No. 00-050-2241), (**c**) CF/syntAg₂S1, (**d**) CF/syntAg₂S2, and (**e**) CF/syntAg₂S3.

3.4. Functional Group Analysis of CF/comAg$_2$S and CF/syntAg$_2$S

According to Figure 6, the commercial (Figure 6a–d) and synthesized (Figure 6g–j) samples had comparable FTIR patterns along the observation scanning range. However, in terms of peak intensity, the sample with the synthesized Ag$_2$S had a sharp and narrow peak compared to the commercial sample. The spectra illustrated the same pattern among the thin films that were either impregnated by the commercial or synthesized Ag$_2$S. The broad bands that were found from wavenumber 3336.5 cm^{-1} to 3419.8 cm^{-1} were designated to the hydroxyl (O-H) group stretching vibration. The shifted peak among these bands was attributed to the interaction of the hydrogen bond of the pristine cellulose with glycerol and PVA. These displacements were also found by [71,85], in which the high content of glycerol and PVA in the cellulose matrix impacted the sharpening and shifting to the higher wavenumber. The spectral band located from 2916.5 cm^{-1} to 2912.59 cm^{-1}, as well as a weak peak at 897.00 cm^{-1} were attributed to the characteristic of symmetrical stretching of C-H from the alkyl group [32] and glycosidic CH deformation [72], respectively.

Thereafter, the transmittance band at the area of 1753.70 cm^{-1} to 1752.13 cm^{-1} was associated with the stretching vibration of the C=O ester carbonyl group and the medium peak at the range of 1238 cm^{-1} was denoted as C-O-C asymmetric stretching [86]. The authentic peak of cellulose was found at the region around 1059 cm^{-1} and remarked as C-O-C stretching vibrations of aliphatic primary and secondary alcohols in cellulose [87,88]. The peak of C-H bending was observed at the regions of 1374 cm^{-1} to 1373 cm^{-1}. The study by Cazón et al. announced that the chemical content of glycerol and PVA in the cellulose matrix may contribute to the displacement of the observed peak. Nevertheless, Figure 6e,f,k,l indicates that the loadings of metal silver sulfide into the cellulose can reduce the intensity of the peak with an oxygen-containing functional group. The more concentrated the Ag$_2$S, the lower the intensity of the peak-contained oxygen functional group. According to Kumar et al., it was confirmed that the strong electrostatic linkages between Ag$_2$S and the functional groups of cellulose were responsible for the lower intensity of the transmittance peak into the oxygen-containing functional group [89]. This means that the different intensity of the peak, particularly in the range of oxygen-containing group regions, was a characteristic that showed the success of Ag$_2$S deposited into the cellulose matrix.

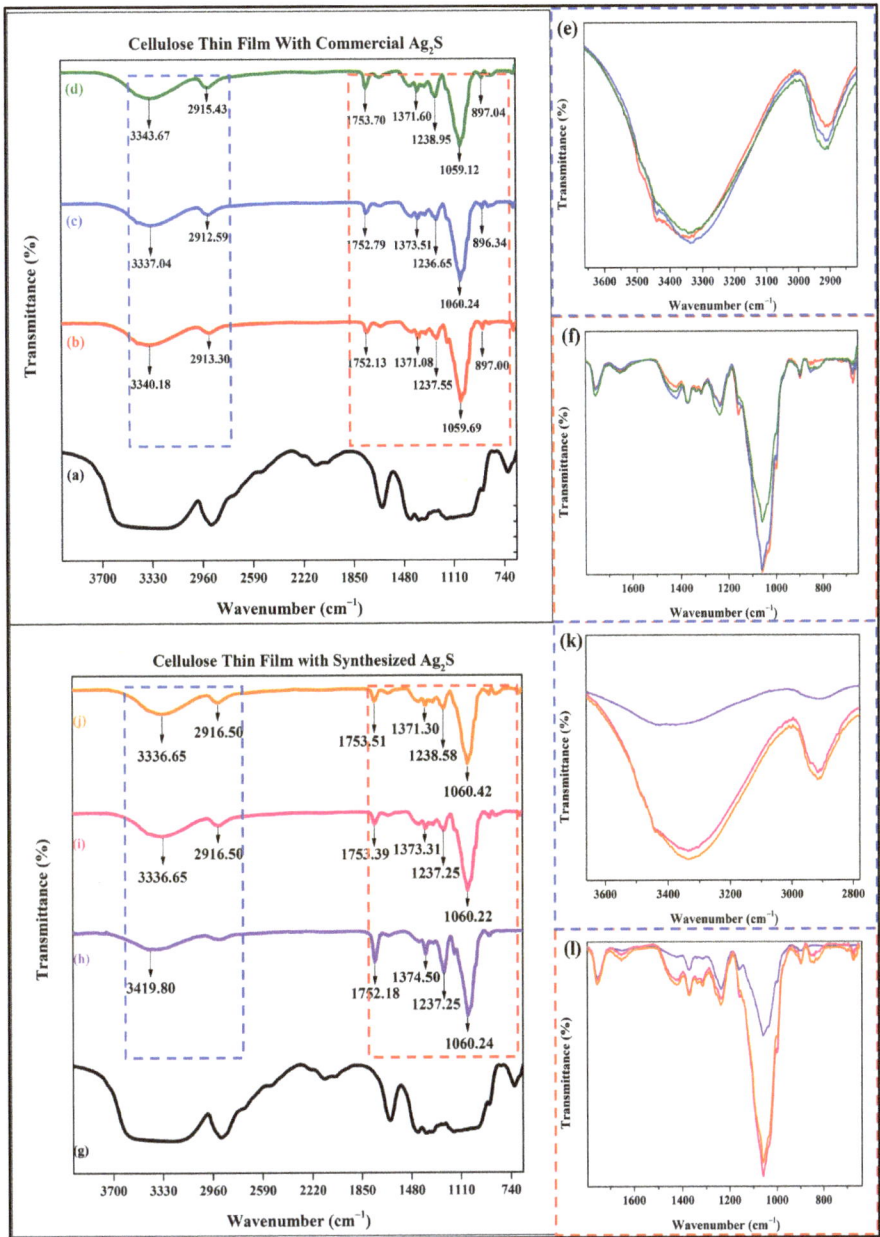

Figure 6. ATR—FTIR spectra of: (**a,g**) cellulose, (**b**) CF/comAg$_2$S1, (**c**) CF/comAg$_2$S2, (**d**) CF/comAg$_2$S3, (**e,f**) inset CF/comAg$_2$S graph, (**h**) CF/syntAg$_2$S1, (**i**) CF/syntAg$_2$S2, (**j**) CF/syntAg$_2$S3, and (**k,l**) inset CF/syntAg$_2$S graph.

3.5. Photocatalytic Degradation Mechanism of Methylene Blue Using the Cellulose/Ag$_2$S Films

The degradation of the methylene blue (MB) dye solution with sunlight exposure and without exposure was investigated. The results are shown in Table 5. For the non-exposure samples, the degradation rate of both synthesized and commercial Ag$_2$S incorporated in the cellulose film was only 35% and 33% (due to the adsorption by cellulose film), respectively, even after 5 h of processing. Compared to the same samples that were exposed to sunlight, the degradation rate of MB reached 98–100% within 2 h. This result showed that solar energy seemed to have a great influence on the photocatalytic activity of the Ag_2S-containing samples, since this semiconductor catalyst absorbed in the NIR region. Nevertheless, according to the direct observation of photocatalytic activity under sunlight exposure, there were two kinds of reactions that might have occurred: firstly, the reaction between the cellulose surface and the MB solution; secondly, the reaction among deposited Ag_2S in the cellulose film and the MB solution. Figure 7 shows the reaction mechanism that might have occurred between the cellulose and MB solution during the process.

Table 5. Comparison of the photocatalytic efficiency of the prepared samples.

Exposure Time (Min)	Degradation Ratio (%)					
	Non-Exposure		Ag_2S Powder	Pristine Cellulose	With Exposure	
	CF/comAg$_2$S	CF/syntAg$_2$S			CF/comAg$_2$S2	CF/syntAg$_2$S1
0	0	0	0	0	0	0
5	3.59	0.66	19.78	44.14	30.99	31.49
10	5.97	1.60	23.37	53.09	40.77	44.42
15	7.84	3.09	26.85	63.43	52.10	71.71
20	10.44	4.31	30.22	71.10	58.56	83.54
25	12.21	5.64	32.82	75.91	64.92	86.69
30	14.42	6.46	34.53	81.66	69.78	89.45
60	22.76	12.43	39.39	90.88	98.95	93.98
90	27.40	16.24	45.41	94.92	99.01	97.29
120	29.78	21.60	48.18	96.19	100	98.56
150	31.33	25.03	52.98	99.56	100	99.78
180	31.93	28.18	54.92	99.56	100	99.78
210	32.32	30.17	59.28	99.83	100	100
240	32.49	33.81	62.49	99.97	100	100
270	32.71	34.75	66.74	100	100	100
300	33.04	35.08	75.03	100	100	100

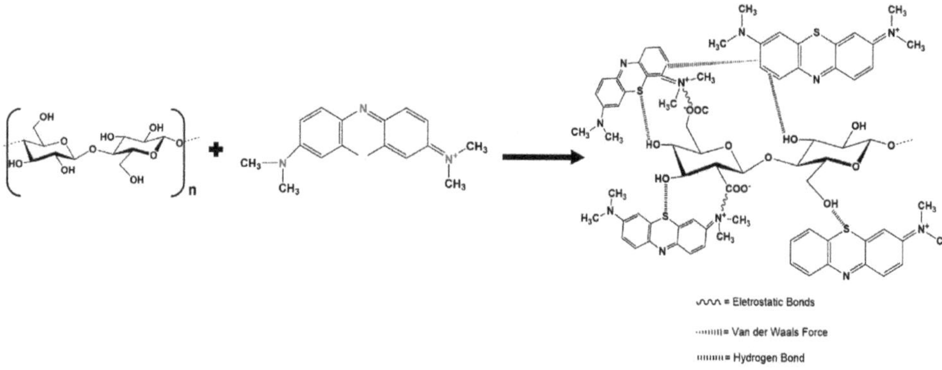

Figure 7. Predicted mechanism reaction between methylene blue and cellulose.

The cellulose interacts with MB through several bonds, such as the electrostatic bond between the ion N^+ in MB and ion O^- in cellulose. The electrostatic attraction that was involved in the adsorption mechanism of MB on cellulose led to the enhancement of the MB molecules to quickly fill the adsorption sites on the surface of the cellulose film, resulting in a high rate of MB adsorption [90]. The Van der Waals force (C=C) and hydrogen bond also formed during this process [91]. These bonds made it possible for the cellulose to experience the photolysis reaction when exposed to the solar source.

On the other hand, under solar illumination, the existence of the catalyst Ag_2S in the cellulose film established the photocatalytic activity due to the configuration of the reducing and oxidizing (electron–hole pairs) ions on the surface of the catalyst, which may degrade the concentration of the MB dyes solution until it reached 100% discoloration (see Table 5). During the process of solar irradiation, photons of solar light with an energy that was either equal to or higher than the band gap of Ag_2S (~1.06 eV) [92] were degraded by Ag_2S. Upon the process of photon absorption, an electron (e^-) from the valence band (VB) was stimulated up to the conduction band (CB). In addition, it was linked to the development of a hole (h^+) in the valence band, which led to the formation of electron–hole pairs that took part in the process of reduction and oxidation. In this process, the electrons in the surface reacted with the oxygen (O_2) that was dissolved in the aqueous solution, which then resulted in the production of anionic superoxide radicals ($\bullet O_2^-$) [93]. Simultaneously, the photogenerated holes also reacted with water to create hydroxyl radicals, which further oxidized the dye molecules. The produced anionic superoxide radicals ($\bullet O_2^-$) and hydroxyl radicals oxidized the dye molecules that can decompose the MB dyes into CO_2 or H_2O and their intermediates [94]. Overall, this phenomenon can be expressed through the following chemical reactions:

$$Ag_2S + hv \rightarrow e^- + h^+ \qquad (7)$$

$$e^- + O_2 \rightarrow \bullet O_2^- \qquad (8)$$

$$h^+ + H_2O \rightarrow \bullet OH + H^+ \qquad (9)$$

$$\text{Methylene Blue} \rightarrow \overbrace{\bullet O_2^- \bullet OH}^{\text{free radicals}} \rightarrow \text{intermediates} + CO_2 + H_2O \qquad (10)$$

However, without light energy, the processes involving reduction–oxidation reactions in the photocatalysis process will not be formed. As a result, the MB dyes wastewater will only interact with the cellulose film as described above.

The incorporation of Ag_2S into cellulose films served primarily to increase the rate of dye degradation by creating a massive contact surface. The catalyst, Ag_2S, was distributed well toward the cellulose films, leading to the films outperforming the powder counterparts in terms of activity because they effectively absorbed light and perhaps underwent internal scattering within the cellulose/Ag_2S film, inducing higher charge carrier formation and hence better photocatalytic efficiency. Table 5 shows the photocatalytic efficiency for all samples based on the MB degradation rate (%). From the table, when Ag_2S powder was used, the degradation rate of MB only reached about 75% even after 5 h of sunlight exposure. For pristine cellulose, the degradation of MB reached 100% after 4 h exposure. The high rate of MB adsorption by pristine cellulose might be due to the electrostatic attraction that was involved in the adsorption mechanism of the MB molecules to quickly fill the adsorption sites on the surface of the cellulose [90]. For CF/com Ag_2S, 100% of the MB was degraded in 2 h of exposure time compared to the CF/syntAg_2S sample with 98.6% degradation. The deterioration in the treatment that applied the catalyst was compared to the degradation in the photolysis experiment that was also conducted without the presence of a catalyst in the aqueous solution of methylene blue. By observing the direct photolysis of MB, it was clear that there were no appreciable color changes over the 5 h of exposure time. The degradation rate also reached its lowest value, ranging from 0% to 1.66%, as shown in Figure 8a,c.

Based on Figure 8b,d, the concentration of the dye solution was elevated to more than 90% in just 60 min for the pristine cellulose, CF/comAg_2S and CF/syntAg_2S samples, while for the Ag_2S powder, less than 40% of the MB was degraded. This showed that the distribution of Ag_2S powders in regenerated cellulose matrices can enhance the photocatalytic activity up to 100% efficiency due to higher surface area, and the affinity of the photocatalyst films that react with the dye molecules during the sunlight irradiation [95].

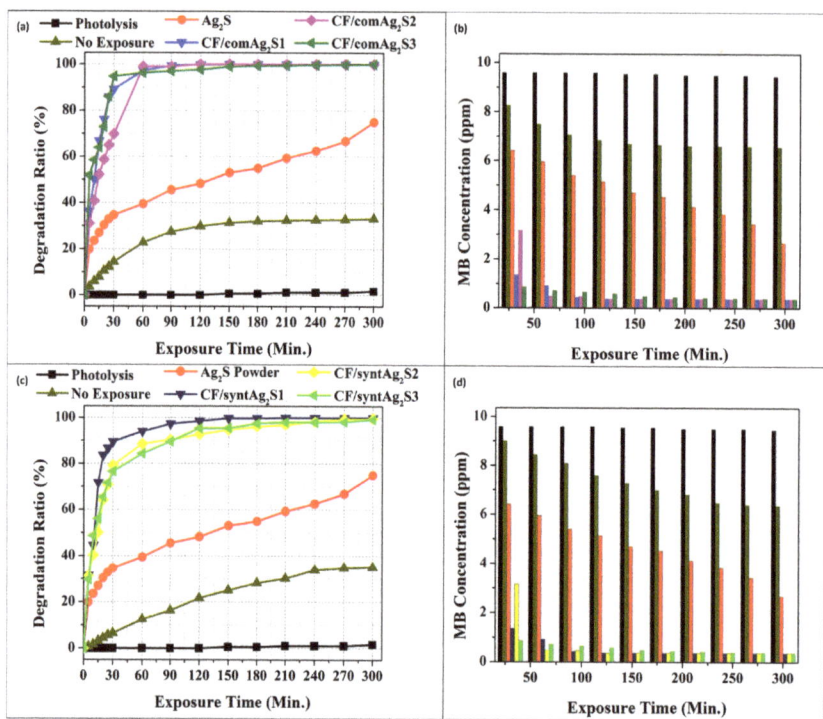

Figure 8. (**a**) Degradation ratio of MB with Ag$_2$S commercial as a catalyst; (**b**) The concentration of MB after photocatalytic activity using CF/comAg$_2$S samples; (**c**) Degradation ratio of MB with synthesized Ag$_2$S as a catalyst; (**d**) The concentration of MB after photocatalytic activity employing CF/syntAg$_2$S samples.

In addition, the surface of the cellulose film acted as a host during the process of dye adsorption, which in turn enabled the donor and acceptor molecules to interact with one another. This process could delay the recombination of charge carriers under direct sunlight. Charge recombination must be avoided during this process since it can bring down the efficiency of the photocatalyst [96–98]. In terms of the degradation ratio, it can be seen that there were differences in the photodegradation characterization of the two types of samples that were synthesized. The samples with commercial Ag$_2$S doping will have an increasing ability with increasing Ag$_2$S loading, where the higher content that can be accepted by cellulose film is 0.1 g. Meanwhile, the other types of samples (CF/syntAg$_2$S) showed the opposite nature. This may occur due to overload on the sample (Figure 3d–f) with loading that might exceed the limit, resulting in poor interfacial charge carrier migration and an increase in the recombination rate of photoexcited electron–hole pairs due to an increase in the concentration of Ag sources [73,74]. Thus, it was found that the CF/comAg$_2$S samples had the efficiency to degrade 100% of the MB content compared to the CF/syntAg$_2$S (98.6%) in 2 h.

4. Conclusions

It has been successfully demonstrated that silver sulfide (Ag$_2$S) particles, which were incorporated into regenerated cellulose, outperformed its properties in the powder state. This was because the distribution of Ag$_2$S particles on the cellulose matrix could increase its photocatalytic activity as the contact surface between the catalyst and dyes became larger. Furthermore, the molecules contained in the cellulose were found to have the ability to interact with molecules in the dyes, especially the methylene blue (MB) dyes. The samples with the commercial Ag_2S showed an increase in photocatalytic activity up to 100% degradation after 120 min of sunlight exposure with the maximum of 0.1 g Ag_2S loading onto the cellulose film, while other types of samples (CF/syntAg$_2$S) showed 98.6%. This situation might be due to the overload of synthesized Ag_2S particles on the sample that exceeded the photocatalyst limit on the film. As a result, poor interfacial charge carrier migration and an increase in the recombination rate of photoexcited electron–hole pairs occurred. Thus, it can be concluded that the photocatalytic efficiency of the photocatalyst film originating from the commercial Ag_2S particles was the highest, with 100% of the degradation rates in 2 h. For the non-exposure sunlight samples, the degradation rates were only 33–35%, showing the importance of the NIR semiconductor Ag_2S catalyst used.

Author Contributions: Conceptualization, Z.R.M., N.M. and M.N.N.; methodology, I.S.M., M.M.A.B.A. and K.B.; software, A.V.S.; validation, Z.R.M., N.M. and P.V.; formal analysis, M.N.N.; investigation, I.S.M., M.M.A.B.A. and K.B.; data curation, K.B., M.N. and M.S.B.; writing—original draft preparation, Z.R.M., N.M., M.N.N. and I.S.M.; writing—review and editing, K.B., M.N. and A.V.S.; visualization, M.M.A.B.A.; supervision, P.V.; project administration, M.M.A.B.A.; funding acquisition, M.S.B. All authors have read and agreed to the published version of the manuscript.

Funding: This work was financially supported by the Fundamental Research Grant Scheme FRGS/1/2021/TK0/UNIMAP/02/33 from the Ministry of Education Malaysia (MOE) and by Gheorghe Asachi Technical University of Iasi—TUIASI Romania, Scientific Research Funds, FCSU-2022.

Conflicts of Interest: The authors declare that they have no known competing financial interest or personal relationships that could have appeared to influence the work reported in this paper.

References

1. Ziarani, G.M.; Moradi, R.; Lashgari, N.; Kruger, H.G. Introduction and Importance of Synthetic Organic Dyes. In *Metal-Free Synthetic Organic Dyes*; Elsevier: Amsterdam, The Netherlands, 2018; pp. 1–7. ISBN 978-0-12-815647-6.
2. Fleck, B.; Cabral, I.; Souto, A.P. Eco Printing of Linen and Tencel Substrates with Onion Skins and Red Cabbage. *Mater. Circ. Econ.* **2022**, *4*, 2. [CrossRef]
3. Kusumastuti, A.; Anis, S.; Fardhyanti, D.S. Production of natural dyes powder based on chemo-physical technology for textile application. *IOP Conf. Ser. Earth Environ. Sci.* **2019**, *258*, 012028. [CrossRef]
4. Elsahida, K.; Fauzi, A.M.; Sailah, I.; Siregar, I.Z. Sustainability of the use of natural dyes in the textile industry. *IOP Conf. Ser. Earth Environ. Sci.* **2019**, *399*, 012065. [CrossRef]
5. Hasan, K.M.F.; Wang, H.; Mahmud, S.; Islam, A.; Habib, M.A.; Genyang, C. Enhancing mechanical and antibacterial performances of organic cotton materials with greenly synthesized colored silver nanoparticles. *IJCST*, **2022**; *ahead-of-print*. [CrossRef]
6. Bahuguna, A.; Singh, S.; Bahuguna, A.; Sharma, S.; Dadarwal, B. Physical method of Wastewater treatment-A review. *J. Res. Environ. Earth Sci.* **2021**, *7*, 2348–2532.
7. Berradi, M.; Hsissou, R.; Khudhair, M.; Assouag, M.; Cherkaoui, O.; El Bachiri, A.; El Harfi, A. Textile finishing dyes and their impact on aquatic environs. *Heliyon* **2019**, *5*, e02711. [CrossRef]
8. Kosswattaarachchi, A.M.; Cook, T.R. Repurposing the Industrial Dye Methylene Blue as an Active Component for Redox Flow Batteries. *ChemElectroChem* **2018**, *5*, 3437–3442. [CrossRef]
9. Lawagon, C.P.; Amon, R.E.C. Magnetic rice husk ash "cleanser" as efficient methylene blue adsorbent. *Environ. Eng. Res.* **2019**, *25*, 685–692. [CrossRef]
10. Zhou, M.; Chen, J.; Hou, C.; Liu, Y.; Xu, S.; Yao, C.; Li, Z. Organic-free synthesis of porous CdS sheets with controlled windows size on bacterial cellulose for photocatalytic degradation and H_2 production. *Appl. Surf. Sci.* **2019**, *470*, 908–916. [CrossRef]
11. Wang, H.; Li, G.; Fakhri, A. Fabrication and structural of the Ag_2S-MgO/graphene oxide nanocomposites with high photocatalysis and antimicrobial activities. *J. Photochem. Photobiol. B Biol.* **2020**, *207*, 111882. [CrossRef]
12. Hammood, Z.A.; Chyad, T.F.; Al-Saedi, R. Adsorption Performance of Dyes Over Zeolite for Textile Wastewater Treatment. *Ecol. Chem. Eng. S* **2021**, *28*, 329–337. [CrossRef]
13. Januário, E.F.D.; Vidovix, T.B.; Calsavara, M.A.; Bergamasco, R.; Vieira, A.M.S. Membrane surface functionalization by the deposition of polyvinyl alcohol and graphene oxide for dyes removal and treatment of a simulated wastewater. *Chem. Eng. Process.—Process Intensif.* **2022**, *170*, 108725. [CrossRef]
14. Khan, M.M.; Pradhan, D.; Sohn, Y. (Eds.) Nanocomposites for Visible Light-induced Photocatalysis. In *Springer Series on Polymer and Composite Materials*; Springer International Publishing: Cham, Germany, 2017; ISBN 978-3-319-62445-7.
15. Popa, N.; Visa, M. New hydrothermal charcoal TiO_2 composite for sustainable treatment of wastewater with dyes and cadmium cations load. *Mater. Chem. Phys.* **2021**, *258*, 123927. [CrossRef]

16. Singh, R.; Munya, V.; Are, V.N.; Nayak, D.; Chattopadhyay, S. A Biocompatible, pH-Sensitive, and Magnetically Separable Superparamagnetic Hydrogel Nanocomposite as an Efficient Platform for the Removal of Cationic Dyes in Wastewater Treatment. *ACS Omega* **2021**, *6*, 23139–23154. [CrossRef]
17. Azimi, B.; Abdollahzadeh-Sharghi, E.; Bonakdarpour, B. Anaerobic-aerobic processes for the treatment of textile dyeing wastewater containing three commercial reactive azo dyes: Effect of number of stages and bioreactor type. *Chin. J. Chem. Eng.* **2021**, *39*, 228–239. [CrossRef]
18. Haddad, M.; Abid, S.; Hamdi, M.; Bouallagui, H. Reduction of adsorbed dyes content in the discharged sludge coming from an industrial textile wastewater treatment plant using aerobic activated sludge process. *J. Environ. Manag.* **2018**, *223*, 936–946. [CrossRef]
19. Idris, I.; Rahmadhani, I.; Sudiana, I.M. Feasibility of Aspergillus keratitidis InaCC1016 for synthetic dyes removal in dyes wastewater treatment. *IOP Conf. Ser. Earth Environ. Sci.* **2020**, *439*, 012027. [CrossRef]
20. Ikram, M.; Zahoor, M.; El-Saber Batiha, G. Biodegradation and decolorization of textile dyes by bacterial strains: A biological approach for wastewater treatment. *Z. Für Phys. Chem.* **2021**, *235*, 1381–1393. [CrossRef]
21. Shen, L.; Jin, Z.; Xu, W.; Jiang, X.; Shen, Y.; Wang, Y.; Lu, Y. Enhanced Treatment of Anionic and Cationic Dyes in Wastewater through Live Bacteria Encapsulation Using Graphene Hydrogel. *Ind. Eng. Chem. Res.* **2019**, *58*, 7817–7824. [CrossRef]
22. Ben Ayed, S.; Azam, M.; Al-Resayes, S.I.; Ayari, F.; Rizzo, L. Cationic Dye Degradation and Real Textile Wastewater Treatment by Heterogeneous Photo-Fenton, Using a Novel Natural Catalyst. *Catalysts* **2021**, *11*, 1358. [CrossRef]
23. Dong, Y.-Y.; Zhu, Y.-H.; Ma, M.-G.; Liu, Q.; He, W.-Q. Synthesis and characterization of Ag@AgCl-reinforced cellulose composites with enhanced antibacterial and photocatalytic degradation properties. *Sci. Rep.* **2021**, *11*, 3366. [CrossRef]
24. Hussin, N.A.M.; Abidin, C.Z.A.; Fahmi; Ibrahim, A.H.; Ahmad, R.; Singa, P.K. Optimization of Anthraquinone Dye Wastewater Treatment using Ozone in the Presence of Persulfate Ion in a Semi-batch Reactor. *IOP Conf. Ser. Earth Environ. Sci.* **2021**, *920*, 012019. [CrossRef]
25. Saeed, M.; Adeel, S.; Muneer, M.; ul Haq, A. Photo Catalysis: An Effective Tool for Treatment of Dyes Contaminated Wastewater. In *Bioremediation and Biotechnology, Vol 2*; Bhat, R.A., Hakeem, K.R., Dervash, M.A., Eds.; Springer International Publishing: Cham, Germany, 2020; pp. 175–187. ISBN 978-3-030-40332-4.
26. Brienza, M.; Katsoyiannis, I. Sulfate Radical Technologies as Tertiary Treatment for the Removal of Emerging Contaminants from Wastewater. *Sustainability* **2017**, *9*, 1604. [CrossRef]
27. Crini, G.; Lichtfouse, E. Advantages and disadvantages of techniques used for wastewater treatment. *Environ. Chem. Lett.* **2019**, *17*, 145–155. [CrossRef]
28. Sengupta, S.; Pal, C.K. Chemistry in Wastewater Treatment. In *Advanced Materials and Technologies for Wastewater Treatment*; CRC Press: Boca Raton, FL, USA, 2021; ISBN 978-1-00-313830-3.
29. Riente, P.; Noël, T. Application of metal oxide semiconductors in light-driven organic transformations. *Catal. Sci. Technol.* **2019**, *9*, 5186–5232. [CrossRef]
30. Gisbertz, S.; Pieber, B. Heterogeneous Photocatalysis in Organic Synthesis. *ChemPhotoChem* **2020**, *4*, 456–475. [CrossRef]
31. Al-Mayyahi, A.; Al-Asadi, A.A. Advanced oxidation processes (aops) for wastewater treatment and reuse: A brief review. *Asian J. Appl. Sci. Technol* **2018**, *2*, 13.
32. Kumar, N.; Mittal, H.; Reddy, L.; Nair, P.; Ngila, J.C.; Parashar, V. Morphogenesis of ZnO nanostructures: Role of acetate (COOH$^-$) and nitrate (NO$_3^-$) ligand donors from zinc salt precursors in synthesis and morphology dependent photocatalytic properties. *RSC Adv.* **2015**, *5*, 38801–38809. [CrossRef]
33. Zailan, S.N.; Bouaissi, A.; Mahmed, N.; Abdullah, M.M.A.B. Influence of ZnO Nanoparticles on Mechanical Properties and Photocatalytic Activity of Self-cleaning ZnO-Based Geopolymer Paste. *J. Inorg. Organomet. Polym.* **2020**, *30*, 2007–2016. [CrossRef]
34. Subramaniam, M.N.; Goh, P.-S.; Lau, W.-J.; Ng, B.-C.; Ismail, A.F. Chapter 3—development of nanomaterial-based photocatalytic membrane for organic pollutants removal. In *Advanced Nanomaterials for Membrane Synthesis and its Applications*; Lau, W.-J., Ismail, A.F., Isloor, A., Al-Ahmed, A., Eds.; Micro and Nano Technologies; Elsevier: Amsterdam, The Netherlands, 2019; pp. 45–67. ISBN 978-0-12-814503-6.
35. Ishchenko, O.M.; Rogé, V.; Lamblin, G.; Lenoble, D. Tio2- and zno-based materials for photocatalysis: Material properties, device architecture and emerging concepts. In *Semiconductor Photocatalysis—Materials, Mechanisms and Applications*; Cao, W., Ed.; InTech: London, UK, 2016; ISBN 978-953-51-2484-9.
36. Kanakkillam, S.S.; Krishnan, B.; Guzman, S.S.; Martinez, J.A.A.; Avellaneda, D.A.; Shaji, S. Defects rich nanostructured black zinc oxide formed by nanosecond pulsed laser irradiation in liquid. *Appl. Surf. Sci.* **2021**, *567*, 150858. [CrossRef]
37. Haounati, R.; El Guerdaoui, A.; Ouachtak, H.; El Haouti, R.; Bouddouch, A.; Hafid, N.; Bakiz, B.; Santos, D.M.F.; Labd Taha, M.; Jada, A.; et al. Design of direct Z-scheme superb magnetic nanocomposite photocatalyst Fe$_3$O$_4$/Ag$_3$PO$_4$@Sep for hazardous dye degradation. *Sep. Purif. Technol.* **2021**, *277*, 119399. [CrossRef]
38. Hu, X.; Li, Y.; Tian, J.; Yang, H.; Cui, H. Highly efficient full solar spectrum (UV-vis-NIR) photocatalytic performance of Ag$_2$S quantum dot/TiO$_2$ nanobelt heterostructures. *J. Ind. Eng. Chem.* **2017**, *45*, 189–196. [CrossRef]
39. Dias, P.; Mendes, A. Hydrogen Production from Photoelectrochemical Water Splitting. In *Encyclopedia of Sustainability Science and Technology*; Meyers, R.A., Ed.; Springer: New York, NY, USA, 2018; pp. 1–52. ISBN 978-1-4939-2493-6.
40. Danish, M.S.S.; Bhattacharya, A.; Stepanova, D.; Mikhaylov, A.; Grilli, M.L.; Khosravy, M.; Senjyu, T. A Systematic Review of Metal Oxide Applications for Energy and Environmental Sustainability. *Metals* **2020**, *10*, 1604. [CrossRef]

41. Sadovnikov, S.I.; Gusev, A.I. Recent progress in nanostructured silver sulfide: From synthesis and nonstoichiometry to properties. *J. Mater. Chem. A* **2017**, *5*, 17676–17704. [CrossRef]
42. Shafi, A.; Ahmad, N.; Sultana, S.; Sabir, S.; Khan, M.Z. Ag$_2$S-sensitized NiO–ZnO heterostructures with enhanced visible light photocatalytic activity and acetone sensing property. *ACS Omega* **2019**, *4*, 12905–12918. [CrossRef]
43. Yu, W.; Yin, J.; Li, Y.; Lai, B.; Jiang, T.; Li, Y.; Liu, H.; Liu, J.; Zhao, C.; Singh, S.C.; et al. Ag$_2$S Quantum Dots as an Infrared Excited Photocatalyst for Hydrogen Production. *ACS Appl. Energy Mater.* **2019**, *2*, 2751–2759. [CrossRef]
44. Badawi, A. Effect of the non-toxic Ag$_2$S quantum dots size on their optical properties for environment-friendly applications. *Phys. E Low-Dimens. Syst. Nanostructures* **2019**, *109*, 107–113. [CrossRef]
45. Shahri, N.N.M.; Taha, H.; Hamid, M.H.S.A.; Kusrini, E.; Lim, J.-W.; Hobley, J.; Usman, A. Antimicrobial activity of silver sulfide quantum dots functionalized with highly conjugated Schiff bases in a one-step synthesis. *RSC Adv.* **2022**, *12*, 3136–3146. [CrossRef]
46. Jiang, W.; Wu, Z.; Yue, X.; Yuan, S.; Lu, H.; Liang, B. Photocatalytic performance of Ag$_2$S under irradiation with visible and near-infrared light and its mechanism of degradation. *RSC Adv.* **2015**, *5*, 24064–24071. [CrossRef]
47. Yuan, L.; Lu, S.; Yang, F.; Wang, Y.; Jia, Y.; Kadhim, M.; Yu, Y.; Zhang, Y.; Zhao, Y. A facile room-temperature synthesis of three-dimensional coral-like Ag$_2$S nanostructure with enhanced photocatalytic activity. *J. Mater. Sci.* **2019**, *54*, 3174–3186. [CrossRef]
48. Al-Shehri, B.M.; Shkir, M.; Bawazeer, T.M.; AlFaify, S.; Hamdy, M.S. A rapid microwave synthesis of Ag$_2$S nanoparticles and their photocatalytic performance under UV and visible light illumination for water treatment applications. *Phys. E Low-Dimens. Syst. Nanostructures* **2020**, *121*, 114060. [CrossRef]
49. Cui, C.; Li, X.; Liu, J.; Hou, Y.; Zhao, Y.; Zhong, G. Synthesis and Functions of Ag$_2$S Nanostructures. *Nanoscale Res. Lett.* **2015**, *10*, 431. [CrossRef] [PubMed]
50. Khan, M.M. Metal oxide powder photocatalysts. In *Multifunctional Photocatalytic Materials for Energy*; Elsevier: Amsterdam, The Netherlands, 2018; pp. 5–18. ISBN 978-0-08-101977-1.
51. Nalajala, N.; Patra, K.K.; Bharad, P.A.; Gopinath, C.S. Why the thin film form of a photocatalyst is better than the particulate form for direct solar-to-hydrogen conversion: A poor man's approach. *RSC Adv.* **2019**, *9*, 6094–6100. [CrossRef] [PubMed]
52. Garusinghe, U.M.; Raghuwanshi, V.S.; Batchelor, W.; Garnier, G. Water Resistant Cellulose—Titanium Dioxide Composites for Photocatalysis. *Sci. Rep.* **2018**, *8*, 2306. [CrossRef] [PubMed]
53. Katheresan, V.; Kansedo, J.; Lau, S.Y. Efficiency of various recent wastewater dye removal methods: A review. *J. Environ. Chem. Eng.* **2018**, *6*, 4676–4697. [CrossRef]
54. Samsami, S.; Mohamadizaniani, M.; Sarrafzadeh, M.-H.; Rene, E.R.; Firoozbahr, M. Recent advances in the treatment of dye-containing wastewater from textile industries: Overview and perspectives. *Process Saf. Environ. Prot.* **2020**, *143*, 138–163. [CrossRef]
55. Wandiga, S.O.; Masese, F.; Mbugua, S.N.; Macharia, J.W.; Otieno, M.A. Challenges and solutions to water problems in Africa. In *Handbook of Water Purity and Quality*; Elsevier: Amsterdam, The Netherlands, 2021; pp. 35–56. ISBN 978-0-12-821057-4.
56. Zhu, C.; Zhang, X.; Zhang, Y.; Li, Y.; Wang, P.; Jia, Y.; Liu, J. Ultrasonic-Assisted Synthesis of CdS/Microcrystalline Cellulose Nanocomposites with Enhanced Visible-Light-Driven Photocatalytic Degradation of MB and the Corresponding Mechanism Study. *Front. Chem.* **2022**, *10*, 892680. [CrossRef]
57. Liu, X.; Sayed, M.; Bie, C.; Cheng, B.; Hu, B.; Yu, J.; Zhang, L. Hollow CdS-based photocatalysts. *J. Mater.* **2021**, *7*, 419–439. [CrossRef]
58. Nawrot, K.C.; Wawrzyńczyk, D.; Bezkrovnyi, O.; Kępiński, L.; Cichy, B.; Samoć, M.; Nyk, M. Functional CdS-Au Nanocomposite for Efficient Photocatalytic, Photosensitizing, and Two-Photon Applications. *Nanomaterials* **2020**, *10*, 715. [CrossRef]
59. Haghighatzadeh, A.; Kiani, M.; Mazinani, B.; Dutta, J. Facile synthesis of ZnS–Ag$_2$S core–shell nanospheres with enhanced nonlinear refraction. *J. Mater. Sci. Mater. Electron.* **2020**, *31*, 1283–1292. [CrossRef]
60. Zhang, Y.; Zhou, W.; Jia, L.; Tan, X.; Chen, Y.; Huang, Q.; Shao, B.; Yu, T. Visible light driven hydrogen evolution using external and confined CdS: Effect of chitosan on carriers separation. *Appl. Catal. B Environ.* **2020**, *277*, 119152. [CrossRef]
61. Nalajala, N.; Salgaonkar, K.N.; Chauhan, I.; Mekala, S.P.; Gopinath, C.S. Aqueous Methanol to Formaldehyde and Hydrogen on Pd/TiO$_2$ by Photocatalysis in Direct Sunlight: Structure Dependent Activity of Nano-Pd and Atomic Pt-Coated Counterparts. *ACS Appl. Energy Mater.* **2021**, *4*, 13347–13360. [CrossRef]
62. Thamer, N.H.; Hatem, O.A. Synthesis and Characterization of TiO$_2$–Ag-Cellulose Nanocomposite and evaluation of photo catalytic degradation efficiency for Congo Red. *IOP Conf. Ser. Earth Environ. Sci.* **2022**, *1029*, 012026. [CrossRef]
63. Mohamed, M.A.; Salleh, W.N.W.; Jaafar, J.; Ismail, A.F.; Mutalib, M.A.; Sani, N.A.A.; Asri, S.E.A.M.; Ong, C.S. Physicochemical characteristic of regenerated cellulose/N-doped TiO$_2$ nanocomposite membrane fabricated from recycled newspaper with photocatalytic activity under UV and visible light irradiation. *Chem. Eng. J.* **2016**, *284*, 202–215. [CrossRef]
64. Verbruggen, S.; Tytgat, T.; Passel, S.; Martens, J.; Lenaerts, S. Cost-effectiveness analysis to assess commercial TiO$_2$ photocatalysts for acetaldehyde degradation in air. *Chem. Pap.* **2014**, *68*, 1273–1278. [CrossRef]
65. Asha, R.C.; Vishnuganth, M.A.; Remya, N.; Selvaraju, N.; Kumar, M. Livestock Wastewater Treatment in Batch and Continuous Photocatalytic Systems: Performance and Economic Analyses. *Water Air Soil Pollut.* **2015**, *226*, 132. [CrossRef]
66. Akbarzadeh, R.; Ibrahim, Q.; Jen, T.-C. Photocatalysis and Energy Cost Analysis of Vanadia Titania Thin Films Synthesis. In Proceedings of the 2019 International Conference on Power Generation Systems and Renewable Energy Technologies (PGSRET), Istanbul, Turkey, 26–27 August 2019; pp. 1–3.

67. Mehrjouei, M.; Müller, S.; Möller, D. Catalytic and photocatalytic ozonation of tert-butyl alcohol in water by means of falling film reactor: Kinetic and cost–effectiveness study. *Chem. Eng. J.* **2014**, *248*, 184–190. [CrossRef]
68. Ibrahim, N.A.; Salleh, K.M.; Fudholi, A.; Zakaria, S. Drying Regimes on Regenerated Cellulose Films Characteristics and Properties. *Membranes* **2022**, *12*, 445. [CrossRef]
69. Yang, Q.; Fukuzumi, H.; Saito, T.; Isogai, A.; Zhang, L. Transparent Cellulose Films with High Gas Barrier Properties Fabricated from Aqueous Alkali/Urea Solutions. *Biomacromolecules* **2011**, *12*, 2766–2771. [CrossRef]
70. Cazón, P.; Vázquez, M.; Velazquez, G. Cellulose-glycerol-polyvinyl alcohol composite films for food packaging: Evaluation of water adsorption, mechanical properties, light-barrier properties and transparency. *Carbohydr. Polym.* **2018**, *195*, 432–443. [CrossRef]
71. Geng, H.; Yuan, Z.; Fan, Q.; Dai, X.; Zhao, Y.; Wang, Z.; Qin, M. Characterisation of cellulose films regenerated from acetone/water coagulants. *Carbohydr. Polym.* **2014**, *102*, 438–444. [CrossRef]
72. Cazón, P.; Velazquez, G.; Vázquez, M. Novel composite films from regenerated cellulose-glycerol-polyvinyl alcohol: Mechanical and barrier properties. *Food Hydrocoll.* **2019**, *89*, 481–491. [CrossRef]
73. Holi, A.M.; Zainal, Z.; Ayal, A.K.; Chang, S.-K.; Lim, H.N.; Talib, Z.A.; Yap, C.-C. Ag_2S/ZnO Nanorods Composite Photoelectrode Prepared by Hydrothermal Method: Influence of Growth Temperature. *Optik* **2019**, *184*, 473–479. [CrossRef]
74. Holi, A.M.; Zainal, Z.; Al-Zahrani, A.A.; Ayal, A.K.; Najm, A.S. Effect of Varying $AgNO_3$ and $CS(NH_2)_2$ Concentrations on Performance of Ag_2S/ZnO NRs/ITO Photoanode. *Energies* **2022**, *15*, 2950. [CrossRef]
75. Haounati, R.; Alakhras, F.; Ouachtak, H.; Saleh, T.A.; Al-Mazaideh, G.; Alhajri, E.; Jada, A.; Hafid, N.; Addi, A.A. Synthesized of Zeolite@Ag_2O Nanocomposite as Superb Stability Photocatalysis Toward Hazardous Rhodamine B Dye from Water. *Arab. J. Sci. Eng.* **2022**, *207*, 112157. [CrossRef]
76. Hajizadeh, Z.; Taheri-Ledari, R.; Asl, F.R. Identification and analytical methods. In *Heterogeneous Micro and Nanoscale Composites for the Catalysis of Organic Reactions*; Elsevier: Amsterdam, The Netherlands, 2022; pp. 33–51. ISBN 978-0-12-824527-9.
77. Medronho, B.; Romano, A.; Miguel, M.G.; Stigsson, L.; Lindman, B. Rationalizing cellulose (in) solubility: Reviewing basic physicochemical aspects and role of hydrophobic interactions. *Cellulose* **2012**, *19*, 581–587. [CrossRef]
78. Xiong, B.; Zhao, P.; Cai, P.; Zhang, L.; Hu, K.; Cheng, G. NMR spectroscopic studies on the mechanism of cellulose dissolution in alkali solutions. *Cellulose* **2013**, *20*, 613–621. [CrossRef]
79. Mohamed, M.A.; Salleh, W.N.W.; Jaafar, J.; Ismail, A.F.; Abd Mutalib, M.; Jamil, S.M. Incorporation of N-doped TiO_2 nanorods in regenerated cellulose thin films fabricated from recycled newspaper as a green portable photocatalyst. *Carbohydr. Polym.* **2015**, *133*, 429–437. [CrossRef]
80. Zeng, J.; Liu, S.; Cai, J.; Zhang, L. TiO_2 Immobilized in Cellulose Matrix for Photocatalytic Degradation of Phenol under Weak UV Light Irradiation. *J. Phys. Chem. C* **2010**, *114*, 7806–7811. [CrossRef]
81. Desai, J.A.; Adhikari, N.; Kaul, A.B. Chemical exfoliation efficacy of semiconducting WS_2 and its use in an additively manufactured heterostructure graphene–WS_2–graphene photodiode. *RSC Adv.* **2019**, *9*, 25805–25816. [CrossRef]
82. Ma, Y.; Zhao, Z.; Xian, Y.; Wan, H.; Ye, Y.; Chen, L.; Zhou, H.; Chen, J. Highly Dispersed Ag_2S Nanoparticles: In Situ Synthesis, Size Control, and Modification to Mechanical and Tribological Properties towards Nanocomposite Coatings. *Nanomaterials* **2019**, *9*, 1308. [CrossRef]
83. Wang, Q.; Cai, J.; Zhang, L. In situ synthesis of Ag_3PO_4/cellulose nanocomposites with photocatalytic activities under sunlight. *Cellulose* **2014**, *21*, 3371–3382. [CrossRef]
84. Wu, J.; Zhao, N.; Zhang, X.; Xu, J. Cellulose/silver nanoparticles composite microspheres: Eco-friendly synthesis and catalytic application. *Cellulose* **2012**, *19*, 1239–1249. [CrossRef]
85. Rouhi, M.; Razavi, S.H.; Mousavi, S.M. Optimization of crosslinked poly (vinyl alcohol) nanocomposite films for mechanical properties. *Mater. Sci. Eng. C* **2017**, *71*, 1052–1063. [CrossRef] [PubMed]
86. Abderrahim, B.; Abderrahman, E.; Aqil, M. Ouassini Krim Kinetic Thermal Degradation of Cellulose, Polybutylene Succinate and a Green Composite: Comparative Study. *Sci. Educ. Publ.* **2015**, *3*, 95–110. [CrossRef]
87. Dutta, D.; Hazarika, R.; Dutta, P.D.; Goswami, T.; Sengupta, P.; Dutta, D.K. Synthesis of Ag–Ag_2S Janus nanoparticles supported on an environmentally benign cellulose template and their catalytic applications. *RSC Adv.* **2016**, *6*, 85173–85181. [CrossRef]
88. Fujii, Y.; Imagawa, K.; Omura, T.; Suzuki, T.; Minami, H. Preparation of Cellulose/Silver Composite Particles Having a Recyclable Catalytic Property. *ACS Omega* **2020**, *5*, 1919–1926. [CrossRef]
89. Kumar, P.; Agnihotri, R.; Wasewar, K.L.; Uslu, H.; Yoo, C. Status of adsorptive removal of dye from textile industry effluent. *Desalination Water Treat.* **2012**, *50*, 226–244. [CrossRef]
90. Tan, K.B.; Abdullah, A.Z.; Horri, B.A.; Salamatinia, B. Adsorption Mechanism of Microcrystalline Cellulose as Green Adsorbent for the Removal of Cationic Methylene Blue Dye. *J. Chem. Soc. Pak.* **2016**, *38*, 651–664.
91. Manna, S.; Roy, D.; Saha, P.; Gopakumar, D.; Thomas, S. Rapid methylene blue adsorption using modified lignocellulosic materials. *Process Saf. Environ. Prot.* **2017**, *107*, 346–356. [CrossRef]
92. Zamiri, R.; Abbastabar Ahangar, H.; Zakaria, A.; Zamiri, G.; Shabani, M.; Singh, B.; Ferreira, J.M.F. The structural and optical constants of Ag_2S semiconductor nanostructure in the Far-Infrared. *Chem. Cent. J.* **2015**, *9*, 28. [CrossRef]
93. Andrade-Guel, M.; Díaz-Jiménez, L.; Cortés-Hernández, D.; Cabello-Alvarado, C.; Ávila-Orta, C.; Bartolo-Pérez, P.; Gamero-Melo, P. Microwave assisted sol–gel synthesis of titanium dioxide using hydrochloric and acetic acid as catalysts. *Boletín De La Soc. Española De Cerámica Y Vidr.* **2019**, *58*, 171–177. [CrossRef]

94. Isac, L.; Cazan, C.; Enesca, A.; Andronic, L. Copper Sulfide Based Heterojunctions as Photocatalysts for Dyes Photodegradation. *Front. Chem.* **2019**, *7*, 694. [CrossRef] [PubMed]
95. Fan, J.; Yu, D.; Wang, W.; Liu, B. The self-assembly and formation mechanism of regenerated cellulose films for photocatalytic degradation of C.I. Reactive Blue 19. *Cellulose* **2019**, *26*, 3955–3972. [CrossRef]
96. Fan, G.-Z.; Wang, Y.-X.; Song, G.-S.; Yan, J.-T.; Li, J.-F. Preparation of microcrystalline cellulose from rice straw under microwave irradiation. *J. Appl. Polym. Sci.* **2017**, *134*, 44901. [CrossRef]
97. Ahmaruzzaman, M. Biochar based nanocomposites for photocatalytic degradation of emerging organic pollutants from water and wastewater. *Mater. Res. Bull.* **2021**, *140*, 111262. [CrossRef]
98. Tavker, N.; Gaur, U.K.; Sharma, M. Agro-waste extracted cellulose supported silver phosphate nanostructures as a green photocatalyst for improved photodegradation of RhB dye and industrial fertilizer effluents. *Nanoscale Adv.* **2020**, *2*, 2870–2884. [CrossRef]

Disclaimer/Publisher's Note: The statements, opinions and data contained in all publications are solely those of the individual author(s) and contributor(s) and not of MDPI and/or the editor(s). MDPI and/or the editor(s) disclaim responsibility for any injury to people or property resulting from any ideas, methods, instructions or products referred to in the content.

Article

Effectiveness of Dimple Microtextured Copper Substrate on Performance of Sn-0.7Cu Solder Alloy

Siti Faqihah Roduan [1], Juyana A. Wahab [1,*], Mohd Arif Anuar Mohd Salleh [1,2], Nurul Aida Husna Mohd Mahayuddin [1], Mohd Mustafa Al Bakri Abdullah [1,2], Aiman Bin Mohd Halil [3], Amira Qistina Syamimi Zaifuddin [3], Mahadzir Ishak Muhammad [3], Andrei Victor Sandu [4,5], Mădălina Simona Baltatu [4,*] and Petrica Vizureanu [4,6,7]

[1] Faculty of Chemical Engineering and Technology, Kompleks Pusat Pengajian Jejawi 2, Universiti Malaysia Perlis (UniMAP), Arau 02600, Malaysia
[2] Geopolymer and Green Technology, Centre of Excellence (CEGeoGTech), Universiti Malaysia Perlis (UniMAP), Arau 02600, Malaysia
[3] Faculty of Mechanical and Automotive Engineering Technology, Universiti Malaysia Pahang, Pekan 26600, Malaysia
[4] Department of Technologies and Equipments for Materials Processing, Faculty of Materials Science and Engineering, Gheorghe Asachi Technical University of Iași, Blvd. Mangeron, No. 51, 700050 Iasi, Romania
[5] Romanian Inventors Forum, Str. Sf. P. Movila 3, 700089 Iasi, Romania
[6] National Institute for Research and Development in Environmental Protection INCDPM, Splaiul Independentei 294, 060031 Bucharest, Romania
[7] Technical Sciences Academy of Romania, Dacia Blvd. 26, 030167 Bucharest, Romania
* Correspondence: juyana@unimap.edu.my (J.A.W.); cercel.msimona@yahoo.com (M.S.B.)

Abstract: This paper elucidates the influence of dimple-microtextured copper substrate on the performance of Sn-0.7Cu solder alloy. A dimple with a diameter of 50 μm was produced by varying the dimple depth using different laser scanning repetitions, while the dimple spacing was fixed for each sample at 100 μm. The dimple-microtextured copper substrate was joined with Sn-0.7Cu solder alloy using the reflow soldering process. The solder joints' wettability, microstructure, and growth of its intermetallic compound (IMC) layer were analysed to determine the influence of the dimple-microtextured copper substrate on the performance of the Sn-0.7Cu solder alloy. It was observed that increasing laser scan repetitions increased the dimples' depth, resulting in higher surface roughness. In terms of soldering performance, it was seen that the solder joints' average contact angle decreased with increasing dimple depth, while the average IMC thickness increased as the dimple depth increased. The copper element was more evenly distributed for the dimple-micro-textured copper substrate than its non-textured counterpart.

Keywords: laser surface texturing; dimple microtexture; lead-free solder; wettability; intermetallic compound

1. Introduction

Biomimetics, also known as biomimicry, is the imitation of nature's movements, materials, and processes. Nature observance and the knowledge gained can be utilised to form new ideas. Engineers have utilised biomimicry to develop and improve new products by imitating the characteristics of plants and animals, and biomimicry products are now everywhere, including everyday items.

An example of a nature-inspired product is Velcro, which owed its genesis to the behaviour of cockleburs sticking to animals and clothes. Cockleburs possess tiny hooks that allow them to adhere tightly to the loops of fur or fabric. Other well-known products are window glass, self-cleaning exterior paint, and umbrella fabric. These products were mimicked from the lotus flower plant, which is made up of tiny bumps covered by waxy crystals to produce a self-cleaning mechanism [1]. Most natural organisms possess characteristic micro/nano-texture designed to ensure their survival and adaptation. These microtextures have been researched extensively in various applications and, when possible, exploited to enhance materials' performance, specifically tribology, friction, and wear [2–4].

Textures are usually tailored per their intended effects. Some typical surface textures include grids, riblets, grooves, and dimples. An example is a golf ball with dimples on its surface. The dimples are designed to enhance its aerodynamic performance [5] by producing a thin, turbulent boundary layer of air that clings to the surface. This permits the air to flow around the ball's surface, culminating at the rear side of the ball, which minimises the air resistance that opposes the golf ball's flight direction.

A study in brazing application has found that the convex platform and groove-textured surfaces fabricated on the Ti_3SiC_2 ceramic surface effectively decreased the wetting angle of the brazing filler metal from 59.6° to 25.7° [6]. According to the research, the brazing filler metal fills the microstructural groove via a capillary action mechanism. Another reported observation was that the Ti3SiC2 edge's diffusion layer width was more significant when it had grooves on its surface. This was because grooves facilitate the migration of Ag and Cu atoms into the Ti3SiC2 ceramic and directly improve the diffusion mechanism during the brazing process.

Li et al. [7] manufactured micro-grooves on Ti_6Al_4V substrates and reported that the wettability was enhanced only when the groove spacing exceeded 0.25 mm. Textured surfaces are commonly used to improve the adhesion between solid substrate surfaces.

Zhiyang Liu et al. [8] fabricated micro/nano-ripples, micro-grooves, and micro-pits on stainless steel using an ultrafast laser to study the wetting and spreading behaviours of aluminium-silicon (Al-Si). They reported that the fabricated micro/nano-ripples successfully decreased the contact angle from 10.8° to 8.7°, enhancing the Al-Si's wettability. They also reported that the thickness of the reaction layer increased due to the presence of the microtexture, which increased the overall surface area. This process significantly improved the interfacial diffusion reaction, thickening the interfacial IMCs layer.

The literature confirms that the presence of microtextures enhances materials' performance in various applications. However, although there have been many studies involving the development of microtextures on material surfaces, minimal information is available on the advantages of microtextured surfaces in soldering.

A microtextured surface can affect solder alloy's spreading area and wettability. The surface texture increases the diffusion area for the copper atom, which increases its migration to the solder alloy. Due to the diffusion and migration processes, an increased concentration of surface reaction will occur during the soldering process, which affects the microstructure and intermetallic compound (IMC) layer formation. The microstructure and IMC formation in a solder joint depends on the type of solder alloy and surface finish of the substrate [9].

The literature posits that the Sn-Cu family of alloys is preferred over other proposed Pb-free solder alloys, and Sn-0.7Cu is the best option due to its low cost and excellent performance [10,11]. The performance of the solder is important in electronic packaging, in which the solder acts as the main interconnect between a printed circuit board (PCB) and electronic components. They also provide electrical, thermal, and mechanical continuity in electronic assemblies. Simply put, the integrated circuit (IC) chip needs to be packaged in a way that it can work with bigger surroundings and function as a single system in an electronic device.

However, other researchers reported several performance problems using the Sn-0.7Cu solder alloy. For example, Gourlay et al. [12] stated that as Sn-0.7Cu solidifies, its flow ability and wettability decrease [13,14], while Mohd Salleh et al. [15] reported that Sn-0.7Cu suffers from poor microstructure refinement, which affected the solder joint's mechanical properties.

Many researchers attempted to address these issues by incorporating alloying elements into the solder or altering its surface finish [16–18]. For example, Jaffery et al. [19] studied the effect of adding Iron (Fe) and Bismuth (Bi) in Sn-0.7Cu solder alloy via the resulting solder's oxidation and wetting characteristics. They reported that the wetting properties of Sn-0.7Cu improved due to the addition of Bi and Fe into the Sn-0.7Cu solder alloy. Teoh et al. [20] reported that the mechanical strength of Sn-0.7Cu improved by adding Bi into the solder alloy. Hanim et al. [21] reported that the electroless nickel immersion silver (ENIAg) surface finish significantly enhanced the shear strength of the solder joint, as it provided a more stable metallurgical bond of the solder joint. Despite the abovementioned studies and others involving the improvement of the performance of Sn-0.7Cu, studies on the modification of the substrate used in the soldering process remain limited.

Therefore, this study looks into the fabrication of a microtexture on a substrate's surface to improve the performance of Pb-free solder in soldering applications. Surface texturing was used to create a suitable microtexture capable of generating excellent interfacial reaction between solder alloy and copper substrate.

2. Materials and Methods

2.1. Raw Materials

A flat, square-shaped, pure copper pad with a purity of 99.9% measuring 15 mm × 15 mm × 1 mm was used as a substrate, while the Sn-0.7Cu ingot was used as a solder material.

2.2. Fabrication of Dimple Micro-Texture on Copper Substrate

The copper pad sample was ground using various grit sizes of silicon carbide (SiC) paper, then polished using 1 μm alumina polishing paste to remove uneven surfaces, scratches, and impurities. Next, a closed-pore dimple was developed on the polished copper surface using LST. An Ytterbium Fibre Laser Marking Machine (Herolaser; Shenzhen, China) produced a dimple micro-texture with a diameter, d, of 50 μm and a dimple spacing, S, of 100 μm. Finally, 3 samples were fabricated by varying the dimple depth, D, using different scan repetitions set at 1 to 3. The LST process parameters are summarised in Table 1, while the dimple microtexture on the copper substrate is shown in Figure 1.

Table 1. Laser surface texturing parameters.

Parameter	Value
Name	Ytterbium Fiber Laser Marking Machine
Model	ML-MF-A01
Wavelength (nm)	1064
Scanning speed (mm/s)	4000
Laser power (W)	70
Frequency (kHz)	20

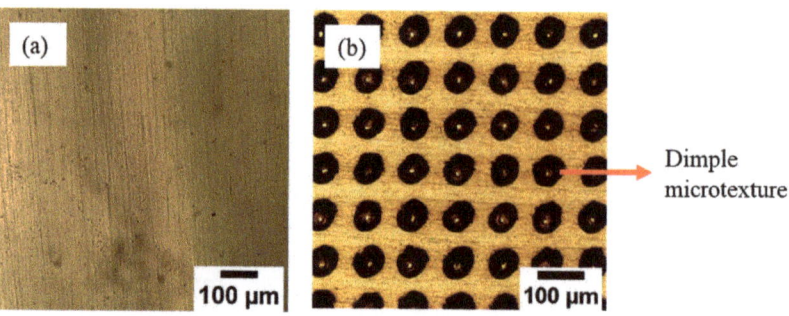

Figure 1. The surface of the copper substrate, (**a**) before surface texturing and (**b**) after surface texturing.

2.3. Fabrication of Sn-0.7Cu Solder Joint

To prepare the solder ball, the Sn-0.7Cu solder ingots were melted at 350 °C in a solder pot. The ingot was cold-rolled to produce thin solder sheets, then cut into small pieces, each weighing ~0.4 g. The solder sheet was placed on a Pyrex sheet with a small amount of rosin mildly activated (RMA) flux and reflowed in a reflow oven at 250 °C and N2 gas flow. Then, the solder balls were cleaned and rinsed thoroughly using acetone and placed on the copper substrate's surface with a small flux. Next, the reflow soldering process was used to fabricate the solder joint, using an F4N Pb-free reflow oven via a lead (Pb)-free reflow profile, per Figure 2. The solder joint was then mounted, using a mixture of epoxy resin and hardener, and cross-sectioned before the metallographic step.

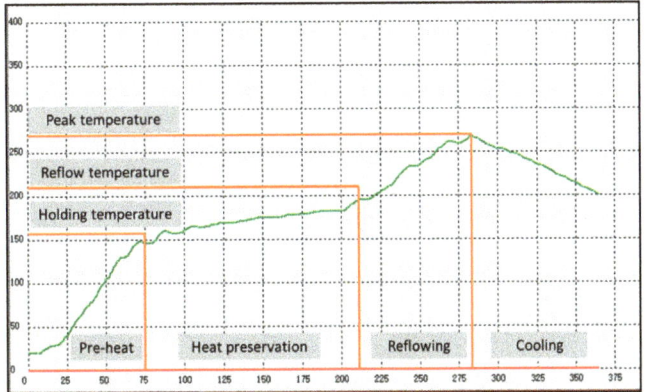

Figure 2. Reflow profile during the reflow soldering process.

2.4. Testing and Characterisation

2.4.1. Microstructure Characterisation of the Textured Copper Substrate

The dimple-microtextured copper substrate was polished using 1 μm polishing paste for 1 min at 50 rpm to remove any resolidified material on the copper substrate. The surface morphology and roughness of the dimple-microtextured surface copper substrate were then imaged using a 3D measuring laser microscope (Model Olympus OLS5000, Shinjuku, Japan).

2.4.2. Wettability of Sn-0.7Cu Solder Joint

The wettability of the solder joint was evaluated using contact angle measurement between the molten solder and copper substrate. First, the cross-sectional area of the solder joint was imaged using an optical microscope (OM), and then the contact angle was measured using Image-J.

2.4.3. Microstructure Characterisation of Sn-0.7Cu Solder Joint

The solder joint's microstructure analysis included the bulk solder's microstructure and the thickness of the intermetallic compound (IMC) layer. The microstructural imaging of the solder substrate was carried out using OM to image the microstructure at the bulk solder. Additionally, the area along the IMC layer was measured, and its thickness was averaged per Equation (1).

$$IMC = Area\ (A)/Length\ (L), \tag{1}$$

2.4.4. Distribution of Copper Element in Solder Joint

Synchrotron Micro-X-ray Fluorescence spectroscopy and Imaging (SR-μ-XRF) was used to determine the distribution of the copper element in the solder joint. The thickness of the mounted sample was 5 mm, while its diameter was 25 mm. The samples were thoroughly cleaned using acetone in an ultrasonic cleaner to remove any impurities. The Synchrotron µ-XRF was performed at the BL6b beamline at Synchrotron Light Research Institute (SLRI), Thailand. At the beamline, a continuous synchrotron was produced from the bending magnet with an energy range of up to 12 keV and a beam size of 30 µm. The exposure time was 20 s for each spot with a step size of 0.05 mm in a helium (He) atmosphere. The data obtained were analysed using PyMca software (version 5.5.3).

2.4.5. Lap Shear Strength of Sn-0.7Cu Solder Joint

A single-lap shear test was carried out using an Instron Universal Tensile Testing Machine, based on the ASTM D1002 with Cu substrate specifications of 101.6 mm × 25.4 mm × 1.5 mm. First, ~1 g of Sn-0.7Cu solder sheets were sandwiched between the two copper substrates, as illustrated in Figure 3. Next, the samples were subjected to a reflow soldering process using a tabletop Pb-free reflow oven model F4N. The samples were shear tested at a shear speed of 2 mm/min with a strain rate of $5 \times 10^{-4} \text{ s}^{-1}$, then analysed using Scanning Electron Microscope (SEM) for fracture analyses.

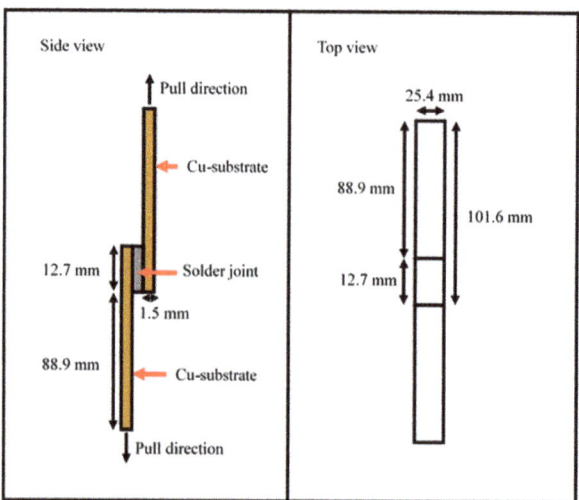

Figure 3. Schematic diagram of Sn-0.7Cu solder joint sample for lap-shear test.

3. Results and Discussions

3.1. Surface Profile of the Copper Substrate

Figure 4 shows the 2D and 3D images of the dimple-microtextured copper substrates. It can be seen that a well-defined dimple shape was developed and arranged evenly on the copper surface, most likely due to the high precision control of the LST process [22]. The LST process involves heat transfer, where the material absorbs heat from laser radiation. Once the material absorbs the heat, the thermalisation process occurs, increasing the surface temperature, which promotes the melting of the copper substrates, and the melted part is subsequently ejected onto the surface. This process alters the materials' surface morphology and topography [23].

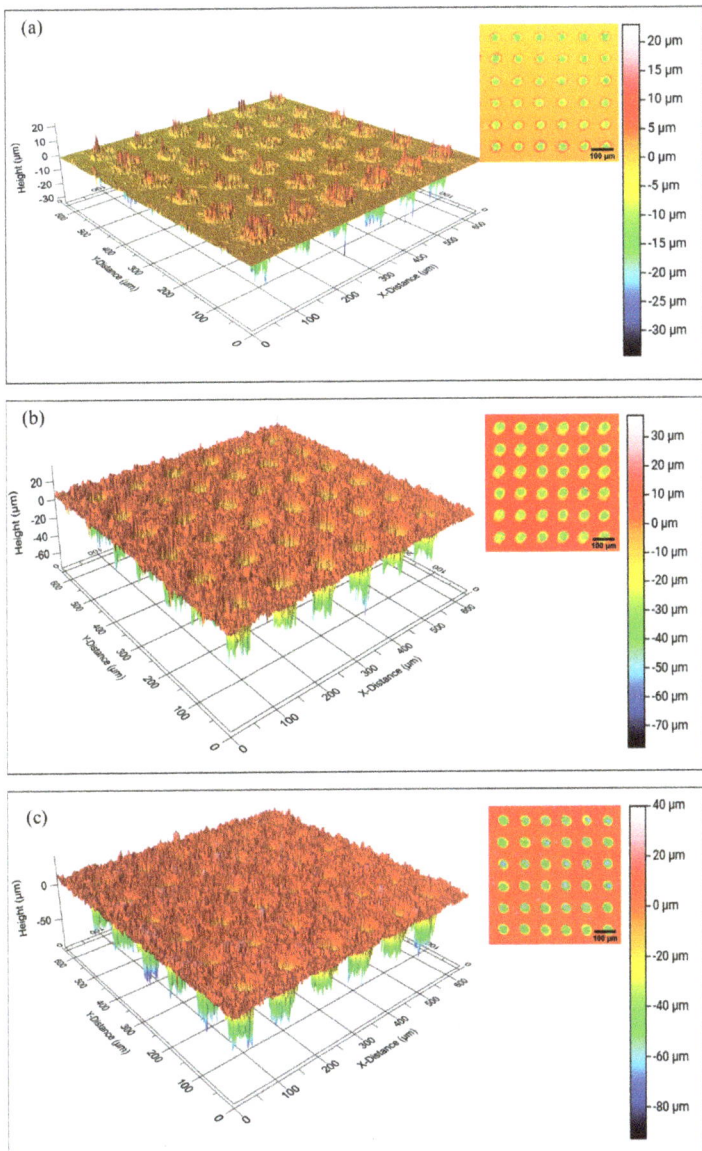

Figure 4. Images of the copper substrate with different depths of dimple, (**a**) 10 μm, (**b**) 30 μm, and (**c**) 50 μm.

The number of laser scan repetitions influenced the dimple depth, and it can be seen that the higher number of scan repetitions increases the dimple's depth. For example, the average depth of a dimple for 1 scan repetition was 10 ± 5 μm, while for 2 scan repetitions, it was 30 ± 5 μm. The deepest dimple depth was for 3 scan repetitions, which was 50 ± 5 μm. During the LST process, a higher number of laser scan repetitions compelled the laser beam to stay in contact longer than a smaller number of laser scan repetitions. As the laser beam was in contact with the material longer, the material was able to absorb more laser energy, which meant more material melted and ejected to the surface, deepening the dimple depth.

When a higher number of laser scan repetitions with low pulse energy were used, the ablation threshold decreased, resulting in an increase in material removal by one laser pulse [24]. As this process occurs, the increased dimple depth directly increases the surface area, thus increasing the surface roughness of the copper substrate. The surface profiles of the dimple-microtextured copper substrate with different dimple depths are shown in Figure 5.

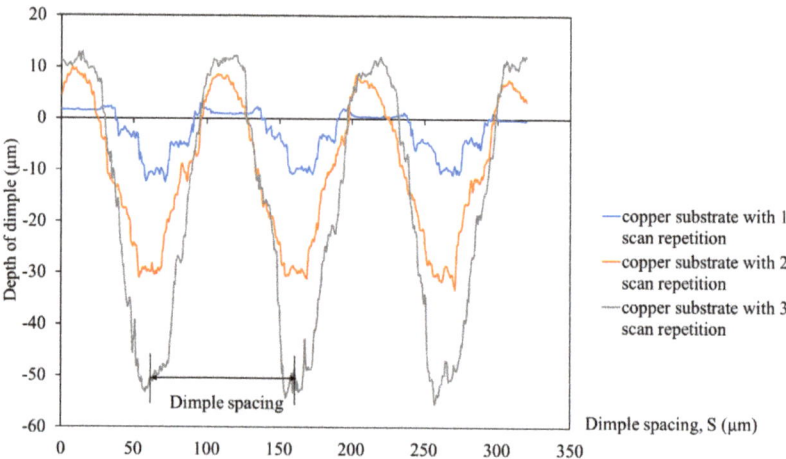

Figure 5. Surface profiles of the dimple–microtextured copper substrate with a different number of laser scan repetitions.

3.2. Solderability of Sn-0.7Cu on Copper-Textured Surfaces

The performance of the dimple-microtextured copper substrate in soldering applications was investigated. The average contact angle of the solder joint and the copper substrates' surface roughness is shown in Figure 6. It can be seen that the highest contact angle is for the sample with a dimple depth of 10 μm, which is 36.5°, while the lowest contact angle is for the sample with a dimple depth of 50 μm, which is 35.1°. It can also be seen that an increase in dimple depth decreased the average contact angle of the Sn-0.7Cu solder, which enhances the wettability of the Sn-0.7Cu solder. During wetting, the molten solder flowed into the valleys via capillary action [25], and Way et al. [26] stated that when the soldering process occurs, the molten solder will start melting and be attracted to the copper substrate, and capillary action pulls the solder into the dimples. However, this phenomenon disturbs the wetting process, as the molten solder must overcome the dimples' energy barriers before thoroughly wetting the copper's surface [27]. Therefore, as the dimple depth increases, a high volume of molten solder is needed to fill in the dimple, leading to a low volume of molten solder remaining on the surface, which decreases the contact angle.

The surface roughness of the microtextured copper substrate affects the solder joint's average contact angle. It can be seen in Figure 6 that the highest surface roughness is for samples with a dimple depth of 50 μm, which is 8.07 μm, while the lowest surface roughness is for the samples with a dimple depth of 10 μm, which is 1.06 μm. Therefore, it can be surmised that a higher number of laser scan repetitions results in deeper dimples produced on the copper substrate. This can be attributed to the formation of dimples modifying the surface profile of the copper substrate and providing additional surface area, which increases surface roughness [28,29]. This observation is consistent with Pratap and Patra [30] and Zhang et al. [31], who reported that an increase in microtexture depth increases surface roughness.

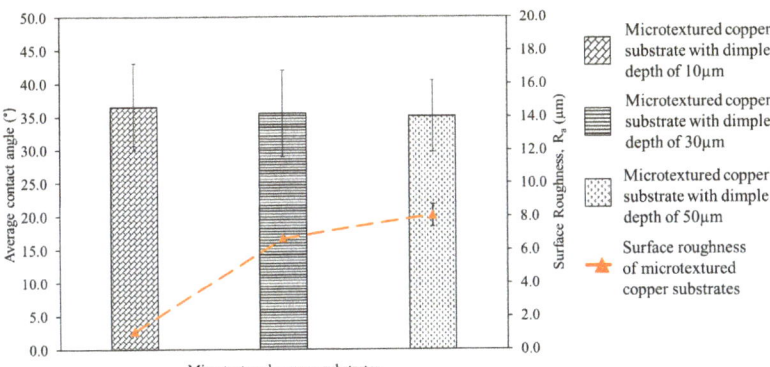

Figure 6. The average contact angle of the solder joint and surface roughness of microtextured copper substrates.

It was observed that the surface roughness is inversely proportional to the contact angle of the solder joint, per Yulong et al. [32], who reported that the contact angle of Sn-35Bi-1Ag solder decreased when the surface roughness of the copper substrates increased, indicating that the solder wettability improved. The result is also consistent with Wu et al. [33], who reported that the substrate's increased surface roughness significantly enhances the wetting properties.

3.3. Microstructure Analysis of Sn-0.7Cu Solder

The analytical results of the bulk microstructure of Sn-0.7Cu solder are shown in Figure 7. During the reflow soldering process, the copper element from the substrate tends to diffuse in the solder alloy and react with tin (Sn) due to copper diffusion. From the reactions taking place during the process, it was observed that two phases formed on the bulk area. Typically, the microstructure of Sn-0.7Cu solder alloy that forms on the substrate consists of a primary β-Sn phase or dendrites surrounded by eutectic, which is a combination of β-Sn and intermetallic Cu6Sn5 [34]. The microstructure of the solder is composed of two visible regions: eutectic and β-Sn phases. The light region represents the β-Sn phase, while the dark region represents the eutectic phase. As seen in the figure, the primary IMC increased with the dimple depth. Furthermore, it was found that the β-Sn area becomes refined and finely dispersed in the bulk microstructure. This phenomenon occurs due to the high amount of copper that dissolves in the solder alloy and reduces the β-Sn area [35]. Shen et al. [36] stated that during the solidification, the Cu atom reacts with Sn, forming intermetallic Cu6Sn5 particles and refining the grain size of the Bi-rich phase, simultaneously acting as heterogeneous nucleation sites. This finding is also similar to Hung et al. [37], who reported that the β-Sn was refined and the primary IMC increased as the copper content increased.

Figure 8 shows the micrograph of the interfacial IMC layer of the Sn-0.7Cu solder on different dimple depths of the microtextured copper substrate. It can be seen that a continuous IMC layer made up of Cu6Sn5 is present at the interface between the Sn-0.7Cu solder and copper substrates. It is also evident that the continuous layer of IMC gradually thickened as the dimple depth increased. This IMC layer will form when the solder alloy reacts with the substrate during the reflow soldering process. When the reflow soldering process begins, the element from the substrate will dissolve into the molten solder. The molten solder is then concentrated with the metals, creating a layer of IMC between the interface of the solder alloy and the substrate. The morphology of the IMC layer consists of pointed and shallow scallops, with more of the latter compared to the former in the IMC layer with increasing dimple depth. The formation of pointed scallops in the IMC layer is undesirable, as it will induce a brittle fracture at the solder joint interface compared to the shallow scallop structure [38].

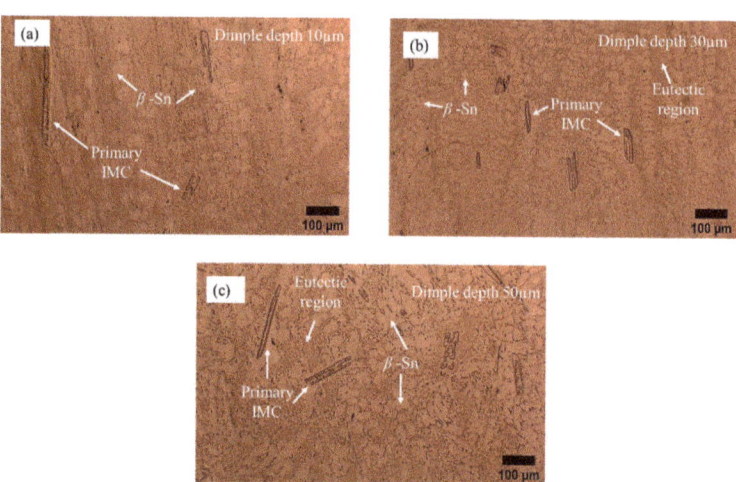

Figure 7. Microstructure analysis of Sn-0.7Cu solder for bulk microstructure on a microtextured copper substrate with dimple depth of (**a**) 10 μm, (**b**) 30 μm, and (**c**) 50 μm.

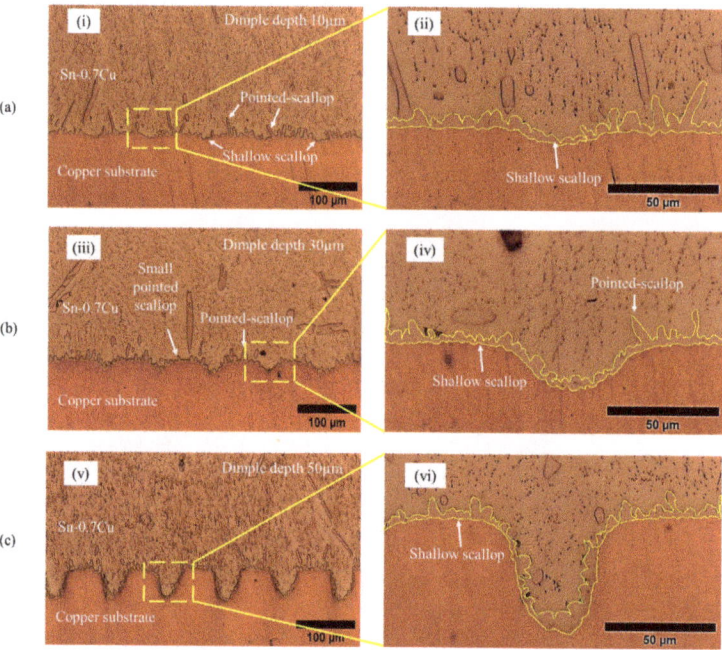

Figure 8. Micrograph of interfacial IMC layer on microtextured copper substrate with a dimple depth of (**a**) 10 μm, (**i**) magnification of 10×, (**ii**) magnification of 40×; (**b**) 30 μm, (**iii**) magnification of 10×, (**iv**) magnification of 40×; and (**c**) 50 μm, (**v**) magnification of 10×, (**vi**) magnification of 40×.

The IMC thickness was averaged to determine the difference in the growth of IMC layers among dimple-microtextured copper substrates, and the results are shown in Figure 9.

It can be seen that the highest average IMC is for the sample with a dimple depth of 50 μm, which is 8.8 μm, while the lowest average is for the sample with a dimple depth of 10 μm, which is 4.3 μm. It is also evident that the average thickness of the interfacial IMC layer substantially increases as the dimple depth increases. The increment in the thickness of the IMC layer for the increased dimple depth can be attributed to the influence of the surface area of the copper substrate. During wetting, chemical reactions and diffusion processes occur where the molten solder dissolves the substrate metal in a process called reactive wetting [39]. As the Sn-0.7Cu solder alloy reflows on the dimple-microtextured copper substrate, more copper diffused into the solder alloy due to the dimple microtextures on the surface providing an additional surface area that promotes diffusion and chemical reaction at the interface of the solder joint [7]. As per Dong et al. [40], due to the presence of peaks and valleys on the surface, a larger surface area was obtained between the interface of the solder joint. This stage results in the faster diffusion of the copper atoms into the solder alloy. Hence, a thick IMC layer was formed at the interface of the solder substrate. This result is supported by the distribution of copper elements in the solder joint, per Figure 10, and is discussed in the following section.

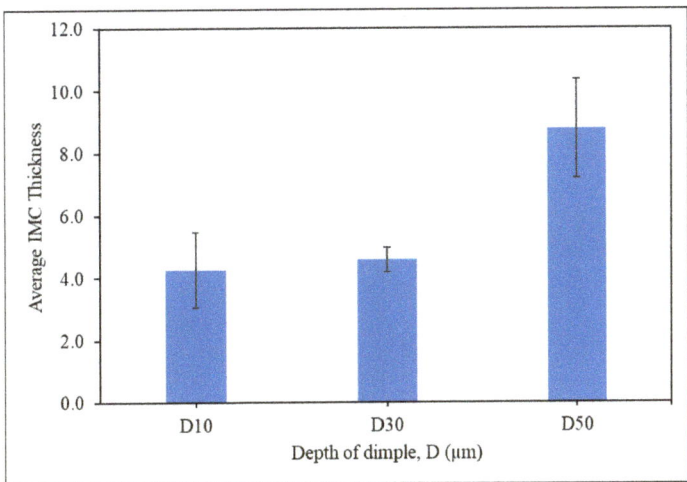

Figure 9. Average IMC thickness of solder joint.

3.4. Distribution of Copper Element in Solder Joint

The Synchrotron Micro-X-ray Fluorescence (SR-μ-XRF) mapping was carried out to determine the distribution of copper elements in the Sn-0.7Cu solder joint. Figure 10 shows the results of SR-μ-XRF mapping. The red and blue markings on the map indicate the highest and lowest concentrations of the copper element, respectively. It can be seen that more copper elements were distributed to the solder area for the dimple-microtextured copper substrate compared to the non-textured copper substrate. The copper element, represented in red, is more evenly distributed throughout the solder balls area for the dimpled copper substrate. The highest copper concentration level in the solder ball area is shown in Figure 10b. These results prove that the dimple developed on the surface provides more surface area, increasing copper dissolution from substrates. According to Chen et al. [6], the grooves' surface microstructure successfully enhances the diffusion mechanism where the grooves promote silver and copper atoms diffusing into titanium silico-carbide (Ti3SiC2) ceramics [41,42].

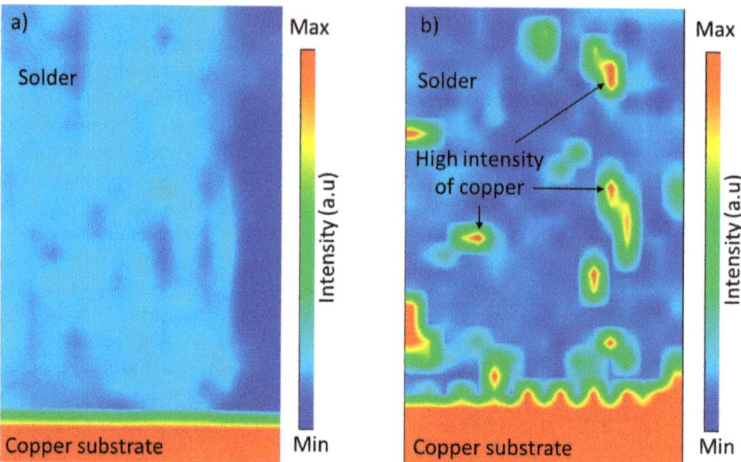

Figure 10. Synchrotron micro-XRF results from Cu mapping distribution of (**a**) non-textured copper substrate, (**b**) dimple-microtextured copper substrate with a dimple depth of 30 µm.

3.5. Lap Shear Strength of Sn-0.7Cu Solder Joint

The strength and the fracture behaviour of the Sn-0.7Cu Pb-free solder on dimple-microtextured copper substrate were determined. The shear strength of Sn-0.7Cu solder on the copper-textured surface is shown in Figure 11. It can be seen that the highest strength is for the sample with a dimple depth of 50 µm, which is 21.93 MPa, while the sample with a dimple depth of 10 µm reported the lowest strength, which is 20.93 MPa. The strength of the solder joint increases with increasing dimple depth due to the dimples on the copper substrates, which affords a greater mechanical interlocking that improves the joint bonding strength [43,44]. As a result, the higher dimple depth needed extra energy to pull the microtextures out during the debonding of the joint [45].

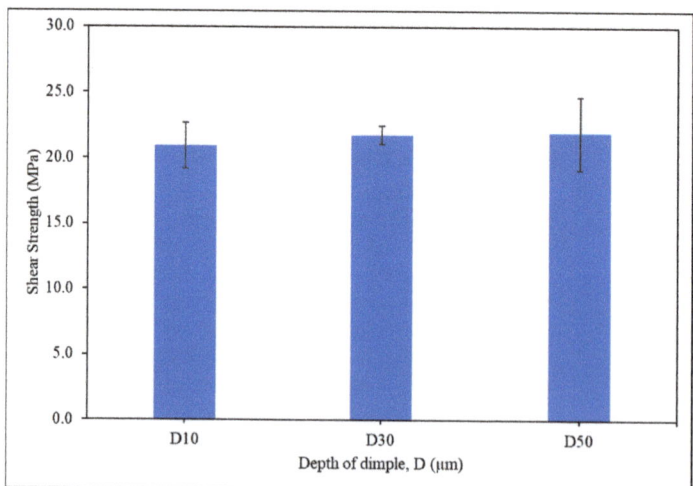

Figure 11. Shear strength of Sn-0.7Cu solder joint.

Figure 12 shows the fractured surface morphology obtained from SEM images for the Sn-0.7Cu Pb-free solder on a microtextured copper substrate. Based on the images, it can be surmised that the Sn-0.7Cu solder alloy has a ductile fracture mode, per the circle shape seen in the figure. The dimples are also evident in the Sn-0.7Cu solder, confirming the material's ductile mode [46]. It can also be confirmed that there is no significant difference in the fracture behaviour of the solder joint.

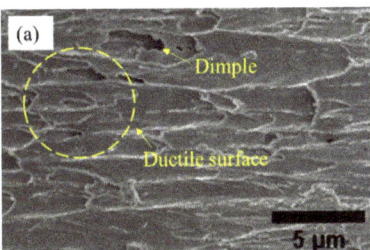

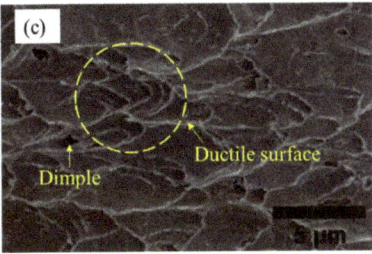

Figure 12. SEM images of fracture surface Sn-0.7Cu lead-free solder on different dimple depths of (**a**) 10 μm, (**b**) 30 μm, and (**c**) 50 μm.

4. Conclusions

A dimple microtexture was successfully developed on a copper substrate to enhance the surface characteristics. It was observed that increased laser scanning repetitions resulted in increased dimple depth. The deepest dimple, 50 μm, was obtained using 3 scan repetitions. In the case of the soldering application, it was confirmed that the increased dimple depth decreased the contact angle of the solder joint from 36.5° to 35.1°. It was also confirmed that the surface roughness of the copper substrates increased with increasing dimple depth. The microstructure at the bulk solder showed that more intermetallic compound Cu_6Sn_5 was present in the bulk solder area per increasing dimple depth. In the case of the interfacial IMC, it was revealed that the deepest dimple, which was 50 μm, has the thickest IMC layer at 8.8 μm. Through copper element mapping, the dimple microtexture on the copper substrate increased the diffusion area for the copper atom to migrate into the solder alloy. The highest strength of the Sn-0.7Cu solder alloy was reported for the sample with a dimple depth of 50 μm, which was obtained at 21.93 MPa.

Author Contributions: Conceptualization, S.F.R., J.A.W. and M.A.A.M.S.; methodology, M.M.A.B.A., N.A.H.M.M. and P.V.; software, A.V.S.; validation, A.B.M.H., A.Q.S.Z. and M.S.B.; formal analysis, M.I.M., A.V.S. and M.S.B.; investigation, S.F.R., J.A.W., M.A.A.M.S. and M.M.A.B.A.; visualization, M.M.A.B.A., N.A.H.M.M. and A.B.M.H.; supervision, M.M.A.B.A.; project administration, A.V.S. All authors have read and agreed to the published version of the manuscript.

Funding: This research was funded by the Ministry of Education (MoE), Malaysia, under the Fundamental Research Grant Scheme (FRGS/1/2019/TK05/UNIMAP/03/3). The paper was also supported by an innovations grant of the TUIASI, project number MedTech_8/2022.

Institutional Review Board Statement: Not applicable.

Informed Consent Statement: Not applicable.

Data Availability Statement: Not applicable.

Acknowledgments: The authors gratefully acknowledge the Faculty of Chemical Engineering and Technology, Universiti Malaysia Perlis (UniMAP), for supporting this research effort through materials and facilities. The authors also acknowledge Nihon Superior Co. Ltd. for the support given throughout the research project. The μ-XRF trace element mapping technique was performed at the BL6b beamline of Synchrotron Light Research Institute (SLRI), Thailand.

Conflicts of Interest: The authors declare no conflict of interest.

References

1. Cohen, Y.H.; Reich, Y.; Greenberg, S. What can we learn from biological systems when applying the law of system completeness? *Procedia Eng.* **2015**, *131*, 104–114. [CrossRef]
2. Mao, B.; Siddaiah, A.; Liao, Y.; Menezes, P.L. Laser surface texturing and related techniques for enhancing tribological performance of engineering materials: A review. *J. Manuf. Process.* **2020**, *53*, 153–173. [CrossRef]
3. Singh, A.; Patel, D.S.; Ramkumar, J.; Balani, K. Single step laser surface texturing for enhancing contact angle and tribological properties. *Int. J. Adv. Manuf. Technol.* **2019**, *100*, 1253–1267. [CrossRef]
4. Arumugaprabu, V.; Ko, T.J.; Thirumalai Kumaran, S.; Kurniawan, R.; Uthayakumar, M. A brief review on importance of surface texturing in materials to improve the tribological performance. *Rev. Adv. Mater. Sci.* **2018**, *53*, 40–48. [CrossRef]
5. Sullivan, M.J. Golf Ball Dimples. U.S. Patent 6,569,038 B2, 27 May 2003.
6. Chen, H.Y.; Wang, X.C.; Fu, L.; Feng, M.Y. Effects of surface microstructure on the active element content and wetting behavior of brazing filler metal during brazing Ti_3SiC_2 ceramic and Cu. *Vacuum* **2018**, *156*, 256–263. [CrossRef]
7. Li, H.; Li, L.; Huang, R.; Tan, C.; Yang, J.; Xia, H.; Chen, B.; Song, X. The effect of surface texturing on the laser-induced wetting behavior of AlSi5 alloy on Ti6Al4V alloy. *Appl. Surf. Sci.* **2021**, *566*, 150630. [CrossRef]
8. Liu, Z.; Yang, J.; Li, Y.; Li, W.; Chen, J.; Shen, L.; Zhang, P.; Yu, Z. Wetting and spreading behaviors of Al-Si alloy on surface textured stainless steel by ultrafast laser. *Appl. Surf. Sci.* **2020**, *520*, 146316. [CrossRef]
9. Adawiyah, M.A.R.; Azlina, O.S. Interfacial reaction between SAC305 lead–free solders and ENImAg surface finish and bare copper: Grenzflächenreaktion zwischen bleifreien Lötmetallen SAC305 und ENImAg-Oberflächenfinish und blankem Kupfer. *Materwiss. Werksttech.* **2017**, *48*, 235–240. [CrossRef]
10. Qu, D.; Li, C.; Bao, L.; Kong, Z.; Duan, Y. Structural, electronic, and elastic properties of orthorhombic, hexagonal, and cubic Cu_3Sn intermetallic compounds in Sn–Cu lead-free solder. *J. Phys. Chem. Solids* **2020**, *138*, 109253. [CrossRef]
11. Han, P.; Lu, Z.; Zhang, X. Sn-0.7Cu-10Bi Solder Modification Strategy by Cr Addition. *Metals* **2022**, *12*, 1768. [CrossRef]
12. Gourlay, C.M.; Nogita, K.; Read, J.; Dahle, A.K. Intermetallic formation and fluidity in Sn-Rich Sn-Cu-Ni alloys. *J. Electron. Mater.* **2010**, *39*, 56–69. [CrossRef]
13. Yang, W.; Lv, Y.; Zhang, X.; Wei, X.; Li, Y.; Zhan, Y. Influence of graphene nanosheets addition on the microstructure, wettability, and mechanical properties of Sn-0.7Cu solder alloy. *J. Mater. Sci. Mater. Electron.* **2020**, *31*, 14035–14046. [CrossRef]
14. Zeng, G.; Xue, S.; Zhang, L.; Gao, L. Recent advances on Sn-Cu solders with alloying elements: Review. *J. Mater. Sci. Mater. Electron.* **2011**, *22*, 565–578. [CrossRef]
15. Mohd Salleh, M.A.A.; McDonald, S.D.; Gourlay, C.M.; Belyakov, S.A.; Yasuda, H.; Nogita, K. Effect of Ni on the Formation and Growth of Primary Cu_6Sn_5 Intermetallics in Sn-0.7 wt.% Cu Solder Pastes on Cu Substrates During the Soldering Process. *J. Electron. Mater.* **2016**, *45*, 154–163. [CrossRef]
16. Amli, S.F.M.; Salleh, M.A.A.M.; Ramli, M.I.I.; Aziz, M.S.A.; Yasuda, H.; Chaiprapa, J.; Nogita, K. Effects of immersion silver (ImAg) and immersion tin (ImSn) surface finish on the microstructure and joint strength of Sn-3.0Ag-0.5Cu solder. *J. Mater. Sci. Mater. Electron.* **2022**, *33*, 14249–14263. [CrossRef]
17. Ramli, M.I.I.; Yusof, M.S.S.; Mohd Salleh, M.A.A.; Said, R.M.; Nogita, K. Influence of Bi addition on wettability and mechanical properties of Sn-0.7Cu solder alloy. *Solid State Phenom.* **2018**, *273*, 27–33. [CrossRef]
18. Kao, C.L.; Chen, T.C. Ball impact responses of Sn-1Ag-0.5Cu solder joints at different temperatures and surface finishes. *Microelectron. Reliab.* **2018**, *82*, 204–212. [CrossRef]
19. Jaffery, S.H.A.; Sabri, M.F.M.; Rozali, S.; Hasan, S.W.; Mahdavifard, M.H.; AL-Zubiady, D.A.A.S.; Ravuri, B.R. Oxidation and wetting characteristics of lead-free Sn-0.7Cu solder alloys with the addition of Fe and Bi. *Microelectron. Reliab.* **2022**, *139*, 114802. [CrossRef]
20. Teoh, A.L.; Mohd Salleh, M.A.A.; Halin, D.S.C.; Foo, K.L.; Abdul Razak, N.R.; Yasuda, H.; Nogita, K. Microstructure, thermal behavior and joint strength of Sn-0.7Cu-1.5Bi/electroless nickel immersion gold (ENIG). *J. Mater. Res. Technol.* **2021**, *12*, 1700–1714. [CrossRef]
21. Hanim, M.A.A.; Kamil, N.M.; Wei, C.K.; Dele-Afolabi, T.T.; Azlina, O.S. Microstructural and shear strength properties of RHA-reinforced Sn-0.7Cu composite solder joints on bare Cu and ENIAg surface finish. *J. Mater. Sci. Mater. Electron.* **2020**, *31*, 8316–8328. [CrossRef]
22. Ibatan, T.; Uddin, M.S.; Chowdhury, M.A.K. Recent development on surface texturing in enhancing tribological performance of bearing sliders. *Surf. Coat. Technol.* **2015**, *272*, 102–120. [CrossRef]
23. Roduan, S.F.; Wahab, J.A.; Mohd Salleh, M.A.A.; Mohd Mahayuddin, N.A.H.; Halil, A.M.; Ishak, M. Effect of LST parameter on surface morphology of modified copper substrates. *J. Tribol.* **2020**, *26*, 84–91.

24. Schille, J.; Ebert, R.; Loeschner, U.; Scully, P.; Goddard, N.; Exner, H. High repetition rate femtosecond laser processing of metals. *Front. Ultrafast Opt. Biomed. Sci. Ind. Appl. X* **2010**, *7589*, 758915. [CrossRef]
25. Satyanarayan; Prabhu, K.N. Spreading behavior and evolution of IMCs during reactive wetting of SAC solders on smooth and rough copper substrates. *J. Electron. Mater.* **2013**, *42*, 2696–2707. [CrossRef]
26. Way, M.; Willingham, J.; Goodall, R. Brazing filler metals. *Int. Mater. Rev.* **2020**, *65*, 257–285. [CrossRef]
27. Chen, Y.Y.; Duh, J.G.; Chiou, B.S. Effect of substrate surface roughness on the wettability of Sn-Bi solders. *J. Mater. Sci. Mater. Electron.* **2000**, *11*, 279–283. [CrossRef]
28. Jain, A.; Kumari, N.; Jagadevan, S.; Bajpai, V. Surface properties and bacterial behavior of micro conical dimple textured Ti6Al4V surface through micro-milling. *Surf. Interfaces* **2020**, *21*, 100714. [CrossRef]
29. Zaifuddin, A.Q.; Ishak, M. Influence of Laser Surface Texturing (LST) Parameters on the Surface Characteristics of Ti6Al4V and the Effects Influence of Laser Surface Texturing (LST) Parameters on the Surface Characteristics of Ti6Al4V and the Effects Thereof on Laser Heating. *Lasers Eng.* **2021**, *51*, 355–367.
30. Pratap, T.; Patra, K. Fabrication of micro-textured surfaces using ball-end micromilling for wettability enhancement of Ti-6Al-4V. *J. Mater. Process. Technol.* **2018**, *262*, 168–181. [CrossRef]
31. Zhang, J.; Rosenkranz, A.; Zhang, J.; Guo, J.; Li, X.; Chen, X.; Xiao, J.; Xu, J. Modified Wettability of Micro-structured Steel Surfaces Fabricated by Elliptical Vibration Diamond Cutting. *Int. J. Precis. Eng. Manuf.-Green Technol.* **2022**, *9*, 1387–1397. [CrossRef]
32. Li, Y.; Wang, Z.; Li, X.; Lei, M. Effect of temperature and substrate surface roughness on wetting behavior and interfacial structure between Sn-35Bi-1Ag solder and Cu substrate. *J. Mater. Sci. Mater. Electron.* **2020**, *31*, 4224–4236. [CrossRef]
33. Wu, M.; Chang, L.; Zhang, L.; He, X.; Qu, X. Effects of roughness on the wettability of high temperature wetting system. *Surf. Coat. Technol.* **2016**, *287*, 145–152. [CrossRef]
34. Harcuba, P.; Janeček, M. Microstructure changes and physical properties of the intermetallic compounds formed at the interface between Sn-Cu solders and a Cu substrate due to a minor addition of Ni. *J. Electron. Mater.* **2010**, *39*, 2553–2557. [CrossRef]
35. Said, R.M.; Mohamad Johari, F.H.; Mohd Salleh, M.A.A.; Sandu, A.V. The Effect of Copper Addition on the Properties of Sn-0.7Cu Solder Paste. *IOP Conf. Ser. Mater. Sci. Eng.* **2018**, *318*, 012062. [CrossRef]
36. Shen, J.; Pu, Y.; Yin, H.; Luo, D.; Chen, J. Effects of minor Cu and Zn additions on the thermal, microstructure and tensile properties of Sn-Bi-based solder alloys. *J. Alloys Compd.* **2014**, *614*, 63–70. [CrossRef]
37. Hung, F.Y.; Lui, T.S.; Chen, L.H.; He, N.T. Resonant characteristics of the microelectronic Sn-Cu solder. *J. Alloys Compd.* **2008**, *457*, 171–176. [CrossRef]
38. Ramli, M.I.I.; Mohd Salleh, M.A.A.; Derman, M.N.; Said, R.M.; Saud, N. Influence of Activated Carbon Particles on Intermetallic Compound Growth Mechanism in Sn-Cu-Ni Composite Solder. *MATEC Web Conf.* **2016**, *78*, 01064. [CrossRef]
39. Warren, J.A.; Boettinger, W.J.; Roosen, A.R. Modeling reactive wetting. *Acta Mater.* **1998**, *46*, 3247–3264. [CrossRef]
40. Dong, C.; Shang, M.; Ma, H.; Wang, Y.; Ma, H. Effect of substrate surface roughness on interfacial reaction at Sn-3.0Ag/(001)Cu interface. *Vacuum* **2022**, *197*, 110816. [CrossRef]
41. Baltatu, M.S.; Vizureanu, P.; Sandu, A.V.; Munteanu, C.; Istrate, B. Microstructural analysis and tribological behavior of Ti-based alloys with a ceramic layer using the thermal spray method. *Coatings* **2020**, *10*, 1216. [CrossRef]
42. Spataru, M.C.; Cojocaru, F.D.; Sandu, A.V.; Solcan, C.; Duceac, I.A.; Baltatu, M.S.; Voiculescu, I.; Geanta, V.; Vizureanu, P. Assessment of the effects of si addition to a new TiMoZrTa system. *Materials* **2021**, *14*, 7610. [CrossRef] [PubMed]
43. Shukla, P.; Waugh, D.G.; Lawrence, J.; Vilar, R. *Laser Surface Structuring of Ceramics, Metals and Polymers for Biomedical Applications: A Review*; Elsevier Ltd.: Amsterdam, The Netherlands, 2016; ISBN 9780081009420.
44. Liu, J.; Dai, Y.; Shi, Y.; Cui, W.; Jiang, T. Effect of Surface Texture on Tensile Shear Strength of 1060Al-PET Welding Joints. *Chin. J. Mech. Eng. (Engl. Ed.)* **2021**, *34*, 134. [CrossRef]
45. Jiang, D.; Long, J.; Han, J.; Cai, M.; Lin, Y.; Fan, P.; Zhang, H.; Zhong, M. Comprehensive enhancement of the mechanical and thermo-mechanical properties of W/Cu joints via femtosecond laser fabricated micro/nano interface structures. *Mater. Sci. Eng. A* **2017**, *696*, 429–436. [CrossRef]
46. Zaimi, N.S.M.; Salleh, M.A.A.M.; Abdullah, M.M.A.B.; Nadzri, N.I.M.; Sandu, A.V.; Vizureanu, P.; Ramli, M.I.I.; Nogita, K.; Yasuda, H.; Sandu, I.G. Effect of Kaolin Geopolymer Ceramics Addition on the Microstructure and Shear Strength of Sn-3.0Ag-0.5Cu Solder Joints during Multiple Reflow. *Materials* **2022**, *15*, 2758. [CrossRef] [PubMed]

Disclaimer/Publisher's Note: The statements, opinions and data contained in all publications are solely those of the individual author(s) and contributor(s) and not of MDPI and/or the editor(s). MDPI and/or the editor(s) disclaim responsibility for any injury to people or property resulting from any ideas, methods, instructions or products referred to in the content.

Article

Influence of Polyformaldehyde Monofilament Fiber on the Engineering Properties of Foamed Concrete

Md Azree Othuman Mydin [1,*], Mohd Mustafa Al Bakri Abdullah [2,3], Mohd Nasrun Mohd Nawi [4], Zarina Yahya [2], Liyana Ahmad Sofri [2], Madalina Simona Baltatu [5,*], Andrei Victor Sandu [5,6,7] and Petrica Vizureanu [5,8]

1 School of Housing, Building and Planning, Universiti Sains Malaysia, Gelugor 11800, Penang, Malaysia
2 Faculty of Chemical Engineering & Technology, Universiti Malaysia Perlis, Arau 01000, Perlis, Malaysia
3 Centre of Excellence Geopolymer and Green Technology (CEGeoGTech), Universiti Malaysia Perlis (UniMAP), Arau 01000, Perlis, Malaysia
4 Disaster Management Institute (DMI), School of Technology Management and Logistics, Universiti Utara Malaysia, Sintok 06010, Kedah, Malaysia
5 Department of Technologies and Equipments for Materials Processing, Faculty of Materials Science and Engineering, Gheorghe Asachi Technical University of Iași, Blvd. Mangeron, No. 51, 700050 Iasi, Romania
6 Romanian Inventors Forum, Str. Sf. P. Movila 3, 700089 Iasi, Romania
7 National Institute for Research and Development in Environmental Protection INCDPM, Splaiul Independentei 294, 060031 Bucharest, Romania
8 Technical Sciences Academy of Romania, Dacia Blvd 26, 030167 Bucharest, Romania
* Correspondence: azree@usm.my (M.A.O.M.); cercel.msimona@yahoo.com (M.S.B.)

Abstract: Foamed concrete is considered a green building material, which is porous in nature. As a result, it poses benefits such as being light in self-weight, and also has excellent thermal insulation properties, environmental safeguards, good fire resistance performance, and low cost. Nevertheless, foamed concrete has several disadvantages such as low strength, a large amount of entrained air, poor toughness, and being a brittle material, all of which has restricted its usage in engineering and building construction. Hence, this study intends to assess the potential utilization of polypropylene fibrillated fiber (PFF) in foamed concrete to enhance its engineering properties. A total of 10 mixes of 600 and 1200 kg/m^3 densities were produced by the insertion of four varying percentages of PFF (1%, 2%, 3%, and 4%). The properties assessed were splitting tensile, compressive and flexural strengths, workability, porosity, water absorption, and density. Furthermore, the correlations between the properties considered were also evaluated. The outcomes reveal that the foamed concrete mix with 4% PFF attained the highest porosity, with approximately 13.9% and 15.9% for 600 and 1200 kg/m^3 densities in comparison to the control specimen. Besides, the mechanical properties (splitting tensile, compressive and flexural strengths) increased steadily with the increase in the PFF percentages up to the optimum level of 3%. Beyond 3%, the strengths reduced significantly due to poor PFF dispersal in the matrix, leading to a balling effect which causes a degraded impact of scattering the stress from the foamed concrete vicinity to another area of the PFF surface. This exploratory investigation will result in a greater comprehension of the possible applications of PFF in LFC. It is crucial to promote the sustainable development and implementation of LFC materials and infrastructures.

Keywords: foamed concrete; polypropylene fibrillated fiber; compression; flexural; tensile; porosity; water absorption

Citation: Mydin, M.A.O.; Abdullah, M.M.A.B.; Mohd Nawi, M.N.; Yahya, Z.; Sofri, L.A.; Baltatu, M.S.; Sandu, A.V.; Vizureanu, P. Influence of Polyformaldehyde Monofilament Fiber on the Engineering Properties of Foamed Concrete. *Materials* 2022, 15, 8984. https://doi.org/10.3390/ma15248984

Academic Editor: José Barroso de Aguiar

Received: 25 November 2022
Accepted: 13 December 2022
Published: 15 December 2022

Publisher's Note: MDPI stays neutral with regard to jurisdictional claims in published maps and institutional affiliations.

Copyright: © 2022 by the authors. Licensee MDPI, Basel, Switzerland. This article is an open access article distributed under the terms and conditions of the Creative Commons Attribution (CC BY) license (https://creativecommons.org/licenses/by/4.0/).

1. Introduction

The construction industry around the world has become the industry that utilizes the most energy, recording approximately fifty percent of the total amount of energy used, as urbanization has progressed, and more people have moved into cities. In the context of global climate change, the construction industry is mulling over an alternative for normal-strength concrete as a means of mitigating the extreme self-weight of the material as well as the substantial carbon discharge connected with the production of cement [1]. As a result, the implementation of environmentally friendly and energy-saving building materials as an alternative to conventional building materials has become the industry standard [2]. The porous nature of foamed concrete, which is a green building material, confers upon it several advantages, including its low cost, excellent thermal insulation properties, environmental protections, and excellent fire resistance performance [3]. Compared to normal-strength concrete, foamed concrete has a higher strength-to-weight ratio and a bulk density that ranges from 500 to 1850 kg/m^3 [4]. This distinction results in a decrease in the total dead load of the structural components, as well as a decrease in the costs of manufacturing and labor during the transportation and construction processes [5].

Foamed concrete is broadly applied in the construction industry as a filling material, as a non-loadbearing wall constituent, and as insulation material for exterior panels [6]. To deliver this porous cellular form of foamed concrete, small air bubbles of differing sizes are first inserted into the newly blended materials [7]. This is typically accomplished through the use of a foaming method that is either chemical or mechanical [8]. In its most basic form, the material component that goes into the production of foamed concrete is identical to that of normal-strength concrete, which consists of Portland cement, aggregate, and water [9]. This is despite the fact that the fabrication of foamed concrete only requires the use of fine sand. In addition, lightweight filler or reinforcement such as rice husk ash or polypropylene fiber can be combined with the foamed concrete base mix to expand its range of characteristics [10].

In spite of this, it is commonly believed that foamed concrete has poor physical, mechanical, and durability properties, which is principally due to its high porosity and void connectivity, both of which allow harmful substances to enter foamed concrete media [11]. In addition, foamed concrete has some disadvantages, including low strength, a high volume of entrapped air, poor toughness, and brittle material, which have limited its use in engineering and building construction [12]. These restrictions have decreased the use of foamed concrete [13]. As a result, it was determined that foamed concrete should not be used for principal load-bearing structural elements. Less than five percent of global use statistics were attributable to the construction industry's utilization of foamed concrete. This study investigated the use of PFF to improve the engineering qualities of foamed concrete in order to counteract its negative tendencies. Steel fiber-reinforced concrete is extensively utilized in a wide range of civil engineering projects [14] due to its low cost, high performance, and ease of production.

Despite this, a few studies have demonstrated that the uneven inclusion of steel fiber in concrete can have a negative effect on the workability of the fresh mix, leading to porous concrete and a reduction in the strength parameter [15,16]. By incorporating fibrous components into the mix, the characteristics of the concrete can be improved. The tensile and shear capacities of concrete can be enhanced by adding polypropylene fiber, according to past studies [17,18]. It is feasible for polypropylene to raise the strength of a material without increasing its dry bulk density [19]. According to some evidence, the application of propylene fiber membrane marginally increases the engineering qualities of concrete, specifically its tensile strength [20]. Compared to thick fibers, a membrane constructed of propylene fiber plays a key function in reducing plastic shrinkage [21].

Polypropylene fibrillated fiber (PFF), one of the synthetic fibers, has the potential to be utilized as an additive in foamed concrete to improve its qualities. All the PFF membrane's exceptional mechanical qualities, alkali resistance, and thermal properties contribute to its superior performance. In addition, the PFF membrane is acknowledged as a fiber that may be incorporated into a cement-based composite material which has a high resistance to cracking. The incorporation of PFF into concrete permits the formation of a three-dimensional dispersion network, which effectively suppresses the development of microcracks in cement-based materials.

On the basis of the preceding review, it can be concluded that the influence of PFF inclusion in foamed concrete on the enhancement of engineering qualities has not been explored and discussed in detail. Consequently, the fundamental objective of this study is to determine the engineering characteristics of foamed concrete, reinforced with a PFF membrane. By altering the weight percentages of PFF in the production process, densities of 600 kg/m^3 and 1200 kg/m^3 were achieved.

2. Materials and Methods

2.1. Mix Constituents

The primary materials used in this study were fine river sand that had been sieved through a 600-micron sieve with a specific gravity of 2.59 g/cm^3 and particle sizes ranging from 0.15 to 2.36 mm, as well as Portland cement that complied with the specifications of the British Standards Institution (1996). The entire required fine river sand quantity underwent a sieve examination to fit the coarse aggregate standard specification, as per what was stated in ASTM-C33. Figure 1 shows the result of the sand grading curve. Besides, using a protein-based surfactant, stable foam with a density of 70 kg/m^3 was produced. It was attenuated in the mortar slurry at a ratio of 1:34 and then aerated using a TM-1 foam generator. Then, four different weight fractions of polypropylene fibrillated fiber (Figure 2) in a range from 1% to 4% were added to the foamed concrete mixtures. Table 1 displays the physical and mechanical characteristics of polypropylene fibrillated fiber (PFF).

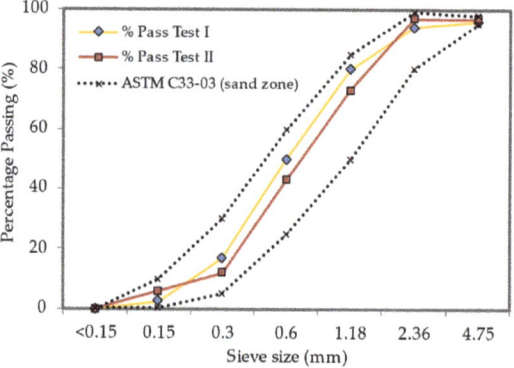

Figure 1. Sand grading curve of fine river sand with ASTM-C33 upper and lower limits.

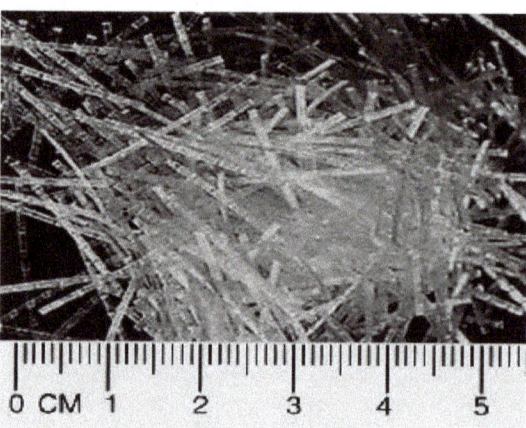

Figure 2. Polypropylene fibrillated fiber (PFF).

Table 1. Physical and mechanical attributes of PFF.

Properties	Value
Elastic modulus (GPa)	7.25
Tensile strength (MPa)	915
Elongation at failure (%)	19.3
Thermal conductivity (W/mK)	0.255
Specific heat capacity (J/kgK)	1335
Melting temperature (°C)	160
Specific weight (g/cm^3)	0.89
Thickness (mm)	0.25
Length (mm)	19

2.2. Mix Proportions

In this investigation, densities of 600 and 1200 kg/m^3 were cast. The cement-to-sand ratio was held at 1:1.5, and the water-to-cement proportion was maintained at 0.45 for all mixtures. Five distinct PFF weight fractions of 1%, 2%, 3% and 4% were selected for addition to foamed concrete mixtures. Table 2 displays the proportions of the created foamed concrete mixture.

Table 2. Mixture proportions of foamed concrete.

Density (kg/m^3)	PFF (%)	Cement (kg/m^3)	Sand (kg/m^3)	Water (kg/m^3)	Foam (kg/m^3)	PFF (kg/m^3)
600	0	230.2	345.4	103.6	43.2	0.0
600	1	230.2	345.4	103.6	43.2	7.2
600	2	230.2	345.4	103.6	43.2	14.4
600	3	230.2	345.4	103.6	43.2	21.7
600	4	230.2	345.4	103.6	43.2	28.9
1200	0	446.9	670.4	201.1	24.4	0.0
1200	1	446.9	670.4	201.1	24.4	13.4
1200	2	446.9	670.4	201.1	24.4	26.9
1200	3	446.9	670.4	201.1	24.4	40.3
1200	4	446.9	670.4	201.1	24.4	53.7

3. Results

3.1. Flow Table Test

This experiment was carried out to find out how adding PFF affected the foamed concrete flow and workability. Using an extended cylinder, spreadability was used to determine the workability (diameter of 76.2 mm × height of 152.4 mm). This apparatus was used following Brewer's open-ended cylinder spread procedure [22,23]. Once the foamed concrete stopped flowing, the average spread diameter of the mixed foamed concrete was calculated. The measurement of foamed concrete's spreadability is shown in Figure 3.

3.2. Porosity Test

The number of pores in foamed concrete is commonly stated as a percentage of volume. The porosity of foamed concrete affects many different aspects of its mechanical, thermal, and durability properties. Additionally, it allows several dangerous substances into the foamed concrete. Salt will corrode the reinforcing bar in reinforced foamed concrete, and rust will grow and cause the foamed concrete to crack. Determining the porosity of foamed concrete with a polyformaldehyde monofilament membrane is therefore critical. As shown in Figure 4, the vacuum-saturated method was used to conduct the porosity test [24]. The cylinder specimens used in casting had diameters of 45 mm and heights of 50 mm. On day 28, this test was executed by placing the foamed concrete samples inside a vacuum desiccator.

Figure 3. Measurement of the spreadability of foamed concrete.

Figure 4. Vacuum saturation procedure.

3.3. Water Absorption Test

The ability of foamed concrete to absorb water is directly related to its ability to repel water, something which is essential in many weakening mechanisms and prevents many harmful factors from entering the environment. The movement of oxygen, chlorine, and carbon dioxide, which start the corrosion of the reinforcing steel in foamed concrete, is made possible by the water absorption capacity of foamed concrete. The water absorption test was performed in this study in conformity with the BS 1881-122 specification [25]. Concrete cylinder specimens, 100 mm in height × 75 mm in diameter, were created.

3.4. Compression Test

The purpose of the axial compression test was to determine the potential strength of the foamed concrete mixtures from which the samples were taken. It evaluates the capacity of a foamed concrete specimen to withstand a load before failing. It permits the verification of whether the correct mix proportions of various foamed concretes and various ingredients were used to achieve the requisite strength. In this investigation, the axial compression test was conducted at a constant rate of 0.03 mm/s in accordance with the BS 12390-3 standard [26]. Cubic specimens with a dimension of 100 × 100 × 100 mm were prepared. The compression test was carried out on days 7, 28, and 56. The result was derived from the mean of three samples of foamed concrete. Figure 5 depicts the axial compression test configuration.

Figure 5. Setup for compression test.

3.5. Flexural Test

Flexural testing was performed to confirm the rigidity of foamed concrete and determine the force required to bend a prism. In a flexural test, ultimate stress represents the last stress level before failure. A three-point flexural test, as indicated in Figure 6, was undertaken in conformity with the BS EN 12390-5 standard [27] to assess the influence of PFF inclusion on the flexural strength of foamed concrete. Each flexural strength test is performed with 3 prisms that are each 100 mm × 100 mm × 500 mm in size and made from foamed concrete. Standard cure times of 7, 28, and 56 days were used for the flexural tests.

Figure 6. Apparatus for three-point bending test.

3.6. Splitting Tensile Test

Foamed concrete's splitting tensile strength is one of the essential and crucial qualities that significantly affects the depth and breadth of cracks in the material. Because of its brittleness and poor tensile strength, a foamed concrete specimen is not often anticipated to bear direct tension. In this research, the splitting tensile strength test was carried out in line with BS EN 12390-6 specifications [28]. The test was performed on a cylinder of foamed concrete, 100 mm in diameter by 200 mm in height, on days 7, 28, and 56. The splitting tensile strength test setup is displayed in Figure 7.

Figure 7. Splitting tensile test apparatus.

4. Discussion

4.1. Spreadability

Figure 8 displays the results of the spreadability for both densities examined in this study (the dataset can be found in Appendix A, Table A2). Figure 8 shows that the self-flowing ability of all the foamed concrete mixes was excellent, with a spreadability greater than 187 mm. Spreadabilities of 255 mm and 230 mm were recorded for the 600 and 1200 kg/m^3 densities of the control specimens, respectively. Spreadability was reduced by adding PFF to foamed concrete, and this effect was proportional to the amount of PFF added. Foamed concrete mixes with 4% PFF had the lowest spreadability when compared to the other foamed concrete mixes. Spreadabilities were measured to be 206 mm for 600 kg/m^3 and 187 mm for 1200 kg/m^3. Because PFF tends to absorb water, there is a consistent decrease in spreadability when it is are present; the external segment of PFF presents significant porosity, which benefits the bond to the matrices, and thus the spreadability of the foamed concrete decreases. In addition, the cementitious matrices aggregate on the PFF's large specific surface area, increasing the foamed concrete's viscosity and causing a decrease in spreadability at greater PFF weight fractions. As the air bubbles are forced out of the cementitious matrix, the slump diameter of the foamed concrete will decrease as a result of the free flow of the intermittent stage above the capacity of the stable spreading form. Mortar must additionally cover the smooth PFF membrane in addition to the fine filler (sand). The spreadability of foamed concrete was reduced as the percentage of PFF was raised from 1% to 4%, indicating that greater quantities filling mortar was required to cover the additional zone of PFF. Furthermore, PFF increases the inner abrasion between foamed concrete components, which in turn causes more cement matrix to diminish the inner resistance, hence reducing the workability of foamed concrete. A similar finding, that increasing the fiber's weight fraction in concrete reduced its workability, was previously described by Mohseni et al. [29].

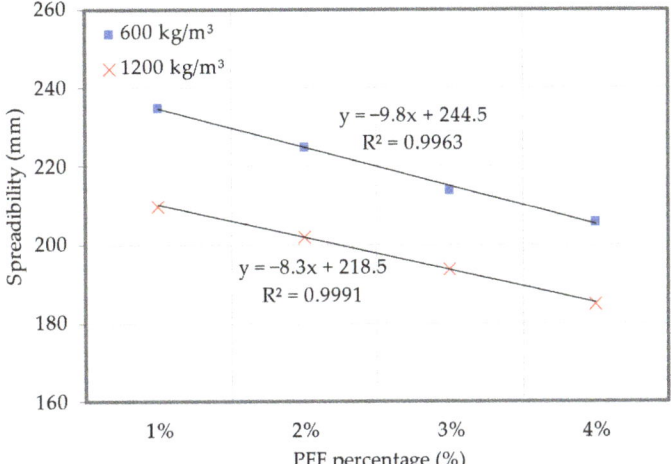

Figure 8. Spreadability of foamed concrete with varying percentages of PFF.

4.2. Density

Differences in the dry density of foamed concrete as a function of PFF weight % are shown in Figure 9 (the dataset can be found in Appendix A, Table A2). According to the findings, the dry density of foamed concrete decreases from that of the control foamed concrete across all three densities when the weight percentage of PFF increases from 1% to 4%. It was discovered that the foamed concrete dry density was highest when PFF was included at a rate of 0% (control specimen), and lowest when PFF was included at a rate of 4%. The decrease in dry density of LFC, combined with larger weight fractions of PF, was caused by difficulties in the compaction process. These resulted in absorbent LFC, resulting in a lower dry density of LFC as linked to the control specimen. Final dry densities were acceptable, falling within the range of 50 kg/m^3 for all foamed concrete mixes that included PFF in different percentages. For instance, for mixes PF0%, PF1%, PF2%, PF3% and PF4%, the discrepancies between final dry densities and planned dry densities of 600 kg/m^3 density were ±2 kg/m^3, ±5 kg/m^3, ±10 kg/m^3, ±19 kg/m^3 and ±24 kg/m^3, respectively. The foamed concrete properties are entirely reliant on the dry density [30].

4.3. Porosity

The porosity of foamed concrete with increasing PFF percentages is represented graphically in Figure 10 (the dataset can be found in Appendix A, Table A2). Generally, the inclusion of PFF in foamed concrete results in a gradual increase in the material's porosity, which reaches the highest value with a 4% addition of PFF. When contrasted with the control foamed concrete sample, the foamed concrete mix with 4% PFF obtained the optimal values of porosity, with around 13.9% and 15.9% decreases in porosity for 600 and 1200 kg/m^3 densities, respectively. It is possible that this is because of the high packing capacity that PFF has in the cementitious matrix of foamed concrete. A porosity value of 64.9% was observed for the foamed concrete mix, with a density of 600 kg/m^3 and a percentage of PFF of 4%, whereas a porosity value of 75.4% was recorded for the control sample. Microcracks formed on the surface of the foamed concrete while it was still in its fresh state condition. At the same time, the surface moisture was rapidly evaporating, which resulted in significant dry shrinkage. When PFF is added to foamed concrete mixes, the segregation can be reduced, and this also helps to reduce the amount of water that is lost through evaporation. In addition to this, the use of PFF has been shown to effectively stop the spread of cracks in foamed concrete that start on the surface and move inward [31]. The morphology of the control foamed concrete specimen, which has a density of 1200 kg/m³, is shown in Figure 11a. It was clear that there were a great number of huge pores that were attached to one another, which resulted in a high porosity value. With the presence of 4% PFF, the compactness of the foamed concrete is improved, and the number of pores that are both large and interconnected is reduced significantly. As the PF was incorporated into the cementitious matrix, as seen in Figure 11b, the internal structure became denser and a homogenous microstructure was achieved, all while decreasing the foamed concrete porosity value.

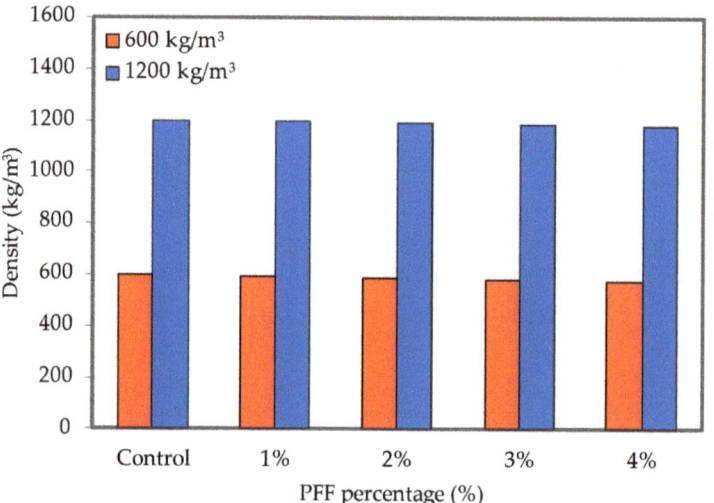

Figure 9. Density of foamed concrete with varying percentages of PFF.

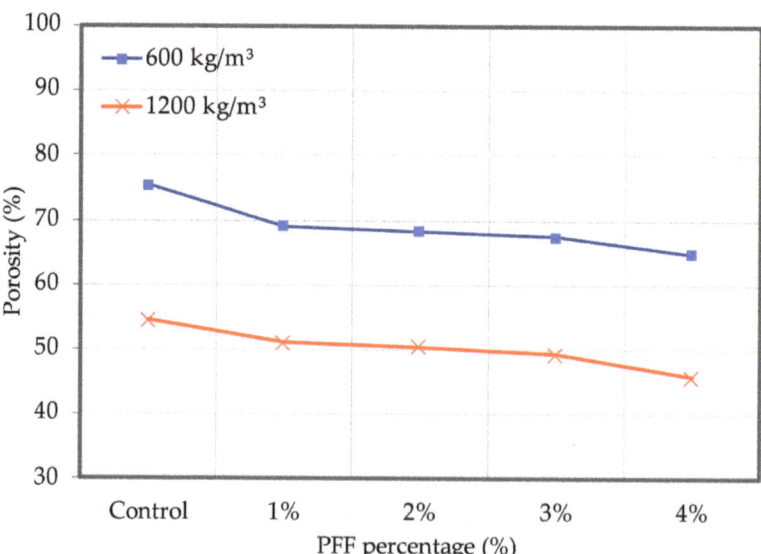

Figure 10. Porosity of foamed concrete with varying percentages of PFF.

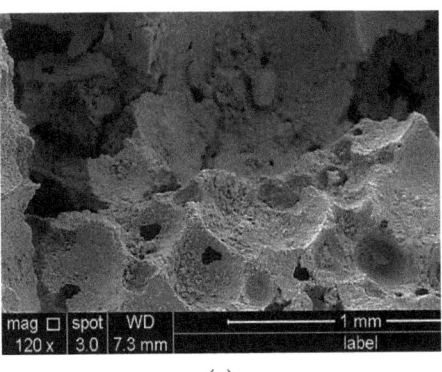 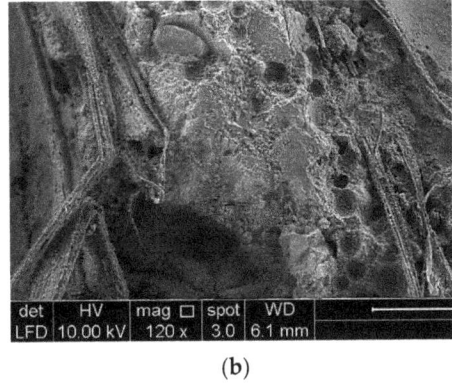

(a) (b)

Figure 11. Morphology of (**a**) control specimen; (**b**) foamed concrete with 4% PFF.

4.4. Water Absorption

The findings of foamed concrete's water absorption with various PFF percentages are shown in Figure 12 (the dataset can be found in Appendix A, Table A2). It was clear that as PFF percentages increased from 1% to 4%, the foamed concrete's ability to absorb water increased as well. For both densities examined in this study, the inclusion of 4% of PFF resulted in the highest water absorption. It should be noted that less cracking occurs in foamed concrete when PFF is added to the mix, and the cracks that do form are smaller and finer than they are in foamed concrete without PFF. This suggests that the presence of PFF can significantly increase foamed concrete's capacity to absorb water and significantly reduce the risk of microcrack merging [32]. Ramezanianpour et al. [33] explored the impact of polypropylene fiber percentages ranging from 0.5% to 4.0% on concrete water permeability. They discovered that adding polypropylene fiber decreased the extent of concrete water penetration. The effect of multi-size polypropylene fibers on the impermeability of concrete was analyzed by Guo et al. [34]. They discovered that, whereas coarse fibers had more obvious macropore inhibition effects on concrete, small fibers had noticeable inhibition on micropores. Additionally, the impenetrability of foamed concrete made with coarse and fine polypropylene fiber is higher than that of foamed concrete made with single-diameter polypropylene fiber. Behfarnia and Behravan [35] evaluated the concrete's ability to absorb water between 0.4% and 0.8% steel fiber and polypropylene fiber. They discovered that the presence of polypropylene fiber diminished the concrete's capacity to absorb water by up to 45%, indicating that concrete's impermeability was greatly increased.

4.5. Water Absorption—Porosity Relationship

The correlation between foamed concrete water absorption capacity and its level of porosity is seen in Figure 13. When the percentage of PFF changes, a linear relationship develops between the foamed concrete water absorption and porosity. This relationship states that the foamed concrete porosity will expand in tandem with the foamed concrete water absorption as the water absorption of the foamed concrete increases. The outward area of the foamed concrete is where the dispersal of free water first begins, eventually making its way into the foamed concrete matrix's deepest segment. In addition, the amount of water that is absorbed internally by foamed concrete may have a marginal impact on the porosity of the material. The fact that there is a definite linear association between the water-absorbing capacity of foamed concrete and its porosity can be seen from the fact that its R-squared values are 0.9805 and 0.9366 for 1200 and 600 kg/m^3 densities, correspondingly.

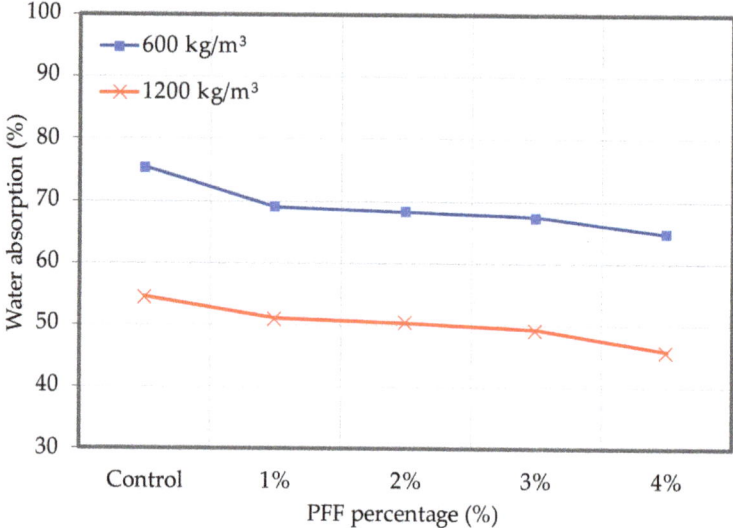

Figure 12. Water absorption of foamed concrete with varying percentages of PFF.

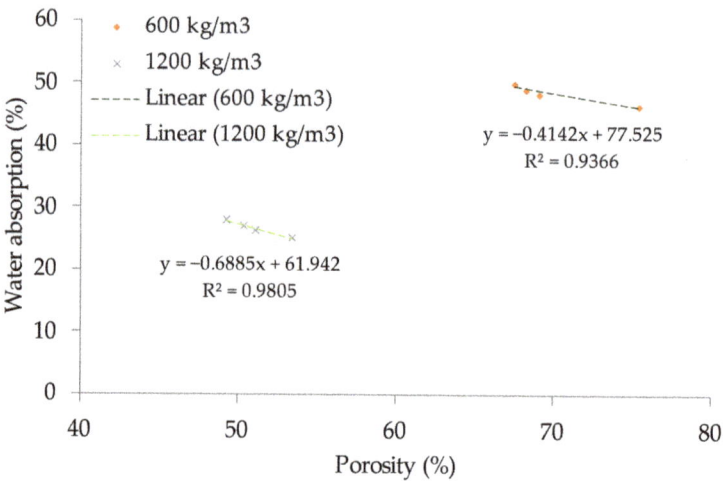

Figure 13. Relationship between water absorption and porosity.

4.6. Compressive Strength

The compressive strength results of 600 and 1200 kg/m³ densities, corresponding with the insertion of various percentages of PFF, are shown in Figures 14 and 15 (the dataset can be found in Appendix A, Table A1). Generally, adding PFF to foamed concrete limited in the growth of the material's compressive strength. For both densities, the optimal percentage of PFF was 3%. PFF was added to foamed concrete, which resulted in a decrease in the amount of entrapped air voids, capillary pores, and entrained air voids. All three of these factors contribute to a rise in the foamed concrete compressive strength. The compressive strengths of the foamed concrete on days 7, 28, and 56 with the inclusion of PFF were greater than the control specimen strength, regardless of the density of the foamed concrete. In comparison to the control specimen, which only achieved compressive strengths of 1.37 MPa (600 kg/m³) and 4.34 MPa (1200 kg/m³), the optimal compressive strengths achieved on day 56 were 2.26 MPa and 7.83 MPa, with the inclusion of a 3%

and 4% weight fraction of PFF for the 600 and 1200 kg/m³ densities, respectively. When an optimal percentage of PFF is evenly scattered in foamed concrete cement paste, the hydrated products of cement amass around the PFF. This is due to their superior surface energy, as their surface behaves as a nucleation site. As the foamed concrete contracts, the PFF membrane absorbs tensile energy through the boundary between the PFF and the foamed concrete cementitious matrix. It then transfers this energy to the neighboring matrix, thereby reducing the amount of concentrated tensile stress and increasing the foamed concrete's resistance to cracking [36]. Poor PFF dispersal in the foamed concrete cementitious matrix results in a balling effect when the percentage of PFF in the foamed concrete exceeds 3%. This creates a deteriorated impact by scattering the tensile stress from the vicinity of the foamed concrete to another place on the PFF surface. This explanation lends credence to the idea that a decline in compressive strength occurred when the weight fractions of PFF increased to a level greater than 3%.

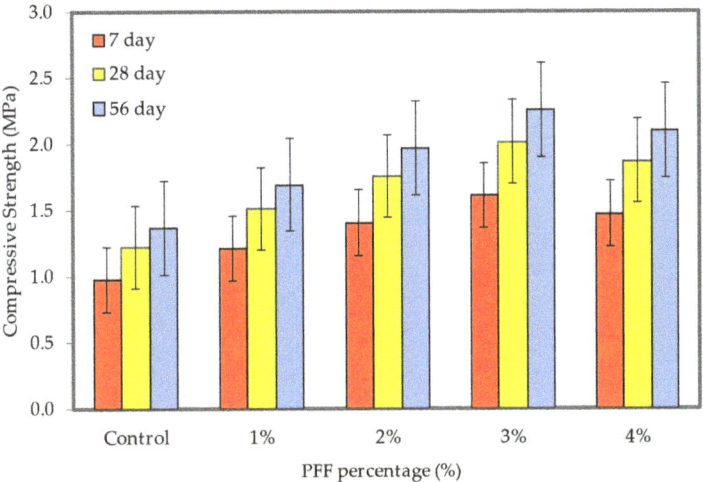

Figure 14. Compressive strength of foamed concrete of 600 kg/m³ density with varying PFF percentages.

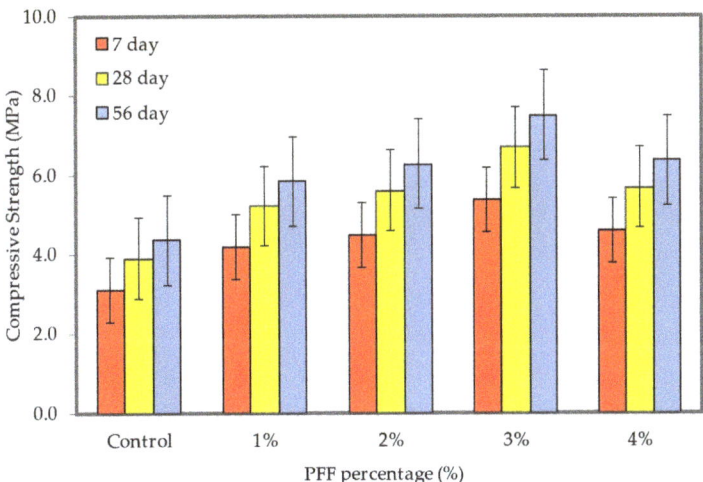

Figure 15. Compressive strength of foamed concrete of 1200 kg/m³ density with varying PFF percentages.

4.7. Flexural Strength

The results of the flexural strength tests at densities of 600 and 1200 kg/m^3 are shown in Figures 16 and 17, respectively (the dataset can be found in Appendix A, Table A1). These densities correspond with different weight fractions of PFF. It is clear from looking at Figures 16 and 17 that the inclusion of PFF in foamed concrete led to a rise in flexural strength. For both densities considered in this study, the best weight fraction of PFF was 3%. The flexural strengths of the foamed concrete with the presence of PFF on days 7, 28, and 56 were greater than the flexural strength of the control specimen, regardless of the density of the foamed concrete. In contrast to the control specimen, which only achieved flexural strengths of 0.38 MPa (600 kg/m^3) and 1.04 MPa (1200 kg/m^3), the ideal flexural strengths that were attained at day 56 were 0.63 MPa and 1.66 MPa, with the addition of a 3% of PFF for the 600 kg/m^3 and 1200 kg/m^3. PFF is a hydrophilic substance, and as a result, it possesses great adhesion when combined with cement paste. After the PFF percentage was increased to 4%, the flexural strength of foamed concrete drastically dropped for both densities that were taken into consideration for this investigation. Because it is difficult to disperse the PFF evenly and because it might cause agglomeration, the flexural strength will be lowered if the percentage of PFF is too high. This is because PFF can cause agglomeration. During the process of the crack spreading out, the PFF will gradually become separated from the matrix until the bond strength is completely exceeded. This will continue until the fracture has completely spread out. Although the matrix is damaged, it is still capable of maintaining its fundamental form. The presence of PFF in foamed concrete plays a significant role in both the strengthening of the foamed concrete cementitious matrix, and the modification of the material's physical characteristics from a brittle state to a ductile one. Both of these effects are the result of the material's transition from a brittle to a ductile state. Because an appropriate percentage of PFF can bond with the hydration products and unhydrated constituents in the foamed concrete matrix to build a three-dimensional grid structure, incorporating an appropriate weight fraction of PFF can efficiently increase the flexural strength of foamed concrete. This is because the three-dimensional network structure can create a subsidiary effect and boost the foamed concrete flexural strength [37].

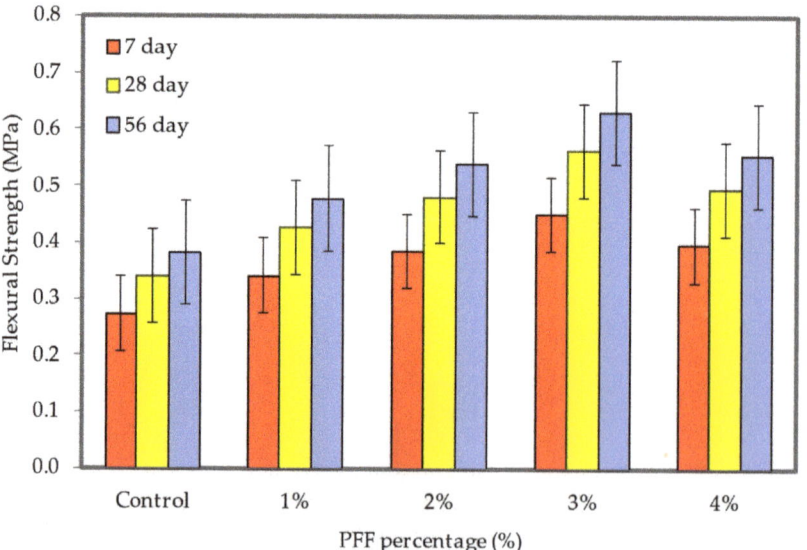

Figure 16. Flexural strength of foamed concrete of 600 kg/m^3 density with varying PFF percentages.

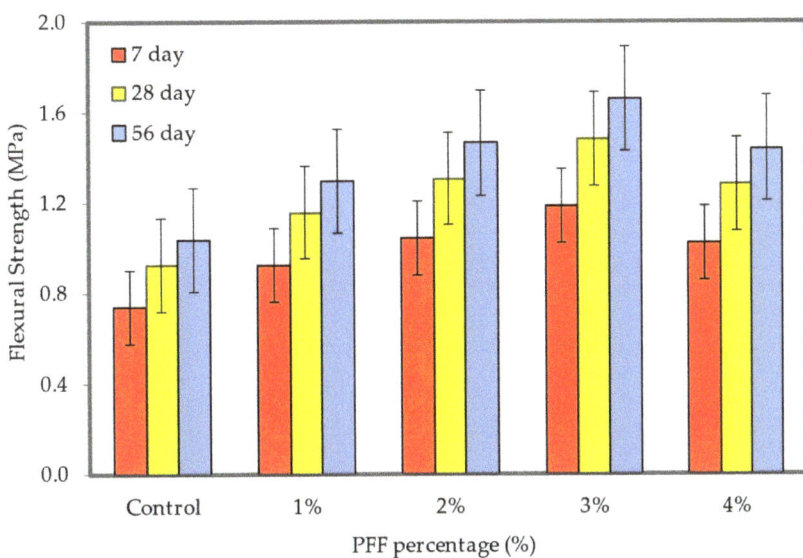

Figure 17. Flexural strength of foamed concrete of 1200 kg/m³ density with varying PFF percentages.

4.8. Splitting Tensile Strength

Figures 18 and 19 illustrate the influence that different percentages of PFF have on the splitting tensile strength with densities of 600 and 1200 kg/m³ (the dataset can be found in Appendix A, Table A1). It is clear from the whole trend that with the rise in PF percentages, the splitting tensile strength of both foamed concrete densities at different ages all exhibit an upward trend. This is the case because the PFF percentages are increasing as the trend continues. In most cases, the splitting tensile strength progressively increases with the growth in the percentages of PFF up to 3%. In comparison to the control specimen, which achieved compressive strengths of 0.24 MPa (600 kg/m³) and 0.64 MPa (1200 kg/m³), the optimal splitting tensile strengths attained at day 56 were 0.39 MPa and 0.93 MPa, with the inclusion of a 3% PFF for the 600 and 1200 kg/m³ densities. Because of the enhanced foamed concrete robustness, helped along by the presence of PFF, the enhancement of splitting tensile strengths was accomplished for both materials that were taken into consideration throughout this investigation. PFF will gradually form a fiber grid skeleton inside the foamed concrete as it goes through the process of hardening. This can improve the brittle state of the foamed concrete matrix by governing the growth and propagation of cracks when it is subjected to external tensile stress, preventing the cracks from leading to an explosive failure. It became apparent that the PFF could be uniformly disseminated when the PFF percentage was 3%, which resulted in a growth in the bonding strength between the cement matrix and the PFF. The PFF is dispersed throughout the foamed concrete matrix in a manner that is nearly homogeneous and does not show any significant signs of buildup. In the presence of a tensile load, foamed concrete demonstrates elastic linear tension prior to the onset of plastic deformation, which occurs after the appearance of the first crack in the material. The addition of PFF membrane to foamed concrete, on the other hand, causes the membrane to take on the role of fastening when the foamed concrete cracks. This ensures that the matrix elastic modulus does not immediately drop to zero when the direct boundary strain is reached [38–40]. When cracks appear in the PFF membrane, the membrane will bear all the tension and then gradually transmit it to the matrix [8,41–45].

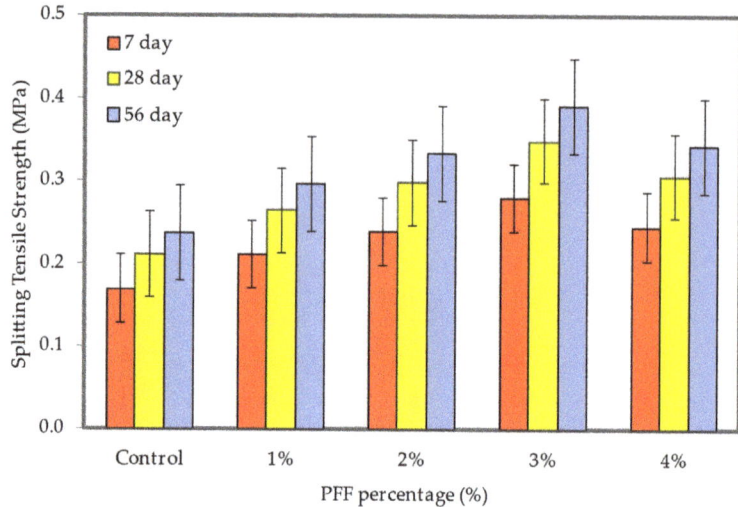

Figure 18. Splitting tensile strength of foamed concrete of 600 kg/m³ density with varying PFF percentages.

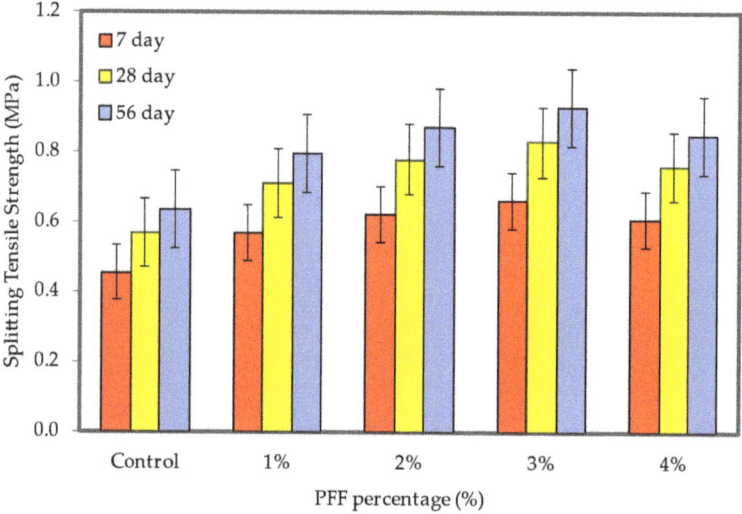

Figure 19. Splitting tensile strength of foamed concrete of 1200 kg/m³ density with varying PFF percentages.

4.9. Compressive—Flexural Strengths Relationship

The density of 600 kg/m³ of foamed concrete was used in the execution of a correlation between the compressive and flexural strengths of the material. For the purpose of this investigation, all curing periods were taken into account. The connection between the flexural and compressive strengths of foamed concrete with various percentages of PFF is demonstrated in Figure 20. It seems that they found out that there is a direct expanding, relationship that can be differentiated in the compressive strength and flexural strength of foamed concrete. From Figure 20, it can be seen that an R-squared value of 0.9588 was obtained, which indicates a highly linear relationship between the two strength parameters. This implies that variations in the predictors are interrelated to deviations in the response variable and that the achieved prediction models explain a significant portion of the response's inherent variability. Because of this relationship, it is clear that the flexural strength of foamed concrete increases as the compressive strength of the concrete increases. This regression model makes it possible to approximate the flexural strength of foamed concrete, based on its axial compressive strength, for the range of values that was investigated in this exploration.

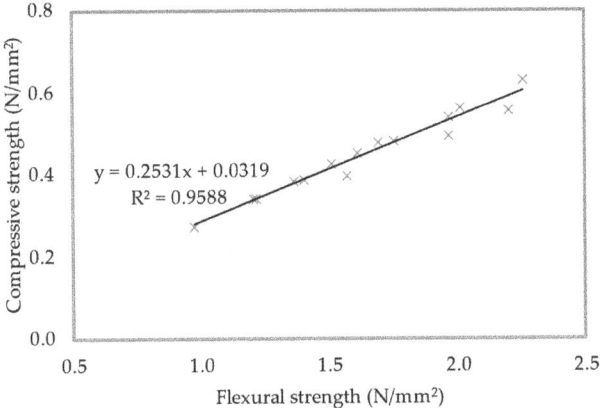

Figure 20. Flexural-compressive strengths relationship for 600 kg/m^3 density.

4.10. Compressive—Splitting Tensile Strengths Relationship

The relationship between the splitting tensile and compressive strengths of 600 kg/m^3 density foamed concrete with different PFF percentages is shown in Figure 21. The compressive strength was mapped against the foamed concrete's splitting tensile strength. According to Figure 21, data dissemination supports the existence of a strong correlation between the splitting tensile and compressive strengths of foamed concrete. In a similar trend, the splitting tensile strengths rose with increasing compressive strength for all curing periods. With an R-squared value of 0.9646, a strong linear connection is evident. The splitting tensile strength was approximately 15% of its strength under compression conditions for the total specimens evaluated in this investigation.

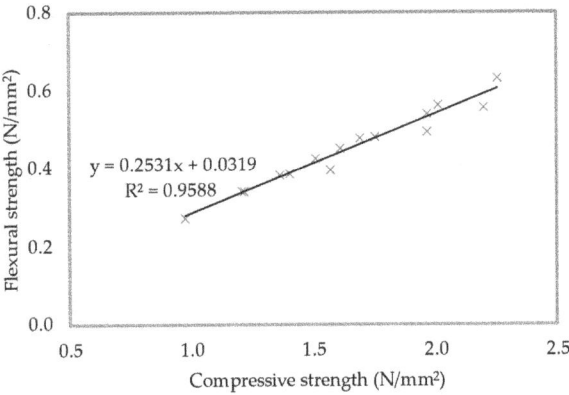

Figure 21. Compressive-tensile strengths relationship for 600 kg/m^3 density.

5. Conclusions

This laboratory investigation aims to evaluate the possible use of polypropylene fibrillated fiber (PFF) to enhance the foamed concrete engineering properties. With the inclusion of four different PFF percentages (1%, 2%, 3%, and 4%), ten mixes of 600 and 1200 kg/m^3 densities were made. Compressive, splitting tensile and flexural strengths were assessed. Additionally, the workability, porosity, water absorption, and density of products were evaluated as well. Additionally, the correlations between the considered properties were analyzed. When PFF is added to foamed concrete, spatial networks are formed and cement paste is incinerated to cover the PFF, resulting in reduced spreadability versus foamed concrete without fibers. However, all foamed concrete mixes exhibited spreadabilities greater than 187 mm, indicating a significant capacity for self-flow. In contrast to the control specimen, the dry density of the foamed concrete decreases as the PFF weight fractions rise from 1% to 4%. The control specimen had the highest dry density of foamed concrete, whereas the 4% PFF inclusion product had the lowest density. Due to the intricacy of the compaction process, which produces porous foamed concrete, the dry density decreased at larger weight fractions of PFF. With the presence of PFF, the foamed concrete porosity grows gradually up to 4%. Foamed concrete mixes containing 4% PFF achieved the optimal porosity with a reduction of around 13.9% and 15.9% for 600 and 1200 kg/m^3 densities correspondingly. This is most likely because foamed concrete's cement matrix has strong PFF packing capabilities. As the PFF percentages rose from 1% to 4%, foamed concrete water absorption was increased. With the addition of 4% PFF, the highest water absorption capacity was achieved. When PFF is added to foamed concrete, the fissures are less noticeable and finer than they are in foamed concrete that does not contain PFF. The foamed concrete flexural, compressive, and splitting tensile strengths were increased by the addition of PFF. The ideal percentage of PFF for both densities was 3%. The foamed concrete's compressive, flexural, and breaking tensile strengths considerably decreased above the 3% PFF. It is difficult to spread the PFF uniformly, and agglomeration is caused if the PFF weight fractions are too large.

Author Contributions: Conceptualization, M.A.O.M., A.V.S., Z.Y. and M.M.A.B.A.; methodology, M.A.O.M., L.A.S., P.V., M.N.M.N. and M.M.A.B.A.; formal analysis, M.A.O.M., P.V. and M.M.A.B.A.; Technology development, M.A.O.M., M.M.A.B.A., P.V. and M.N.M.N.; resources, M.A.O.M., Z.Y. and M.M.A.B.A.; data curation, M.A.O.M., M.N.M.N. and M.M.A.B.A.; writing—original draft preparation, M.A.O.M., M.N.M.N. and M.M.A.B.A.; writing— review and editing, M.A.O.M., A.V.S., L.A.S. and M.M.A.B.A.; funding acquisition, M.A.O.M. and M.S.B. All authors have read and agreed to the published version of the manuscript.

Funding: The authors gratefully thank the support from the Ministry of Higher Education (MOHE) through the Fundamental Research Grant Scheme (FRGS) (FRGS/1/2022/TK01/USM/02/3). Besides, this paper was financially supported by the Project "Network of excellence in applied research and innovation for doctoral and postdoctoral programs/InoHubDoc", project co-funded by the European Social Fund financing agreement no. POCU/993/6/13/153437. This paper was also supported by "Gheorghe Asachi" Technical University from Iaşi (TUIASI), through the Project "Performance and excellence in postdoctoral research 2022".

Conflicts of Interest: The authors declare no conflict of interest.

Appendix A

Table A1. Compressive, flexural, and splitting tensile strengths experimental dataset.

Specimen			600 kg/m^3			1200 kg/m^3		
			7 Day	28 Day	56 Day	7 Day	28 Day	56 Day
Compressive Strength (MPa)		Control	0.98	1.22	1.37	3.11	3.88	4.35
		1%	1.21	1.51	1.69	4.16	5.20	5.83
		2%	1.40	1.76	1.97	4.47	5.59	6.26
		3%	1.61	2.02	2.26	5.34	6.68	7.48
		4%	1.47	1.87	2.10	4.57	5.67	6.35

Table A1. Cont.

Specimen		600 kg/m³			1200 kg/m³		
		7 Day	28 Day	56 Day	7 Day	28 Day	56 Day
Flexural Strength (MPa)	Control	0.27	0.34	0.38	0.74	0.93	1.04
	1%	0.34	0.43	0.48	0.93	1.16	1.30
	2%	0.38	0.48	0.54	1.04	1.31	1.46
	3%	0.45	0.56	0.63	1.19	1.48	1.66
	4%	0.40	0.49	0.55	1.02	1.28	1.44
Splitting Tensile (MPa)	Control	0.17	0.21	0.24	0.46	0.57	0.64
	1%	0.21	0.26	0.30	0.57	0.71	0.80
	2%	0.24	0.30	0.33	0.62	0.78	0.87
	3%	0.28	0.35	0.39	0.66	0.83	0.93
	4%	0.25	0.31	0.34	0.61	0.76	0.85

Table A2. Slump, density, porosity, and water absorption experimental dataset.

	Specimen	600 kg/m³	1200 kg/m³
Workability (mm)	Control	255	230
	1%	235	210
	2%	225	202
	3%	214	194
	4%	206	185
Density (kg/m³)	Control	600	1200
	1%	594	1195
	2%	589	1190
	3%	580	1185
	4%	574	1181
Porosity (%)	Control	75.4	54.5
	1%	69.1	51.1
	2%	68.3	50.4
	3%	67.5	49.3
	4%	64.9	45.8
Water Absorption (%)	Control	46.4	25.9
	1%	48.4	26.5
	2%	49.2	27.3
	3%	50.0	28.1
	4%	50.7	29.2

References

1. Huang, Z.; Zhang, T.; Wen, Z. Proportioning and characterization of Portland cement-based ultra-lightweight foam concretes. *Constr. Build. Mater.* **2015**, *79*, 390–396. [CrossRef]
2. Mohamad, N.; Abdul Samad, A.A.; Lakhiar, M.T.; Othuman Mydin, M.A.; Jusoh, S.; Sofia, A.; Efendi, S.A. Effects of Incorporating Banana Skin Powder (BSP) and Palm Oil Fuel Ash (POFA) on Mechanical Properties of Lightweight Foamed Concrete. *Int. J. Integr. Eng.* **2018**, *10*, 169–176. [CrossRef]
3. Ganesan, S.; Othuman Mydin, M.A.; Sani, N.M.; Che Ani, A.I. Performance of polymer modified mortar with different dosage of polymeric modifier. *MATEC Web Conf.* **2014**, *15*, 01039. [CrossRef]
4. Nensok, M.H.; Othuman Mydin, M.A.; Awang, H. Investigation of Thermal, Mechanical and Transport Properties of Ultra Lightweight Foamed Concrete (ULFC) Strengthened with Alkali Treated Banana Fibre. *J. Adv. Res. Fluid Mech. Therm. Sci.* **2021**, *86*, 123–139. [CrossRef]
5. Othuman Mydin, M.A.; Phius, A.F.; Sani, N.M.; Tawil, N.M. Potential of Green Construction in Malaysia: Industrialised Building System (IBS) vs. Traditional Construction Method. *E3S Web Conf.* **2014**, *3*, 01009. [CrossRef]
6. Serri, E.; Othuman Mydin, M.A.; Suleiman, M.Z. The effects of oil palm shell aggregate shape on the thermal properties and density of concrete. *Adv. Mater. Res.* **2014**, *935*, 172–175. [CrossRef]
7. Kudyakov, A.I.; Steshenko, A.B. Cement foamed with low shrinkage. *Adv. Mater. Res.* **2015**, *1085*, 245–249. [CrossRef]

8. Othuman Mydin, M.A.; Mohamed Shajahan, M.F.; Ganesan, S.; Sani, N.M. Laboratory investigation on compressive strength and micro-structural features of foamed concrete with addition of wood ash and silica fume as a cement replacement. *MATEC Web Conf.* **2014**, *17*, 01004. [CrossRef]
9. Jones, M.R.; Zheng, L.; Ozlutas, K. Stability and instability of foamed concrete. *Mag. Concr. Res.* **2016**, *68*, 542–549. [CrossRef]
10. Awang, H.; Othuman Mydin, M.A.; Roslan, A.F. Effects of fibre on drying shrinkage, compressive and flexural strength of lightweight foamed concrete. *Adv. Mater. Res.* **2012**, *587*, 144–149. [CrossRef]
11. Wee, T.H.; Daneti, S.B.; Tamilselvan, T. Effect of w/c ratio on air-void system of foamed concrete and their influence on mechanical properties. *Mag. Concr. Res.* **2011**, *63*, 583–595. [CrossRef]
12. Mohamad, N.; Iman, M.A.; Othuman Mydin, M.A.; Samad, A.A.A.; Rosli, J.A.; Noorwirdawati, A. Mechanical properties and flexure behaviour of lightweight foamed concrete incorporating coir fibre. *IOP Conf. Ser. Earth Environ. Sci.* **2018**, *140*, 012140. [CrossRef]
13. Tambichik, M.A.; Abdul Samad, A.A.; Mohamad, N.; Mohd Ali, A.Z.; Othuman Mydin, M.A.; Mohd Bosro, M.Z.; Iman, M.A. Effect of combining Palm Oil Fuel Ash (POFA) and Rice Husk Ash (RHA) as partial cement replacement to the compressive strength of concrete. *Int. J. Integr. Eng.* **2018**, *10*, 61–67. [CrossRef]
14. Zheng, Y.; Cai, Y.; Zhang, G.; Fang, H. Fatigue property of basalt fiber-modified asphalt mixture under complicated environment. *J. Wuhan Univ. Technol. Mater. Sci. Ed.* **2014**, *29*, 996–1004. [CrossRef]
15. Jaivignesh, B.; Sofi, A. Study on mechanical properties of concrete using plastic waste as an aggregate. *IOP Conf. Ser.: Earth Environ. Sci.* **2017**, *80*, 012016. [CrossRef]
16. Akca, A.H.; Özyurt, N. Effects of re-curing on residual mechanical properties of concrete after high temperature exposure. *Constr. Build. Mater.* **2018**, *159*, 540–552. [CrossRef]
17. Ahmad, J.; Manan, A.; Ali, A.; Khan, M.W. A study on mechanical and durability aspects of concrete modified with steel fibers (SFs). *Civ. Eng. Archit.* **2020**, *8*, 814–823. [CrossRef]
18. Shafigh, P.; Mahmud, H.; Jumaat, M.Z. Effect of steel fiber on the mechanical properties of oil palm shell lightweight concrete. *Mater. Des.* **2011**, *32*, 3926–3932. [CrossRef]
19. Yew, M.K.; Mahmud, H.B.; Ang, B.C.; Yew, M.C. Influence of different types of polypropylene fibre on the mechanical properties of high-strength oil palm shell lightweight concrete. *Constr. Build. Mater.* **2015**, *90*, 36–43. [CrossRef]
20. Yap, S.P.; Alengaram, U.J.; Jumaat, M.Z. Enhancement of mechanical properties in polypropylene and nylon–fibre reinforced oil palm shell concrete. *Mater. Des.* **2013**, *49*, 1034–1041. [CrossRef]
21. Abadel, A.; Abbas, H.; Almusallam, T.; Al-Salloum, Y.; Siddiqui, N. Mechanical properties of hybrid fibre-reinforced concrete–analytical modelling and experimental behaviour. *Mag. Concr. Res.* **2016**, *68*, 823–843. [CrossRef]
22. Jones, M.R.; McCarthy, M.J.; McCarthy, A. Moving fly ash utilization in concrete forward: A UK perspective. In Proceedings of the 2003 International Ash Utilisation Symposium, Center for Applied Energy Research, University of Kentucky, Lexington, KY, USA, 20–22 October 2003; pp. 20–22.
23. Kearsley, E.P.; Mostert, H.F. Designing mix composition of foamed concrete with high fly ash contents. In Proceedings of the International Conference, University of Dundee, Scotland, UK, 5 July 2005; pp. 29–36.
24. Kearsley, E.P.; Wainwright, P.J. Porosity and permeability of foamed concrete. *Cem. Concr. Res.* **2001**, *31*, 805–812. [CrossRef]
25. BS 1881-122; Testing Concrete. Method for Determination of Water Absorption. British Standards Institute: London, UK, 1983.
26. BS 12390-3; Testing Hardened Concrete. Compressive Strength of Test Specimens. British Standards Institute: London, UK, 2011.
27. BS EN 12390-5; Testing Hardened Concrete. Flexural Strength of Test Specimens. British Standards Institute: London, UK, 2019.
28. BS EN 12390-6; Testing Hardened Concrete. Tensile Splitting Strength of Test Specimens. British Standards Institute: London, UK, 2009.
29. Mohseni, E.; Yazdi, M.A.; Miyandehi, B.M.; Zadshir, M.; Ranjbar, M.M. Combined effects of metakaolin, rice husk ash, and polypropylene fiber on the engineering properties and microstructure of mortar. *J. Mater. Civ. Eng.* **2017**, *29*, 04017025. [CrossRef]
30. Zamzani, N.M.; Mydin, M.A.O.; Ghani, A.N.A. Effectiveness of 'cocos nucifera linn' fibre reinforcement on the drying shrinkage of lightweight foamed concrete. *ARPN J. Eng. Appl. Sci.* **2019**, *14*, 3932–3937.
31. Kochova, K.; Gauvin, F.; Schollbach, K.; Brouwers, H.J.H. Using alternative waste coir fibres as a reinforcement in cement fibre composites. *Constr. Build Mater.* **2020**, *231*, 117121. [CrossRef]
32. Elkatatny, S.; Gajbhiye, R.; Ahmed, A.; Mahmoud, A.A. Enhancing the cement quality using polypropylene fiber. *J. Pet. Explor. Prod. Technol.* **2019**, *10*, 1097–1107. [CrossRef]
33. Ramezanianpour, A.A.; Esmaeili, M.; Ghahari, S.A.; Najafi, M.H. Laboratory study on the effect of polypropylene fiber on durability, and physical and mechanical characteristic of concrete for application in sleepers. *Constr. Build. Mater.* **2013**, *44*, 411–418. [CrossRef]
34. Guo, Z.Q. Experimental Study on the Impermeability of Multi-Scale Polypropylene Fibre Concrete. Master's Thesis, Chongqing University, Chongqing, China, 2018.
35. Behfarnia, K.; Behravan, A. Application of high performance polypropylene fibers in concrete lining of water tunnels. *Mater. Des.* **2014**, *55*, 274–279. [CrossRef]
36. Ferreira, S.R.; de Andrade Silva, F.; Lima, P.R.L.; Filho, R.D.T. Effect of hornification on the structure, tensile behavior and fiber matrix bond of sisal, jute and curaua' fiber cement based composite systems. *Constr. Build. Mater.* **2017**, *139*, 551–561. [CrossRef]

37. Othuman Mydin, M.A.; Nawi, M.N.M.; Odeh, R.A.; Salameh, A.A. Durability Properties of Lightweight Foamed Concrete Reinforced with Lignocellulosic Fibers. *Materials* **2022**, *15*, 4259. [CrossRef]
38. Othuman Mydin, M.A. The effect of raw mesocarp fibre inclusion on the durability properties of lightweight foamed concrete. *ASEAN J. Sci. Technol. Dev.* **2021**, *38*, 59–66. [CrossRef]
39. Ariffin, N.; Abdullah, M.M.A.B.; Zainol, M.R.R.M.A.; Baltatu, M.S.; Jamaludin, L. Effect of Solid to Liquid Ratio on Heavy Metal Removal by Geopolymer-Based Adsorbent. *IOP Conf. Ser. Mater. Sci. Eng.* **2018**, *374*, 012045. [CrossRef]
40. Aziz, I.H.; Abdullah, M.M.A.; Salleh, M.A.A.M.; Yoriya, S.; Abd Razak, R.; Mohamed, R.; Baltatu, M.S. The investigation of ground granulated blast furnace slag geopolymer at high temperature by using electron backscatter diffraction analysis. *Arch. Metall. Mater.* **2022**, *67*, 227–231. [CrossRef]
41. Othuman Mydin, M.A.; Mohd Nawi, M.N.; Mohamed, O.; Sari, M.W. Mechanical Properties of Lightweight Foamed Concrete Modified with Magnetite (Fe_3O_4) Nanoparticles. *Materials* **2022**, *15*, 5911. [CrossRef] [PubMed]
42. Othuman Mydin, M.A.; Nawi, M.N.M.; Odeh, R.A.; Salameh, A.A. Potential of Biomass Frond Fiber on Mechanical Properties of Green Foamed Concrete. *Sustainability* **2022**, *14*, 7185. [CrossRef]
43. Serudin, A.M.; Othuman Mydin, M.A.; Ghani, A.N.A. Investigating the load carrying capacities of lightweight foamed concrete strengthen with fiber mesh. *Int. J. Integr. Eng.* **2022**, *14*, 360–376. [CrossRef]
44. Suhaili, S.S.; Othuman Mydin, M.A.; Awang, H. Influence of Mesocarp Fibre Inclusion on Thermal Properties of Foamed Concrete. *J. Adv. Res. Fluid Mech. Therm. Sci.* **2021**, *87*, 1–11. [CrossRef]
45. Nensok, M.H.; Othuman Mydin, M.A.; Awang, H. Optimization of mechanical properties of cellular lightweight concrete with alkali treated banana fiber. *Rev. De La Constr.* **2021**, *20*, 491–511. [CrossRef]

Article

Study on the Effect of Ultraviolet Absorber UV-531 on the Performance of SBS-Modified Asphalt

Li Liu *, Leixin Liu, Zhaohui Liu, Chengcheng Yang, Boyang Pan and Wenbo Li

National Engineering Laboratory for Highway Maintenance Technology, School of Traffic and Transportation Engineering, Changsha University of Science & Technology, Changsha 410114, China
* Correspondence: liuli@csust.edu.cn; Tel.: +86-137-8715-5595

Abstract: Asphalt pavements at high altitudes are susceptible to aging and disease under prolonged action of UV light. To improve their anti-ultraviolet aging performance, UV-531/SBS-modified asphalts with UV-531 dopings of 0.4%, 0.7%, and 1.0% were prepared by the high-speed shear method, and the effect of UV-531 on the conventional performance of SBS-modified asphalt before aging was studied by needle penetration, softening point and 5 °C ductility tests. The high- and low-temperature rheological properties of UV-531/SBS-modified asphalt before and after aging were also analyzed by high temperature dynamic shear rheology test and low-temperature glass transition temperature test. Finally, the effect of UV-531 on the anti-aging performance of SBS-modified asphalt was evaluated by three methods, including rutting factor ratio, viscosity aging index, and infrared spectroscopy. The results show that with the increase of UV-531 doping, the needle penetration and 5 °C ductility show an increasing trend, but the effect on the softening point is small. The high temperature stability of SBS-modified asphalt is not much affected by the addition of UV-531, and the low-temperature stability is improved, and when 0.7% UV absorber is added, SBS-modified asphalt shows better low-temperature performance. The results of all three evaluation methods show that the addition of UV-531 significantly improved the anti-UV aging performance of SBS-modified asphalt, with the amount of 0.7% providing the asphalt with the best anti-UV aging performance. The results of the study can provide an important reference for improving the anti-ultraviolet aging performance of SBS-modified asphalt.

Keywords: road engineering; UV aging; UV absorber; high and low-temperature performance; infrared spectrum

Citation: Liu, L.; Liu, L.; Liu, Z.; Yang, C.; Pan, B.; Li, W. Study on the Effect of Ultraviolet Absorber UV-531 on the Performance of SBS-Modified Asphalt. *Materials* 2022, 15, 8110. https://doi.org/10.3390/ma15228110

Academic Editor: Andrei Victor Sandu

Received: 19 October 2022
Accepted: 14 November 2022
Published: 16 November 2022

Publisher's Note: MDPI stays neutral with regard to jurisdictional claims in published maps and institutional affiliations.

Copyright: © 2022 by the authors. Licensee MDPI, Basel, Switzerland. This article is an open access article distributed under the terms and conditions of the Creative Commons Attribution (CC BY) license (https://creativecommons.org/licenses/by/4.0/).

1. Introduction

When using asphalt mixes, the process of ultraviolet radiation induced by asphalt UV aging cannot be neglected, since asphalt mixes are prone to aging under heat or natural conditions [1,2]. Since old asphalt is more susceptible to spalling, potholes, and other pavement diseases, it is imperative to find solutions to improve its performance. Because of their exceptional performance at both high and low temperatures and their resilience to fatigue, styrene butadiene styrene (SBS)-modified asphalt binders are employed in a significant portion of asphalt pavement construction projects across the world [3].

Although research on the current asphalt thermal aging mechanism is fairly advanced [4], because road workers do not pay attention to asphalt ultraviolet aging, research progress is slow. This is because the thermal aging of asphalt and ultraviolet aging incentives are entirely different. Because the UV aging study of asphalt plays an important guiding role in building roads in western China, several road researchers have recently begun to pay attention to the effects of UV aging on asphalt mixes and asphalt [5]. Several researchers have found that many factors affect the properties of asphalt during aging [6–9]. Yuanyuan Li et al. pointed out that UV aging has a significant effect on the service life of asphalt pavements due to the fact that the increased penetration of UV light during

UV aging increases the UV aging depth [10]. Zhang Hailin found that when studying the changes in elemental composition and rheological properties of asphalt before and after thermal and UV aging using elemental analysis and DSR tests, UV aging resulted in greater changes in indices and more aging [11]. Lixing Ma et al. found that the water stability performance of asphalt mixes decayed as the UV aging time increased [12]. Mona Nobakht et al. demonstrated that intense UV exposure reduces the shear strength of asphalt, and that the low-temperature cracking and fatigue properties of asphalt decrease with aging time [13]. Several people have begun to work on the mechanism of UV aging of asphalt and improvement methods, since the risks of UV aging have been made clear through the tests of some road researchers. Zheng et al. used permeability, viscosity, and ductility to develop a nonlinear equation for the decay of asphalt properties after UV aging [14]. Min. Xiao simulated the dynamic behavior of asphalt microstructure during UV aging using Materials Studio software, a process different from the internal thermal aging mechanism of asphalt [15]. Feng Zhengang et al. found that the addition of UV absorber UV-531 can improve the high and low temperature performance of asphalt in addition to effectively improving the anti-UV aging ability of asphalt [16]. Lingling Hong found that adding UV-531 to SBS-modified asphalt with block ratios of 40/60 and 30/70 could effectively improve its high- and low-temperature properties [17]. UV-531 provides superior performance, is the most widely used ultraviolet absorber, can powerfully absorb the wavelength of 300–345 nm ultraviolet light, and is one of the most used UV absorbers by road researchers to improve the UV aging performance of asphalt materials. Therefore, finding the right dosage is essential to improve the performance and economy of asphalt pavements by improving the UV aging resistance of asphalt.

According to relevant studies, the UV-531 content that can improve the UV aging resistance of asphalt is in the range of 0.2% to 1.2% [15]. Hence, the paper chose three UV-531 doping levels—0.4%, 0.7%, and 1.0%. UV-531 was added to SBS-modified asphalt by melt blending. Needle penetration, softening point, and 5 °C ductility tests were chosen to assess the effect of UV-531 on the conventional properties of SBS-modified asphalt. UV aging of treated asphalt was performed using DSR and DMA tests to evaluate the effect of UV-531 on the high- and low-temperature performance of SBS-modified asphalt. Finally, the optimum dose was derived after evaluating the UV-531/SBS-modified asphalt for UV aging resistance using rutting factor ratio, viscosity index, and infrared spectroscopy.

2. Materials and Methods

2.1. Testing Raw Materials

UV absorber: UV-531 (2-Hydroxy-4-n-Octyloxy benzophenone) is a light yellow powder that absorbs ultraviolet rays from sunlight as a light stabilizer. The property indicators are shown in Table 1.

Table 1. Performance index of UV-531.

Indicators	Technical Requirements	Indicators	Technical Requirements
Density/g/cm^3	1.068	Boiling/°C	457.9
Melting point/°C	47~49	Flash point/°C	155.1

The basic properties of SBS-modified asphalt are shown in Table 2.

Table 2. Fundamental characteristics of SBS-modified asphalt.

Test Items		SBS-Modified Asphalt	
		Real Test	Technical Requirements
Ductility (5 °C, 5 cm/min)/cm Not less than		28.0	20
Softening point (Global Law)/°C Not less than		74.2	60
Needle penetration (25 °C, 100 g, 5 s)/0.1 mm		53.0	40~60
Resilient Recovery (25 °C)/% Not less than		94.0	75
Flash Point/°C Not less than		300	230
Needle penetration index PI Not less than		0.26	0
Kinematic viscosity (135 °C)/Pa·s No greater than		2.55	3.0
Solubility (trichloroethylene)/% Not less than		99.7	99
After RTFOT test	Quality change/% No greater than	−0.008	±1.0
	Ductility (5 °C)/cm Not less than	19.7	15
	Needle penetration ratio (25 °C)/% Not less than	83.2	65

2.2. UV-531/SBS-Modified Asphalt Preparation

The high-speed shear method was used to prepare UV-531- and SBS-modified asphalts to integrate them fully. The specific preparation procedure is as follows.

(1) Put the SBS-modified asphalt into 160 °C ovens for 30 min and then take it out.
(2) 0%, 0.4%, 0.7%, and 1.0% of UV-531 (external admixture) were added to SBS-modified asphalt and sheared using a high-speed shear rate of 3000 r/min and a shear temperature controlled at 165 °C to 170 °C.
(3) UV-531- and SBS-modified asphalt was mixed and sheared at high speed for 40 min to prepare UV-531/SBS-modified asphalt, blended into the amount of 0% of UV-531 produced that is SBS-modified asphalt (not involving UV absorber).

Figure 1 depicts the UV-531/SBS-modified asphalt preparation procedure. In order to more clearly identify between the four distinct UV-531/SBS-modified asphalt dosages, the produced modified asphalts were given numbers. The numbering outcomes are displayed in Table 3.

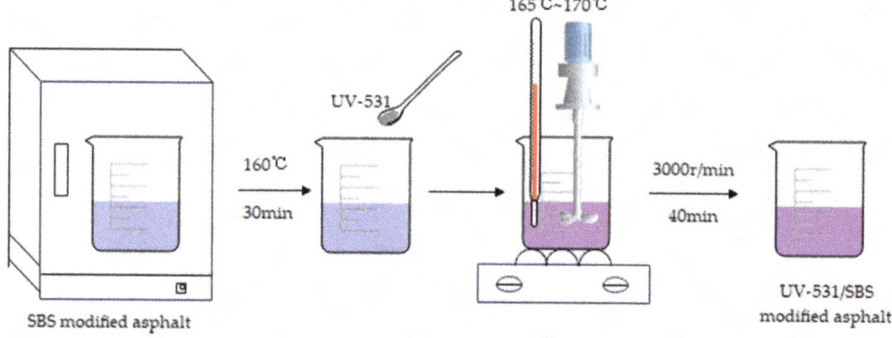

Figure 1. Flow chart of UV-531/SBS asphalt preparation.

Table 3. Modified Asphalt Number.

Modified Asphalt	Number
SBS-Modified Asphalt	SBS
0.4%UV-531 + SBS-Modified Asphalt	A
0.7%UV-531 + SBS-Modified Asphalt	B
1.0%UV-531 + SBS-Modified Asphalt	C

2.3. Asphalt UV Aging Test Process

The UV aging tests on the prepared modified asphalts were performed in a UV infrared chamber, as shown in Figure 2. According to the existing study [18], asphalt specimens should have a thickness of 1 mm during UV aging; thus, in order to effectively simulate the asphalt film thickness of the in-situ asphalt mixture during the test, the sample container area and the asphalt film thickness (1 mm) were used to calculate the mass of the asphalt UV aging specimens.

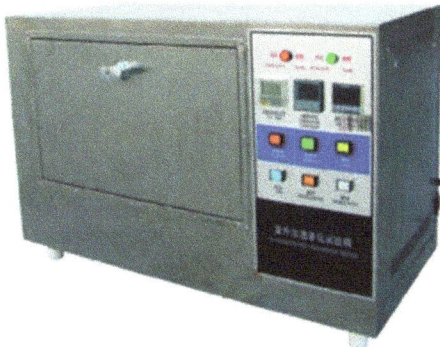

Figure 2. UV aging environment box for asphalt.

The diameter of the glass disc used to hold the asphalt during the test was 14 cm and the thickness of the asphalt film was 1 mm, so 15.4 g of asphalt was weighed into the glass disc. The discs were warmed in the oven at 170 °C until the asphalt was uniformly laid on the surface of the glass discs. Finally, the asphalt aging specimens were placed in the UV aging environmental chamber for 7 d. The UV-aged asphalt specimens are shown in Figure 3, and the process of the asphalt UV aging test is shown in Figure 4.

Figure 3. Aging asphalt sample.

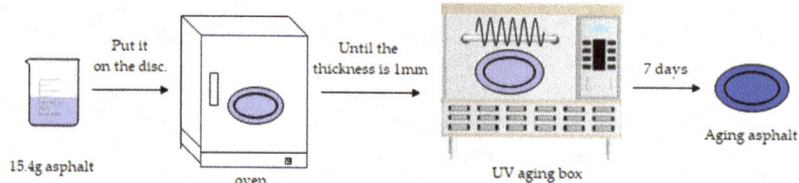

Figure 4. Ultraviolet aging experiment diagram.

2.4. Tests and Methods

2.4.1. Needle Penetration

The needling values of four SBS-modified asphalts were tested at 25 °C, 100 g, and 5 s using a fully automatic needling instrument. Testing method: dry needle penetration sample dishes are prepared, the asphalt is placed in the vessels cooled to ambient temperature, and then the asphalt vessels are placed in a constant temperature water bath at 25 °C for 1.5 h. Remove the test sample after the water bath, then evaluate the needle penetration of the test sample using the needle penetration tester to determine the needle penetration of the measured data to take the average of three calculations. The resulting value is the needle penetration of modified asphalt.

2.4.2. Softening Point

In this paper, the effect of UV-531 on the high temperature performance of SBS-modified asphalt was evaluated by testing the softening point values of four SBS-modified asphalts using a fully automatic softening point tester. The test procedure is as follows: take a certain amount of modified asphalt and place it in a standard mold, cool the mold together with the internal asphalt to room temperature, and then scrape the test sample using a hot scraper. Place the test sample in a constant temperature bath of 5 ± 0.5 °C for 30 min. Use the softening point tester to test the samples until the test sample wrapped inside the steel ball just falls on the receiving plate when stopped, then record the temperature of the test sample. The temperature data is the softening point of the modified asphalt.

2.4.3. 5 °C Ductility

The authors chose a temperature of 5 °C for the ductility test. The test was performed by slowly pouring the modified asphalt into the mold from one end to the other, allowing the mold and the asphalt inside to cool to room temperature before scraping the test sample with a hot scraper. The test sample was placed in a water bath at a temperature of 5 ± 0.5 °C for 1.5 h. Using a ductility tester, the sample was tested until the asphalt was pulled off and the scale was read, which provided the ductility of the modified asphalt.

2.4.4. High-Temperature Dynamic Shear Rheology Test

Temperature scanning tests were performed on four asphalt specimens using a dynamic shear rheometer (DSR) with a temperature range of 30–100 °C and a frequency of 10 ± 0.1 rad/s. The DSR test method is accomplished by measuring the viscous and elastic properties of a thin asphalt binder specimen sandwiched between an oscillating plate and a fixed plate [19].

The high temperature PG index is determined by the DSR of the unaged asphalt, as reflected in the rutting factor curve, at the temperature corresponding to $G^*/\sin\delta \leq 1.00$ kpa [20].

2.4.5. Low-Temperature Glass Transition Temperature Test

The modulus of a viscoelastic material can be tested using the Dynamic Mechanical Analysis (DMA) test as a function of time, temperature, or frequency. Only a tiny sample is needed to ascertain the material's dynamic mechanical properties over a large range of temperatures or frequencies. Figures 5 and 6 display the clamp and dynamic mechanical analyzer.

Figure 5. Dynamic Mechanical Analysis.

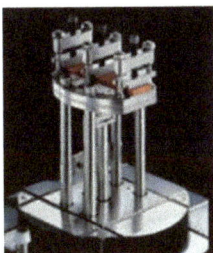

Figure 6. The clamp of Dynamic Mechanical Analysis.

The ideal modulus–temperature curve of modified asphalt typically has four regions: the glassy region, the glassy–rubbery transition region, the rubbery plateau region, and the flow region. The service temperature of asphalt mixtures undergoes the entire process from the glassy to the viscous flow state, and the mechanical condition of the asphalt mixture is essentially a macroscopic depiction of molecular mobility. The glass transition temperature can describe the asphalt mixture's low-temperature performance. When the glass transition temperature is lower than the minimum service temperature and the asphalt can perform as intended, this is the perfect state for the asphalt mixture. This gives the asphalt mixture good deformability throughout the service temperature and can relax the temperature stresses caused by the lowering of the temperature, reducing the generation of low-temperature cracks.

2.4.6. Rutting Factor Ratio

The rutting factor ratio is the ratio of the asphalt rutting factor after aging and before aging. The formula for calculating the rutting factor ratio is shown in Equation (1).

$$TR = \frac{(G_2^* / \sin \delta_2)}{(G_1^* / \sin \delta_1)} \tag{1}$$

where: TR is the rutting factor ratio; δ_1 and δ_2 are the phase angles before and after aging, respectively; G_1^* is the complex modulus before aging, and G_2^* is the complex modulus after aging.

2.4.7. Viscosity Aging Index

Asphalt aging, viscosity increases, and the viscosity curve move up a distance, and this distance can be called the viscosity aging index. A lower viscosity aging index indicates superior anti-aging performance, and the viscosity aging index calculation formula is shown in Formula (2).

$$C = \lg\lg(\eta_a \times 10^3) - \lg\lg(\eta_0 \times 10^3) \qquad (2)$$

where: C is the viscosity aging index, η_a is the viscosity of the asphalt after aging, and η_0 is the viscosity of the original asphalt.

2.4.8. Infrared Spectroscopy

Infrared spectroscopy is based on the information of atomic vibrations and rotations inside molecules to analyze and determine substance composition and molecular structure [21]. Infrared spectrograms can be applied to the determination of the molecular structure of compounds, identification of unknowns, and analysis of mixture composition [22]. In other words, changes in the compositional makeup of asphalt prior to and subsequent to aging can be studied by infrared spectroscopy.

3. Results

3.1. Effect of UV-531 on the Conventional Performance of SBS-Modified Asphalt

The effect of UV-531 on the conventional performance of SBS-modified asphalt was investigated by adding UV-531 at 0.4%, 0.7%, and 1.0% to SBS-modified asphalt and testing the needle penetration, softening point, and 5 °C ductility.

3.1.1. Needle Penetration Results

Needle penetration measures asphalt viscosity and is strongly related to asphaltene content; the higher the asphaltene content and the lower the aromatic fraction content, the lower the needle penetration index [23]. Figure 7 depicts the test results.

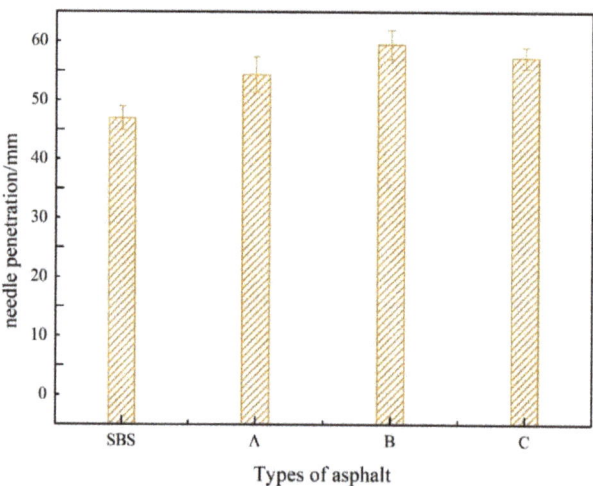

Figure 7. Needle Penetration of Modified Asphalt at 25 °C.

As can be seen from Figure 7, with the increase of UV-531 admixture, the needle penetration first increased and then decreased, at 0.7% admixture, increased by 24%, overall UV-531/SBS-modified asphalt needle penetration than SBS-modified asphalt slightly increased. Mainly due to the molten state, between the heterogeneous polymer molecules break up and convective agitation, coupled with the strong shear effect of mixing equipment, the formation of a part of the graft or block copolymer, so that the original asphalt intermolecular force is reduced, mobility increased, the needle penetration increased.

3.1.2. Softening Point Results

The softening point of asphalt characterizes its high-temperature performance; the higher the softening point, the better the asphalt's high-temperature performance. The test results are shown in Figure 8.

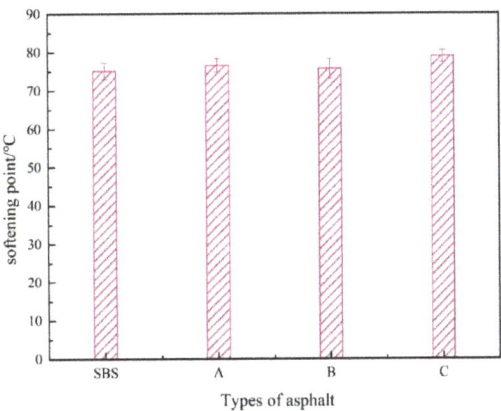

Figure 8. Softening Point of Modified Asphalt.

Figure 8 indicates that the softening point tended upward when the concentration of UV-531 rose. Although there is a small increase of 4.9% at 1.0% compared to SBS-modified bitumen, this indicates that UV-531 has little effect on the high-temperature properties of SBS-modified bitumen.

3.1.3. 5 °C Ductility Results

Ductility characterizes the low temperature cracking resistance of asphalt. The higher the ductility value, the greater the elastic component and the lower the viscous component, the better the low-temperature crack tolerance of the asphalt. In this study, 5 °C ductility was used to investigate the impact of UV-531 on the low-temperature properties of SBS-modified asphalt, and the test is shown in Figure 9.

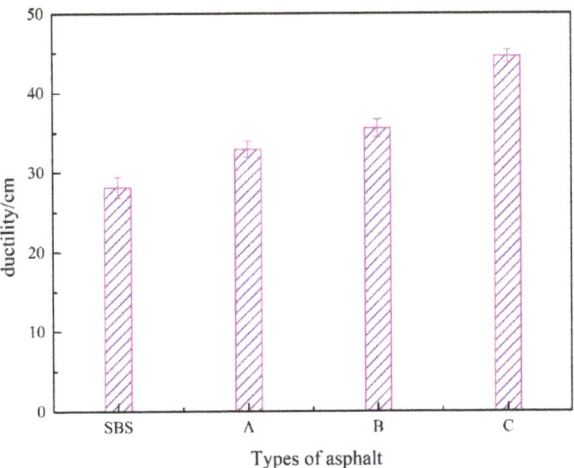

Figure 9. Ductility of Modified Asphalt at 5 °C.

Figure 9 illustrates how various UV-531 concentrations significantly impacted the ductility of the SBS-modified asphalt. From 0.4% to 1.0% of UV-531, the ductility of SBS-modified asphalt gradually increased, with an increase of 58.3% at 1.0%, showing that UV-531 improved the ability of SBS-modified asphalt to withstand low-temperature distortion.

3.2. Effect of UV-531 on the High and Low-Temperature Performance of SBS-Modified Asphalt before and after UV Aging

3.2.1. High-Temperature Dynamic Shear Rheology Test Results

To evaluate the effect of UV-531 on the high-temperature properties of asphalt, we measured the curves of composite shear modulus and phase angle variation with temperature. The test outcomes are shown in Figures 10–13.

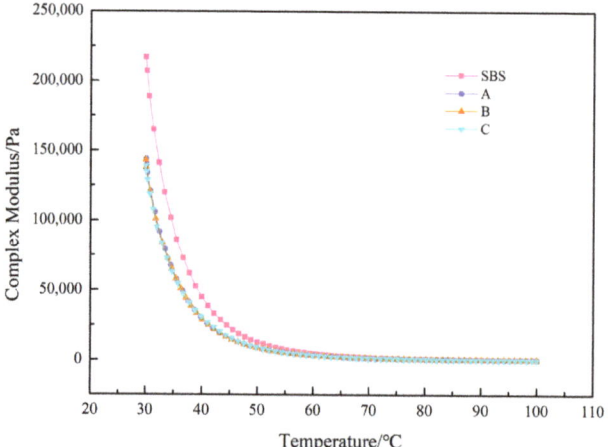

Figure 10. Modified asphalt complex modulus before UV aging.

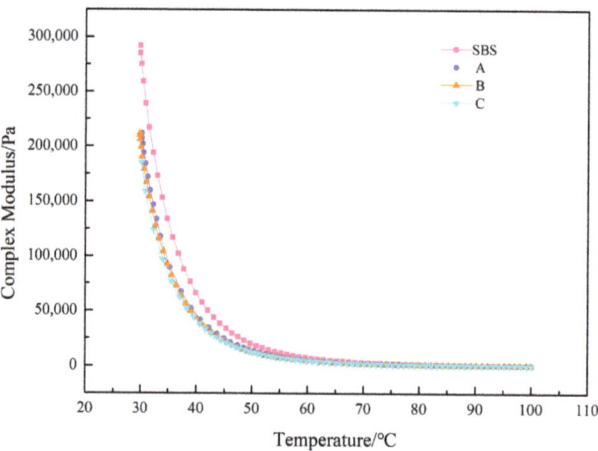

Figure 11. Modified asphalt complex modulus after UV aging.

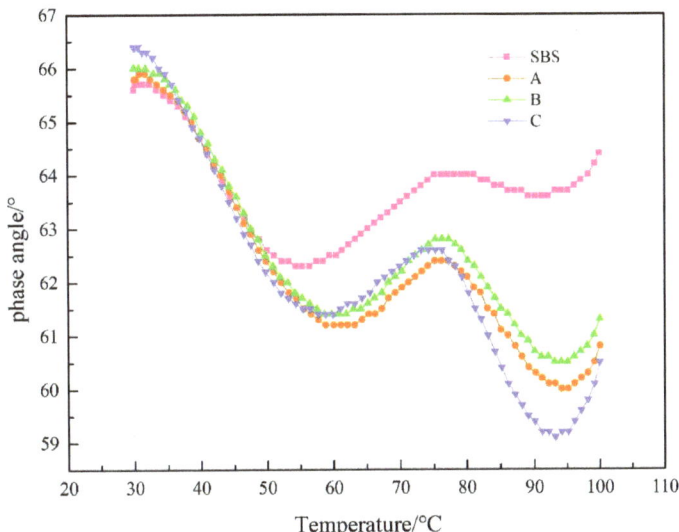

Figure 12. The phase angle of modified asphalt before UV aging.

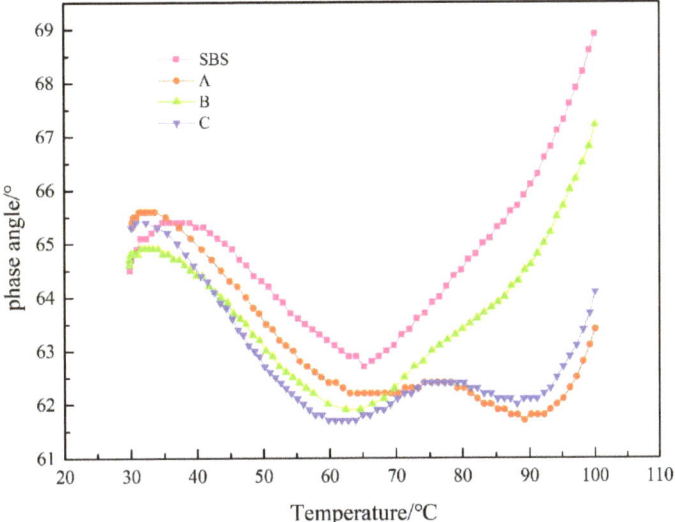

Figure 13. The phase angle of modified asphalt after UV aging.

Figure 10 shows that the composite modulus curves before UV-aging vary widely until 50 °C, and the composite modulus decreases slightly with increasing UV-531 incorporation and is essentially close after 50 °C. As shown in Figure 11, the composite shear modulus for the four asphalts increased to different degrees after UV aging in comparison with the pre-aging period. After UV aging, pitch molecules absorbed radiation energy, bond energy was broken, free radicals were generated, combined with oxygen, and compound modulus inwardly increased.

The phase angle δ reflects the proportion of the viscous component in the composite modulus [24,25]. A larger δ indicates a higher viscous proportion, and vice versa, a higher elastic proportion. From Figure 12, it can be seen that the addition of UV-531 before aging has a small effect on the stage angle of the low temperature section, and after 50 °C, the stage angle tends to drop with the amount of admixture, indicating that UV-531 can increase elasticity of SBS-modified asphalt. Observing Figure 13, it is evident that the warmer the post-aging temperature is, the lower the phase gradient of UV-531/SBS-modified asphalt is with the rising amount of admixture, compared with the increase of elastic component of SBS-modified asphalt, indicating that the anti-aging properties of SBS-modified asphalt can be enhanced by UV-531. Comparing Figures 12 and 13, it is observed that the overall phase-angle increases upon aging, but the phase-angle change of UV-531/SBS-modified asphalt is smaller than that of SBS-modified asphalt, which means that the viscosity ratio of the four asphalts after aging increases and the incorporation of UV-531 suppresses the increase of the viscous component to some extent.

The results of the high temperature PG test are shown in Figure 14, as can be seen, the PG grade decreases after adding UV-531. The analysis is due to the fact that UV-531 particles added to SBS asphalt are dispersed between asphalt molecules, which hinders the relative movement of asphalt molecular chains and increases the intermolecular reactions. The temperature range of PG ranges between 70–82 °C and the phase angle of the four asphalts is about 60°. The A, B, C three asphalt phase angles are less than the SBS-modified asphalt, the complex modulus is basically the same, and the smaller the phase angle, the smaller the rutting factor. It is consistent with the complex modulus curve and phase angle curve.

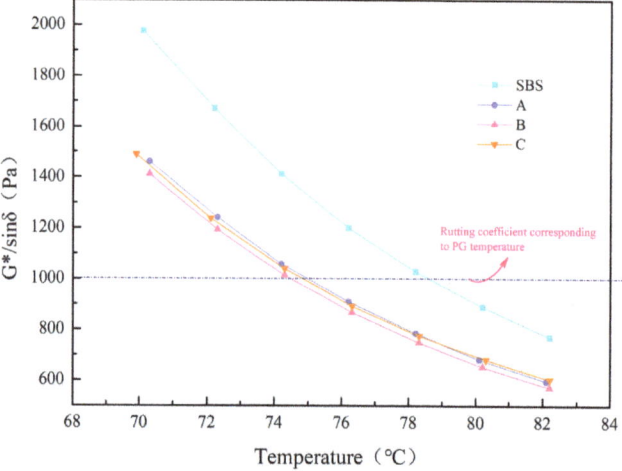

Figure 14. Rutting coefficient of each sample at different temperatures.

3.2.2. Low-Temperature Glass Transition Temperature Test Results

The peak value of the loss modulus determines how hot the asphalt material is at the glass transition temperature of the loss modulus–temperature curve. Mamuye, Y. et al. indicate that the glass transition temperature test method is simple, has a clear physical meaning, correlates well with the measured properties of asphalt mixes, and can be used as an evaluation criterion for the good or bad low-temperature properties of asphalt [26]. DiWang et al. investigated the effects of the glass transition temperature Tg and the use of the modulus shift factor in measuring the rheological properties of an asphalt binder at low temperatures, and the glass transition temperature has been widely used [27]. As a result, we can determine how modified asphalt performs at low-temperatures by comparing the modified asphalt's glass transition temperature prior to and subsequent to UV aging.

From Figure 15, the glass transition temperature of SBS asphalt was reduced by each admixture before aging, and it was reduced by 54.74% at 1.0% admixture, indicating that the incorporation of UV-531 increased aromatic structure of asphalt, and the aromatic content increased with the admixture, improving the overall low temperature resistance by SBS asphalt. The overall temperature after aging decreases, and reaches its lowest at 0.7%, when the glass transition temperature is −12.29 °C, indicating that the increase of UV-531 doping has resisted the increase of glass transition temperature to a certain extent. At this time, UV-531 is likely to capture the free radicals so that the free radicals cannot combine with oxygen and inhibit the oxidation reaction, thus slowing down the aging. Comparing the glass transition temperature before and after aging, it was discovered that the glass transition temperature of asphalt has increased with aging, indicating that UV-531 and SBS asphalt underwent addition and oxidation reactions during aging, resulting in long-chain compounds and oxides, aromatic and resin turning into asphaltene, and worsening low-temperature performance. After aging at 0.7% doping, the curve's inflection point demonstrates that a more significant dose is not better. In general, the increase of UV-531 has an inhibitory effect on both the chain breaking and oxidation reactions that occur during asphalt aging.

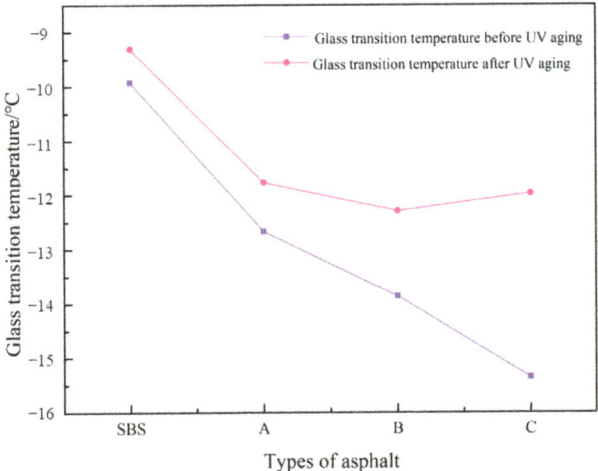

Figure 15. Glass Transition Temperature of Modified Asphalt.

4. Evaluation of UV-531 on the Anti-UV Aging Performance of SBS-Modified Asphalt

4.1. Evaluation of the Aging Behavior of Modified Asphalt Based on Rutting Factor Ratio

The size of the rutting factor ratio reflects the extent of asphalt hardening due to aging; the more significant its value, the worse the asphalt's resistance to short-term aging performance. The paper chose rutting coefficients of 70–88 °C to investigate the aging properties of the four asphalts. Table 4 displays the calculated results. The rutting factor change curve is shown in Figure 16.

Table 4. The G*/sinδ ratio before and after aging of SBS/A/B and C.

Temperature (°C)	SBS			A			B			C		
	$G_1^*/\sin\delta_1$	$G_2^*/\sin\delta_1$	TR	$G_1^*/\sin\delta_1$	$G_2^*/\sin\delta_1$	TR	$G_1^*/\sin\delta_1$	$G_2^*/\sin\delta_1$	TR	$G_1^*/\sin\delta_1$	$G_2^*/\sin\delta_1$	TR
70	1977	3714	1.87	1462	2249	1.53	1413	2536	1.79	1490	1980	1.32
76	1201	2048	1.71	910	1399	1.54	869	1245	1.43	893	1207	1.35
82	771	1204	1.56	598	905	1.51	573	751	1.31	604	783	1.29
88	515	753	1.46	412	607	1.47	395	488	1.23	428	525	1.22

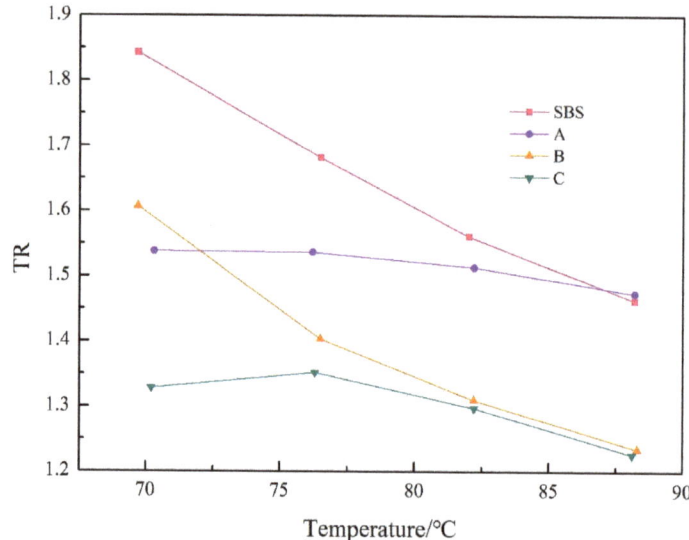

Figure 16. The G*/sinδ ratio before and after aging of SBS/A/B and C.

From Figure 16, the same trends were observed for all four asphalts, which decreased with increasing temperature, but to different degrees. At the same temperature, the rutting factor ratio decreases with increasing doping, indicating that UV-531 incorporation improves the aging resistance. Among the three curves of A, B, and C, the TR decreases with the increase of doping amount, but after a doping amount of 0.7%, there is almost no change in TR, which indicates that the higher doping amount is not better. Comprehensive analysis shows that the addition of UV-531 can inhibit asphalt aging, and the optimal dose is 0.7%.

4.2. Evaluation of Asphalt Aging by Viscosity Aging Index

The aging resistance index of asphalt in this experiment uses the 60 °C viscosity ratio of pre-aging and post-aging asphalt. The aging indices of SBS-modified asphalt and A, B, and C asphalts are listed in Table 5.

Table 5. The aging index of SBS, A, B and C modified asphalt at 60 °C.

Asphalt Type	Original η_a	Aging η_a	C
SBS-modified asphalt	773.918	1275.485	0.01571
A	608.138	910.510	0.01297
B	596.864	767.58	0.00814
C	610.044	822.675	0.00964

From Table 5, there is evidence that prior to aging, the viscosities of A, B, and C are less than those of SBS-modified asphalt, and the reason for this analysis is that UV-531 has little effect upon the branched chain fat composition of road asphalt, but it increases the aromatic composition of asphalt [28], so the viscosity decreases. The overall viscosity of the aged asphalt rises due to the simultaneous oxidative hardening of the matrices and the aging decomposition of the SBS polymer [29], the decrease of the aromatic content, the increase of the carbon-based index, the increase of ashaltene [30], and the increase of molecular volume of asphalt, which causes the asphalt to harden and the visibility to increase. Aging index can be found that UV-531/SBS-modified asphalt is less than SBS-modified asphalt, meaning that UV-531 improves asphalt aging resistance. The aging index at 0.7% admixture is the smallest, indicating that UV-531 admixture has the best content. In general, UV-531 was favorable to improve the aging resistance of asphalt.

4.3. Analysis of the Degree of Aging Using Infrared Spectroscopy

Fusong Wang et al. indicated that some double bonds in SBS copolymers degrade during aging and oxidize to oxygenated groups, such as hydroxyl, carbonyl group, and ether bonds [31], carbonyl absorption peak size can reflect the aging degree, the larger the peak, the deeper the aging. This experiment tested SBS-modified asphalt with UV-531 dosed at 1.0% (i.e., C), Figures 17 and 18 show the test results.

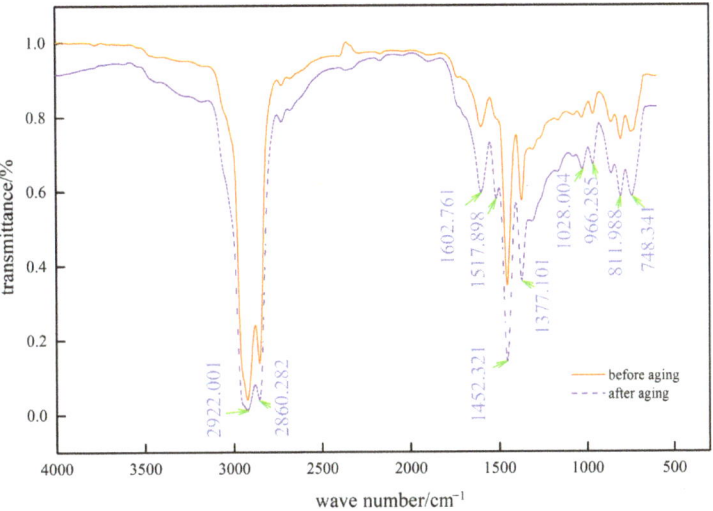

Figure 17. Infrared spectra of SBS-modified asphalt before and after UV aging.

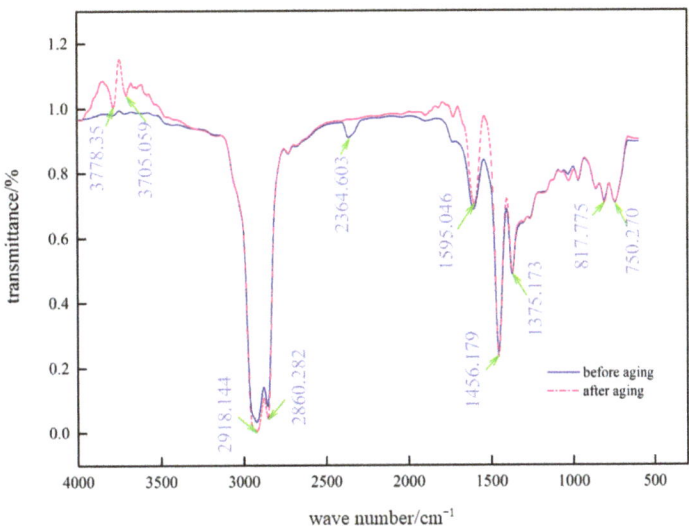

Figure 18. Infrared spectra of C asphalt before and after UV aging.

As seen in Figures 17 and 18, the characteristic peaks of the infrared spectra changed more obviously in the first region (4000–2800), the second region (2800–1900), and the third region (1900–1300), while the characteristic peaks in the fourth region (1300–500) did not change; only the absorption intensity appeared to be different.

Table 6 shows the wave numbers corresponding to some characteristic peaks at wave numbers 1595.046, 2364.603, and 3705.059 for the -C=O group, -C≡C group, and -OH group, respectively. After UV aging, an apparent -OH (conjoined) absorption peak was observed in the first region near position 3750, indicating the occurrence of chemical reactions that resulted in the production of alcohols, phenols, or organic acids, which should be a component of long-chain compounds in the aging process of chain-breaking reactions, primarily oxidation reactions. At position 2364.603 in the second area, the disappearance of the -C≡C group was discovered, showing that the -C≡C group carbon element was involved in the chemical reaction that produced R'≡R symmetric molecules without an infrared spectral band; the number of carbonyl groups in the aromatic rings of the two asphalts have changed significantly as a result of UV aging, as shown by the third region in 1595.046 position -C=O absorption peak, which becomes more prominent. A larger carbonyl absorption peak denotes more severe UV aging, thus, the SBS-modified asphalt and C asphalt exhibit UV aging ranging from severe to mild.

Table 6. Characteristic peak distribution table.

Wave Number/cm^{-1}	1595.046	2364.603	3705.059
Feature Peak	-C=O	-C≡C	-OH group of alcohols and phenols

5. Conclusions

In this paper, the needle penetration, softening point, and 5 °C ductility of UV-531/SBS asphalt for the purpose of assessing the effect of UV-531 on the conventional properties of SBS-modified asphalt were measured. DSR and DMA tests were performed to estimate their effects on the high and low temperature properties of SBS-modified bitumen. Three methods were chosen to investigate the aging behavior of modified asphalt UV-531/SBS, such as rutting factor ratio, viscosity aging index, and infrared spectroscopy. In a comprehensive view, the addition of UV-531 can enhance the UV aging resistance of asphalt effectively with an optimal dose of 0.7%. The main conclusions are listed below.

(1) The ductility at 5 °C and needle penetration of SBS-modified asphalt were improved with the increase of UV-531 addition, while the softening point was less affected. This means that the addition of UV-531 improved the low temperature properties and viscosity of SBS-modified asphalt, while the high temperature properties were less affected.

(2) The DSR test revealed that UV-531 enhanced the high temperature rheological properties of SBS-modified asphalt to some degree. It was shown in the DMA test that the increase of UV-531 could reduce the glass transition temperature by a maximum of 54.74%, which led to a significant improvement in the low temperature anti-cracking properties of asphalt after ultraviolet light aging.

(3) Three evaluation methods were selected to evaluate the aging. In the evaluation method based on the rutting coefficient ratio, the rutting coefficient ratio reduced as the amount of UV-531 doping increased, demonstrating that UV-531 inhibits UV aging. In the evaluation, a method based on viscosity aging index was used. As the viscosity before aging decreased with increasing dosing, the viscosity aging index also decreased, reaching a minimum at a dosing of 0.7%, indicating that UV-531 can increase the aromatic structure of asphalt. According to the analysis of the degree of aging by infrared spectra, the characteristic peaks of infrared spectra vary more obviously in the first, second and third regions, and the analysis results were that the UV aging degree of UV-531/SBS asphalt with 0.7% admixture was significantly lower than that of SBS-modified asphalt.

Author Contributions: Conceptualization, L.L. (Li Liu) and L.L. (Leixin Liu); Data curation, L.L. (Leixin Liu) and C.Y.; Formal analysis, L.L. (Leixin Liu); Funding acquisition, L.L. (Li Liu) and Z.L.; Investigation, L.L. (Leixin Liu), C.Y., B.P. and W.L.; Methodology, L.L. (Leixin Liu); Resources, L.L. (Li Liu); Software, L.L. (Leixin Liu); Supervision, Z.L.; Validation, L.L. (Li Liu); Writing—original draft, L.L. (Leixin Liu); Writing—review & editing, L.L. (Li Liu) and C.Y. All authors have read and agreed to the published version of the manuscript.

Funding: This research was funded by National Key R&D Program of China, grant number 2021YFB2601000.

Institutional Review Board Statement: Not applicable.

Informed Consent Statement: Not applicable.

Data Availability Statement: Not applicable.

Conflicts of Interest: The authors declare no conflict of interest.

References

1. Liu, H.; Zhang, Z.; Tian, Z.; Lu, C. Exploration for UV Aging Characteristics of Asphalt Binders based on Response Surface Methodology: Insights from the UV Aging Influencing Factors and Their Interactions. *Constr. Build. Mater.* **2022**, *347*, 128460. [CrossRef]
2. Liu, L.; Liu, Z.; Hong, L.; Huang, Y. Effect of ultraviolet absorber (UV-531) on the properties of SBS-modified asphalt with different block ratios. *Constr. Build. Mater.* **2020**, *234*, 117388. [CrossRef]
3. Zhang, Y.; Leng, Z. Quantification of bituminous mortar ageing and its application in ravelling evaluation of porous asphalt wearing courses. *Mater. Des.* **2017**, *119*, 1–11. [CrossRef]
4. Liu, L.; Liu, Z.; Liu, J. Effects of Silane-Coupling Agent Pretreatment on Basalt Fibers: Analyzing the Impact on Interfacial Properties and Road Performance. *J. Mater. Civ. Eng.* **2020**, *32*, 04020041. [CrossRef]
5. Mohammad Asib, A.S.; Rahman, R.; Romero, P.; Hoepfner, M.P.; Mamun, A. Physicochemical characterization of short and long-term aged asphalt mixtures for low-temperature performance. *Constr. Build. Mater.* **2022**, *319*, 30–38. [CrossRef]
6. Behnood, A.; Modiri Gharehveran, M. Morphology, rheology, and physical properties of polymer-modified asphalt binders. *Eur. Polym. J.* **2019**, *112*, 766–791. [CrossRef]
7. Behnood, A.; Olek, J. Rheological properties of asphalt binders modified with styrene-butadiene-styrene (SBS), ground tire rubber (GTR), or polyphosphoric acid (PPA). *Constr. Build. Mater.* **2017**, *151*, 464–478. [CrossRef]
8. Li, Y.; Feng, J.; Wu, S.; Chen, A.; Kuang, D.; Bai, T.; Gao, Y.; Zhang, J.; Li, L.; Wan, L.; et al. Review of ultraviolet ageing mechanisms and anti-ageing methods for asphalt binders. *J. Road Eng.* **2022**, *2*, 137–155. [CrossRef]
9. Liao, M.; Liu, Z.; Gao, Y.; Liu, L.; Xiang, S.J.C.; Materials, B. Study on UV aging resistance of nano-TiO_2/montmorillonite/styrene-butadiene rubber composite modified asphalt based on rheological and microscopic properties. *Constr. Build. Mater.* **2021**, *301*, 124108. [CrossRef]
10. Li, Y.; Feng, J.; Chen, A.; Wu, S.; Bai, T.; Liu, Q.; Zhu, H. Development Mechanism of Aging Depth of Bitumen with Increasing UV Aging Time. *Case Stud. Constr. Mater.* **2022**, *2022*, e01057. [CrossRef]
11. Hai-lin, Z. Comparative study of multi-scale characteristics of hard-Grade asphalt and SBS modified asphalt under thermal oxygen and ultraviolet aging. *J. China Foreign Highw.* **2020**, *40*, 305–310. [CrossRef]
12. Ma, L.; Wang, F.; Cui, P.; Yunusa, M.; Xiao, Y. Effect of aging on the constitutive models of asphalt and their mixtures. *Constr. Build. Mater.* **2021**, *272*, 121611. [CrossRef]
13. Nobakht, M.; Sakhaeifar, M.S. Dynamic modulus and phase angle prediction of laboratory aged asphalt mixtures. *Constr. Build. Mater.* **2018**, *190*, 740–751. [CrossRef]
14. Zheng, N.; Ji, X.; Hou, Y. Nonlinear Prediction of Attenuation of Asphalt Performance after Ultraviolet Aging. *J. Highw. Transp. Res. Dev.* **2009**, *26*, 33–37.
15. Xiao, M.M.; Fan, L. Ultraviolet aging mechanism of asphalt molecular based on microscopic simulation. *Constr. Build. Mater.* **2022**, *319*, 126157. [CrossRef]
16. Zhen-gang, F.; Hao, Z.; Xin-jun, L.; Pei-long, L.; Li-ke, Z. Effect of ultraviolet absorber UV531 on morphology of SBS modified bitumen. *J. Wuhan Univ. Technol.* **2015**, *37*, 53–58.
17. Ling-ling, H.; Zhao-hui, L.; Guo-jie, F.; Sheng, L.; Li, L. Effect of Ultraviolet Absorber on Properties of SBS Modified Bitumen with Different Block Ratios. *J. Wuhan Univ. Technol.* **2017**, *39*, 28–36. [CrossRef]
18. Li, Q.; Zeng, X.; Wang, J.; Luo, S.; Meng, Y.; Gao, L.; Wang, X. Aging performance of high viscosity modified asphalt under complex heat-light-water coupled conditions. *Constr. Build. Mater.* **2022**, *325*, 126314. [CrossRef]
19. Ren, S.; Liu, X.; Fan, W.; Wang, H.; Erkens, S. Rheological Properties, Compatibility, and Storage Stability of SBS Latex-Modified Asphalt. *Materials* **2019**, *12*, 3683. [CrossRef]
20. Ozdemir, D.K. High and low temperature Theological characteristics of linear alkyl benzene sulfonic acid modified bitumen. *Constr. Build. Mater.* **2021**, *301*, 124041. [CrossRef]

21. Yaping, H.; Sanpeng, M.; Guitao, Z.; Wei, J. Application and Development Trend of Infrared Spectroscopy in Asphalt Analysis. *Pet. Asph.* **2021**, *35*, 35–39. [CrossRef]
22. Xing, C.; Li, M.; Zhao, G.; Liu, N.; Wang, M. Analysis of bitumen material test methods and bitumen surface phase characteristics via atomic force microscopy-based infrared spectroscopy. *Constr. Build. Mater.* **2022**, *346*, 128373. [CrossRef]
23. Wang, Z.; Wu, X.; Li, H.; Si, Y. Study on The Correlation between the Four Components of Matrix Asphalt and Conventional Indexes. *Pet. Asph.* **2020**, *34*, 26–32, 36. [CrossRef]
24. Yu, H.; Bai, X.; Qian, G.; Wei, H.; Li, Z.J.P. Impact of Ultraviolet Radiation on the Aging Properties of SBS-Modified Asphalt Binders. *Polymers* **2019**, *11*, 1111. [CrossRef] [PubMed]
25. Dokandari, P.A.; Topal, A.; Ozdemir, D.K. Rheological and Microstructural Investigation of the Effects of Rejuvenators on Reclaimed Asphalt Pavement Bitumen by DSR and AFM. *Int. J. Civ. Eng.* **2021**, *19*, 749–758. [CrossRef]
26. Mamuye, Y.; Liao, M.C.; Do, N.D. Nano-Al_2O_3 composite on intermediate and high temperature properties of neat and modified asphalt binders and their effect on hot mix asphalt mixtures. *Constr. Build. Mater.* **2022**, *331*, 127304. [CrossRef]
27. Investigation on the low temperature properties of asphalt binder: Glass transition temperature and modulus shift factor. *Constr. Build. Mater.* **2020**, *245*, 118351. [CrossRef]
28. Feng, Z.G.; Wang, S.J.; Bian, H.J.; Guo, Q.L.; Li, X.J.J.C. FTIR and rheology analysis of aging on different ultraviolet absorber modified bitumens. *Constr. Build. Mater.* **2016**, *115*, 48–53. [CrossRef]
29. Zhu, H.; Shen, J.N.; Yu, J. Micro and Macro-level Properties of Weathered Asphalt Binders. *J. Mater. Sci. Eng.* **2021**, *39*, 801–807. [CrossRef]
30. Qu, X.; Liu, Q.; Guo, M.; Wang, D.; Oeser, M. Study on the effect of aging on physical properties of asphalt binder from a microscale perspective. *Constr. Build. Mater.* **2018**, *187*, 718–729. [CrossRef]
31. Wang, F.; Zhang, L.; Zhang, X.; Li, H.; Wu, S. Aging mechanism and rejuvenating possibility of SBS copolymers in asphalt binders. *Polymers* **2020**, *12*, 92. [CrossRef] [PubMed]

Article

Influence of Fe$_2$O$_3$, MgO and Molarity of NaOH Solution on the Mechanical Properties of Fly Ash-Based Geopolymers

Brăduț Alexandru Ionescu [1,2], Mihail Chira [1], Horațiu Vermeșan [3,*], Andreea Hegyi [1,*], Adrian-Victor Lăzărescu [1,*], Gyorgy Thalmaier [3], Bogdan Viorel Neamțu [3], Timea Gabor [3,*] and Ioana Monica Sur [3]

1 NIRD URBAN-INCERC Cluj-Napoca Branch, 117 Calea Floresti, 400524 Cluj-Napoca, Romania
2 IOSUD UTCN Doctoral School, Technical University of Cluj-Napoca, 15 Daicoviciu Street, 400020 Cluj-Napoca, Romania
3 Faculty of Materials and Environmental Engineering, Technical University of Cluj-Napoca, 103-105 Muncii Boulevard, 400641 Cluj-Napoca, Romania
* Correspondence: horatiu.vermesan@imadd.utcluj.ro (H.V.); andreea.hegyi@incerc-cluj.ro (A.H.); adrian.lazarescu@incerc-cluj.ro (A.-V.L.); timea.gabor@imadd.utcluj.ro (T.G.)

Abstract: The use of waste from industrial activities is of particular importance for environmental protection. Fly ash has a high potential in the production of construction materials. In the present study, the use of fly ash in the production of geopolymer paste and the effect of Fe$_2$O$_3$, MgO and molarity of NaOH solution on the mechanical strength of geopolymer paste are presented. Samples resulting from the heat treatment of the geopolymer paste were subjected to mechanical tests and SEM, EDS and XRD analyses. Samples were obtained using 6 molar and 8 molar NaOH solution with and without the addition of Fe$_2$O$_3$ and MgO. Samples obtained using a 6 molar NaOH solution where Fe$_2$O$_3$ and MgO were added had higher mechanical strengths compared to the other samples.

Keywords: geopolymer; compressive strength; Fe$_2$O$_3$; MgO

1. Introduction

One of mankind's most important problems is climate change, which is occurring as a result of the significant increase in annual temperatures [1]. Anthropogenic industrial activities release large amounts of greenhouse gases such as carbon dioxide (CO$_2$) into the atmosphere, which adversely affect the climate [2]. The construction industry is one of the main sources of greenhouse gas emissions, accounting for half of global emissions [3].

After water, concrete is the most widely used material in the world, with annual production exceeding 20 billion tonnes [4–7]. Portland cement is the most widely used powder to bind different constituents of concrete. According to a report, about 4.2 billion tonnes of Portland cement were manufactured in 2016 just to meet the high market demand [8]. Energy consumption for Portland cement manufacturing is estimated at 3% of global energy consumption [9]. In addition, one tonne of Portland cement production causes one tonne of CO$_2$ emissions, and the cement industry contributes with 7–8% of global CO$_2$ emissions [10,11]. Due to high emissions and energy consumption, the Portland cement industry is considered one of the main causes of climate change and contributes about 65% to global warming [12]. Approximately 1.5 tonnes of virgin raw materials are used to manufacture one tonne of Portland cement, leading to the depletion of natural resources [13,14]. Climate experts suggest that mankind should reduce emissions to zero by 2050 to limit global warming to 1.5 °C [1]. Therefore, scientists are focusing on limiting Portland cement production by finding environmentally friendly and energy efficient concrete binders. A first step is to substitute a quantity of Portland cement with fly ash and obtain concrete with self-healing properties [15,16].

One of the most important alternative binders to Portland cement are geopolymers also called inorganic polymers, consisting of alternating tetrahedral chains of SiO_4 and AlO_4, connected by a common oxygen atom and balanced by cations [17,18]. Some precursor materials used to produce geopolymers contain large amounts of iron. Although the presence of iron could play an important role in the structure and properties of geopolymers, Al substitutions with Fe have not yet been fully studied, even though they might occur in clays [19–21]. For example, fly ash, with an iron content of about 10%, stands out among these commonly used iron-rich precursor materials and up to 40% for some low-calcium ferric slag materials [22]. The compressive strength of samples obtained with these types of materials ranged from 20 to 80 MPa.

Studies on geopolymers are largely based on traditional precursor materials such as metakaolinite (2% Fe_2O_3 content), fly ash (10% Fe_2O_3 content) and blast furnace slag (0.5% Fe_2O_3 content). However, recent studies have shown that precursors with higher iron content than typically found in fly ash can be activated in alkaline mediums [20,21] with engineering applications. The presence of iron trioxide in the heat-treated geopolymer paste results in the formation of a ferro-silicate geopolymer [-Fe-O-Si-O-Al-O]. The amount of substituted Fe atoms can vary between 5% and 50% of the total amount of Fe_2O_3 contained in the geopolymer binder [23].

Additionally, several studies [23–27] indicate that binders and concrete obtained from alkali-activated slag show high mechanical strength and good performance to chemical attack, freeze–thaw cycles and high temperatures.

However, previous research [28–31] has shown that alkali-activated slag mortar and concrete is subject to substantial shrinkage by drying. This is one of the main disadvantages of the definitive use of alkali-activated slag as an alternative to traditional Portland cement binders. There are a number of factors that determine the drying shrinkage of alkali-activated slag, including the type and content of alkali activators [30,32–34], aggregate and slag properties [28,35], and the curing environment [36–39].

In general, sodium silicate-activated slag has higher shrinkage than sodium hydroxide-activated slag, and the drying shrinkage of alkali-activated slag increases with increasing activator dosage as well as with slag fineness [34,40]. In addition, the shrinkage of alkali-activated slag is very sensitive to the curing medium.

The use of magnesium oxide, MgO, as a shrinkage-reducing mineral additive dates back to the mid-1970s. Volume compensation during the drying process was due to the chemical reaction between MgO and water forming brucite ($Mg(OH)_2$), which results in a 118% increase in volume [41]. The effect of MgO in alkaline-activated slag systems has been investigated recently, either in terms of its naturally variable content in different slag compositions [42] or as an additive [43]. Ben Haha et al. [42] investigated the effect of the natural MgO content in different slags on the performance of alkali-activated slag and showed that although the main hydration product is still C-S-H gel, MgO reacts with slag to form hydrotalcite ($Mg_6Al_2(OH)_{16}CO_3 \bullet 4H_2O$), the content of which increases as the MgO content of the slag increases. They also concluded that because these hydrotalcite-like phases are bulkier than C-S-H, they lead to higher strength, therefore the higher the MgO content, the higher the strength. In the work of Fei Jin et al. [44], the effect of adding MgO as a commercial reagent on the drying shrinkage and strength of alkali-activated slag was studied. It was found that MgO with high reactivity accelerated the early hydration of alkali-activated slag, while MgO with medium reactivity had little effect. Drying shrinkage was significantly reduced by highly reactive MgO, but cracking resulted after drying the samples. On the other hand, MgO with medium reactivity caused a reduction in shrinkage only after one month, but cement strength was improved.

In general, as the concentration of NaOH solution increases, the compressive strengths of samples obtained by alkali-activation of fly ash increase, but there are situations where the strength decreases. This variation in the effect of NaOH concentration on compressive strength is probably due to the different nature and type of molecules that form the fly ash particles. These differences between the types of molecules affect the degree of leaching of SiO_2, Al_2O_3, CaO and Fe_2O_3 in the alkaline activator, where SiO_2 leaching is slower than the other components [45]. According to Fernández-Jiménez and Palomo [46], NaOH concentration is responsible for the decomposition of the bonds of the main oxides. Thus, fly ash with higher SiO_2 content requires a higher NaOH concentration to release SiO_2 and the other oxides from fly ash particles to initiate geopolymerization.

The presence of coarser particles in the fly ash reduces the surface area that is exposed to the alkaline activator [47]. This means that a low chemical reaction and partial dissolution may occur on the surface of the coarse particles. As a result, these unreacted particles will be a weak point in the geopolymer matrix, which consequently reduces the compressive strength of the geopolymer specimens.

In terms of the mechanism of geopolymerization reactions, it is currently estimated worldwide that this process is a result of the dissolution of Si_2O_3 and Al_2O_3 oxides into atoms under the influence of the Na^+ and hydroxyl (OH^-) ion supplying alkali activator. These dissolved atomic species of Si and Al, in the presence of water, form a gel in which the atoms move freely, allowing the formation of monomers, followed by poly- and oligomerization, finally leading to the formation of three-dimensional chain networks. In contrast to the hydration-hydrolysis mechanism specific to Portland cement, geo-polymerization expels water in the polymerization/hardening/maturation process. This process is called "dehydroxylation", water having only the function of facilitating the mobilities of the constituent groups in the gel matrix to form the specific bonds, the whole process can be represented in a generalized equation of the form (Equation (1)) [48,49]:

$$[R] - O - Si - Si - (OH)\ [R] - Si - O - [r] + H_2O \quad (1)$$

where:
[R] = atoms connected to -O-Si-OH, (Al or Fe)
[r] = new chain sequences that connect to [R]-Si-O- to form a larger chain
"+H_2O" indicates the expulsion of water for the bonds to form. The geopolymer acquires its strength by creating long networks of three-dimensional chains, leading to the initial use of a large amount of capillary H_2O, followed by its expulsion once a suitable bond can form.

Worldwide, in the general study on the production of geopolymer binders, there are still a number of controversies or insufficiently clarified elements, as their mechanical strengths and other physical-mechanical performances are strongly influenced by the oxide composition of the main raw materials and additives, the type and molarity of the alkali activators, the existence or not of heat treatment. From the point of view of the oxide composition and other characteristics of the raw materials, depending on their origin, there is a great heterogeneity, which is the main difficulty in the production of geopolymer binders, the customization of the mixtures by the mass ratio of raw materials and the molarity of the alkaline activator, from case to case, being essential.

The aim of this paper is to analyse the possibility of producing geopolymer binders and the influence of Fe_2O_3 and MgO additions and the molarity of the NaOH solution on its mechanical strengths, under the conditions of using a local fly ash, specific for Romanian thermal power plants.

2. Materials and Methods

2.1. Materials

The raw materials used in this study for producing alkali-activated geopolymer pastes were selected locally and consisted of fly ash (F.A.), iron trioxide (Fe_2O_3), magnesium oxide (MgO), sodium hydroxide solution (6M, respectively, 8M) and sodium silicate solution Na_2SiO_3 34%.

Fly ash used in the production of the geopolymer binder was obtained from the Rovinari Thermal Power Plant, Romania. Iron trioxide, magnesium oxide, sodium silicate and NaOH in the form of micropearls with 99.7% purity were purchased commercially. The chemical composition of the fly ash is shown in Table 1.

Table 1. Fly ash chemical composition.

Oxides	SiO$_2$	Al$_2$O$_3$	Fe$_2$O$_3$	CaO	MgO	SO$_3$	Na$_2$O	K$_2$O	TiO$_2$	L.O.I
F.A %	46.9	23.8	10.1	10.7	2.7	0.5	0.6	1.7	0.9	2.1

2.2. Synthesis of Geopolymer

Two geopolymer pastes were prepared from fly ash, and an alkaline activating solution by combining NaOH solution (6M and 8M, respectively) with Na$_2$SiO$_3$ solution and two geopolymer pastes to which iron trioxide and magnesium oxide were added in addition to ash. The ratio of sodium silicate solution to Na$_2$SiO$_3$/NaOH sodium hydroxide solution was set to 2. The procedure used in the production of the alkali-activated fly ash based geopolymer binders with added iron trioxide and magnesium oxide is shown in Figure 1. The mixing of fly ash with iron trioxide and magnesium oxide was done for 3 min for complete homogenization using a paddle mixer. Subsequently, after mixing the fly ash with the iron trioxide and magnesium oxide, the alkali activator solution was poured in gradually over 70 s, initially at low speed. Pouring the alkaline activator too abruptly can lead to instant curing effect. The mixing of alkaline activators with fly ash, iron trioxide and magnesium oxide was done for 10 min. After mixing the obtained mixture was poured into rectangular moulds with inner dimensions of 40 × 40 × 160 mm. The geopolymer mixture was kept in the oven for 24h at an activation temperature of 70 °C. After removal from the oven the obtained samples were kept in the climate chamber at a temperature of 23 °C and relative humidity of 50% and their flexural and compressive strengths were measured at 7, 14 and 28 days.

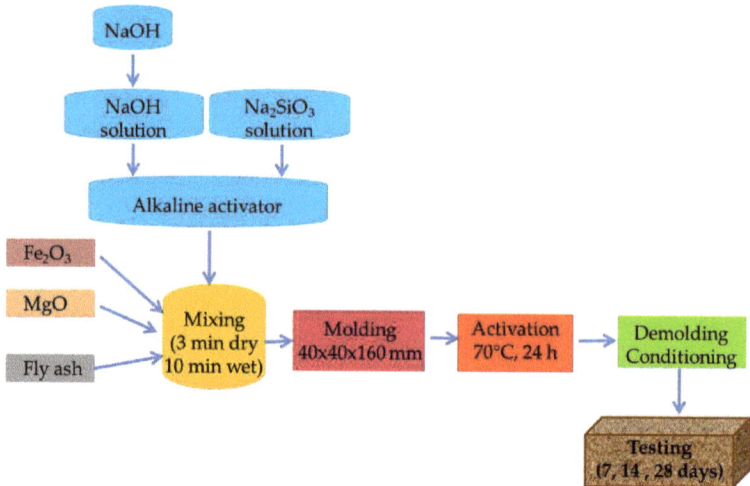

Figure 1. Methodology used for the preparation of the geopolymer paste.

Four geopolymer paste mixtures were prepared and investigated, two with NaOH solution (6M) and two with NaOH solution (8M). The mass ratio of sodium silicate solution to sodium hydroxide solution was set to 2, and the solution mass to dry mass ratio was 0.9. The proportions of substances used for the preparation of the four recipes are shown in Table 2. In order to understand the influence of Fe$_2$O$_3$, MgO and the molarity of the NaOH solution, 1% Fe$_2$O$_3$ and 1% MgO was added to the amount of ash used. This addition of Fe$_2$O$_3$ and MgO was done for both a NaOH solution molarity of 6M and a molarity of 8M.

Table 2. Mass ratio of materials used in the production of the mixtures.

Sample (NaOH conc.)	m_{FA}(g)	$m_{Fe_2O_3}$(g)	m_{MgO}(g)	m_{sol}(g)	$\dfrac{m_{Na_2SiO_3}}{m_{NaOH}}$	$\dfrac{m_{sol}}{m_{dry}}$
6M (6M)	267	-	-	240.30	2	0.9
6MX1 (6M)	267	2.67	2.67	245.10	2	0.9
8M (8M)	267	-	-	240.30	2	0.9
8MX1 (8M)	267	2.67	2.67	245.10	2	0.9

The literature indicates the possibility of using NaOH solution for the preparation of alkaline activator, with various molarities, to obtain geopolymeric materials. For economic and environmental impact reasons, but also to create a basis for preliminary analysis for further research, using NaOH solutions with higher molarity, in this study the 6M and 8M variants were chosen.

Preliminary investigations carried out only with the addition of 1%, 5% and 10% iron trioxide (Fe_2O_3) in relation with mass ratio of fly ash and 6M NaOH solution molar concentration used for the production of the alkaline activator, showed that, in terms of mechanical strengths, as the amount of Fe_2O_3 increased, they decreased by up to 3.3%. Therefore, for further research, 1%, in relation to the amount of ash, was considered as the optimum addition of Fe_2O_3.

Similarly, for the sodium hydroxide solution with 6M molar concentration, the compressive strength of the samples produced using 1% Fe_2O_3 + 1% MgO, 1% Fe_2O_3 + 5% MgO and 1% Fe_2O_3 + 10% MgO were analysed. The experimental results showed a reduction of this parameter by up to 16% (compressive strength of the mixtures with 1% Fe_2O_3 + 10% MgO compared to that of the mixtures with 1% Fe_2O_3 + 1% MgO). Therefore, for further research, 1%, in relation to the amount of ash, was considered as the optimal addition of Fe_2O_3 and 1%, in relation to the amount of ash, as the optimal addition of MgO.

Based on preliminary results obtained for the situation using 6M NaOH solution, the hypothesis of preservation of the trend of evolution of the compressive strength in the variant using 8M NaOH solution was verified. The negative influence of excess addition of oxides was also observed, so the mixtures presented in Table 2 were established in order to analyse the influence of the molarity of the NaOH solution on the geopolymer mechanical performances.

Crystal structure analysis of the layers was performed by X-ray diffraction (XRD) using a high-resolution Brucker D8 diffractometer with copper anode (CuKα1 = 1.54056 Å). X-ray diffraction (XRD) was used to estimate structural and microstructural properties.

Morphological and microstructural characterization was performed using scanning electron microscopy (SEM) and energy dispersive spectroscopy (EDS)). A JEOL JSM 5600 LV (JEOL, Ltd., Tokyo, Japan) high-resolution scanning electron microscope (SEM) equipped also with an electron back-scattered diffraction detector (EBSD) was used in the present work.

The evolution of a three-dimensional shrinkage phenomenon was observed during the 7 days of specimen conditioning, prior to testing in terms of mechanical characteristics, and the shrinkage was evaluated as a volume reduction in relation to the initial volume (40 × 40 × 160 mm). Uniaxial bending and compressive strength were measured using the Advantest 9 testing machine (Advantest Corporation, Tokyo, Japan).

Flexural strength was determined using the three-point bending test (3PB) in accordance with EN 196-1. A digital flexural strength tester suitable for loads up to 10 kN (±10%) with a loading rate of (50 ± 10) N/s was used. The testing machine is provided with a bending device consisting of two steel support rollers with a diameter of (10 ± 0.5) mm, arranged at a distance of (100 ± 0.5) mm from each other and a third load roller, placed centrally.

The compressive strength of the specimens was determined in accordance with EN 196-1, using the compression test of the prismatic specimen halves resulting from the three-point bend test (3PB). The compressive loading rate used was 50 N/s (0.12 MPa/s). Samples were tested at 7, 14 and 28 days of age.

Experimental testing was carried out under laboratory conditions, ensuring compliance with repeatability and reproducibility requirements.

3. Results and Discussions

The experimental results show both the influence of the addition of Fe_2O_3 and MgO oxides and the molarity of the NaOH solution used for the preparation of the alkaline activator on the physical-mechanical performance of the geopolymer material. From the beginning, a volume shrinkage of the tested specimens between 1.5% and 5.0% was recorded. This can be attributed to the water content in the mixture, which is directly influenced by the molarity of the NaOH solution, water that is removed in the heat treatment process.

3.1. Flexural Strength

Analysing Figure 2 shows an increase in flexural strength of all samples from 7 days to 14 days of age. It is also found that the samples containing Fe_2O_3 and added MgO have a higher strength than the control samples. For example, for a 6M molarity of NaOH solution, the sample (6MX1) containing 1% Fe_2O_3 and 1% MgO added to the ash used, has a flexural strength of more than 69% compared to the sample (6M) containing no additional elements, at 7 days, over 110% at 14 days and over 80% at 28 days. Similarly, but less obvious, for a molarity of 8M NaOH solution, we have an increase in the bending strength of the 8MX1 sample of 1%, 7% and 12% at 7, 14 and 28 days, respectively. The flexural strength of sample 6MX1 has the highest value compared to the other samples. Although the molarity of the NaOH solution for sample 8MX1 is higher than that of sample 6MX1 the bending strength of sample 6MX1 is higher by more than 15% at 28 days.

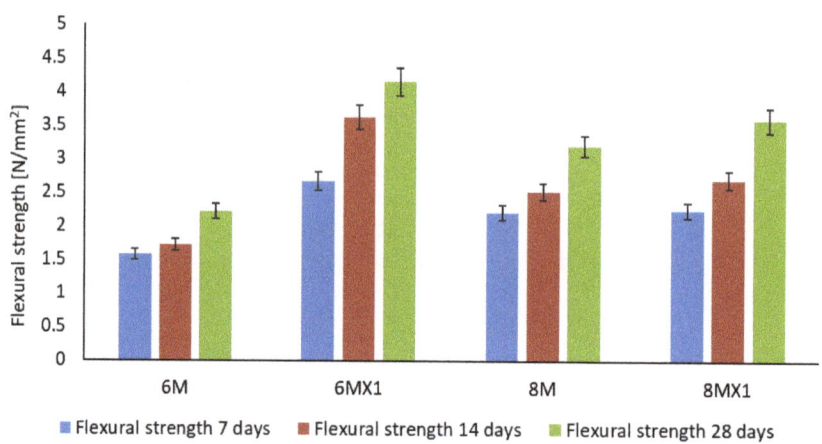

Figure 2. Flexural strength of the alkali-activated geopolymer samples.

3.2. Compressive Strength

The evolution of the compressive strength of the samples obtained at different test intervals is shown in Figure 3.

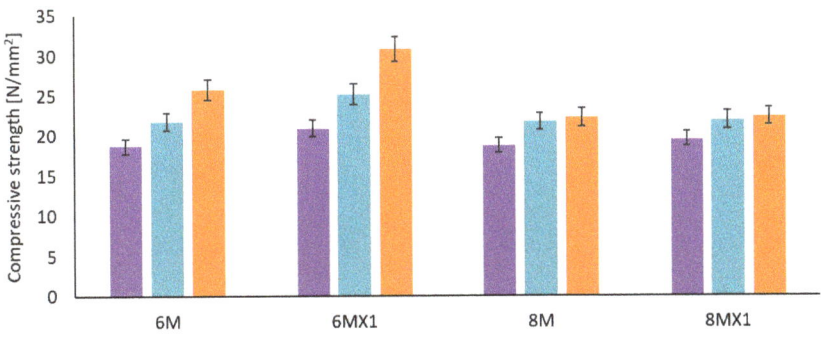

Figure 3. Compressive strength of the alkali-activated geopolymer samples.

It can be seen that the compressive strength for all samples increases with the age of testing. It is also observed that for a molarity of 6M of NaOH solution, the sample (6MX1) containing 1% Fe_2O_3 and 1% MgO added to the ash used has a compressive strength more than 11% and 15%, respectively, 19% higher than the sample (6M) not containing these additional elements, at the age of 7, 14 and 28 days.

The evolution of the compressive strength of the samples obtained using a NaOH solution of molarity 8 is almost identical whether 1% Fe_2O_3 and 1% MgO was added or not (increase of less than 1% for ages 14 days and 28 days after casting). The most significant increase in compressive strength due to the addition of iron and magnesium oxides in the case of 8M NaOH solution molarity is recorded early, at the age of 7 days after casting (4%).

The compressive strength of sample 6MX1 at the test age of 28 days has the highest value being more than 37% higher than samples 8M and 8MX1. The increase in compressive strength can also be attributed to the microcrack filling effect of Fe_2O_3 and MgO acting as inactive granular fillers by the presence of unreacted phases (hematite, forsterite and periclase) in the geopolymer specimens [50].

The increase in mechanical strength, with the increase in molarity of the NaOH solution used in the preparation of the alkaline activator, could be attributed, in addition to the maturation in time and the completion of the geopolymer reactions, to the contribution of Na+ ions and OH- groups provided by the alkaline activator. Thus, in accordance with the literature, with references on the main steps of the geopolymerization mechanism, it is appreciated that both Na+ ions and hydroxyl groups play an important role in the dissolution and hydrolysis processes, breaking the bonds existing in the Si and Al source raw materials, contributing to produce Si-O-Al bonds, bonds known as geopolymer precursors [51]. More specifically, Na+ ions contribute to the balancing of negative charges produced by Si-O-Al formation, while OH ions play an essential role in the hydrolysis process of the geopolymer, producing the geopolymerization reaction and the formation of the aluminosilicate network with stable, stronger bonds, which ultimately lead to better mechanical strength of the material. This assessment is also in agreement with Khale and Chaudhary who appreciate that to obtain good mechanical strengths of the geopolymer binder it is necessary to identify an optimal molarity of NaOH, which, by providing Na+ ions, contributes to balance and optimize the geopolymerization reactions, reactions strongly dependent on the oxidative nature of the raw materials [52]. On the other hand, an excess of Na+ ions or hydroxyl groups, resulting from a too high molarity of the NaOH solution used in the preparation of the alkaline activator, will result in an early precipitation of the aluminosilicate gel, respectively, a reduction of the geopolymer specific bonds and, consequently, a reduction of the mechanical strengths [53–55]. In the present case, the experimentally recorded values indicate that with the addition of iron and magnesium oxides, this positive influence of the NaOH solution size on the mechanical strength is not necessarily maintained. The best situation is identified as sample 6MX1, the influence of the addition of oxides being weaker than the influence of the molarity of the NaOH solution, the microstructural analysis presented below providing elements to support this hypothesis.

Quantitatively analysing the results obtained in terms of the mechanical strengths, it can be said that they are in agreement with some specifications in the literature and even exceed other reported results, Figure 4. This observation can be explained on the one hand based on the contribution of source elements for the geopolymerization reaction, i.e., the oxide composition of fly ash, and on the other hand on the basis of the differences in the grain size of the raw material, an element which by its influence on the specific reactive surface that influences the kinetics of the geopolymerization reactions.

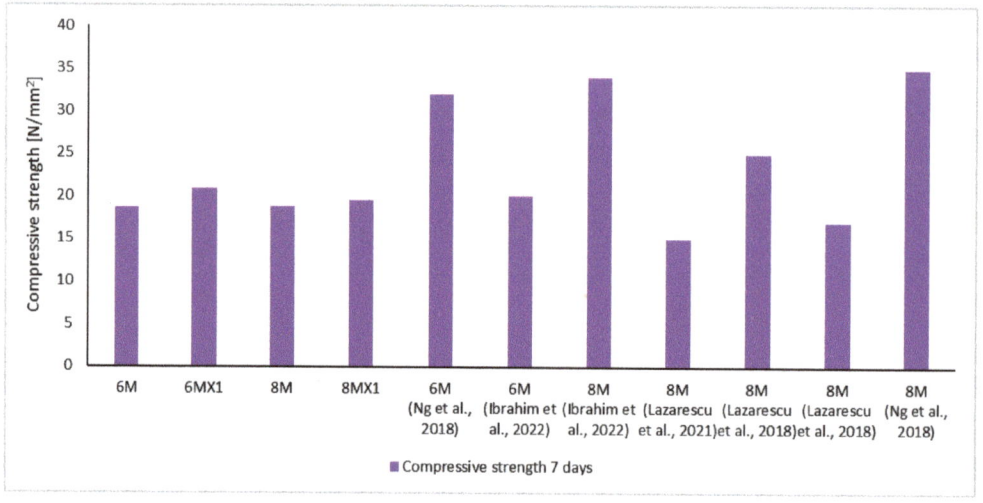

Figure 4. Comparative analysis of experimental results with literature reports.

This is due to the fact that geopolymer materials are very sensitive to the physical and chemical characteristics of the raw material, which is the reason why there are situations when mixtures, which although prepared with the same NaOH solution molarity, but using fly ash with different chemical/oxide composition or grain size (specific characteristics of the raw material source) or situations where the solution is prepared from NaOH flakes or pearls, the physico-mechanical performances are very different (Figure 4).

As can be seen from Figure 5, from the point of view of the evolution of the mechanical strengths, with the increase of the age of the specimens, there is also an increase of these parameters. It can also be seen that, in the case of the flexural tensile strength, this improvement is more evident in the 7–28 days period, while the compressive strength increases to a lesser extent in this period, a sign that, according to the literature, in the first days after preparation geopolymer binders tend to reach even 90% of the final compressive strength.

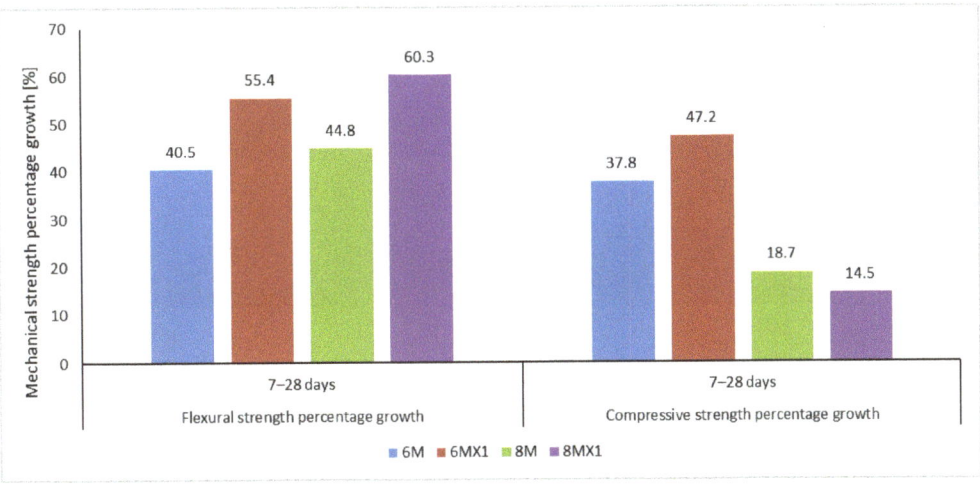

Figure 5. Percentage increase in flexural strength and compressive strength over 7–28 days after casting.

3.3. Scanning Electron Microscopy and Energy Dispersive Spectroscopy Analysis

Figures 6–9 represent the SEM micrographs for the studied samples after 28 days of curing. These show the presence of pores (1), partially reacted raw materials (3), microcracks (2) and compact areas where the raw materials have fully reacted (4). The difference between the microstructure of samples 6M and 6MX1 is the portion of the geopolymer matrix and the amount of unreacted fly ash (Figures 6 and 7). It can be seen that sample 6MX1 has a more compact geopolymer structure than 6M. The larger pore size of sample 6M and the larger amount of partially reacted fly ash were part of the reasons why its mechanical strength was lower than that of sample 6MX1.

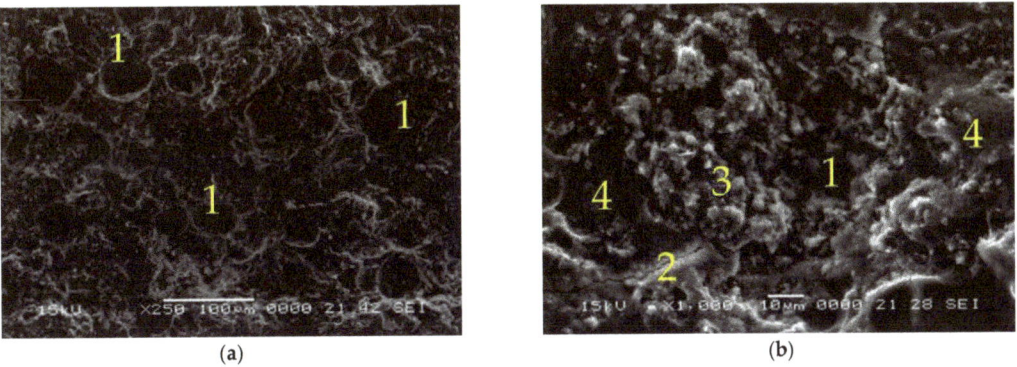

Figure 6. SEM micrographs of: (**a**,**b**) Sample 6M, ×250, respectively, ×1000 magnification (1—pores, 2—microcracks, 3—partially reacted fly ash and 4—dense zone of reacted fly ash).

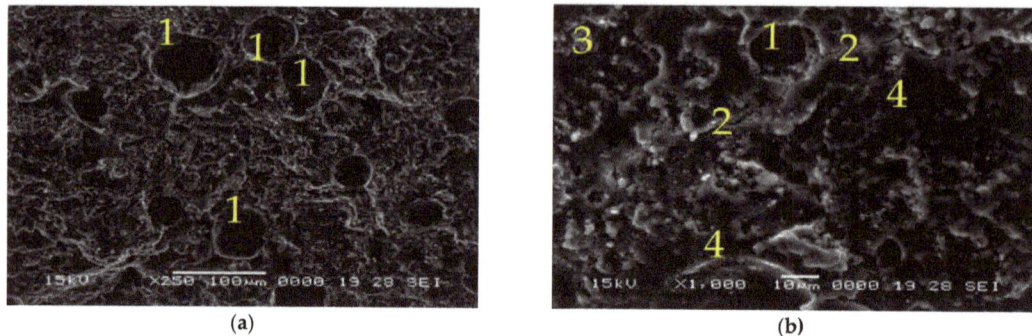

Figure 7. SEM micrographs of: (**a**,**b**) Sample 6MX1, ×250, respectively, ×1000 magnification (1—pores, 2—microcracks, 3—partially reacted fly ash and 4—dense zone of reacted fly ash).

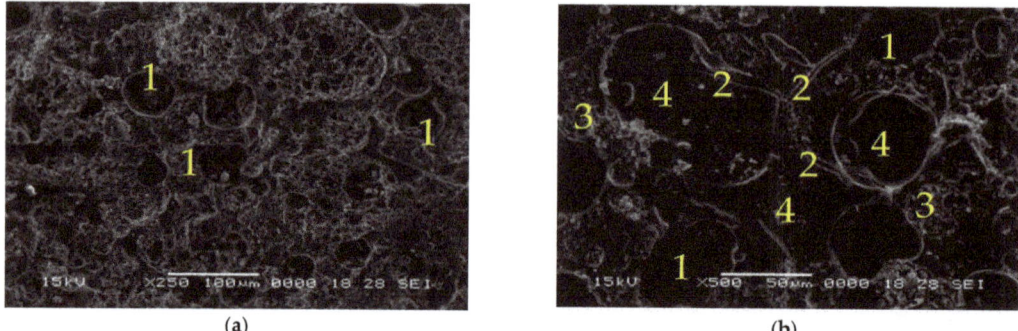

Figure 8. SEM micrographs of: (**a**,**b**) Sample 8M, ×250, respectively, ×1000 magnification (1—pores, 2—microcracks, 3—partially reacted fly ash and 4—dense zone of reacted fly ash).

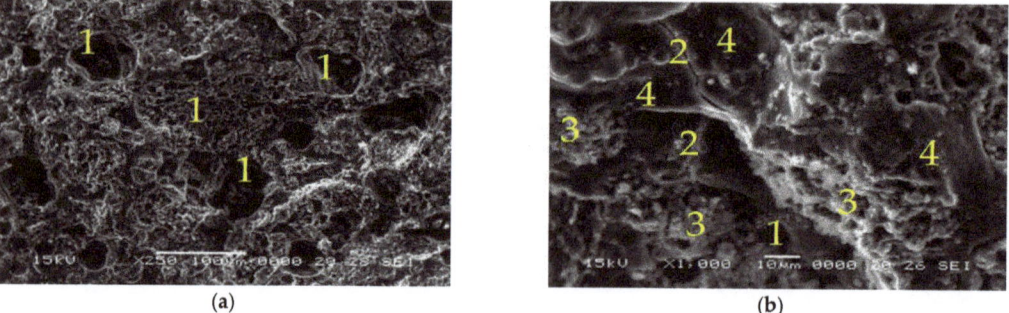

Figure 9. SEM micrographs of: (**a**,**b**) Sample 8MX1, ×250, respectively, ×1000 magnification (1—pores, 2—microcracks, 3—partially reacted fly ash and 4—dense zone of reacted fly ash).

The SEM micrographs for samples 8M and 8MX1 shown in Figures 8 and 9 reveal that the pore number and microcracks size are higher compared to samples 6M and MX1. The higher number and size of pores may also be due to the higher molarity of the NaOH solution leading to an energetic reaction between the alkaline activator and the ash. Microcracks may be due to shrinkage during curing [56,57], exothermic dissolution of activators [58] and heat treatment. Although the molarity of 8M and 8MX1 is higher than that of 6M and 6MX1 samples, the presence of high pore numbers and high amount of unreacted ash has the effect of lowering the mechanical strengths of these samples.

Figures 10 and 11 show the EDS images with the distribution of elements in the selected areas for the studied samples, and Table 3 gives the percentages of elements present in the samples.

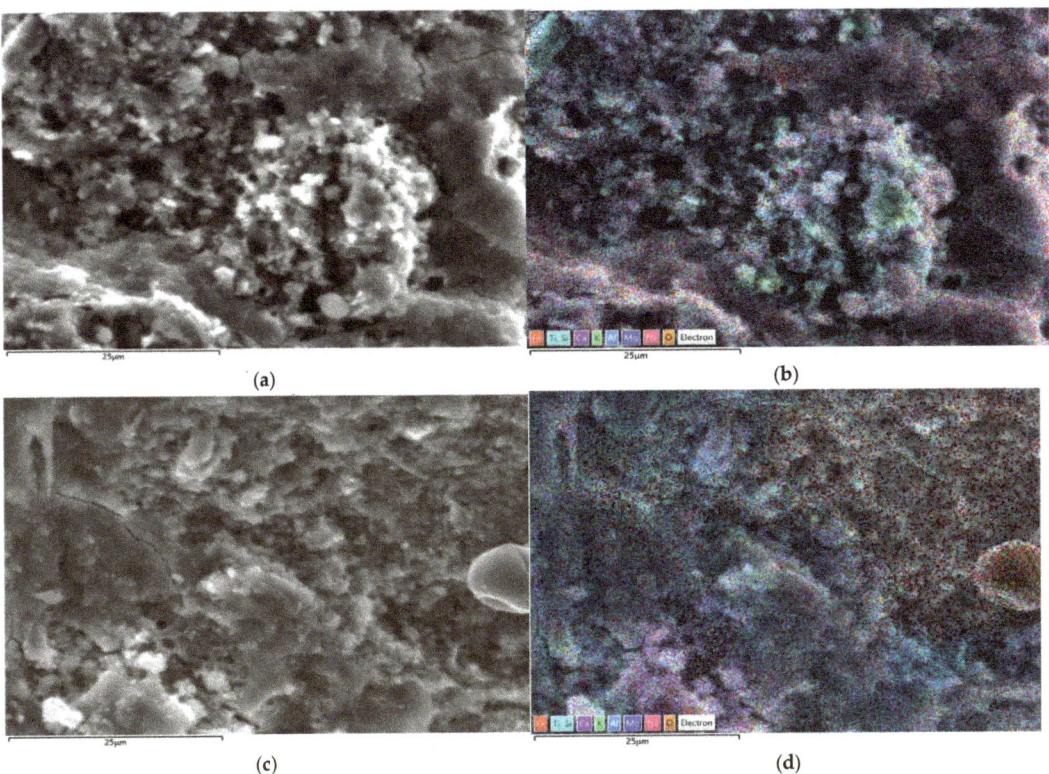

Figure 10. EDS analysis for sample 6M: (**a**) SEM image of selected area and (**b**) EDS stratified image of selected area; for sample 6MX1: (**c**) SEM image of selected area and (**d**) EDS stratified image of selected area.

Table 3. EDS data for selected zones corresponding to samples 6M, 6MX1, 8M and 8MX1.

		O	Si	Al	Na	Fe	Ca	K	Mg	Ti
6M	Element									
	Weight%	44.9	27.4	10.1	7.9	4.1	2.4	1.5	1.2	0.5
	σ	0.2	0.1	0.1	0.1	0.1	0.1	0.0	0.0	0.1
6MX1	Element									
	Weight%	40.1	27.1	9.6	8.1	5.7	4.6	2.3	1.8	0.7
	σ	0.2	0.2	0.1	0.1	0.2	0.1	0.1	0.1	0.1
8M	Element									
	Weight%	42.0	28.5	9.4	9.2	4.8	2.0	2.1	1.2	0.7
	σ	0.2	0.2	0.1	0.1	0.1	0.1	0.1	0.0	0.1
8MX1	Element									
	Weight%	41.7	25.2	8.9	8.1	6.5	3.8	2.7	2.8	0.4
	σ	0.2	0.2	0.1	0.1	0.2	0.1	0.1	0.1	0.1

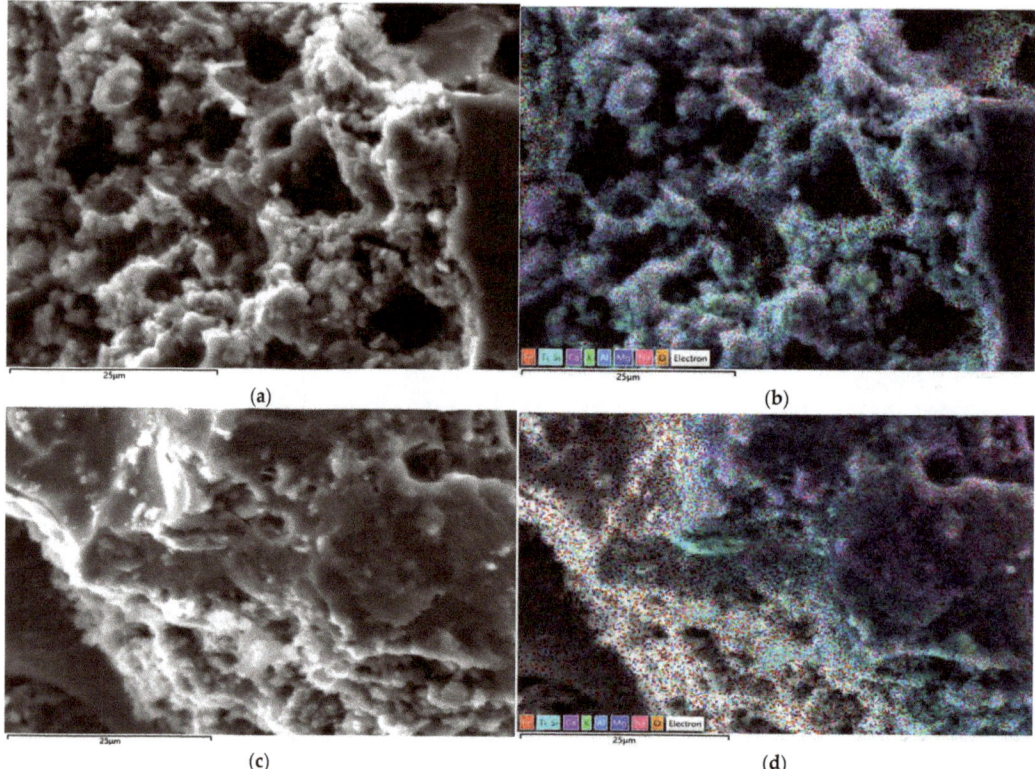

Figure 11. EDS analysis for sample 8M: (**a**) SEM image of selected area and (**b**) EDS layered image of selected area; for sample 8MX1: (**c**) SEM image of selected area and (**d**) EDS layered image of selected area.

The literature states that in EDS analysis for Si/Al ratio between 2.39 and 2.84 and for Na/Al between 0.34 and 0.53, respectively, the major reaction product is formed by alkaline hydroaluminosilicates of the zeolitic type in which Ca incorporation is also assumed. Additionally, for sufficiently high Si/Al ratio of 2.49 and a fairly high Na/Al ratio of 2.27, a carbonate variety of sodium and calcium hydroalumino-silicates is obtained [59]. Based on these clarifications and in accordance with the experimental results obtained and presented in Table 3, it can be stated that in the case of the studied samples (6M, 6MX1, 8M and 8MX1) the majority reaction product consists of zeolitic type hydroalumino-silicates and with slight tendencies to form carbonate variety of sodium and calcium hydroalumino-silicates.

Analysing Figures 10d and 11d, areas where Fe_2O_3 is unreacted are noticed, which functions as a crack filler, while this could be a possible explanation for both the improvement in compressive strength and the macroscopic reddish colour identified in the tested specimens.

Analysing the ratio of the identified concentration of Fe and Al, respectively, Na, according to the values presented in Table 3, it is observed that the Fe/Al ratio varies in the range 0.4–0.7, and the Fe/Na ratio varies in the range 0.5–0.8, always the ratio characteristic of the mixture recipe with the addition of Fe_2O_3 and MgO being higher, both for both molarity of the NaOH solution used in the preparation of the alkaline activator. This trend is also considered to be a possible explanation for the higher mechanical strengths in the samples recorded for samples prepared with the addition of oxides.

3.4. X-Ray Diffraction (XRD) Analysis

Analysing the X-ray spectra for samples 6M and 6MX1 in Figure 12, the presence of quartz, feldspar, calcite and mullite is observed, and for samples 8M and 8MX1 in Figure 13, the presence of quartz and feldspar in the geopolymer is observed.

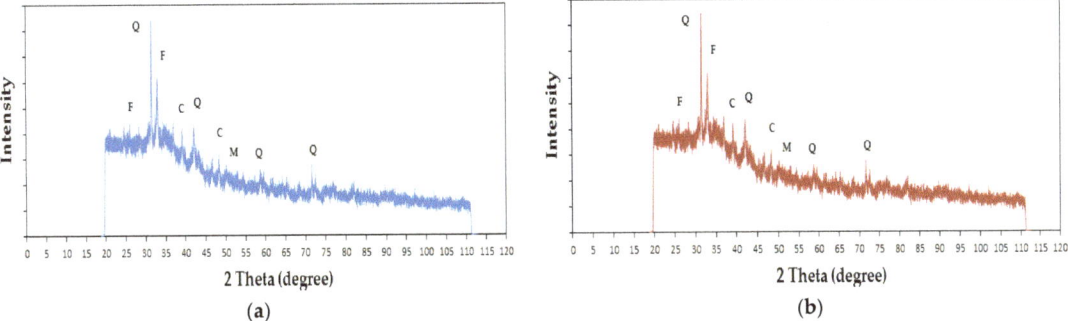

Figure 12. XRD analysis of geopolymer binders for (**a**) sample 6M and (**b**) sample 6MX1, where Q represents quartz, F—feldspar, C—calcite and M—mullite.

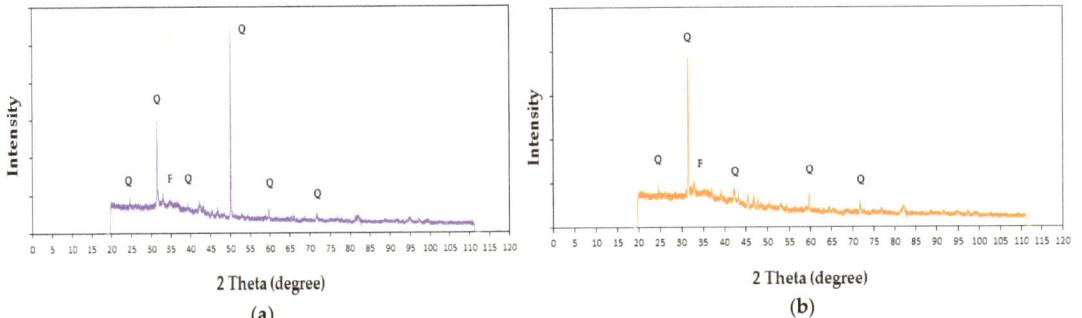

Figure 13. XRD analysis of geopolymer binders for (**a**) sample 8M and (**b**) sample 8MX1, where Q represents quartz, F—feldspar.

Analysing Figures 12a and 13a, it can be seen that the peaks with maximum intensities were for quartz, with the peak for the highest intensity 2θ maximum at 32° for sample 6M, respectively, at 50° for sample 8M, under the conditions of maintaining also an obvious peak at the 32° angle in the case of this sample characterized by a higher molarity of the NaOH solution used in the preparation of the alkaline activator. This displacement of the maximum 2θ angle is considered to be an indicator for preferential directions of crystallization, depending on the molarity of the NaOH solution used to prepare the alkaline activator. Comparing Figures 12b and 13b, it is observed that the same major peak is identified for quartz at the maximum 2θ at 32° angle, but with a much higher intensity for the case of sample 8MX1, suggesting again the influence of the molarity of the NaOH solution used in the preparation of the alkaline activator on the crystallization mechanism. Comparing the samples preprepared with 6M NaOH solution, with and without the addition of Fe_2O_3 and MgO, no major differences in terms of the crystallization angles are identified (Figure 12). On the other hand, with increasing molarity of the NaOH solution to 8M, between the characteristic spectra of the samples prepared without, respectively, with addition of Fe_2O_3 and MgO, it is observed the maintenance of the characteristic quartz angle, 2θ, at 32°, but of a much higher intensity for the sample prepared with addition of oxides, concomitant with the maintenance of the characteristic feldspar peaks. In the literature it is stated that for a higher amount of iron trioxide and magnesium oxide added to fly ash the presence of hematite (Fe_2O_3), periclase (MgO) and forsterite ($MgFeSiO_4$) mineral phases is observed in the X-ray spectrum [60], which in the present cases has not been confirmed.

From the study, it can be stated that iron trioxide, magnesium oxide and the molarity of the sodium hydroxide solution used in the preparation of fly ash-based geopolymer paste influence the physico-mechanical properties of the obtained heat-treated samples.

The bending tensile strength and compressive strength of the 6MX1 sample (containing iron trioxide, magnesium oxide and for which a molarity of 6M sodium hydroxide solution was used) had higher values compared to the other samples. This observation can be interpreted as a signal that for the specific case of fly ash with the oxide composition shown in Table 1, the most favourable case for obtaining the geopolymer binder would be the use of an alkaline activator prepared with 6M NaOH solution.

This increase in mechanical strengths can be explained by the action of iron trioxide which causes the formation of ferro-sialate groups and by the action of magnesium oxide which reduces the shrinkage of the sample. Additionally, the lower molarity of the hydroxide solution results in a less energetic reaction and fewer pores in the sample.

SEM micrographs reveal areas with fewer pores and fewer cracks for samples obtained with lower molarity of sodium hydroxide solution and smaller pore size, provided, however, that sufficient Na+ and OH− ions are available to allow good dissolution and extraction of Al and Si atoms from the raw material.

XRD analysis shows the presence of quartz, calcite, feldspar and mullite in samples obtained with a molarity of 6 of sodium hydroxide solution, and quartz and feldspar in samples obtained with a molarity of 8M. The formation of these elements is also influenced by the type of ash used and its chemical composition, and the addition of Fe_2O_3 and MgO leads to a preferential crystallization directive especially for quartz.

Identifying quartz (hardness 7, trigonal crystallization system), mullite (hardness 6/7.5, orthorhombic crystallization system), feldspar (hardness 6/6.5, tri- or monoclinic crystallization system) and calcite (hardness 3, trigonal or triclinic crystallization system), the following is estimated:

- The hardness of the crystallites as well as the specific crystallization system directly influences the compressive strength of the material;
- In the crystallite contact zone, for the crystallite combinations identified in the geopolymer material prepared with 6M NaOH solution, the cohesive energy would be higher than the cohesive energy specific to the intercrystallite contact zone of the geopolymer prepared with 8M NaOH solution.

The layered EDS images and the provided data reveal that in the case of the studied samples (6M, 6MX1, 8M and 8MX1) the majority reaction product is formed by zeolitic-type hydroaluminosilicates and with slight tendencies to form carbonate varieties of sodium and calcium hydroaluminosilicates. Iron trioxide and magnesium oxide are also observed to have a microcrack filling effect, i.e., they act as inactive granular fillers.

The mechanical strengths of the samples obtained, comparable to Portland cement, justify the use of these geopolymer pastes in the production of geopolymer concretes and in the production of precast concrete. Results obtained in the current study are in accordance with results previously obtained in the literature, while completing the knowledge about the production of alkaline-activated geopolymer materials [61–75].

4. Conclusions

The aim of this experimental study was to investigate the influence of the addition of Fe_2O_3 and MgO, respectively, and the influence of the molarity of the NaOH solution used in the preparation of the alkaline activator, on the mechanical strengths of the geopolymer binder prepared using locally sourced fly ash. Based on the obtained results, the following conclusions can be drawn:

1. The compressive and flexural strength of the 6MX1 specimen is higher than the other specimens (6M, 8M and 8MX1).
2. SEM micrographs reveal areas with fewer pores and fewer cracks for samples obtained with lower molarity of sodium hydroxide solution and smaller pore size.
3. XRD analysis shows the presence of quartz, calcite, feldspar and mullite in samples obtained with a molarity of 6M of sodium hydroxide solution, and quartz and feldspar in samples obtained with a molarity of 8M.
4. The EDS data show that the major reaction product is formed of zeolitic-type hydroaluminosilicates with slight tendencies to form carbonate varieties of sodium and calcium hydroaluminosilicates.
5. The addition of Fe_2O_3 and MgO to a geopolymer improves its physico-mechanical properties.

This paper contributes to the research developed so far worldwide on alkali-activated geopolymer materials with the following:

- The chemical, oxidic and mineralogical composition of the raw material used (flz ash) is specific only to the main source from which was provided and, according to the literature, has a major influence on the physico-mechanical characteristics of the geopolymer matrix;
- The NaOH solution used to prepare the alkaline liquid was prepared with local raw materials;
- Although some specifications in the literature analyse the influence of Fe and Mg oxides on the performance of geopolymer materials, in this case, these oxides do not represent the input of the basic raw material (fly ash), but are introduced as a controlled addition;
- The mix-design ratio and production technology are obtained following the analysis of literature but customized to the availability of resources and equipment. It is known that reproducibility is strongly influenced by the particularities of the materials and production techniques.

All these specific elements represented both challenges and risks, but also elements of novelty in the development of the experimental programme.

In the future, it is important to determine the optimum molar concentration of the NaOH solution used for the preparation of the alkaline activator and the optimal temperature range for obtaining samples with higher mechanical strengths.

Author Contributions: Conceptualization, B.A.I., M.C. and A.-V.L.; methodology, B.A.I., M.C., A.H. and A.-V.L.; formal analysis, H.V., A.-V.L. and T.G.; investigation, B.A.I., M.C., H.V., A.H., A.-V.L., G.T., B.V.N., T.G. and I.M.S.; resources, B.A.I., M.C. and H.V.; data curation, M.C., H.V., G.T., B.V.N., T.G. and I.S; writing—original draft preparation, B.A.I., M.C. and A.H.; writing—review and editing, H.V. and A.-V.L.; visualization, B.A.I., M.C., H.V., A.H., A.-V.L., G.T., B.V.N., T.G. and I.M.S.; supervision, H.V., A.-V.L., G.T., B.V.N., T.G. and I.M.S. All authors have read and agreed to the published version of the manuscript.

Funding: This research received no external funding.

Institutional Review Board Statement: Not applicable.

Informed Consent Statement: Not applicable.

Data Availability Statement: Not applicable.

Acknowledgments: This paper was financially supported by the Project "Entrepreneurial competences and excellence research in doctoral and postdoctoral programs—ANTREDOC", project co-funded by the European Social Fund financing agreement no. 56437/24.07.2019. Partial support was received from Programme Research for sustainable and ecological integrated solutions for space development and safety of the built environment, with advanced potential for open innovation—"ECOSMARTCONS", Programme code: PN 19 33 04 02: "Sustainable solutions for ensuring the population health and safety within the concept of open innovation and environmental preservation", financed by the Romanian Government.

Conflicts of Interest: The authors declare no conflict of interest.

References

1. Shahmansouri, A.A.; Yazdani, M.; Ghanbari, S.; Akbarzadeh Bengar, H.; Jafari, A.; Farrokh Ghatte, H. Artificial neural network model to predict the compressive strength of eco-friendly geopolymer concrete incorporating silica fume and natural zeolite. *J. Clean. Prod.* **2021**, *279*, 123697. [CrossRef]
2. Kabirifar, K.; Mojtahedi, M.; Wang, C.; Tam, V.W.Y. Construction and demolition waste management contributing factors coupled with reduce, reuse, and recycle strategies for effective waste management: A review. *J. Clean. Prod.* **2020**, *263*, 121265. [CrossRef]
3. Khasreen, M.M.; Banfill, P.F.; Menzies, G.F. Life-cycle assessment and the environmental impact of buildings: A review. *Sustainability* **2009**, *1*, 674–701. [CrossRef]
4. Fernando, S.; Gunasekara, C.; Law, D.W.; Nasvi, M.C.M.; Setunge, S.; Dissanayake, R. Life cycle assessment and cost analysis of fly ash–rice husk ash blended alkaliactivated concrete. *J. Environ. Manag.* **2021**, *295*, 113140. [CrossRef]
5. Kazmi, S.M.S.; Munir, M.J.; Wu, Y.-F. Application of waste tire rubber and recycled aggregates in concrete products: A new compression casting approach. *Resour. Conserv. Recycl.* **2021**, *167*, 105353. [CrossRef]
6. Munir, M.J.; Kazmi, S.M.S.; Wu, Y.-F.; Patnaikuni, I.; Zhou, Y.; Xing, F. Stress strain performance of steel spiral confined recycled aggregate concrete. *Cem. Concr. Compos.* **2020**, *108*, 103535. [CrossRef]
7. Munir, M.J.; Kazmi, S.M.S.; Wu, Y.-F.; Patnaikuni, I.; Wang, J.; Wang, Q. Development of a unified model to predict the axial stress–strain behavior of recycled aggregate concrete confined through spiral reinforcement. *Eng. Struct.* **2020**, *218*, 110851. [CrossRef]
8. Jamora, J.B.; Gudia, S.E.L.; Go, A.W.; Giduquio, M.B.; Loretero, M.E. Potential CO_2 reduction and cost evaluation in use and transport of coal ash as cement replacement: A case in the Philippines. *Waste Manag.* **2020**, *103*, 137–145. [CrossRef]
9. Damtoft, J.S.; Lukasik, J.; Herfort, D.; Sorrentino, D.; Gartner, E.M. Sustainable development and climate change initiatives. *Cem. Concr. Res.* **2008**, *38*, 115–127. [CrossRef]
10. Juenger, M.C.G.; Winnefeld, F.; Provis, J.L.; Ideker, J.H. Advances in alternative cementitious binders. *Cem. Concr. Res.* **2011**, *41*, 1232–1243. [CrossRef]
11. Davidovits, J. Global warming impact on the cement and aggregates industries. *World Resour. Rev.* **1994**, *6*, 263–278.
12. Zhou, W.; Yan, C.; Duan, P.; Liu, Y.; Zhang, Z.; Qiu, X.; Li, D. A comparative study of high- and low-Al_2O_3 fly ash based-geopolymers: The role of mix proportion factors and curing temperature. *Mater. Des.* **2016**, *95*, 63–74. [CrossRef]
13. Rashad, A.M. A comprehensive overview about the influence of different admixtures and additives on the properties of alkali-activated fly ash. *Mater. Des.* **2014**, *53*, 1005–1025. [CrossRef]
14. Dehghani, A.; Aslani, F.; Ghaebi Panah, N. Effects of initial SiO_2/Al_2O_3 molar ratio and slag on fly ash-based ambient cured geopolymer properties. *Constr. Build. Mater.* **2021**, *293*, 123527. [CrossRef]
15. Toader, T.P.; Mircea, A.C. Self-healing concrete mix-design based on engineered cementitious composites principles. *Proceedings* **2020**, *63*, 5.
16. Toader, T.P.; Mircea, A.C. Designing concrete with self-healing properties using engineered cementitious composites as a model. *IOP Conf. Ser. Mater. Sci. Eng.* **2020**, *877*, 012035.
17. Davidovits, J. Ancient and modern concretes: What is the real difference? *Concr. Int.* **1984**, *9*, 23–35.
18. Palomo, A.; Lopez de la Fuente, J.I. Alkali-activated cementitious materials: Alternative matrices for the immobilisation of hazardous wastes. part I: Stabilisation of boron. *Cem. Concr. Res.* **2003**, *33*, 281–288. [CrossRef]
19. Bland, W.; Roll, D. *Weathering: An Introduction to the Scientific Principles*; Arnolds: London, UK, 1998.
20. Gomes, K.C.; Torres, S.M.; De Barros, S.; Barbosa, N.P. Geopolymer Bonded Steel Plates. In Proceedings of the ETDCM8—8th Seminar on Experimental Techniques and Design in Composite Materials, Sant'Elmo Beach Hotel, Castiadas, Costa Rei, Sardinia, Italy, 3–6 October 2007.
21. Gomes, K.C.; Torres, S.M.; de Barros, S.S.; Barbosa, N.P. Solid Mechanics in Brazil 09. In *Associação Brasileira de Engenharia e Ciências Mecânicas*; Mattos, H.S.C., Alves, M., Eds.; ABCM: Rio de Janeiro, Brazil, 2009; Volume 2.
22. Komnitsas, K.; Zaharaki, D. *Structure, Processing, Properties and Industrial Applications PART II: Manufacture and Properties of Geopolymers*; Provis, J., van Deventer, J.S.J., Eds.; CRC Press: Boca Raton, FL, USA; Woodhead Publishing Ltd.: Oxford, UK, 2009.
23. Rashad, A.M. A comprehensive overview about the influence of different additives on the properties of alkali-activated slag—A guide for civil engineer. *Constr. Build. Mater.* **2015**, *47*, 29–55. [CrossRef]
24. Palacios, M.; Puertas, F. Effect of shrinkage-reducing admixtures on the properties of alkali-activated slag mortars and pastes. *Cem. Concr. Res.* **2007**, *37*, 691–702. [CrossRef]
25. Collins, F.; Sanjayan, J.G. Effect of pore size distribution on drying shrinking of alkali-activated slag concrete. *Cem. Concr. Res.* **2000**, *30*, 1401–1406. [CrossRef]
26. Lou, Z.; Ye, Q.; Chen, H.; Wang, Y.; Shen, J. Hydration of MgO in clinker and its expansion property. *J. Chin. Ceram. Soc.* **1998**, *26*, 430–436.
27. Gao, P.-W.; Wu, S.-X.; Lu, X.-L.; Deng, M.; Lin, P.-H.; Wu, Z.-R.; Tang, M.-S. Soundness evaluation of concrete with MgO. *Constr. Build. Mater.* **2007**, *21*, 132–138. [CrossRef]

28. Cincotto, M.A.; Melo, A.A.; Repette, W.L. Effect of different activators type and dosages and relation to autogenous shrinkage of activated blast furnace slag cement. In Proceedings of the 11th International Congress on the Chemistry of Cement, Durban, South Africa, 11–16 May 2003; Grieve, G., Owens, G., Eds.; pp. 1878–1888.
29. Collins, F.; Sanjayan, J.G. Workability and mechanical properties of alkali activated slag concrete. *Cem. Concr. Res.* **1999**, *29*, 455–458. [CrossRef]
30. Bakharev, T.; Sanjayan, J.G.; Cheng, Y.B. Alkali activation of Australian slag cements. *Cem. Concr. Res.* **1999**, *29*, 113–120. [CrossRef]
31. Puligilla, S.; Mondal, P. Role of slag in microstructural development and hardening of fly ash-slag geopolymer. *Cem. Concr. Res.* **2013**, *43*, 70–80. [CrossRef]
32. Shi, C. Strength, pore structure and permeability of alkali-activated slag mortars. *Cem. Concr. Res.* **1996**, *26*, 1789–1799. [CrossRef]
33. Shi, C.; Day, R.L. Some factors affecting early hydration of alkali-slag cements. *Cem. Concr. Res.* **1996**, *26*, 439–447. [CrossRef]
34. Krizan, D.; Zivanovic, B. Effects of dosage and modulus of water glass on early hydration of alkali-slag cements. *Cem. Concr. Res.* **2002**, *32*, 1181–1188. [CrossRef]
35. Andersson, R.; Gram, H.E. Properties of alkali-activated slag. In *Alkali-Activated Slag*; Sweedish Cement and Concrete Institute Report 1.88; Anderson, R., Gram, H.E., Malolepszy, J., Deja, J., Eds.; Sweedish Cement and Concrete Institute: Stockholm, Sweden, 1988; pp. 9–65.
36. Hikkinen, T. The influence of slag content on the microstructure, permeability and mechanical properties of concrete Part 1 Microstructural studies and basic mechanical properties. *Cem. Concr. Res.* **1993**, *23*, 407–421. [CrossRef]
37. Kutti, T.; Berntsson, L.; Chandra, S. Shrinkage of cements with high content of blast-furnace slag. In Proceedings of the Fourth CANMET/ACI International Conference on Fly Ash, Silica Fume, Slag and Natural Pozzolans in Concrete, Istanbul, Turkey, 3–8 May 1992; pp. 615–625.
38. Douglas, E.; Bilodeau, A.; Malhotra, V.M. Properties and durability of alkali-activated slag concrete. *ACI Mater. J.* **1992**, *89*, 509–516.
39. Li, Y.; Sun, Y. Preliminary study on combined-alkali–slag paste materials. *Cem. Concr. Res.* **2000**, *30*, 963–966. [CrossRef]
40. Melo Neto, A.A.; Cincotto, M.A.; Repette, W. Drying and autogenous shrinkage of pastes and mortars with activated slag cement. *Cem. Concr. Res.* **2008**, *38*, 565–574. [CrossRef]
41. Holt, E.E. *Early Age Autogenous Shrinkage of Concrete*; Technical Research Center of Finland: Espoo, Finland, 2001; p. 446.
42. Ben Haha, M.; Lothenbach, B.; Le Saout, G.; Winnefeld, F. Influence of slag chemistry on the hydration of alkali-activated blast-furnace slag—Part I: Effect of MgO. *Cem. Concr. Res.* **2011**, *41*, 955963.
43. Shen, W.; Wang, Y.; Zhang, T.; Zhou, M.; Li, J.; Cui, X.J. Magnesia modification of alkali-activated slag fly ash cement. *J. Wuhan Univ. Technol. Mater. Sci. Ed.* **2011**, *26*, 121. [CrossRef]
44. Jin, F.; Gu, K.; Al-Tabbaa, A. Strength and drying shrinkage of reactive MgO modified alkali-activated slag paste. *Constr. Build. Mater.* **2014**, *51*, 395–404. [CrossRef]
45. Marjanović, N.; Komljenović, M.; Baščarević, Z.; Nikolić, V. Improving reactivity of fly ash and properties of ensuing geopolymers through mechanical activation. *Constr. Build. Mater.* **2014**, *57*, 151–162. [CrossRef]
46. Fernández-Jiménez, A.; Palomo, A. Composition and microstructure of alkali activated fly ash binder: Effect of the activator. *Cem. Concr. Res.* **2005**, *35*, 1984–1992. [CrossRef]
47. Kumar, S.; Kristály, F.; Mucsi, G. Geopolymerisation behaviour of size fractioned fly ash. *Adv. Powder Technol.* **2015**, *26*, 24–30. [CrossRef]
48. Davidovits, J. *Geopolymer Chemistry and Applications*, 3rd ed.; Institute Geopolymer: Saint-Quentin, France, 2011.
49. Abullah, M.M.; Hussin, K.; Bnhussain, M.; Ismail, K.; Ibrahim, N. Mechanism and chemical reaction of fly ash geopolymer cement—A review. *Int. J. Pure Appl. Sci. Technol.* **2011**, *6*, 35–44.
50. Lăzărescu, A.; Szilagyi, H.; Baeră, C.; Hegyi, A. Alternative concrete—Geopolymer concrete. Emerging research and opportunities. *Mater. Res. Found.* **2021**, *109*, 138.
51. Tahir, M.F.M.; Abdullah, M.M.A.B.; Rahim, S.Z.A.; Mohd Hasan, M.R.; Sandu, A.V.; Vizureanu, P.; Ghazali, C.M.R.; Kadir, A.A. Mechanical and durability analysis of fly ash based geopolymer with various compositions for rigid pavement applications. *Materials* **2022**, *15*, 3458. [CrossRef] [PubMed]
52. Khale, D.; Chaudhary, R. Mechanism of geopolymerization and factors influencing its development: A review. *J. Mater. Sci.* **2007**, *42*, 729–746. [CrossRef]
53. Shukla, A.; Chaurasia, A.K.; Mumtaz, Y.; Pandey, G. Effect of Sodium Oxide on Physical and Mechanical properties of Fly-Ash based geopolymer composites. *Indian J. Sci. Technol.* **2020**, *13*(38), 3994–4002. [CrossRef]
54. Chindaprasirt, P.; Jaturapitakkul, C.; Chalee, W.; Rattanasak, U. Comparative study on the characteristics of fly ash and bottom ash geopolymers. *Waste Manag.* **2009**, *29*, 539–543. [CrossRef] [PubMed]
55. Arafa, T.A.; Ali, A.Z.M.; Awal, A.S.M.A.; Loon, L.Y. Optimum mix of fly ash binder based on workability and compressive strength. *IOP Conf. Ser. Earth Environ. Sci.* **2018**, *140*, 01215.
56. Dong, M.; Elchalakani, M.; Karrech, A. Development of high strength one-part geopolymer mortar using sodium metasilicate. *Constr. Build. Mater.* **2020**, *236*, 117611. [CrossRef]
57. Yousefi Oderji, S.; Chen, B.; Ahmad, M.R.; Shah, S.F.A. Fresh and hardened properties of one-part fly ash-based geopolymer binders cured at room temperature: Effect of slag and alkali activators. *J. Clean. Prod.* **2019**, *225*, 1–10. [CrossRef]

58. Askarian, M.; Tao, Z.; Samali, B.; Adam, G.; Shuaibu, R. Mix composition and characterisation of one-part geopolymers with different activators. *Constr. Build. Mater.* **2019**, *225*, 526–537. [CrossRef]
59. Georgescu, M.; Cătănescu, I.; Voicu, G. Microstructure of some fly ash based geopolimer binders. *Rev. Romana De Mater.* **2011**, *41*, 183–191.
60. Kaya, M.; Koksal, F.; Gencel, O.; Munir, M.J.; Kazmi, S.M.S. Influence of micro Fe_2O_3 and MgO on the physical and mechanical properties of the zeolite and kaolin based geopolymer mortar. *J. Build. Eng.* **2022**, *52*, 104443. [CrossRef]
61. Lăzărescu, A.-V.; Szilagyi, H.; Baeră, C.; Ioani, A. Parameters affecting the mechanical properties of fly ash-based geopolymer binders—Experimental results. *IOP Conf. Ser. Mat. Sci. Eng.* **2018**, *374*, 012035. [CrossRef]
62. Ng, H.T.; Heah, C.Z.; Liew, Y.M. The effect of various molarities of NaOH soluton on fly ash geopolymer paste. *AIP Conf. Proc. 2045* **2018**, *2018*, 020098.
63. Hardjito, D.; Wallah, S.E.; Sumajouw, D.M.J.; Rangan, B.V. On the development of fly ash-based geopolymer concrete. *Mater. J.* **2004**, *101*, 467–472.
64. Ibrahim, W.M.W.; Abdullah, M.M.A.B.; Ahmad, R.; Sandu, A.V.; Vizureanu, P.; Benjeddou, O.; Rahim, A.; Ibrahim, M.; Sauffi, A.S. Chemical distributions of different sodium hydroxide molarities on fly ash/dolomite-based geopolymer. *Materials* **2022**, *15*, 6163. [CrossRef] [PubMed]
65. Nordin, N.; Abdullah, M.M.B.B.; Tahir, M.F.M.; Sandu, A.V.; Hussin, K. Utilization of fly ash waste as construction material. *Int. J. Conserv. Sci.* **2016**, *7*, 161–166.
66. Burduhos Negris, D.D.; Abdullah, M.; Vizureanu, P.; Tahir, M.F.M. Geopolymers and their uses. *IOP Conf. Ser. Mat. Science Eng.* **2018**, *374*, 012019. [CrossRef]
67. Bouaissi, A.; Li, L.Y.; Moga, L.M.; Sandu, I.G.; Abdullah, M.M.A.; Sandu, A.V. A review on fly ash as a raw cementitious material for geopolymer concrete. *Rev. Chim.* **2018**, *69*, 1661–1667. [CrossRef]
68. Shahedan, N.F.; Abdullah, M.M.A.; Hussin, K.; Sandu, I.; Ghazali, C.M.R.; Binhussain, M.; Yahya, Z.; Sandu, A.V. Characterization and design of alkali activated binder for coaling application. *Technology* **2014**, *51*, 258–262.
69. Razak, R.A.; Abdullah, M.M.A.; Hussin, K.; Ismail, K.N.; Sandu, I.G.; Hardjito, D.; Yahya, Z.; Sandu, A.V. Assessment on the potential of volcano ash as artificial lightweight aggregates using geopolymerisation method. *Rev. Chim.* **2014**, *65*, 828–834.
70. Abdullah, M.M.A.; Tahir, M.F.M.; Hussin, K.; Binhussain, M.; Sandu, I.G.; Yahya, Z.; Sandu, A.V. Fly ash based lightweight geopolymer concrete using foaming agent technology. *Rev. Chim.* **2015**, *66*, 1001–1003.
71. Ibrahim, W.M.W.; Abdullah, M.M.A.; Sandu, A.V.; Hussin, K.; Sandu, I.G.; Ismail, K.N.; Radir, A.A.; Binhussain, M. Processing and characterization of fly ash-based geopolymer bricks. *Rev. Chim.* **2014**, *65*, 1340–1345.
72. Nergis, D.D.B.; Vizureanu, P.; Sandu, A.V.; Nergis, D.P.B.; Bejinariu, C. XRD and TG-DTA study of new phosphate-based geopolymers with coal ash or metakaolin as aluminosilicate source and mine tailings addition. *Materials* **2022**, *15*, 202. [CrossRef] [PubMed]
73. Abdila, S.R.; Abdullah, M.M.A.; Ahmad, R.; Nergis, D.D.B.; Rahim, S.Z.A.; Omar, M.F.; Sandu, A.V.; Vizureanu, P. Potential of soil stabilization using ground granulated blast furnace slag (GGBFS) and fly ash via geopolymerization method: A review. *Materials* **2022**, *15*, 375. [CrossRef] [PubMed]
74. Huang, Y.; Yilmaz, E.; Cao, S. Analysis of strength and microstructural characteristics of mine backfills containing fly ash and desulfurized gypsum. *Minerals* **2021**, *10*, 922. [CrossRef]
75. Li, J.; Cao, S.; Yilmaz, E. Characterization of macro mechanical properties and microstructures of cement-based composites prepared from fly ash, desulfurized gypsum, and steel slag. *Minerals* **2021**, *12*, 6. [CrossRef]

Article

Unexpected Method of High-Viscosity Shear Thickening Fluids Based on Polypropylene Glycols Development via Thermal Treatment

Mariusz Tryznowski [1,*], Tomasz Gołofit [2], Selim Gürgen [3], Patrycja Kręcisz [4] and Marcin Chmielewski [5,6]

1. Faculty of Mechanical and Industrial Engineering, Warsaw University of Technology, Narbutta 85, 02-524 Warsaw, Poland
2. Faculty of Chemistry, Warsaw University of Technology, Noakowskiego 3, 00-664 Warsaw, Poland
3. Department of Aeronautical Engineering, Eskişehir Osmangazi University, Eskişehir 26040, Turkey
4. Faculty of Material Engineering, Warsaw University of Technology, Wołoska 141, 02-507 Warsaw, Poland
5. Institute of Microelectronics and Photonics, Łukasiewicz Research Network, Lotników 32/46, 02-668 Warsaw, Poland
6. National Centre for Nuclear Research, Materials Research Lab, Świerk, 05-400 Otwock, Poland
* Correspondence: mariusz.tryznowski@pw.edu.pl

Citation: Tryznowski, M.; Gołofit, T.; Gürgen, S.; Kręcisz, P.; Chmielewski, M. Unexpected Method of High-Viscosity Shear Thickening Fluids Based on Polypropylene Glycols Development via Thermal Treatment. *Materials* 2022, *15*, 5818. https://doi.org/10.3390/ma15175818

Academic Editors: Andrei Victor Sandu and Béla Iván

Received: 7 July 2022
Accepted: 22 August 2022
Published: 24 August 2022

Publisher's Note: MDPI stays neutral with regard to jurisdictional claims in published maps and institutional affiliations.

Copyright: © 2022 by the authors. Licensee MDPI, Basel, Switzerland. This article is an open access article distributed under the terms and conditions of the Creative Commons Attribution (CC BY) license (https://creativecommons.org/licenses/by/4.0/).

Abstract: This study aimed to analyze the influence of the thermal treatment of shear thickening fluids, STFs, on their viscosity. For this purpose, shear thickening fluids based on polypropylene glycols PPG400 and PPG1000 and Aerosil®200 were developed. The shear thickening behavior of obtained fluids was confirmed by using a parallel-plate rheometer. Next, thermogravimetric (TG) analyses were used to characterized thermal stability and weight loss of the STFs at a constant temperature. Finally, the thermal treatment of the STFs obtained was provided using the apparatus developed for this purpose. The received STFs exhibited a very high maximum viscosity up to 15 kPa. The rheology of the STFs measured after thermal treatment indicated that the proposed method allowed the development of STFs with a very high maximum viscosity. The maximum viscosity of the STFs increased twofold when thermal treatment of the STFs at elevated temperature for 210 min was performed. TG confirmed the convergence of the weight loss in the apparatus. Our results show that controlling the thermal treatment of STFs allows STFs to be obtained with high viscosity and a dilatation jump of the STFs by degradation of the liquid matrix.

Keywords: shear thickening fluids; viscosity; thermal treatment; composite

1. Introduction

Shear thickening fluids, STFs, are intelligent materials that are characterized by the increase in their viscosity with either an increased shear rate or applied stress [1], known as a dilatation jump. This phenomenon can be observed as a transition from a liquid into a solid. Due to their properties, STFs can find several engineering and industrial applications. The most common application of STFs is their usage in armor protection. STFs are used to impregnate aramid fabrics to improve the protection of soft body armor [1–4]. Starch-based STFs show shear thickening properties suitable for soft body armor [5]. Furthermore, the use of STFs in combination with aluminum panels has also been shown in the literature [6]. STFs can be applied in shock-absorber systems for the automotive industry [7,8] or as protection for screen devices [9]. Furthermore, STFs have been applied to promote vibration damping in cutting tools [10] and as a fast, low-cost method for smoothing various surfaces [11,12].

In these materials, the solid powder nanoparticles are dispersed in a liquid matrix or a carrier fluid, forming a ceramic-polymer highly concentrated colloidal suspension. STFs are mainly composed of various inorganic powders, such as SiO_2 [13], TiO_2 [14], or $CaCO_3$ [14] and liquid polymers such as glycerin [15], poly(ethylene glycol) [1,16,17], or poly(propylene glycol) [18–20]. Instead of SiO_2, carbon nanofillers can be used [20]. Furthermore, with the addition of carbonyl iron powder to STFs, a fluid with magnetorheological properties can be developed [7]. Instead of inorganic particles, various polymers can be used. Synthetic polymers such as polystyrene-ethyl acrylate particles as solids can be used for STF development [6,21,22].

STFs are mainly prepared by stepwise addition of solids in nanoparticle size to the liquid matrix. This process requires appropriate mixing, which is limited by the increasing viscosity of the fluid being created [23]. Thus, the procedure involves the use of powerful mixers, and the fluid takes several hours to prepare [24]. Liu et al. prepared SiO_2-poly(ethylene glycols) STFs by dispersing the silica in the liquid poly(ethylene glycol) 200 using a ball mill at room temperature [7]. Unfortunately, this process takes a long time, and the mixing was performed for 24 h. Some authors report the development of STFs by stepwise addition of carrier fluid to powder silica [18]. Furthermore, Mahesh et al. report using mixing and sonification to develop STFs by adding solids to a liquid matrix [21,25]. Furthermore, an emulsion polymerization can be used for the development of STFs [6,21]. Some authors proposed a synthesis of STFs by mixing the polymer matrix and silica in the excess of a solvent, such as ethanol [26]. This procedure needs an evaporation process of the solvent. The development of liquid STFs with a high dilatation method is impossible due to the increase in viscosity and a problem with the proper mixing and dispersing of the solid content in the liquid polymer matrix. However, the shear thickening effect is not fully understood [27]. There are a few studies that discuss the problem of modeling of the shear thickening effect of STFs [28–32].

In this work, we show a path to obtain high-viscosity STFs. The STFs were obtained by the traditional method of stepwise addition of fumed silica into the liquid PPG. Next, the shear thickening behavior was confirmed by rheology measurement. The degradation of the STF was measured with the TG technique at a constant temperature (100 °C, 110 °C, 120 °C, 130 °C, and 140 °C). The TG test was needed to assess the degradation behavior of STF and it was done for the selected STF with the highest dilatation effect. Finally, the thermal treatment of the STFs was performed, and their viscosity was confirmed. The unexpected effect of thermal treatment was an increase in the STFs' viscosity. We suggest that thermal treatment in a mild condition can be a method for the development of high-viscosity STFs. The novelty of this work is focused on showing an easy way to obtain STFs with a high dilatant jump by degradation of the liquid matrix.

2. Materials and Methods

2.1. Materials

Poly(propylene glycol)s PPG400 and PPG1000 (CAS 25322-69-4, Sigma-Aldrich, St. Louis, MO, USA) were used as a liquid phase and Aerosil®200 (CAS 112 945-52-5, Evonik Industries, Hanau-Wolfgang, Germany) hydrophilic-fumed silica as a solid phase for STF preparation, respectively. Table 1 shows the properties of the materials used for STF preparation.

Table 1. Properties of materials used for STF development [1].

Abbreviation	Parameter	Value
PPG400	M_n (g mol^{-1})	~400
	Density (g mL)	1.01
PPG1000	M_n (g mol^{-1})	~1000
	Density (g cm^{-3})	1.005
	Dynamic viscosity (mPa·s)	78.34
Aerosil®200	Specific surface BET (m^2·g^{-1})	200 ± 25

[1] According to the Safety Data Sheet provided by the suppliers.

2.2. STFs Preparation

Table 2 shows the STFs' formulation. The STFs were developed by the method shown in our previous work [27]. The liquid PPG was placed in a 250 mL reactor equipped with a mechanical stirrer (R50D, Ingenieurbüro CAT, Ballrechten-Gottingen, Germany) and a stainless-steel propeller-mixing geometry. Next, the silica was added stepwise in minimal amounts to the mixture. To achieve high-viscosity STFs, the stirring speed (from 200 rpm) was gradually decreased to a very low mixing speed (down to 1 rpm) as the silica was added. Stirring at nearly 1 rpm for 14 days was required to obtain a very viscous liquid (STF1000-24).

Table 2. Composition of STFs.

Abbreviation	Ceramic Powder Content in %	Carrier Fluid
STF400-18	18	PPG400
STF400-24	24	
STF1000-18	18	PPG1000
STF1000-24	24	

2.3. Rheology Measurement

A rotational MCR 102 rheometer (Anton Paar Company, Graz, Austria) equipped with a plate-to-plate (top plate φ 20 mm; bottom plate φ 100 mm; spacing gap 0.7 mm geometry) that applied shear stress (up to 250 s^{-1}) was used to determine the shear thickening response of fluids. The rheological measurements were performed prior to and after the thermal treatment of the STFs. The measurement was repeated twice with a new sample.

2.4. Thermogravimetric Analysis

Thermogravimetric analysis (TG) was carried out using TA Instruments SDT Q600 (New Castle, DE, USA) apparatus. The weight loss was measured at constant temperatures of 100 °C, 110 °C, 120 °C, 130 °C, and 140 °C as a function of time. For TG measurement purposes, a sample of approx. 10 mg was used.

2.5. Thermal Treatment Procedure

In order to develop high-viscosity STFs, a special apparatus was designed. The computer-aided design (CAD) scheme of the apparatus is displayed in Figure 1. The weight loss procedure is schematically shown in Figure 2.

An insulated chamber with a movable carousel was heated by five heaters (Finger Patron Heater, Selfa GE SA, Szczecin, Poland), with dimensions 10 × 60 mm, at 300 W each. The power of the heaters was regulated by a Eurotherm 7100A thyristor controller (Worthing, UK) triggered by a PLC under the control of the TwinCAT 3.1.4024.29 system (Beckhoff Automation GmbH & Co. KG, Verl, Germany). The amount of heat supplied to the chamber was regulated by a PID controller. After the temperature inside the chamber was stabilized, six vessels with the STFs for testing were introduced. After the top insulated cover was closed, the control algorithm responsible for the controlled degradation process was triggered. During the degradation process, the sample carousel was rotated clockwise through an angle of 60° every 120 s (see Figure 2a). This ensured that the temperature of all the degraded samples was homogeneous. The AM3012-0C41 servo motor (Beckhoff Automation GmbH & Co. KG, Verl, Germany) with an absolute encoder and a mechanical gear PLE40-M01-10 with a gear ratio of 1:10, powered by the Beckhoff AX5203 module, was responsible for the correct positioning of the carousel (Beckhoff Automation GmbH & Co. KG, Verl, Germany). The weight loss was measured by a P/N 83020552 strain gauge beam from an Ohaus MB25 moisture analyzer (Parsippany, NJ, USA) with an accuracy of 0.005 g/0.05%. The strain gauge beam with the pan was placed on a linear sliding table (Figure 2b). The YR-GZS90K-100 (Lishui City Yongrun Precision Machinery Co., Ltd., Lishui, China) was equipped with a 57HD4016-01 stepper motor (Dongguan Golden Motor

Co., Ltd., Dongguan, China) powered by a Beckhoff EL7041-1000 module (Figure 2c). Due to the micro-vibrations during the taring and weighing process, the stepper motor was turned off.

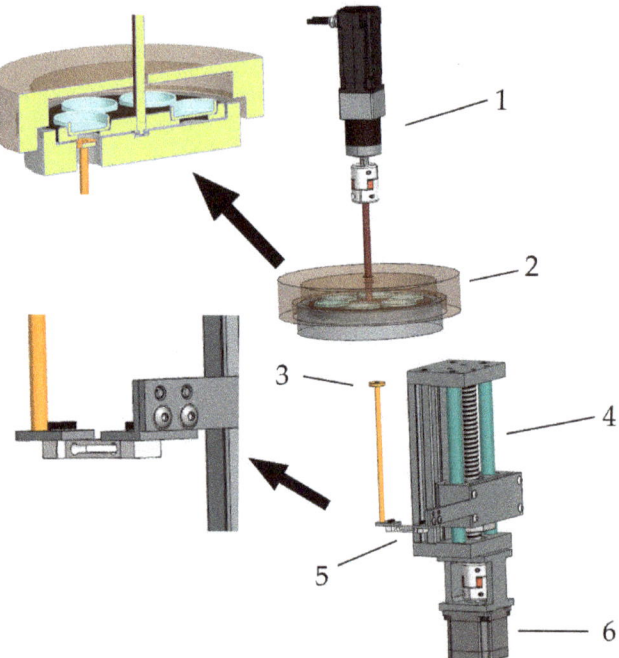

Figure 1. Apparatus scheme for degradation process: 1—servo motor with gearbox and encoder; 2—vessel carousel with heating; 3—beam balance; 4—linear table; 5—load cell; 6—stepper motor.

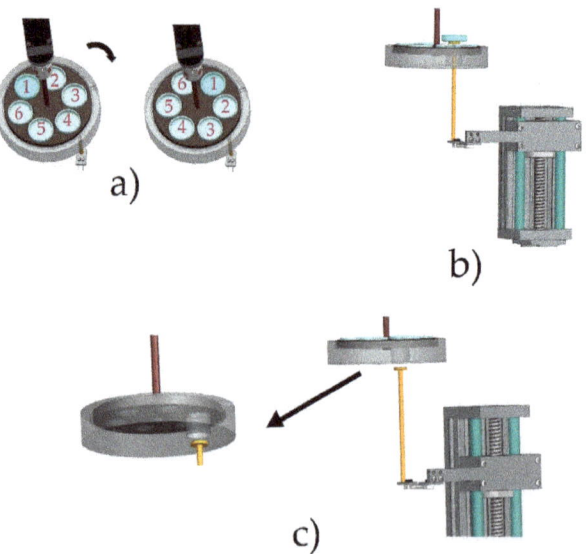

Figure 2. The weight loss measurement procedure: (**a**) clockwise rotation through an angle of 60°; (**b**) weighing of the vessel; (**c**) balance arm retraction and balance adjustment.

Measurement data (vessel weight) were periodically saved by the PLC system to the SQLite database. The sample of approx. 1 g was placed in the vessels (50 mm diameter), and the sample was taken for rheology measurement after a specified period of time. The samples' weight loss was controlled to ensure that the decomposition of the STF samples was similar to that during TGA measurement.

3. Results and Discussion

3.1. STFs Properties

The STFs were developed with poly(propylene glycol)s having a molecular weight of 400 g/mol (PPG400) and 1000 g/mol (PPG1000) and fumed silica Aerosil®200 with a particle size of ~12 nm [33] and a specific surface area of ~200 $m^2 \cdot g^{-1}$. Using PPGs as a carrier fluid instead of PEGs (polyethylene glycols) resulted in higher viscosities at lower shear rates [34]. Furthermore, the melting points of PEGs were much higher than PPGs, and the STFs based on PEGs were temperature-sensitive at low temperatures [35].

Figure 3 presents the viscosity-shear rate dependence of the developed STFs with various concentrations of fumed silica dispersed in poly(propylene)glycol. It can be seen that the developed STFs exhibit typical shear-thickening behavior. It is crucial to obtain STFs with very low viscosity at low shear rates and a very high dilatation jump (or viscosity jump). In other words, the developed STFs should be a liquid, not a paste, and exhibit a high maximum viscosity upon stress. As expected, the increasing silica content resulted in a more significant dilatant effect. Simultaneously, the oligomer's higher molecular weight provided higher viscosity values. The STF1000-24 and STF1000-18 exhibited a dilatant effect with a maximum viscosity of 1539 Pa·s and 830 Pa·s. Furthermore, the STFs based on PPG400 exhibited approx. three times lower values of maximum viscosity: 501 and 269 Pa·s for the STF400-24 and STF400-18, respectively. Comparing the results of the viscosity of the STFs already reported in the literature, it can be seen that the viscosities and the solid content achieved by us were much higher. Fisher et al. developed STFs based on Aerosil®200 and PPG1000 with a solid content of up to 15% w/w [18]. Wierzbicki et al. reported STFs with dilatant effects around 200 Pa·s using PPG400 and fumed silica with a specific surface 200 $m^2 \cdot g^{-1}$ [36]. Arora et al. and Bajya et al. reported the development of STFs with a maximum viscosity reaching 170 Pa s [37,38]. Furthermore, using hydrophobic silica with the same particle size and specific surface area and PPG400 and PPG1000, even when introducing more content of solids, resulted in much lower viscosities (up to 62 Pa·s). Nevertheless, in our previous work, using the same methodology of STF development, we reported fluids with a maximum viscosity exceeding 3000 Pa·s [27].

3.2. Thermal Treatment

In this work, the thermal decomposition of the developed STF1000-24 was performed with TG (thermogravimetric analysis). This method was used to control weight loss changes and to confirm the influence of the temperature on the nature of the degradation. However, it was not possible to compare the TG degradation course and thermal treatment directly, because the sample mass was 100 times larger. Additionally, TG makes it possible to determine a thermal decomposition with a very small sample (approx. 10 mg) and rheological measurement requires much bigger samples (approx. 0.7 g). Hence, we developed an apparatus that allows the thermal degradation of materials in similar conditions to those in TG.

The TG degradation was performed at five various constant temperatures (100 °C, 110 °C, 120 °C, 130 °C, and 140 °C) for the STF1000-24 as shown in Figure S1 in Supplementary Materials. Figure 4 shows the TG curves of STF1000-24 at 140 °C. According to the data (see Table 3, Figures S1 and 4), three steps of thermal degradation could be identified for the STF1000-24 sample, independently of the degradation temperature. The weight loss was accompanied by an exothermic effect. These steps were characterized by various weight loss and heat effects. The first step (I) was related to a slight heat effect connected with a weight loss rate of 0.02% and 0.18% per min at 100 °C and 140 °C, respectively. It might have been attributed to the evaporation of low molecular mass glycols. As shown in Figure 4, no changes in STFs were observed in step I. The second one (II) was characterized by a faster weight loss rate: 0.05% and 0.37% per min at 100 °C and 140 °C, respectively, and it was accompanied by a bigger heat effect. We suspect that this step was related to the vaporization and burning of volatile compounds arising during the degradation of the poly(propylene glycol) molecules. Finally, in the third step (III), the weight loss speed was the highest, leading to the decomposition of the STFs. Step III was related to visible changes in the appearance of the material. The non-dispersed silica appeared on the surface of the fluid. The maximum heat effect of the decomposition was approx. 8 times higher at 140 °C than at 100 °C. Hence, the reaction was exothermic and due to the high maximum heat effect at higher temperatures, during thermal treatment or mixing at elevated temperatures, the mixture could have overheated, the heat could have accumulated, and accelerating decomposition was observed. Furthermore, carrying out the process at an elevated temperature might have led to an uncontrolled explosion of heat. Therefore, the process should be performed at lower temperatures on a large scale.

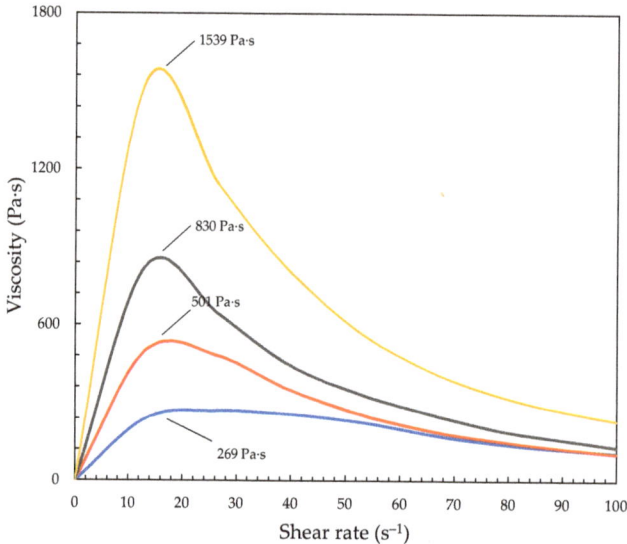

Figure 3. Viscosity vs. shear rate for developed STFs: yellow line—STF1000-24; gray line—STF1000-18; red line—STF400-24; blue line—STF400-18.

The samples of STFs were thermally treated at temperatures of 100 °C, 120 °C, and 140 °C with the apparatus shown in Figure 1. The control lines of weight loss are shown in Figure 5. The rotating carousel made it possible to ensure the homogeneous temperature of all the degraded samples. After a specified period of time, the samples of STFs were taken for viscosity measurement. According to the control lines of thermal treatment, the approx. 25% weight loss was observed after 210 min of thermal treatment.

Figure 6 shows the viscosity of the obtained STF1000-24 after thermal treatment for 210 min at 140 °C. The exposure of the fluids to elevated temperature caused the gradual degradation of the liquid matrix in the STFs, which could be observed as an increase in viscosity. The liquid matrix content decreased by approx. 25%, so the degradation was in the second step of degradation (see Figure 4). A higher reduction in the liquid matrix might have caused irreversible changes in the STFs and total fluid degradation. Hence, the approx. 200 min of thermal treatment at 140 °C seemed to be optimal, allowed to be with the degradation in step II, and prevented the total degradation of the STF and loss of the shear thickening properties of the STF. As shown in Figure 6, the viscosity of STF400-18 and STF1000-18 after thermal treatment increased twofold (blue and gray dashed lines), reaching maximum viscosity of 542 Pa·s and 1632 Pa·s, respectively. It can be seen that the maximum viscosity after thermal treatment had increased.

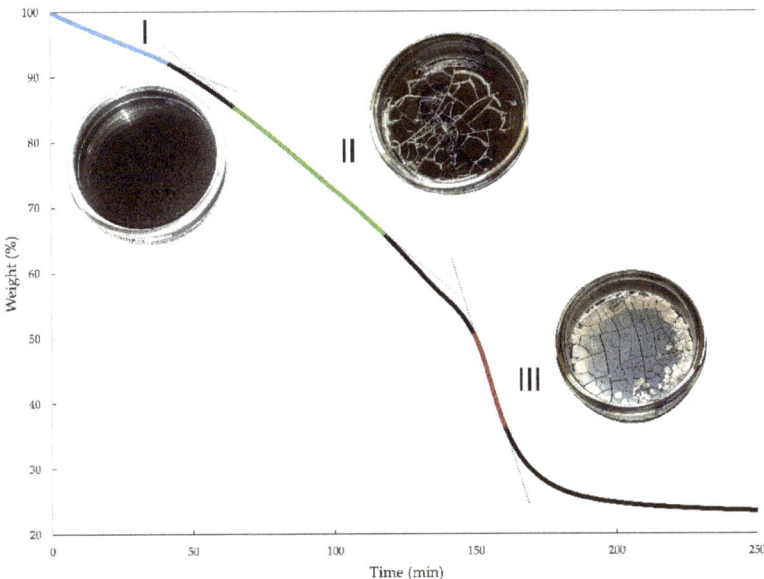

Figure 4. The degradation of STF1000-24 at 140 °C. The degradation steps are highlighted with different colors: I—first step; II—second step, III—third step.

Table 3. Weight loss rate (% per min) and maximum heat effect in step III (W·g^{-1}) after TG curves at various temperatures.

Step	Temperature/°C				
	100	110	120	130	140
I	0.02	0.03	0.05	0.10	0.18
II	0.05	0.09	0.12	0.23	0.37
III	0.05	0.22	0.28	0.56	1.31
Max. heat effect	0.15	0.23	0.35	0.68	1.31

Last but not least, we also thermally treated the STF1000-24 at 140 °C. The process was carried out at a constant temperature until the desired mass was achieved: 15%, 20%, and finally 25% weight loss. The viscosities of the STF1000-24 samples after the specific thermal treatment course are presented in Figure S2. The thermal treatment of STF1000-24 with the degradation of 15% resulted in a maximum viscosity of 6 kPa·s. Further degradation of STF1000-24 revealed a fluid with maximum viscosity at 8.5 kPa·s (degradation of 20% of the matrix) and 14 kPa·s (25% matrix degradation), respectively.

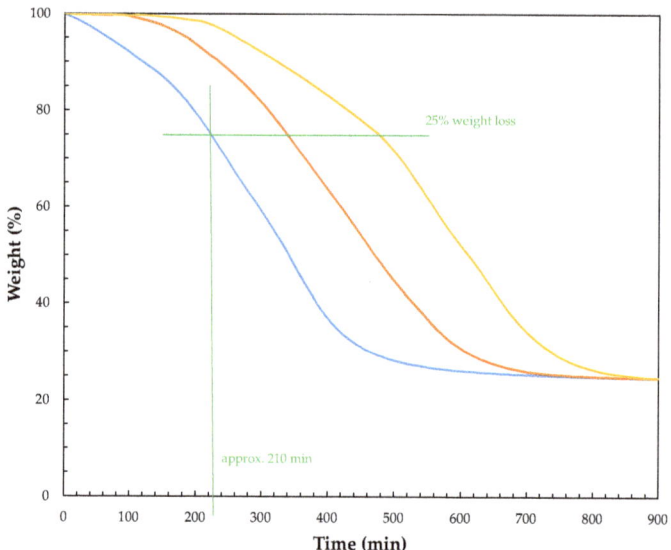

Figure 5. Control lines of weight loss during thermal treatment of STFs at various temperatures: 100 °C (yellow line); 130 °C (orange line); 140 °C (blue line) for the STF1000-24.

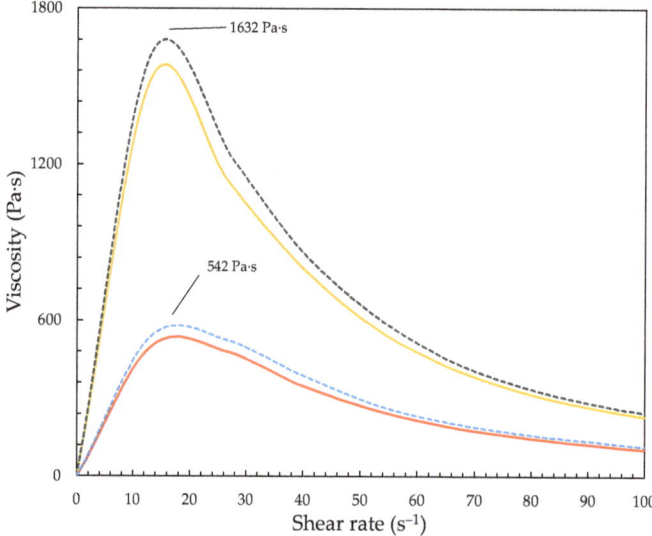

Figure 6. Viscosity vs. shear rate for STFs prior and after thermal treatment: yellow line—STF1000-24; gray dashed line—STF1000-18 after thermal treatment; red line—STF400-24; blue dashed line—STF400-18 after thermal treatment.

4. Conclusions

In this work, we developed four types of STFs based on poly(propylene glycol)s and fumed silica with high solid content (18% and 24% w/w). The developed STFs exhibited high-viscosity properties with a maximum viscosity level at 1539 Pa·s for STFs based on PPG1000 and fumed silica with 24% solid content. The STFs were thermally treated, which allowed a twofold increase in the viscosity properties of the obtained STFs.

We showed that controlled thermal treatment in mild conditions of low viscosity STFs can be used as a tool to obtain STFs characterized by high viscosity and high solid content. Our results reveal the possibility of using thermal treatment at elevated temperature to increase the viscosity of STFs as a novel method of high-viscosity STF preparation. This is a breakthrough paper in the topic of STFs because it shows the importance of controlling the temperature during the development of the fluids. Overheating the STFs during development can lead to complete degradation, but controlled heating can significantly increase the viscosity without mixing problems.

Supplementary Materials: The following supporting information can be downloaded at: https://www.mdpi.com/article/10.3390/ma15175818/s1, Figure S1. TG curves for STF100-24 at constant temperatures: 100 °C (blue line), 110 °C (orange line), 120 °C (grey line), 130 °C (yellow line) and 140 °C (green line); Figure S2. Viscosity vs. shear rate for STF1000-24 after thermal treatment: yellow line—STF1000-24 prior thermal treatment; gray line—approx. 15% weigh loss of STF1000-24; blue line—approx. 20% weigh loss of STF1000-24; orange line—approx. 25% weigh loss of STF1000-24.

Author Contributions: Conceptualization, M.T.; methodology, M.T.; software, M.T.; validation, M.T.; formal analysis, M.T.; investigation, M.T., P.K. and T.G.; resources, M.C.; data curation, M.T.; writing—original draft preparation, M.T.; writing—review and editing, M.T., S.G. and M.C.; visualization, M.T.; project administration, M.T. All authors have read and agreed to the published version of the manuscript.

Funding: Research was funded by POB Technologie Materiałowe of Warsaw University of Technology within the Excellence Initiative: Research University (IDUB) program.

Institutional Review Board Statement: Not applicable.

Informed Consent Statement: Not applicable.

Data Availability Statement: The data presented in this study are available on reasonable request from the corresponding author.

Conflicts of Interest: The authors declare no conflict of interest.

References

1. Ghosh, A.; Majumdar, A.; Butola, B.S. Rheometry of Novel Shear Thickening Fluid and Its Application for Improving the Impact Energy Absorption of P-Aramid Fabric. *Thin-Walled Struct.* **2020**, *155*, 106954. [CrossRef]
2. Domańska, U.; Zołek-Tryznowska, Z. Thermodynamic Properties of Hyperbranched Polymer, Boltorn U3000, Using Inverse Gas Chromatography. *J. Phys. Chem. B* **2009**, *113*, 15312–15321. [CrossRef] [PubMed]
3. Domańska, U.; Zołek-Tryznowska, Z. Temperature and Composition Dependence of the Density and Viscosity of Binary Mixtures of (Hyperbranched Polymer, B-U3000 + 1-Alcohol, or Ether). *J. Chem. Thermodyn.* **2009**, *41*, 821–828. [CrossRef]
4. Pais, V.; Silva, P.; Bessa, J.; Dias, H.; Duarte, M.H.; Cunha, F.; Fangueiro, R. Low-Velocity Impact Response of Auxetic Seamless Knits Combined with Non-Newtonian Fluids. *Polymers* **2022**, *14*, 2065. [CrossRef]
5. Cho, H.; Lee, J.; Hong, S.; Kim, S. Bulletproof Performance of Composite Plate Fabricated Using Shear Thickening Fluid and Natural Fiber Paper. *Appl. Sci.* **2020**, *10*, 88. [CrossRef]
6. Liu, H.; Zhu, H.; Fu, K.; Sun, G.; Chen, Y.; Yang, B.; Li, Y. High-Impact Resistant Hybrid Sandwich Panel Filled with Shear Thickening Fluid. *Compos. Struct.* **2022**, *284*, 115208. [CrossRef]
7. Liu, B.; Du, C.; Deng, H.; Fan, Z.; Zhang, J.; Zeng, F.; Fu, Y.; Gong, X. Mechanical Properties of Magneto-Sensitive Shear Thickening Fluid Absorber and Application Potential in a Vehicle. *Compos. Part A Appl. Sci. Manuf.* **2022**, *154*, 106782. [CrossRef]
8. Sheikhi, M.R.; Gürgen, S. Anti-Impact Design of Multi-Layer Composites Enhanced by Shear Thickening Fluid. *Compos. Struct.* **2022**, *279*, 114797. [CrossRef]
9. Zhao, C.; Gong, X.; Wang, S.; Jiang, W.; Xuan, S. Shear Stiffening Gels for Intelligent Anti-Impact Applications. *Cell Rep. Phys. Sci.* **2020**, *1*, 100266. [CrossRef]
10. Gürgen, S.; Sofuoğlu, M.A. Integration of Shear Thickening Fluid into Cutting Tools for Improved Turning Operations. *J. Manuf. Process.* **2020**, *56*, 1146–1154. [CrossRef]
11. Gürgen, S.; Sert, A. Polishing Operation of a Steel Bar in a Shear Thickening Fluid Medium. *Compos. Part B Eng.* **2019**, *175*, 107127. [CrossRef]
12. Shao, Q.; Duan, S.; Fu, L.; Lyu, B.; Zhao, P.; Yuan, J. Shear Thickening Polishing of Quartz Glass. *Micromachines* **2021**, *12*, 956. [CrossRef] [PubMed]

13. Sharma, S.; Kumar Walia, Y.; Grover, G.; Sanjeev, V.K. Effect of Surface Modification of Silica Nanoparticles with Thiol Group on the Shear Thickening Behaviors of the Suspensions of Silica Nanoparticles in Polyethylene Glycol (PEG). *IOP Conf. Ser. Mater. Sci. Eng.* **2022**, *1225*, 012053. [CrossRef]
14. Yang, H.G.; Li, C.Z.; Gu, H.C.; Fang, T.N. Rheological Behavior of Titanium Dioxide Suspensions. *J. Colloid Interface Sci.* **2001**, *236*, 96–103. [CrossRef] [PubMed]
15. Li, D.; Wang, R.; Liu, X.; Zhang, S.; Fang, S.; Yan, R. Effect of Dispersing Media and Temperature on Inter-Yarn Frictional Properties of Kevlar Fabrics Impregnated with Shear Thickening Fluid. *Compos. Struct.* **2020**, *249*, 112557. [CrossRef]
16. Soutrenon, M.; Michaud, V.; Manson, J.-A.E. Influence of Processing and Storage on the Shear Thickening Properties of Highly Concentrated Monodisperse Silica Particles in Polyethylene Glycol. *Appl. Rheol.* **2013**, *23*, 54865. [CrossRef]
17. Singh, M.; Verma, S.K.; Biswas, I.; Mehta, R. Effect of Molecular Weight of Polyethylene Glycol on the Rheological Properties of Fumed Silica-Polyethylene Glycol Shear Thickening Fluid. *Mater. Res. Express* **2018**, *5*, 55704. [CrossRef]
18. Fischer, C.; Braun, S.A.; Bourban, P.-E.; Michaud, V.; Plummer, C.J.G.; Månson, J.-A.E. Dynamic Properties of Sandwich Structures with Integrated Shear-Thickening Fluids. *Smart Mater. Struct.* **2006**, *15*, 1467–1475. [CrossRef]
19. Żurowski, R.; Antosik, A.; Głuszek, M.; Szafran, M. Shear Thickening Ceramic-Polymer Composite. *Compos. Theory Pract.* **2015**, *15*, 255–258.
20. Nakonieczna-Dąbrowska, P.; Wróblewski, R.; Płocińska, M.; Leonowicz, M. Impact of the Carbon Nanofillers Addition on Rheology and Absorption Ability of Composite Shear Thickening Fluids. *Materials* **2020**, *13*, 3870. [CrossRef]
21. Mahesh, V.; Harursampath, D.; Mahesh, V. An Experimental Study on Ballistic Impact Response of Jute Reinforced Polyethylene Glycol and Nano Silica Based Shear Thickening Fluid Composite. *Def. Technol.* **2022**, *18*, 401–409. [CrossRef]
22. Cao, S.; Pang, H.; Zhao, C.; Xuan, S.; Gong, X. The CNT/PSt-EA/Kevlar Composite with Excellent Ballistic Performance. *Compos. Part B Eng.* **2020**, *185*, 107793. [CrossRef]
23. Chang, C.P.; Shih, C.H.; You, J.L.; Youh, M.J.; Liu, Y.M.; Ger, M. Der Preparation and Ballistic Performance of a Multi-Layer Armor System Composed of Kevlar/Polyurea Composites and Shear Thickening Fluid (Stf)-Filled Paper Honeycomb Panels. *Polymers* **2021**, *13*, 3080. [CrossRef]
24. Caglayan, C.; Osken, I.; Ataalp, A.; Turkmen, H.S.; Cebeci, H. Impact Response of Shear Thickening Fluid Filled Polyurethane Foam Core Sandwich Composites. *Compos. Struct.* **2020**, *243*, 112171. [CrossRef]
25. Liu, X.-Q.; Bao, R.-Y.; Wu, X.-J.; Yang, W.; Xie, B.-H.; Yang, M.-B. Temperature Induced Gelation Transition of a Fumed Silica/PEG Shear Thickening Fluid. *RSC Adv.* **2015**, *5*, 18367–18374. [CrossRef]
26. Hassan, T.A.; Rangari, V.K.; Jeelani, S. Sonochemical Synthesis and Rheological Properties of Shear Thickening Silica Dispersions. *Ultrason. Sonochem.* **2010**, *17*, 947–952. [CrossRef] [PubMed]
27. Żurowski, R.; Tryznowski, M.; Gürgen, S.; Szafran, M.; Świderska, A. The Influence of UV Radiation Aging on Degradation of Shear Thickening Fluids. *Materials* **2022**, *15*, 3269. [CrossRef]
28. Gürgen, S. Numerical Modeling of Fabrics Treated with Multi-Phase Shear Thickening Fluids under High Velocity Impacts. *Thin-Walled Struct.* **2020**, *148*, 106573. [CrossRef]
29. Shende, T.; Niasar, V.J.; Babaei, M. An Empirical Equation for Shear Viscosity of Shear Thickening Fluids. *J. Mol. Liq.* **2021**, *325*, 115220. [CrossRef]
30. Salehin, R.; Xu, R.-G.; Papanikolaou, S.; Lamura, A.; Petukhov, A. V Materials Colloidal Shear-Thickening Fluids Using Variable Functional Star-Shaped Particles: A Molecular Dynamics Study. *Materials* **2021**, *14*, 6867. [CrossRef]
31. Zhang, X.; Yan, R.; Zhang, Q.; Jia, L. The Numerical Simulation of the Mechanical Failure Behavior of Shear Thickening Fluid/Fiber Composites: A Review. *Polym. Adv. Technol.* **2022**, *33*, 20–33. [CrossRef]
32. Lam, L.; Chen, W.; Hao, H.; Li, Z.; Ha, N.S.; Pham, T.M. Numerical Study of Bio-Inspired Energy-Absorbing Device Using Shear Thickening Fluid (STF). *Int. J. Impact Eng.* **2022**, *162*, 104158. [CrossRef]
33. Warren, J.; Offenberger, S.; Toghiani, H.; Pittman, C.U.; Lacy, T.E.; Kundu, S. Effect of Temperature on the Shear-Thickening Behavior of Fumed Silica Suspensions. *ACS Appl. Mater. Interfaces* **2015**, *7*, 18650–18661. [CrossRef] [PubMed]
34. Moriana, A.D.; Tian, T.; Sencadas, V.; Li, W. Comparison of Rheological Behaviors with Fumed Silica-Based Shear Thickening Fluids. *Korea-Aust. Rheol. J.* **2016**, *28*, 197–205. [CrossRef]
35. Prabhu, T.A.; Singh, A. Effect of Carrier Fluid and Particle Size Distribution on the Rheology of Shear Thickening Suspensions. *Rheol. Acta* **2021**, *60*, 107–118. [CrossRef]
36. Wierzbicki, Ł.; Danelska, A.; Chrońska, K.; Tryznowski, M.; Zielińska, D.; Kucińska, I.; Szafran, M.; Leonowicz, M. Shear Thickening Fluids Based on Nanosized Silica Suspensions for Advanced Body Armour. *Compos. Theory Pract.* **2013**, *13*, 241–244.
37. Arora, S.; Majumdar, A.; Butola, B.S. Soft Armour Design by Angular Stacking of Shear Thickening Fluid Impregnated High-Performance Fabrics for Quasi-Isotropic Ballistic Response. *Compos. Struct.* **2020**, *233*, 111720. [CrossRef]
38. Bajya, M.; Majumdar, A.; Butola, B.S.; Verma, S.K.; Bhattacharjee, D. Design Strategy for Optimising Weight and Ballistic Performance of Soft Body Armour Reinforced with Shear Thickening Fluid. *Compos. Part B Eng.* **2020**, *183*, 107721. [CrossRef]

Review

Scientometric Review for Research Patterns on Additive Manufacturing of Lattice Structures

Chiemela Victor Amaechi [1,2,*], Emmanuel Folarin Adefuye [1,3,*], Irish Mpho Kgosiemang [4], Bo Huang [5] and Ebube Charles Amaechi [6]

1. School of Engineering, Lancaster University, Bailrigg, Lancaster LA1 4YR, UK
2. Standards Organisation of Nigeria (SON), 52 Lome Crescent, Wuse Zone 7, Abuja 900287, Federal Capital Territory, Nigeria
3. Department of Mechanical/MetalWork Technology, Federal College of Education [Technical], Akoka 100001, Lagos State, Nigeria
4. Department of Management, University of Central Lancashire (UCLAN), Preston PR1 2HE, UK; mikgosiemang1@uclan.ac.uk
5. School of Civil Engineering, Hunan University of Science and Technology, Xiangtan 411201, China; bohuang@hnust.edu.cn
6. Department of Zoology, University of Ilorin, Ilorin 240003, Kwara State, Nigeria; amaechi.ec@unilorin.edu.ng
* Correspondence: c.amaechi@lancaster.ac.uk (C.V.A.); e.adefuye@lancaster.ac.uk (E.F.A.)

Abstract: Over the past 15 years, interest in additive manufacturing (AM) on lattice structures has significantly increased in producing 3D/4D objects. The purpose of this study is to gain a thorough grasp of the research pattern and the condition of the field's research today as well as identify obstacles towards future research. To accomplish the purpose, this work undertakes a scientometric analysis of the international research conducted on additive manufacturing for lattice structure materials published from 2002 to 2022. A total of 1290 journal articles from the Web of Science (WoS) database and 1766 journal articles from the Scopus database were found using a search system. This paper applied scientometric science, which is based on bibliometric analysis. The data were subjected to a scientometric study, which looked at the number of publications, authorship, regions by countries, keyword co-occurrence, literature coupling, and scientometric mapping. VOSviewer was used to establish research patterns, visualize maps, and identify transcendental issues. Thus, the quantitative determination of the primary research framework, papers, and themes of this research field was possible. In order to shed light on current developments in additive manufacturing for lattice structures, an extensive systematic study is provided. The scientometric analysis revealed a strong bias towards researching AM on lattice structures but little concentration on technologies that emerge from it. It also outlined its unmet research needs, which can benefit both the industry and academia. This review makes a prediction for the future, with contributions by educating researchers, manufacturers, and other experts on the current state of AM for lattice structures.

Keywords: additive manufacturing; lattice structure; 3D printing; research pattern; research trend; scientometric; bibliometric; COVID19; scientific literature review; review; VOSviewer

1. Introduction

Over the past 15 years, interest in polymer additive manufacturing with several engineered materials has dramatically increased [1,2]. With numerous materials, this technique can be applied to quickly design and directly create three-dimensional (3D) objects without adding complexity to the production process. Recent studies into material developments include the microstructures, interfacial behaviour, pore density, layup patterns, layer thickness, and material development [3–7]. With the increasing need for better materials, there are increased techniques and technologies in material processing, materials developments, more customised materials, and newer engineered lattice-structured materials called additive manufactured materials [8–14]. Additive manufacturing (AM) has been practised for over 15 years in numerous manufacturing industries with the aid of 3D printing, referred to as additive manufacturing technology [15–18]. The medical, automotive, aerospace, and materials industries have all benefited from the innovation that additive manufacturing (AM) has brought forth [19–22]. It uses lithographic techniques to join materials, layer-by-layer, on top of an existing structure to create parts from 3D model data [23–25]. A variety of technologies, including rapid prototyping (RP), selective laser melting (SLM), and electron beam melting (EBM) technologies, are used in additive manufacturing (AM) [26–29]. The earlier technology, rapid prototyping, is a concept that refers to the rise of additive manufacturing (AM) and the development of the polymer material used for prototype [30]. One aspect of development of the models is the use of CAD (computer-aided design) to develop the AM designs, like on lattice structure prototyping. Contrary to conventional production methods like casting and machining, additive manufacturing enables designers to quickly prototype and reduce operational costs and material waste in the process [31–33]. For the process, two commonly used methods for building metal components from powder feedstock are the Directed Energy Deposition (DED) and Powder Bed Fusion (PBF) [34]. Another aspect of AM is the use of machine learning (ML), which has been proven in the optimization of the material properties, strength, material mix, and array of the lattice-structured additive manufactured materials [35–37].

Earlier research using scientometric reviews on additive manufacturing reflect that there is an increase in research on other aspects of AM being investigated [38–40]. The current market demand places an ever-increasing emphasis on the efficient use of 3D printing for the production of complicated shapes [41–43]. However, the use of lattice structures in additive manufacturing has seen increasing demand due to their unique applications [44–51]. Lattice structures could be classified as porous and non-porous materials, depending on their applications [52–60]. These lattice structures for additive manufacturing have increasing applications from 3D printing to 4D printing, such as biological and medical applications [61–68]. Practical applications in the biomedical area that utilises lattice structures as additive manufactured materials include the manufacture of prosthetic legs and 3D-printed dental teeth. Conventional materials utilised for different advanced materials, such as ceramics, composites, or metals, can be found in additive manufactured (AM) materials [69–71]. However, AM has a method that spreads quickly in the manufacturing sectors, making AM products usable most of the time [72,73]. A 3D object scanner is used in additive manufacturing (AM), which enables the production of items with accurate geometrical details. In contrast to traditional manufacturing, which frequently necessitates milling or other processes to eliminate superfluous material, these are constructed layer by layer, much like a 3D printing process. There are also more experimental investigations on AM that are used with numerical investigations to further understand engineered lattice structures in AM [74–76]. Additionally, employing stereolithography (SL) for 3D systems, additive manufacturing (AM) technology solidified the thin UV (ultraviolet) layers with light-sensitive liquid polymer through laser operations. Additionally, additive manufacturing (AM) started to advance in the early 1980s when equipment was upgraded from a lower level of operation to a higher level using new conventional equipment as opposed to the previous equipment employed at the time, then in late 1980s to early 1990s, rapid prototyping increased [75,76]. These developments resulted from the use of more sophisticated jigs, medical implants, engineering applications, and tooling on the typical production floor, as earlier illustrated by Graham Tromans (UK) and Terry Wohler (USA) [75,76]. The timeline for the developments on AM showing the past, present, and potential future, including rapid casting (1994), rapid tooling (1995), AM for automotive (2001), aerospace polymers (2004), medical polymers and jigs (2005), medical metal implants (2009), aerospace polymers (2011), nano-manufacturing (2013–2016), architecture (2013–2017), biomedical implants (2013–2018), 3D printing of face shields, masks, ventilators during COVID19 pandemic (2019–2022), lattice structures and 4D Printing for medical organs (2013–2022), in situ bio-manufacturing (2013–2022), and full body organ printing (2013–2032), is represented in Figure 1.

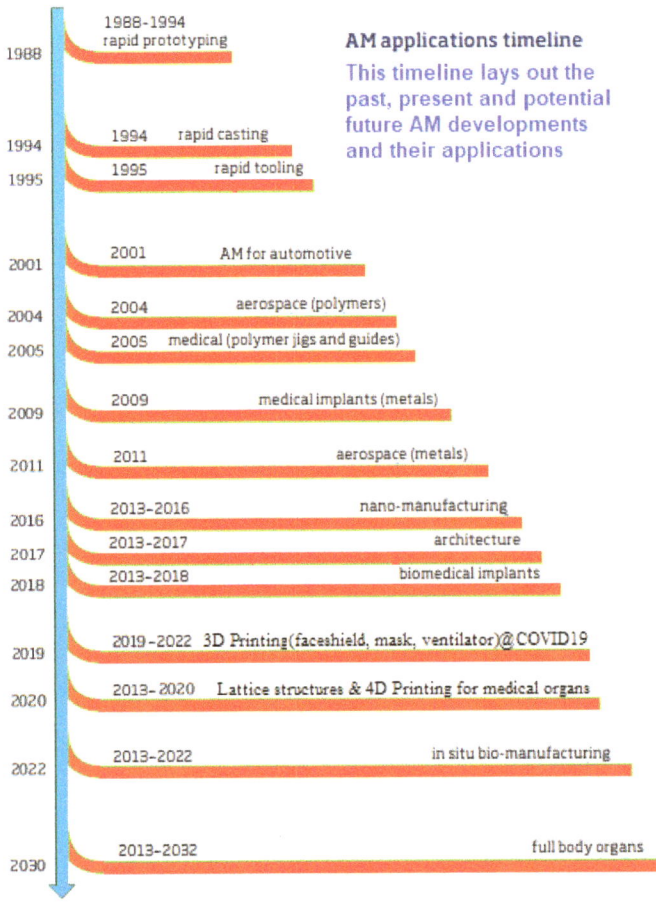

Figure 1. The timeline for the developments on additive manufacturing (AM), showing the past, present, and potential future (Adapted from original image with permission. Courtesy: Graham Tromans of Graham Tromans Associates, London, UK & Terry Wohlers of Wohlers Associates, Youngstown, OH, USA).

These AM processes include material extrusion, material jetting, binder jetting, powder bed fusion, directed energy deposition, photopolymerization, and sheet lamination [76,77]. These processes are all used in additive manufacturing (AM) technologies. They are procedures and techniques for using parts produced through additive manufacturing in production facilities and public spaces. Some state-of-the-art reviews also present advantages of additive manufacturing with related bibliometric analysis, but they did not consider lattice structures [77–79]. Generally, the advantages of additive manufacturing (AM) technology include the use of complex geometries, lighter structures, and the material's ability to allow customization. In addition, it allows manufacturing processes that involve an increase in the geometric complexity of the design or an increment of material volume that leads to a rise in the manufacturing cost or time, thus the need to improve upon this technology, as seen in the trends in both 3D printing and 4D printing [80–86]. Additionally, it is crucial to comprehend the current status of the literature in relation to additive manufacturing procedures and the mechanical properties of 3D printed materials. This technique has been used to produce 3D/4D objects without adding complexity to the production process. This understanding will help to establish a research horizon and create future works on this subject.

Hence the need to conduct this scientometric analysis on additive manufacturing for engineered lattice structures. The majority of the article is structured as follows: Section 1 introduces the subject area of AM. The research methodology is detailed in Section 2. The result and implications of the scientometric analysis are then addressed in Section 3. The implications of the systematic review with discussions on the research trends are detailed in Section 4. Lastly, the summary on the systematic review with recommendations for future research are presented in Section 5.

2. Materials and Methods

In this section, the materials used for the data analysis and the research methodology for the study are presented.

2.1. Data Retrieval

Data collection from the available literature was crucial to this study, notably for the scientometric analysis's result. The data collection for this study followed the already established procedure for bibliometric reviews. Different studies on AM have covered a range of technologies applied [31–34,76–79], hence it is necessary to have a strategy for the selection of the papers. Two criteria were used in the literature collection strategy: (1) contemporary and relevance: all publications from 2002 to 2022 were searched, and the papers were manually screened by carefully reading the keywords and abstracts; and (2) quality assurance: only peer-reviewed papers from journals were included because journal papers typically go through careful reviews to remove errors and mistakes. For the literature review, the database choice was crucial. Due to its large coverage of the subject area in journal publications, the research repositories and academic databases were considered as knowledge domains. Thus, with the diverse range of academic databases cthat are presently available, there was need to decide on the choice of the database(s) to utilise. This decision was achieved by having an initial comparative study between the Scopus database and Web of Science (WoS), as both were considered to obtain the data, and they gave good results. The search was conducted using wildcards. The variations of one keyword were captured using the wildcard character *. The keywords chosen were ("additive manufacturing *" OR "lattice *" AND ("structure *") based on the goal of this research. Searching for terms inside a publication's title, abstract, or keywords turned up all of the available literature on multi-material additive manufacturing of polymers in the Scopus database. The 2002–2022 search window was chosen to reflect the current growth of polymer additive manufacturing using several materials. To restrict the number of papers published in peer-reviewed English journals, a screening procedure was used.

2.2. Research Methodology

In this study, a scientometric analysis is carried out using research database and visualization-mapping tools to investigate research trends and patterns on the subject area. By inflection, scientometrics is used to reveal the research impact of publications, researchers, journals, and research institutions in a particular field of study. By definition, scientometrics also includes the quantitative study of science, science policy, and science communication, which gives an in-depth understanding of the research through the scientific citation and offers a deeper understanding of scientific citations [87–89]. To gain a thorough understanding of the evolution of this research field from 2002 to 2022, this study will conduct a scientometric evaluation and analysis of the papers pertaining to the multi-material additive manufacturing of engineered lattice structures. The scientometric review is also qualitatively validated by comparing the present data with other bibliometric studies on AM [38–40,90–93]. An extensive systematic review is then offered to offer deeper insights into the technology and applications of multi-material additive manufacturing of polymers based on the findings of the scientometrics analysis. In a nutshell, this study used a mixed review methodology, which combines scientometric analysis and systematic review, to examine the state of research on additive manufacturing for lattice-structure materials. By combining subjective research with a robust quantitative description and evaluation using science network mapping techniques, this study's contribution can be seen as extending past review works in this field. The flowchart of the methodology is presented in Figure 2.

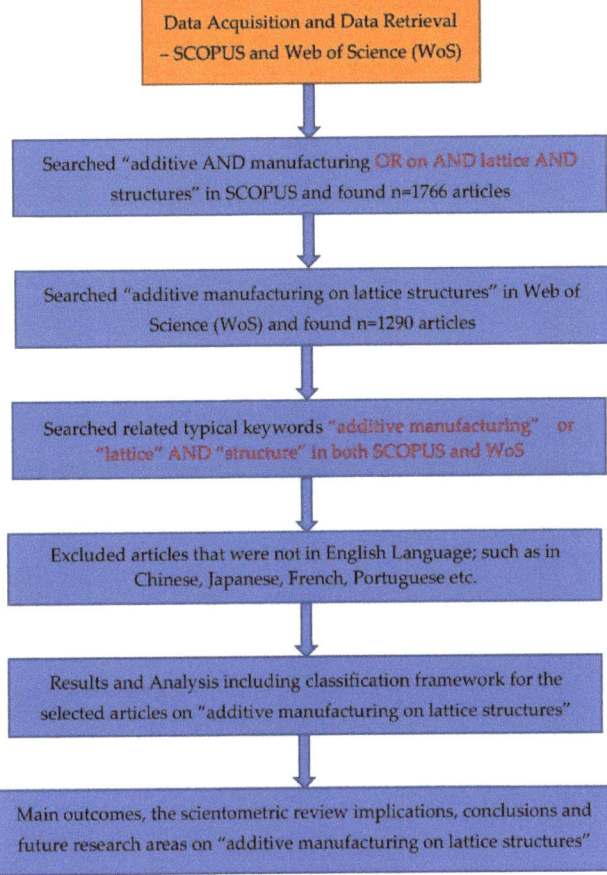

Figure 2. Research methodology on the scientometric study.

2.3. Article Selection

The method taken into account for choosing the academic papers is also a crucial component of the meta-science analysis carried out in this literature review. Finding research trends, threads, and advancements on additive manufacturing is one of the primary goals of this review. As represented in Figure 2, a public database named Scopus has been taken into consideration for this review in order to accomplish this objective. Scopus was accessed through Lancaster University, UK. After certain adjustments and exclusions to make sure the data used fits within the targeted study on additive manufacturing, a total number of papers were taken into account in the meta-analysis. Descriptors in the English language were taken from the Scopus database. Additionally, as the non-English papers were all disqualified, only English-language articles were taken into consideration. Clarification regarding the keywords that were used in this study are mentioned in the keyword search on Figures 2–4, which show the databases used in this study.

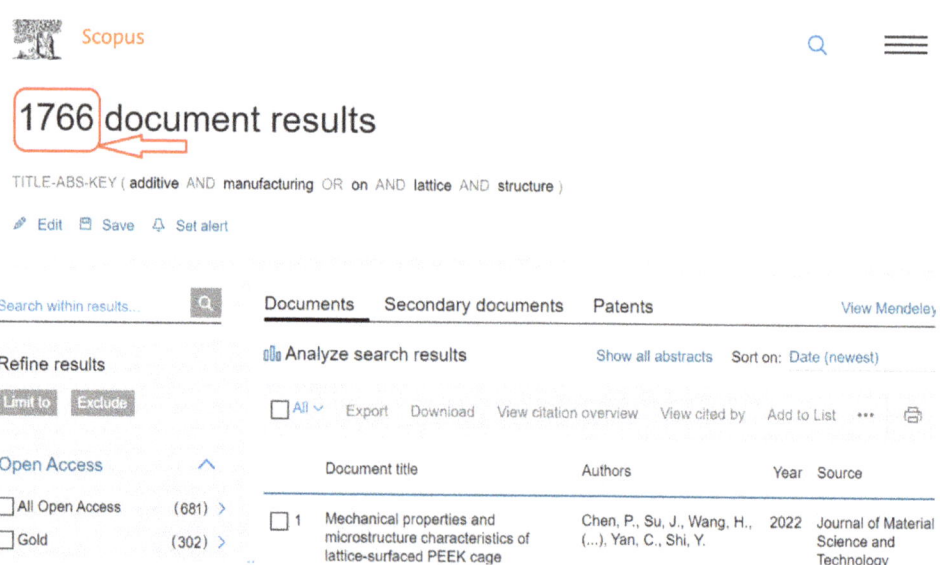

Figure 3. Scopus database supplied by Lancaster University UK, showing keyword "additive manufacturing on lattice structures" for meta-analysis (on 20 June 2022).

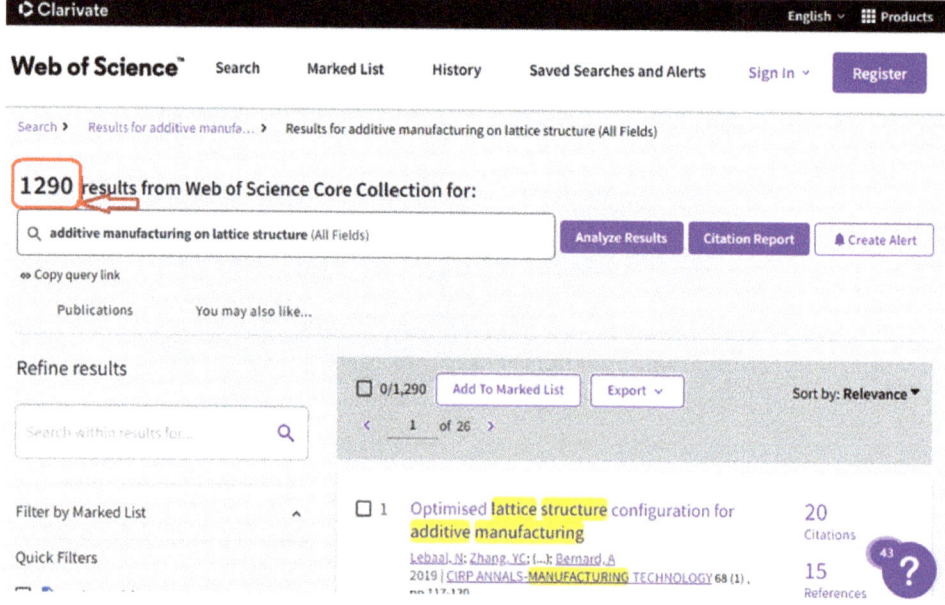

Figure 4. Web of Science (WoS) database supplied by Lancaster University UK, showing keyword "additive manufacturing on lattice structures" for meta-analysis (on 20 June 2022).

Although some comparisons between data from the Scopus database (see Figure 3) and Web of Science database (see Figure 4) were done to determine the trend in development in other forms of the subject area, which remained the main keyword that the research was focused on, it should be noted that the representations in Figures 3 and 4 were used to show different keyword searches used, as each database has a different search structure.

Additionally, these database search images reflect that the search terms used in both show different lines but mean the same thing.

2.4. Research Indicators

The research indicators are key in identifying the significance of any research area. The influence of authorship, co-authorship, regions by countries, affiliations (or institutions), publication sources, and keywords are some of the aspects that are taken into account in the formulation of the scientometric investigation or similar bibliometric reviews [88–90,94–102]. Some mapping was conducted on the publication data retrieved using VoS Viewer [103–112], using standard methods of bibliometric mapping [113–122]. However, the publications were also screened by sampling some data, as the results were too much to check each paper. The trajectory of the analysed subject was tracked by measuring the impact factor and the h-index of the publications selected for the study. The impact factors of the published sources were found by scanning the Clarivate Analytics database [123], Web of Science (WoS) database [124], Scopus database [125], and the SCIMAGO database [126], also available in 2021 Journal Citation Reports [127]. Some studies investigated different databases ranging from PubMed to Scopus and Web of Science databases to conduct bibliometric analysis using different indicators [87–89,127–130]. However, as seen in Figures 3 and 4, the search output obtained from Scopus were 1766 results, whereas the result from Web of Science (WoS) were 1290 results on the same keyword for this scientometric analysis. Hence, most of the data considered were from the Scopus database, whereas data from WoS was used to validate the studies. It also showed that Scopus had higher data collection on the subject area for the time range under consideration in this study. It should be noted that this does not reflect that one database has more collection of publication record than the other. Additionally, it should be noted that H-index is a particular indicator established by JE Hirsch in 2005, which measures each researcher's number of publications and number of citations [128,129,131]. In that study, it was inferred that when a writer has N publications, and those publications have been mentioned at least N times by other writers, then that writer's h-index is equal to N.

2.5. Scientometric Analysis

The knowledge domain structure of the multi-material additive manufacturing of polymers can be clarified by the discovered research clusters; however, the in-depth research problems and research demands cannot be revealed by scientometric analysis. In order to enhance the scientometric analysis in this work, a systematic review was carried out. The systematic review was first split into two parts by the authors: technology and applications. A consensus-based debate on the results of the scientometric review study led to the classification structure of research subjects in these two areas. It was reported that Nalimov and Mulchenko coined the word "Scientometrics" for the first time in 1969 as an evaluation of science [88,89,94,95]. From the second half of the 19th century to the present, scientometrics has been a growing field of study. In the past century, scientometrics research has progressed from the unconscious to consciousness, from qualitative to quantitative research, and from outward description to a thorough examination of the fundamental characteristics of scientific production. Recent research has shown the effectiveness of scientometrics in a variety of fields, including additive manufacturing [38–40], data analysis [94,95], built environment [96,97,128,129], research impact [131], sustainability [132,133], energy [134], project management [135], construction [136,137], water supply [138,139], medical applications [99–101,140,141], visualisation of data [103–108,114–118], and author collaborations [142–144]. Modern scientometric analysis enables researchers to access scientific contributions, map knowledge structures, to access scientific advancement, and identify emerging patterns within a certain study subject from these literature studies. It is quite difficult to describe the total field of multi-material additive manufacturing of polymers using simply systematic analysis due to the large range of research subjects that fall under this umbrella [145,146]. The research field can be understood in depth through systematic analysis, but this method has limitations in terms of subjective interpretation and is open to bias [134–136]. In order to analyse the findings of earlier studies in the field of multi-material additive manufacturing of polymers, a scientometrics analysis method was proposed in this work.

2.6. VOS Viewer

The VOSviewer is an open-source programme and was used in this study's network modelling and visualisation [103–108]. Nees Jan van Eck and Ludo Waltman currently own the VOSviewer. For this study, the version of the software used is VOSviewer version 1.6.18, and it was run with Java version 1.8.0_333 and Microsoft Graph. It is important to note that when organising research subjects for the ensuing systematic review, both the scientific mapping of research communities and themes derived from literature coupling analysis and keyword co-occurrence analysis were taken into account. Numerous analyses were conducted from different angles, including analyses of the countries/regions' activity, authorship, co-occurrence, keyword, literature coupling, and number of publications, as summarised in Figure 5.

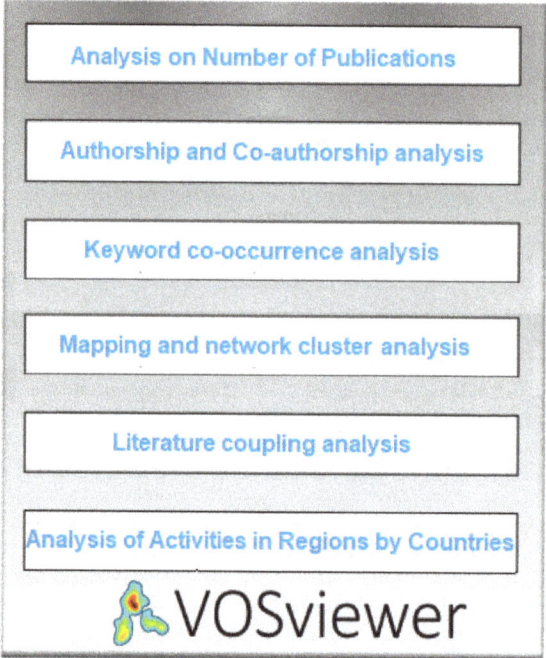

Figure 5. Different analyses conducted using VOSviewer.

3. Results and Analysis

3.1. Publication History

The first aspect of the results for the component meta-analysis is the impact of the research and its breakdown of publication years. Data from the Scopus database was obtained on 20 June 2022, as shown in Figure 6. From the result, the publishing output from 2015–2021 showed a modest trend shift when the most recent articles were taken into account. The output increased from 33 publications in 2014, to 36 publications in 2015, to 80 publications in 2016, before it reduced to 115 publications in 2017. Then it increased to 173 publications in 2018, went up to 265 publications in 2019, increased to 343 publications in 2020, peaked at 436 publications in 2021 while they were producing materials to control the Corona Virus, then decreased to 230 publications by the middle of 2022. As a result, among other things, it may be said that the research is a function of economic activity, as 1766 journal papers were published between 2002 and 2022 using the literature search technique described in Section 3. Figure 6 displays the annual number of journal publications on the subject of additive manufacturing. This statistic shows a general rising trend from 2006 to 2009. Starting in 2010, a burst was noticeable because there were only four publications, which can be ascribed to the global economic crisis. From there, it increased dramatically until the year 2021, rising to nine publications in 2011 and fourteen publications in 2012. The number of publications increased at an astounding rate between 2013 and 2021. Notably, the surge that began in 2013 coincided with an important development in additive manufacturing technology, also summarised in Figure 1. The particular developments seen in recent times from Figure 1 are seen in nano-manufacturing, architecture, engineering of car body parts like brake pedals, engineering of COVID19 control devices like ventilators, personal protective equipment (PPE) like face shields, biomedical implants like prosthetic bones, in situ bio-manufacturing, and full body organs. The growing accessibility of established additive manufacturing technologies may be responsible for the rise in study into the additive manufacture of designed lattice structures in recent years.

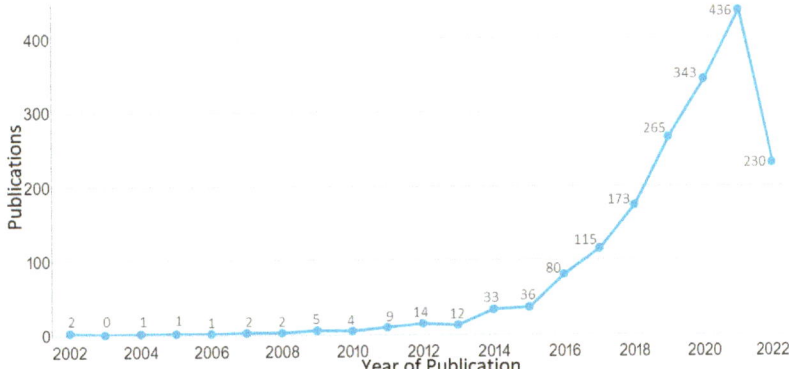

Figure 6. The number of publications against years of publication for publications from 2002 to 2022 (data retrieved from the Scopus database on 20 June 2022).

3.2. Publication Sources

The publishing sources are the subject of the current meta-analysis. Other academic databases were searched as well, though, to verify the information from the Scopus database. It was decided to use the Scopus, despite considering major academic databases like PubMed, Science Direct, DOAJ, Web of Science, Google Scholar, and Scopus. In a detailed, methodical, and scientometric review of scientific scholarly articles (or papers) from journals and conferences, it was possible to make more inferences on the investigation of additive manufacturing. Academic publishers with academic repositories and databases, such as Taylor & Francis, Elsevier, Sage, and Springer Link, were also taken into account, as seen in Figures 7 and 8. Journals that had high significance are specialist journals like Additive Manufacturing, which has a high h-index with an impact factor of 10 and citescore of 11.60, as well as international conferences like ASME, ASCE, ICE, ICCM, ICCS, NIST, ICCS, SAMPE, ISOPE, OTC, etc., were also taken into consideration. As can be seen in Figure 7, the majority of publications on AM were presented in journal papers from two important conference proceedings. However, these publications were less numerous than those that appeared in related Q1 journals. The outcome was subsequently vetted to include the best journals in additive manufacturing. Elsevier's Additive Manufacturing, Elsevier's Materials and Design, MDPI's Materials, MDPI's Polymers, and MDPI's Metals were the journals that appeared the most frequently. The other periodicals are International Journal of Advanced Manufacturing Technology, Rapid Prototyping Journal, Materials Today Proceedings, and Journal of Manufacturing Processes. This was further analysed in Table 1 to show that the highest data was published in Additive Manufacturing journal, especially from the years 2018–2022, where they have high marginal increase.

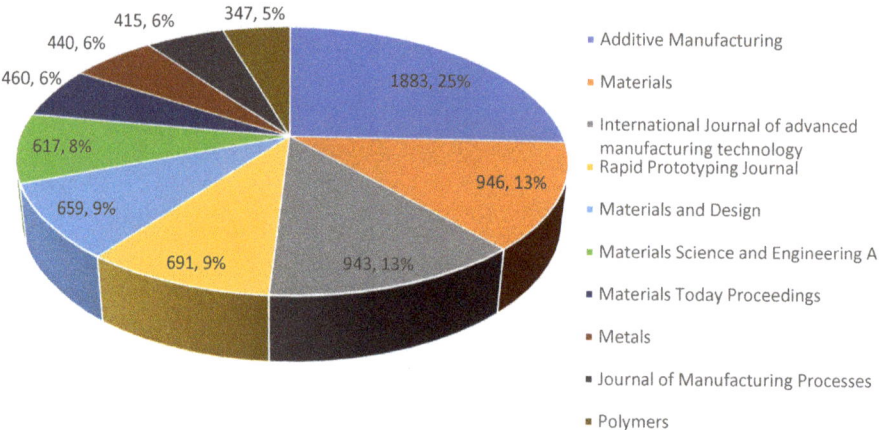

Figure 7. The total number of publications per year by source against percentage of publications from 2001 to 2022 (data retrieved from the Scopus database on 20 June 2022). Note: Each segment shows the number of publications and the corresponding percentages, which is separated by a comma (,) for each publication source.

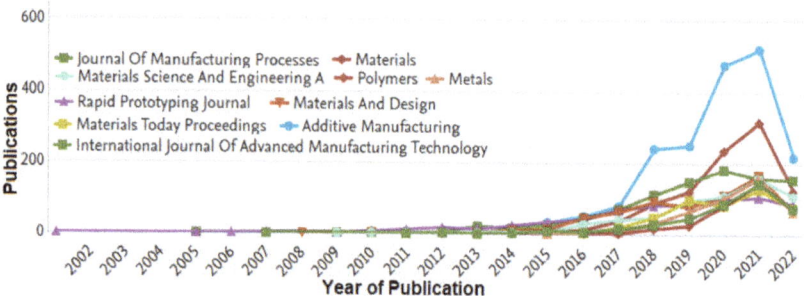

Figure 8. The number of publications per year by source against years of publication for publications from 2001 to 2022 (data retrieved from the Scopus database on 20 June 2022).

Further analysis of the publications per year by source was conducted on this area using data from Web of Science (WoS). It was observed that many publications were available that generally researched on additive manufacturing from 2002–2022. From the data obtained from Scopus, Additive Manufacturing published 1883 articles, Materials published 946 articles, International Journal of advanced manufacturing technology published 940 articles, Rapid Prototyping Journal published 683 articles, Materials and Design published 658 articles, Materials Science and Engineering A published 616 articles, Materials Today Proceedings published 460 articles, Metals published 440 articles, Journal of Manufacturing Processes published 415 articles, and Polymers published 347 articles. However, the sourcing of the data was also conducted using the WoS (Web of Science) database. From the WoS database, 46,821 articles were retrieved, whereas the Scopus database had 43,602 articles. The survey reveals that all of the papers' research output grew from 2013, but Elsevier's Additive Manufacturing had the highest publishing rate. This demonstrates that additive manufacturing researchers have encountered similar problems. These problems primarily revolve around the mechanics of materials with lattice structures, the number of layers, the thickness of the lattice, the material compositions, and the development of standards for lattice structure additive manufacturing. Secondly, it was discovered that, between 2002 and 2022, many patents were published by various inventors as a result of the earliest increased developments in lattice-structured additive produced materials, which were noted as early as in 2002.

Table 1. Publications in Scopus in the top journals for additive manufacturing on lattice structures.

Publication Source	Scopus								
	Total Publications (TP)	2021 Articles	2020 Articles	2019 Articles	2018 Articles	2017 Articles	CiteScore	SJR	SNIP
Additive Manufacturing	1883	521	477	250	241	80	11.6	2.71	2.946
Materials	946	316	236	121	89	37	4.2	0.682	1.261
International Journal of advanced manufacturing technology	943	159	182	149	112	71	5.6	0.946	1.486
Rapid Prototyping Journal	691	106	100	85	83	70	6.0	0.827	1.281
Materials and Design	659	168	111	78	93	64	13.0	1.842	2.264
Materials Science and Engineering A	617	159	109	96	47	42	8.8	1.574	1.973
Materials Today Proceedings	460	128	83	98	48	19	1.8	0.341	0.657
Metals	440	160	102	66	31	8	3.4	0.57	1.062
Journal of Manufacturing Processes	415	144	86	45	27	15	6.6	1.387	2.084
Polymers	347	144	82	25	16	3	4.7	0.77	1.2

3.3. Publication Subjects

The meta-analysis conducted on the scientometric review in this section focuses on the literature search using publication subjects as presented in Figures 9 and 10. They represent the subject-based categorization of papers on additive manufacturing for engineered lattice structures. Engineering-related disciplines accounted for the largest percentage in the 2022 data at 37.5%, followed by Materials Sciences at 27.1%; these two occupied over 50% of the quadrat on publication subjects. It was followed by Physics and Astronomy at 10.6%, then Computer Sciences at 8.2%, then Mathematics at 5.5%, then Chemical Engineering at 2.8%, then Chemistry at 2.0%, then Biochemistry at 1.1%, then Energy at 1.0%. The least was achieved by Business Management at 0.9%, whereas Others, which included minor subgroups, were at 3.1%, which showed that there were other evolving areas that worked on application of additive manufacturing. Furthermore, it was noted that research on Engineering in 2022 data surpassed other areas, which could be seen in the need to develop control materials for the COVID-19 pandemic and systems for manufacturing and the production of oil and gas, among others. These are seen in some of the sampled papers from the screening conducted on the papers used in this study. In other comparable domains, similar transitions were seen using Web of Science data (see Table 2). The data on Table 2 were used to give the best significance of the study, as it has been unified and approximated to be 3 s.f. (significant figures). The tabulated data were also used to have a breakdown of different engineering subjects, such as Engineering Manufacturing, Engineering Mechanical, and Engineering Multidisciplinary. The visualisation treemap used for all publications on additive manufacturing showed that Materials Science Multidisciplinary had 672 publications, followed by Engineering Manufacturing at 337 publications, followed by Engineering Mechanical at 224 publications, followed by Mechanics at 164 publications, followed by Engineering Metallurgy at 160 publications, followed by Applied Physics at 114 publications, followed by Engineering Multidisciplinary at 75 publications, followed by Physical Chemistry at 72 publications, followed by Condensed Matter Physics at 71 publications, and the least was Materials Science Composites at 59 publications. This further demonstrates how interest on additive manufactured materials in engineering subjects has been influenced by their use in full-scale applications, control systems for the COVID-19 pandemic (like face shields), pipeline fabrication, fabrication of machine parts, and deployment on cutting-edge systems.

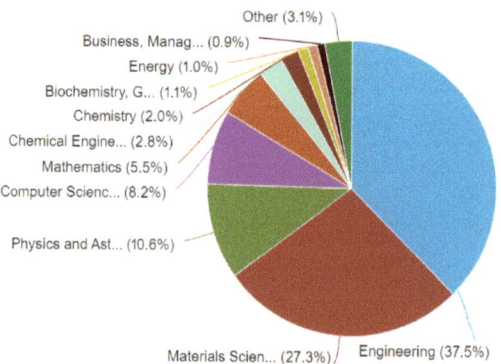

Figure 9. Literature search distribution on the classification of publications by subjects on 'additive manufacturing on lattice structures', (Scopus database on 20 June 2022).

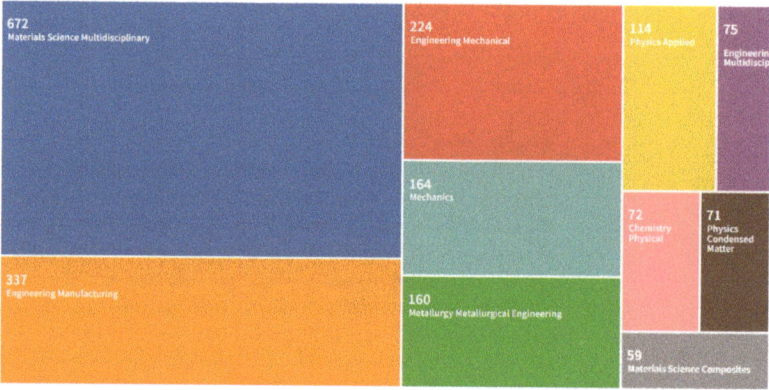

Figure 10. Visualisation treemap chart of different disciplines that published articles on "additive manufacturing on lattice structure" research (data retrieved from the WoS database on 20 June 2022).

Table 2. Data on publication subjects on "additive manufacturing on lattice structure" from WoS.

Web of Science Categories	Record Count	% of 1294
Materials Science Multidisciplinary	672	51.932
Engineering Manufacturing	337	26.043
Engineering Mechanical	224	17.311
Mechanics	164	12.674

Table 2. *Cont.*

Web of Science Categories	Record Count	% of 1294
Metallurgy Metallurgical Engineering	160	12.365
Physics Applied	114	8.81
Engineering Multidisciplinary	75	5.796
Chemistry Physical	72	5.564
Physics Condensed Matter	71	5.487
Materials Science Composites	59	4.56
Engineering Biomedical	56	4.328
Materials Science Biomaterials	48	3.709
Nanoscience Nanotechnology	48	3.709
Automation Control Systems	46	3.555
Materials Science Characterization Testing	42	3.246
Engineering Industrial	35	2.705
Computer Science Interdisciplinary Applications	29	2.241
Mathematics Interdisciplinary Applications	26	2.009
Thermodynamics	26	2.009
Engineering Electrical Electronic	25	1.932
Chemistry Multidisciplinary	22	1.7
Engineering Civil	21	1.623
Instruments Instrumentation	21	1.623
Multidisciplinary Sciences	21	1.623
Polymer Science	21	1.623

3.4. Publication Type

The meta-analysis conducted on the scientometric review in this section focuses on the literature search using publication type that is presented in Figure 11. It represents the type-based categorization of papers on additive manufacturing for lattice structures. Journal papers (or articles) are seen to be the highest, with 69.3% having 1226 publications, followed by conference papers, at 24.1% having 427 publications. Next are review papers, at 3.7% producing 65 documents, then book chapters, at 1.6% producing 29 documents, followed by conference review, at 0.7% producing 13 documents. The other types including the notes, letters, errata, editorials, and data papers; each produced 0.1%, reflecting two documents from each type. This implies that the research scrutiny on AM is reflected on the volume of publication outputs, which are significantly research articles.

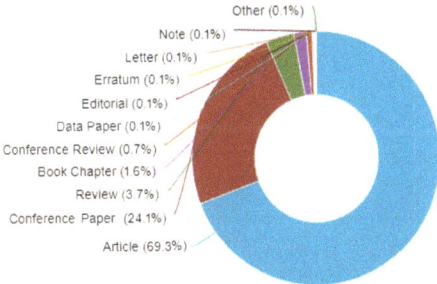

Figure 11. Literature search distribution on the classification of publication type on 'additive manufacturing on lattice structures', (Scopus database on 20 June 2022).

3.5. Publication Keywords

The scientometic analysis on the publication keywords on the search keywords on this investigation. This investigation was initially conducted on the keywords using word cloud, which showed that some words had higher density than others, as seen in Figure 12. The densest keywords are represented with higher font sizes and unique font colours. The keywords are visualized in order using a word cloud generator, which shows the highest to the lowest as boldest to the least bold. The keywords include lattice, structures, additive, manufacturing, design, behaviour, mechanical-properties, optimization, laser, microstructure, structure, melting, mechanical, topology, porous, melting, powder, etc. The word cloud was developed, using text mining via an online Free Word Cloud Generator, to generate two schemes of a word cloud based on different amounts of keywords, as seen in Figure 12a,b. The lesser the number of keywords, the smaller the form of the word cloud, as seen in Figure 12a. However, when more keywords were used, the limit of the word cloud generator had to be increased to develop Figure 12b, but the limit was 100 words. It was then compared with another generator called Voyant tool, which had much larger limit of up to 500 words.

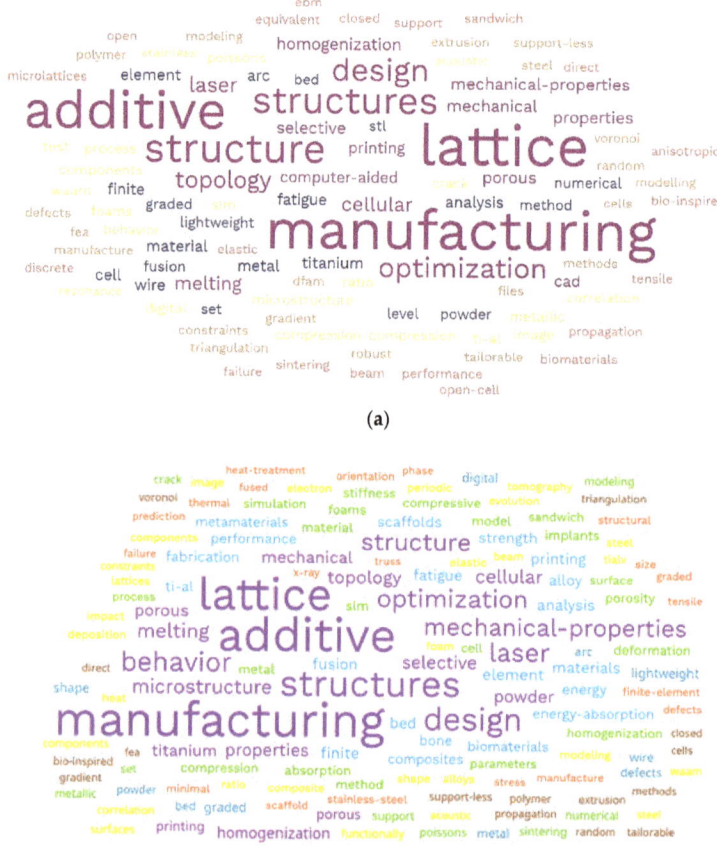

Figure 12. Word cloud on keywords published articles on "additive manufacturing on lattice structure" research (data retrieved from WoS database on 20 June 2022), showing (**a**) scheme 1 and (**b**) scheme 2.

However, the keywords from the scientometric analysis were further post-processed, using VOSviewer version 1.6.18, to obtain the network visualization and density visualization in Figures 13 and 14. The mapped networks showed 36 clusters, showing the co-occurrences of bibliometric items used for the keywords. From this search, there were 3920 items from the results for the component meta-analysis. It was observed that the highest keyword co-occurrence was "element analysis", which shows that a lot of work on this area has been considered based on the different designs for lattice structures used in additive manufacturing. The type of element used has an impact on the research by increasing more micropores, microstructures, and element analysis of the lattices used for the breakdown of publication years from 2002 to 2022. Other keywords that make a mark on this area are: formation, microlattice, minimal surface, porous biomaterial, FE (finite element) result, etc. These range show the diverse research conducted within the scope of additive manufacturing on lattice structures. See the supplementary data for the keyword files used in developing the word clouds and other aspects of this scientometric review.

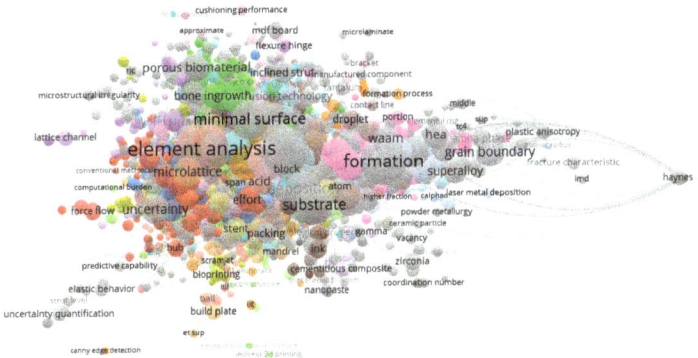

Figure 13. Visualization of a network for keywords published in articles on "additive manufacturing on lattice structure" research (data retrieved from the WoS database on 20 June 2022 and visualized with VOSviewer).

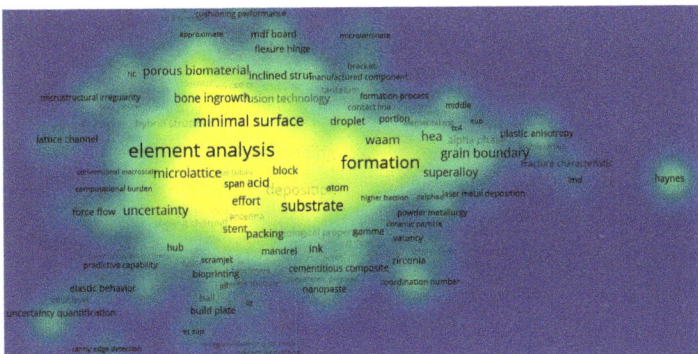

Figure 14. Bibliometric map showing a visualization of the density for keywords published in articles on "additive manufacturing on lattice structure" research (data retrieved from the WoS database on 20 June 2022 and visualized via VOSviewer).

3.6. Publication Affiliation

This sub-section presents the results of the publication affiliation from the bibliometric analysis on the subject area. The results of research output related to the publication in this field are important in understanding the research patterns and the impact of affiliations (institutions and organisations), on the research. A deeper understanding of the support from different affiliations to additive manufacturing on lattice structures is necessary to assess the research impact from the institution or organisation, which is given as a breakdown of publication volume from different departments. Moreso, applications of additive manufacturing on lattice structures have been seen in bioengineering, medical applications, and mechanical engineering. Hence, the outputs seen from the databases were cross-field publications. Currently, different research institutes, polytechnics, universities, and companies have contributed to the scientific literature on additive manufacturing on lattice-manufactured materials. However, there was a recent increase in small-scale research and small AM businesses during the recent COVID-19 pandemic, as detailed in Section 4. AM applications were seen in the control of CoronaVirus for the production of PPEs like face shields, and also in fabricating ventilators [9]. To better understand the influence of affiliations, the analysis of publication affiliations was conducted using data from SCOPUS and WoS databases. To visualise the mapped network, the author's names were further filtered to see publications produced on this subject area per year. This also helps to see the impact of the institution on the research strength in that area. Figure 15 shows the affiliation contributions on the subject matter. There are over 160 institutions that have contributed to research in AM on lattice structures. Each of these institutions have different authors, and some of the publications are sponsored or funded by different funders. However, further analysis on the impact of the funding agencies is presented in the next sub-section. The affiliations have collection of documents, as well as some have only one publication, as AM on lattice structures is still developing in some institutions. Table 3 presents the twenty (20) institutions with the highest publications on the subject area. From this study's data, the Georgia Institute of Technology had the highest publications, as it produced 40 publications on Scopus database while 25 publications on WoS database. The affiliation with the second highest publication is Beijing Institute of Technology which produced 38 publications on both Scopus and WoS databases. The next affiliation is Royal Melbourne Institute of Technology (RMIT) as it produced 37 publications on Scopus database while 38 publications on WoS database. It was also observed that the publications from these top affiliations were also published in high impact journals. Also, the publications from these highest affiliations are among those who have had many years of research experience on the subject area on additive manufacturing. Lastly, the visualized network map in Figure 16 also shows the research connectivity of the different affiliations with 13 clusters.

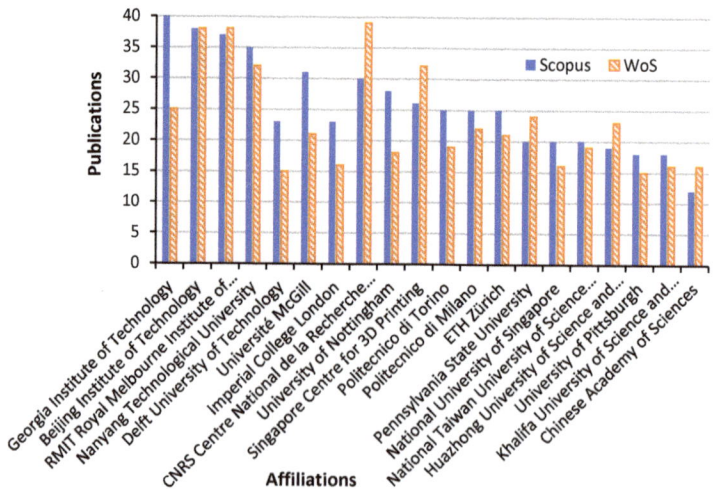

Figure 15. Publication affiliation for "additive manufacturing on lattice structure" research (data retrieved from Scopus and WoS on 20 June 2022).

Table 3. Comparisons on publication affiliation for "additive manufacturing on lattice structure" research (data retrieved from Scopus and WoS on 20 June 2022).

Affiliation	Scopus	WoS	% of 513	% of 465
Georgia Institute of Technology	40	25	7.797	5.376
Beijing Institute of Technology	38	38	7.407	8.172
RMIT Royal Melbourne Institute of Technology	37	38	7.212	8.172
Nanyang Technological University	35	32	6.823	6.882
Delft University of Technology	23	15	4.483	3.226
Université McGill	31	21	6.043	4.516
Imperial College London	23	16	4.483	3.441
CNRS Centre National de la Recherche Scientifique	30	39	5.848	8.387
University of Nottingham	28	18	5.458	3.871
Singapore Centre for 3D Printing	26	32	5.068	6.882
Politecnico di Torino	25	19	4.873	4.086
Politecnico di Milano	25	22	4.873	4.731
ETH Zürich	25	21	4.873	4.516
Pennsylvania State University	20	24	3.899	5.161
National University of Singapore	20	16	3.899	3.441
National Taiwan University of Science and Technology	20	19	3.899	4.086
Huazhong University of Science and Technology	19	23	3.704	4.946
University of Pittsburgh	18	15	3.509	3.226
Khalifa University of Science and Technology	18	16	3.509	3.441
Chinese Academy of Sciences	12	16	2.339	3.441

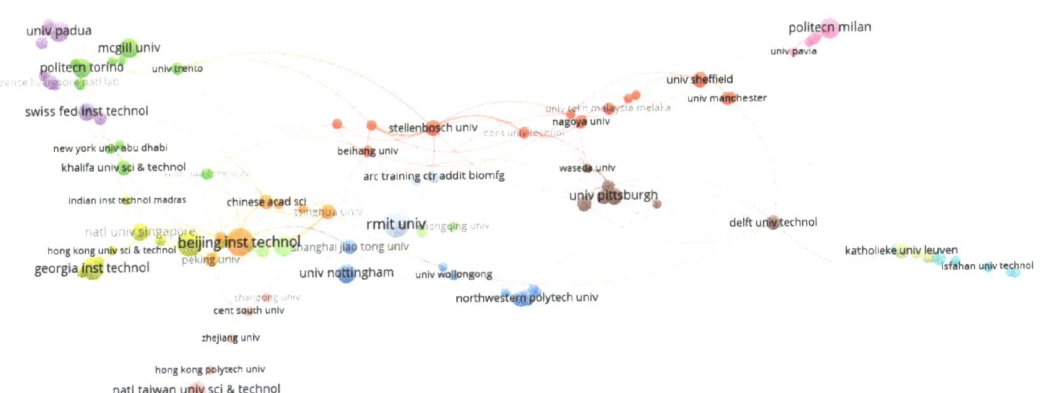

Figure 16. Visualization of network for affiliations on "additive manufacturing on lattice structure" research (data retrieved from WoS database on 20 June 2022 and visualized with VOS Viewer).

3.7. Publication Authors

Another aspect of the investigation of research patterns on "additive manufacturing on lattice structures" is based on the publication authorship. The first aspect of the component meta-analysis is understanding the impact of authorship on the research and its breakdown of publication volume. Different researchers have contributed to the scientific literature on additive manufacturing on lattice-manufactured materials. To visualise the mapped network, the author's names were further filtered to see publications that did not have more than 25 authors per publication. Figure 17 shows the authorship contributions on the subject matter. There are over 2000 authors in the collection of documents, and some of them have only one publication. Table 4 presents the eighteen authors with the highest h-index and highest publications. Additionally, the year the documents were published is shown. The total number of citations since they first published documents and the quantity of references cited for each work were retrieved from academic databases. From this study's data, the author with the highest h-index is Leary, M., followed by Brandt, M., and then next is Zhao, Y.F. The authors who received the most citations per publication were also examined in more detail in the next section. It should be noted that the latest works by authors with the highest h-index are the contributions of the most widely referenced related work. However, authors with the highest h-index are among those who have had many years of research experience on the subject area on additive manufacturing.

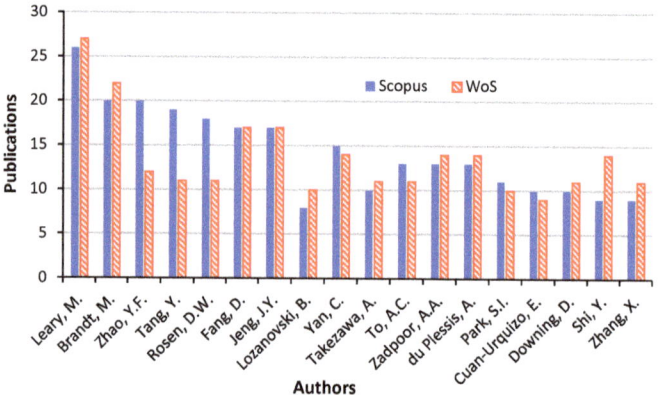

Figure 17. Publication authors for "additive manufacturing on lattice structure" research (data retrieved from Scopus and WoS on 20 June 2022).

Table 4. Comparative results on publication authors for "additive manufacturing on lattice structure" research (data retrieved from Scopus and WoS on 20 June 2022).

Author Name	Scopus	WoS	% of 258	% of 246
Leary, M.	26	27	10.078	10.976
Brandt, M.	20	22	7.752	8.943
Zhao, Y.F.	20	12	7.752	4.878
Tang, Y.	19	11	7.364	4.472
Rosen, D.W.	18	11	6.977	4.472
Fang, D.	17	17	6.589	6.911
Jeng, J.Y.	17	17	6.589	6.911
Lozanovski, B.	8	10	3.101	4.065
Yan, C.	15	14	5.814	5.691
Takezawa, A.	10	11	3.876	4.472
To, A.C.	13	11	5.039	4.472
Zadpoor, A.A.	13	14	5.039	5.691
du Plessis, A.	13	14	5.039	5.691
Park, S.I.	11	10	4.264	4.065
Cuan-Urquizo, E.	10	9	3.876	3.659
Downing, D.	10	11	3.876	4.472
Shi, Y.	9	14	3.488	5.691
Zhang, X.	9	11	3.488	4.472

3.8. Publication Citations

In this sub-section, the scientometric review on publication citations is conducted. The citations in this field against the quantity of publications is one important factor to consider. The citations are used to assess the strength of a research area, the scientific significance, and the impact of the publications in the subject area. One of the indicators used in this assessment is the h-index. The h-index value is based on a list of publications ranked in descending order by the Times Cited count. It can be said that an index of 'h' implies that there are 'h' papers that have each been cited at least 'h' times. Additionally, the h-index is based on the depth of years of the WoS database product subscription and your selected timespan. The source items that are not part of the WoS database product subscription were not factored into the calculation. There were less publications from WoS, whereas there were more publications in Scopus, in a ratio of 1766:1294. It was observed that there was a h-index of 79 and an average citation per publication of 18.71. In the total documents on the subject area, there were also 12,691 citing articles, whereby 11,750 publications were without self-citations. These articles were cited 24,205 times, cumulatively, whereas those publications without self-citations were cited 19,028 times. This is shown in the citation data presented in Figure 18. The number of documents has increased significantly since 2013, whereas the slope of the cumulative publications has barely changed. With the exception of a little decline in the 2012–2013 era, the most substantial changes in the slope of the cumulative publications are shown between 2009–2021. It is important to note there has not been a plateau pattern in recent years, which shows that additive manufacturing research for lattice structures has been relevant recently. Additionally, the drop in the 2021–2022 data shows a drop because it is mid-2022; as such, it is expected to tip higher.

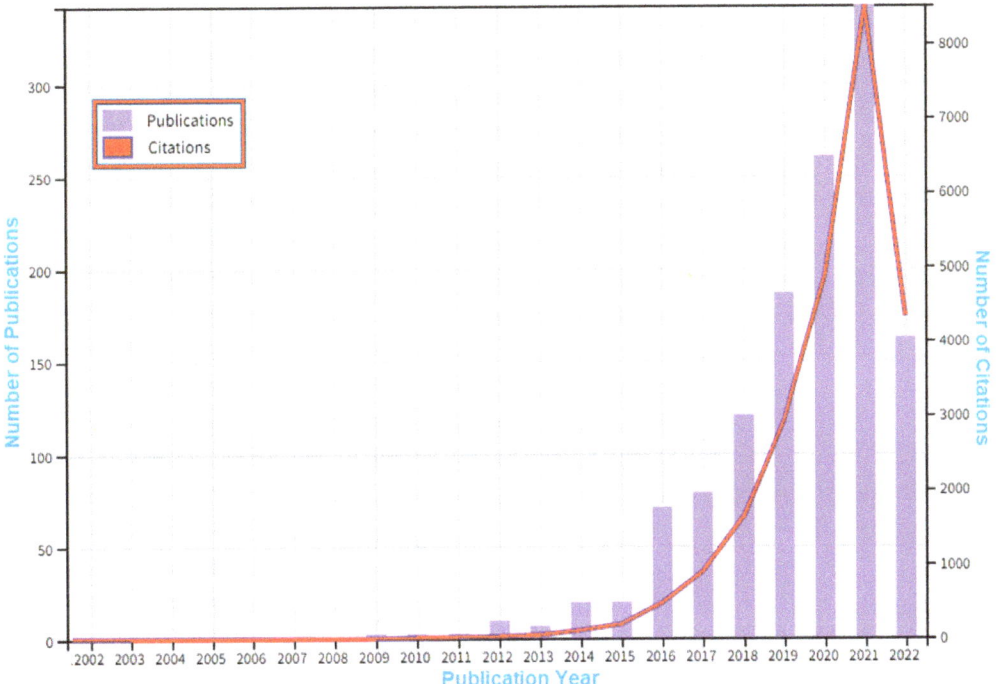

Figure 18. Number of citations and number of publications from 2002 to 2022 for "additive manufacturing on lattice structure" research (data retrieved from WoS on 20 June 2022).

3.9. Publication Collaborations by Co-Authorship

The scientometric analysis on publication authorship was conducted using the data from the publication databases. For the co-authorship analysis, the counting method used was the full counting method, and the publications that had above 25 authors were ignored from this study. The number of documents per author were limited to five, and 92 met the thresholds out of 4475 authors. The authors with the greatest total link strength were used in the selection. The threshold system used to filter the authorship was a minimum of one publication in the area, as 3446 authors met this threshold. In this data clustering, there were two methods considered in the analysis. For the first method, 17 clusters were used for the authorship, as seen in Figures 19 and 20. The highest publications were identified in the green node by cluster 2 showing Seung Ki Moon as the most published author in additive manufacturing, with 23 links, 17 publications, and a total link strength of 45.

Figure 19. Mapping based on co-authorship showing the network visualization for the first method.

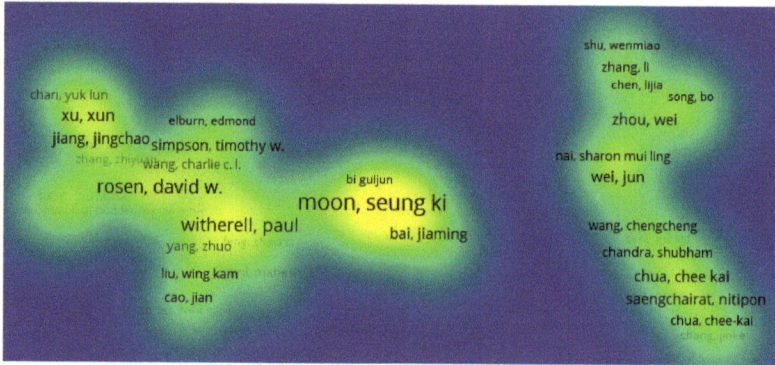

Figure 20. Mapping based on co-authorship showing the density visualization for the first method.

The second method of analysis was conducted using fractionalization to normalize the data. It showed a much wider network of co-occurrences between publications but mapped more links and clusters between the different authors in different locations, as seen in Figure 21. This method also showed the impact of the research, as seen through the authors. The density visualization in Figure 22 also showed the link strength of the authors on this subject area, and the breakdown of publication can be tracked to see the research patterns.

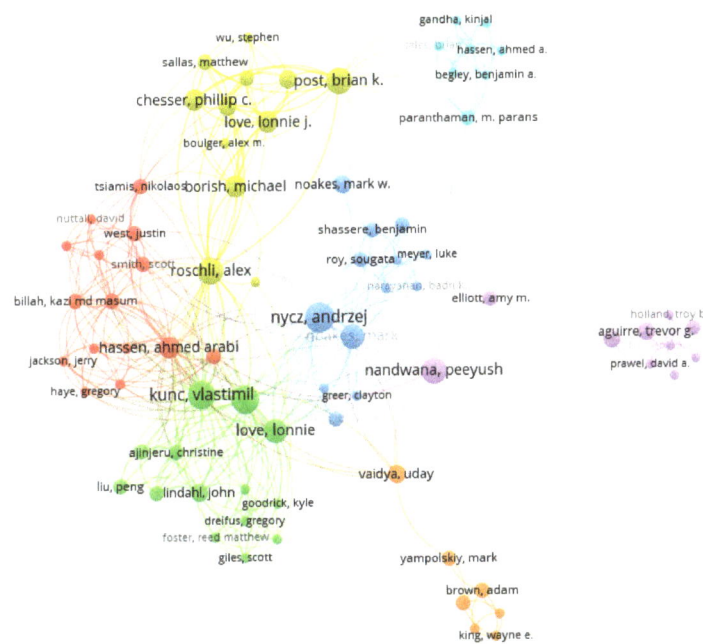

Figure 21. Mapping based on co-authorship showing the network visualization for the second method.

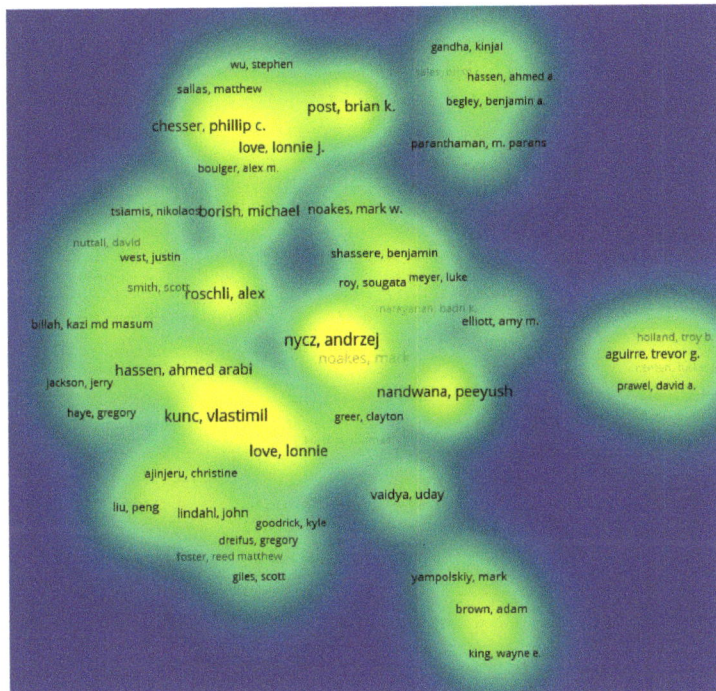

Figure 22. Mapping based on co-authorship showing the density visualization for the second method.

3.10. Publication Countries/Regions

The scientometric analysis on publications conducted in this subject area showed that the researchers from 75 different nations have contributed to the scientific literature on additive manufacturing on lattice structure materials, although only 25 of those nations have more than 10 publications to their names. Figure 23 shows the 25 nations with the most quantity of publications, namely: United States, China, Italy, United Kingdom, Germany, France, Australia, Canada, Singapore, India, Japan, Switzerland, Russian Federation, South Korea, Turkey, Netherlands, Iran, Belgium, South Africa, Taiwan, Sweden, Spain, Malaysia, United Arab Emirates, and Poland. According to the overall number of publications that are not dependent on international collaboration, the USA comes out as the top nation. From 2002–2022, it was observed that different databases reflected close results for each country, as seen in Table 5. The Scopus database showed that the USA produced 397 publications, whereas WoS showed that the USA produced 252 publications. The top four countries with the highest scientific production also include Italy, China, and the UK. In this way, it can also be seen that the regions of Europe and Asia, where there are more than 60 publications, are more interested in research on additive manufacturing with lattice structures. As shown in Figure 23, the USA, China, Italy and the UK, which have the broadest worldwide network of collaboration, are at the forefront of academic engagement. It is hoped that other nations will close the gap in the publication ratio from that of the top countries, such as USA, which almost doubles the third (Italy).

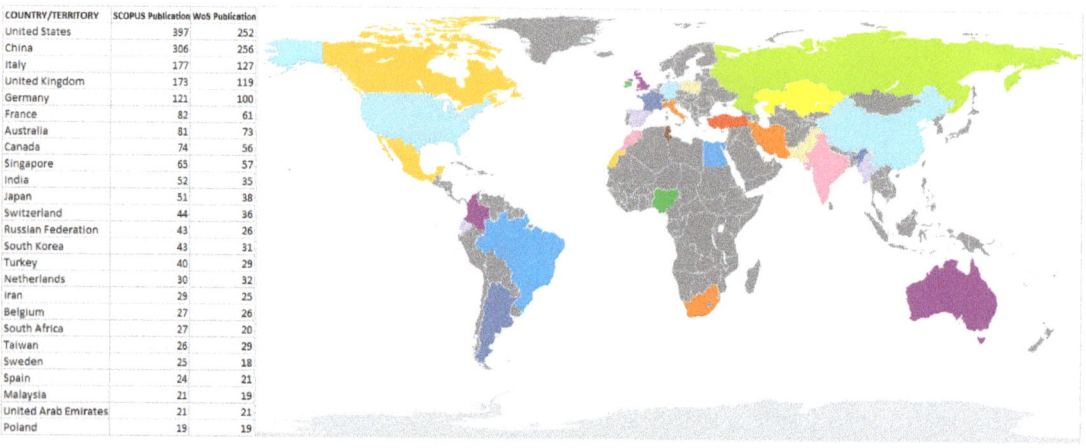

Figure 23. The global research activities showing countries with the highest relevance for "additive manufacturing on lattice structures".

Table 5. The global research activities showing countries with the highest relevance for "additive manufacturing on lattice structures" from the Scopus and WoS databases.

Country/Territory	Scopus	WoS	% of Scopus Sum (1998)	% of WoS Sum (1526)
United States	397	252	19.869	16.514
China	306	256	15.315	16.776
Italy	177	127	8.859	8.322
United Kingdom	173	119	8.659	7.798
Germany	121	100	6.056	6.553
France	82	61	4.104	3.997
Australia	81	73	4.054	4.784

Table 5. Cont.

Country/Territory	Scopus	WoS	% of Scopus Sum (1998)	% of WoS Sum (1526)
Canada	74	56	3.704	3.669
Singapore	65	57	3.253	3.735
India	52	35	2.603	2.294
Japan	51	38	2.553	2.490
Switzerland	44	36	2.202	2.359
Russian Federation	43	26	2.152	1.704
South Korea	43	31	2.152	2.031
Turkey	40	29	2.002	1.900
Netherlands	30	32	1.502	2.097
Iran	29	25	1.451	1.638
Belgium	27	26	1.351	1.704
South Africa	27	20	1.351	1.311
Taiwan	26	29	1.301	1.900
Sweden	25	18	1.251	1.180
Spain	24	21	1.201	1.376
Malaysia	21	19	1.051	1.245
United Arab Emirates	21	21	1.051	1.376
Poland	19	19	0.951	1.245

4. Implications of Trends for Future Research

4.1. Implications of Publication Volume

Due to the significant advantages that polymer-based materials have recently brought to the research and industrial community globally, new studies and technological developments have centred on enhancing levels of multifunctionality in diverse applications. When compared to single homogenous structures, the ability to fabricate bespoke multi-material structures utilising additive manufacturing technology enabled particular material selection and improved various attributes [145–150]. Considering the nature of the topics covered in this review, more discussion has been extended to the limitations of "additive manufacturing on lattice structures" and the challenges of including more aspects of the bibliometric analysis. Hence, further discussions should be looked at based on three (3) very relevant aspects: "type of additive manufacturing technology", "lattice topologies", and "additive manufacturing". Based on the scientometric analysis, the following evaluation of recent multi-material polymers with AM applications in the engineering, biomedical, and information technology sectors is provided.

The literature on lattice-structured materials manufactured using additive processes has grown significantly since 2002 up to 2022, as seen in Figure 6. This pattern not only reflects the advancement of additive manufacturing technology, but it also reflects the rising need for further research. There were over 10 unique high-impact journals that publish in a widely diversified set of publications, which were presented in Table 1. Although journal publications are evenly distributed, the field of additive manufacturing on lattice structures has the most publications overall at 25% (Additive Manufacturing Journal). The field of additive manufacturing, which encompasses a variety of applications, systems, methodologies, techniques, materials, and technologies, is often regarded as having the top journal in the world. Table 1 indicates that the majority of the journals, with the exception of Additive Manufacturing and Rapid Prototyping Journal, concentrate on materials. Therefore, in the process of determining where to publish their papers, researchers working on technologies or processes of additive manufacturing may run into problems. Some of these additive manufacturing studies include different types of lattice structures [150–159]. Typical representations showing the computational model for typical lattice structures, such as (a) the strut-based lattice structure, and (b) the surface-based lattice structure, are seen in Figure 24.

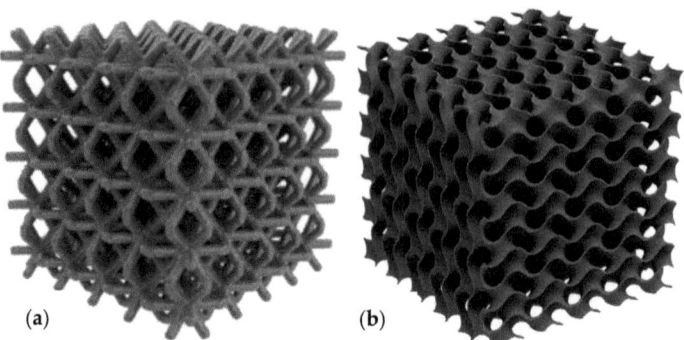

Figure 24. Typical additive manufactured representations showing the computational model (a) a strut-based lattice structure, and (b) a surface-based lattice structure.

4.2. Implications of Additive Manufacturing Processes

The correlations between keywords in articles were taken into account in this analysis. Different studies show that the mechanical properties of both polymer and metal material additive manufacturing received the most attention from researchers [160–168]. Although important for new functional polymers, the applications of fire-resistance, electrical, thermal, bioprinting, electronics, 4D printing, and biocompatible qualities garnered far less research in this discipline. Different processes for AM, including related terminologies on AM, can be seen in the ISO standard [169]. Fused deposition modelling (FDM), which is thought to be the most popular method for multi-material additive manufacturing due to its expanding choice of materials and doable technique, is another hot place from the keywords network. The problems with the FDM technology of additive manufacturing, particularly the weak bond strength between various materials, have not yet been solved, and both academia and business should pay greater attention to this issue.

The purpose of the AM research on lattice structures is to advance 3D/4D printing techniques that fill the gaps between this technology and conventional production processes. Currently, the creation of polymeric compounds opens the door to the flexibility to examine and regulate the characteristics and functionality of the product manufacturing process that are relevant to their actual application. To enhance the quality of the finished product in accordance with its geometry, research into the development of novel materials and investigation of the physical processes involved during the deposition process must be combined. The researcher will be able to determine the composition of the microstructures and the mechanical properties based on the chosen printing parameters. Additionally, future research can be done to evaluate the behaviour during the development of numerical models to assess the temperature variation and the thermal stress sustained by the material.

The scientometric analysis conducted in this study was able to examine different parameters of publications on the field. However, the research findings were only able to quantitatively determine any prospective research gaps and probable future trends. Based on new areas to look into, one possible topic for research in the realm of additive manufacturing is printing efficiency. It is frequently necessary to make a trade-off between printing effectiveness (such as scanning speed) and part quality (such as print resolution) [38–41]. In order to increase printing efficiency, more energy power or quicker scanning speeds can be used, although printing accuracy may suffer as a result. Additionally, lengthy, complicated post-processes lengthen the printing process overall. It is also challenging to expand the printing or production platform, since there are post-processing problems such as with heat treatment and support material removal. Therefore, it is essential to continuously create and enhance efficient post-processing techniques.

4.3. Implications of Policy Documents

The implication of policy documents on the subject area were also looked at based on the scientometric analysis conducted in this study. However, it is suggested that an in-depth analysis be conducted in this area, as policies related to AM are evolving due to the material advances made, developments of techniques for AM, applications, global challenges, and economic values. More study is recommended on the elaboration of AM standards to cover both 3D printing and 4D printing. Additionally, since there are other forms of lattice structures like the planar lattice structure, octet lattice structure, and the BCC lattice structure, it is pertinent that more specialised bibliometric analysis will be conducted that would be confined to particular forms of lattice structures. However, the present scientometric analysis was able to examine different parameters of publications on the field. Additionally, the research findings were only able to quantitatively determine these prospective research gaps and probable future trends.

Another constraint on lattice architectures is the interfacial bonding strength. Engineering the interfaces between objects made of different materials presents one of the challenges in multi-material 3D printing. Determining the proper level of connections for the lattice structure is therefore necessary. Even though different additive manufacturing techniques have made significant advancements and have a great deal of room for growth in the future, the weak bond strength between adjacent printed layers of various materials remains a challenge. This challenge is due to the formation of defects brought on by variations in the physical and chemical properties of the materials, which would ultimately affect the overall mechanical performance of the printed parts [38–40].

Lattice structures benefit from the optimization of additive manufacturing. There are various methods for utilising this solution to improve printing parameters. To get over this problem, this can be achieved using prototype fabrication, numerical simulation, in-situ monitoring, or artificial intelligence. Additionally, anisotropy in the printed object may result from the use of many materials in 3D printing, and each layer's mechanical characteristics may decline as a result of the manufacturing process' temperature gradient. There are two main categories of methods for increasing the mechanical strength of multimaterial 3D parts: processing parameter optimization [149,158,159] and additional external energy input [147,148,156]. To push the limits of multi-material additive manufacturing, fundamental scientific understanding on the inter-layer cohesion mechanisms between incompatible printed materials is required, in addition to technical innovations.

4.4. Implications of the COVID-19 Pandemic

Thus, it is possible to develop more unique materials and applications for this field. An application of response from this field also induces the publication history as recorded in Section 3.1. Two global events have been considered in the publication trend, as the 2016–2017 oil price decline and the COVID-19 pandemic have been highlighted. It is clear from the 2022 data that the sizes are compared to one another. Additionally, it illustrates the impact of several events, such as the COVID-19 pandemic and the global decline in the price of oil per barrel in the years 2016–2017. The authors propose that these two worldwide events have an impact on the volume of publications for additive manufacturing on lattice structures, especially COVID-19 [169–179]. The production of high-caliber research articles has been significantly influenced by the recent worldwide COVID-19 epidemic in 2020–2021 and national lockdowns that happened in over 80% of countries globally for many months due to the challenge of finding COVID-19 controls such as the use of 3D printed face shields [9,120,180–188]. Also, lessons learnt from the recent COVID19 pandemic has shown that challenging situations can lead to advances in AM [189,190].

Although the majority of commercial 3D printers can create macro-scale parts, there are several real-world uses for 3D printed parts at various scales. A promising printing technology for 3D printing parts from the micro to nanoscale of various materials appears to be the hybrid 3D printing platform, which can balance resolution and printing efficiency at the micro-nanometer scale. The next part provides a summary of the future directions that the systematic review suggests, with concluding remarks that are based on the authors' expertise.

5. Conclusions

With the increasing designs and advances in material development, additive manufacturing has not been left behind. Additive manufacturing has aided the manufacture of 3D parts and other multi-material polymer/metal additive manufactured components. Thus, the need for this scientometric study to investigate the research patterns for additive manufacturing on lattice structures. In-depth assessments were conducted on the literature in the area by looking at publication records from the Scopus and WoS databases. The primary research frameworks, articles, and pertinent research subjects were subsequently discovered through analysis of the number of publications, literature coupling, keyword co-occurrence, authorship, and countries/regions activities.

This paper offers a thorough, systematic analysis of the most recent developments in additive manufacturing, both in terms of technique and applications for publications from 2002 to 2022. The scientometric analysis revealed a strong bias in favour of investigating materials in this area but little concentration on emerging technologies. The author keywords from this bibliometric review shows that there are different aspects of lattice structure that are related to the subject area. For instance, 3D printing efficiency, interfacial bonding strength between multiple materials, cross-contamination, scalability, and applications stated above are just a few of the lingering issues that additive manufacturing technology still faces today, seen in the keywords.

To overcome these obstacles, interdisciplinary research and development will be crucial, and developments in additive manufacturing on lattice structures and its innovative uses in new fields will hasten scientific research and technical advancement in a variety of fields. VOSviewer was also used to visualize the items, the co-relationships of the clusters, and the mapped networks needed. This was achieved by using a consistent data format to retrieve the data from both databases. From this study, the USA was the highest region that worked on additive manufacturing for lattice structures. This research also shows patterns of research from authors and their affiliations.

It is recommended that the data from one database be validated by using a second database and comparing the publication records. To gain more complete data, it can be improved by merging several datasets like Scopus, Google Scholar, and Web of Science. Additionally, the scientometric analysis in this study is unable to directly offer or depict the expertise of researchers or publication authors, which may prevent further review. Further areas of scientific reviews can be involved with technical focus groups, whereby the publications can be grouped into many categories. Additionally, the results of the scientometric analysis in each category can be sent to a corresponding expert from technical focus groups. This approach can be used to retrieve technical views from experts in various fields. Findings from the plethora of literature on AM can further present advancements made in the field and the research gaps to broaden the scope of recommendations.

Supplementary Materials: The following supporting information can be downloaded at: Amaechi, Chiemela Victor; Adefuye, Emmanuel (2022), "Supplementary Dataset on Scientometrics of Additive Manufacturing for lattice structure", Mendeley Data, V1, https://doi.org/10.17632/wxd98kfkrp.1 (accessed on 22 July 2022).

Author Contributions: Conceptualization, C.V.A.; methodology, C.V.A.; software, C.V.A.; validation, C.V.A.; formal analysis, C.V.A.; investigation, C.V.A.; resources, C.V.A.; writing—original draft preparation, C.V.A.; writing—reviewing draft, C.V.A., E.F.A., I.M.K., E.C.A. and B.H.; data curation, C.V.A.;

visualization, C.V.A., E.F.A., I.M.K., E.C.A. and B.H.; supervision, C.V.A.; project administration, C.V.A.; funding acquisition, C.V.A. and E.F.A. All authors have read and agreed to the published version of the manuscript.

Funding: The Department of Engineering, Lancaster University, UK and Engineering and Physical Sciences Research Council (EPSRC)'s Doctoral Training Centre (DTC), UK are highly appreciated. In addition, the funding of Overseas Scholarships by Niger Delta Development Commission (NDDC), Nigeria is also appreciated, as well as the support of Standards Organisation of Nigeria (SON), F.C.T Abuja, Nigeria. The funding of Tertiary Education Trust Fund (TETFUND), Nigeria is acknowledged.

Institutional Review Board Statement: Not Applicable.

Informed Consent Statement: Not Applicable.

Data Availability Statement: The raw/processed data required to reproduce these findings have been shared along with supplementary data as used on the present study.

Acknowledgments: The author acknowledges the technical support from Lancaster University Engineering Department and the Library Unit for support with research materials from the databases. The authors also acknowledge the use of the following tools: SCOPUS database, Web of Science database, VOSviewer data mining software and free word cloud generator.

Conflicts of Interest: The authors declare no conflict of interest. The funders had no role in the design of the study; in the collection, analyses, or interpretation of data; in the writing of the manuscript, or in the decision to publish the results.

References

1. Cheng, B.; Chou, K. Geometric consideration of support structures in part overhang fabrications by electron beam additive manufacturing. *Comput. Aided Des.* **2015**, *69*, 102–111. [CrossRef]
2. Bui, T.Q.; Hu, X. A review of phase-field models, fundamentals and their applications to composite laminates. *Eng. Fract. Mech.* **2021**, *248*, 107705. [CrossRef]
3. Amaechi, C.V.; Gillet, N.; Odijie, A.C.; Hou, X.; Ye, J. Composite Risers for Deep Waters Using a Numerical Modelling Approach. *Compos. Struct.* **2019**, *210*, 486–499. [CrossRef]
4. Ye, J.; Cai, H.; Liu, L.; Zhai, Z.; Amaechi, C.V.; Wang, Y.; Wan, L.; Yang, D.; Chen, X.; Ye, J. Microscale intrinsic properties of hybrid unidirectional/woven composite laminates: Part I: Experimental tests. *Compos. Struct.* **2021**, *262*, 113369. [CrossRef]
5. Amaechi, C.V.; Gillet, N.; Ja'e, I.A.; Wang, C. Tailoring the local design of deep water composite risers to minimise structural weight. *J. Compos. Sci.* **2022**, *6*, 103. [CrossRef]
6. Blok, G.; Longana, L.; Yu, H.; Woods, S. An investigation into 3D printing of fibre reinforced thermoplastic composites. *Addit. Manuf.* **2018**, *22*, 176–186. [CrossRef]
7. Amaechi, C.V. Local tailored design of deep water composite risers subjected to burst, collapse and tension loads. *Ocean. Eng.* **2022**, *250*, 110196. [CrossRef]
8. Mirzendehdel, A.M.; Suresh, K. Support structure constrained topology optimization for additive manufacturing. *Comput. Aided Des.* **2016**, *81*, 1–13. [CrossRef]
9. Advincula, R.; Dizon, J.; Chen, Q.; Niu, I.; Chung, J.; Kilpatrick, L.; Newman, R. Additive manufacturing for COVID-19: Devices, materials, prospects, and challenges. *MRS Commun.* **2020**, *10*, 413–427. [CrossRef]
10. Hussein, A.; Hao, L.; Yan, C. Advanced Lattice Support Structures for Metal Additive Manufacturing. *J. Mater. Process. Technol.* **2013**, *213*, 1019–1026. [CrossRef]
11. Ranjan, R.; Samant, R.; Anand, S. Design for Manufacturability in Additive Manufacturing Using a Graph Based Approach. In Proceedings of the ASME 2015 International Manufacturing Science and Engineering Conference, Charlotte, NC, USA, 8–12 June 2015; Volume 1. [CrossRef]
12. Vayre, B.; Vignat, F.; Villeneuve, F. Designing for Additive Manufacturing. *Procedia CIRP* **2012**, *3*, 632–637. [CrossRef]
13. Dong, G.; Tessier, D.; Zhao, Y.F. Design of Shoe Soles Using Lattice Structures Fabricated by Additive Manufacturing. In *Proceedings of the Design Society: International Conference on Engineering Design*; Cambridge University Press: Cambridge, UK, 2019; Volume 1, pp. 719–728. [CrossRef]
14. Diegel, O.; Schutte, J.; Ferreira, A.; Chan, Y.L. Design for additive manufacturing process for a lightweight hydraulic manifold. *Addit. Manuf.* **2020**, *36*, 101446. [CrossRef]
15. Dar, U.A.; Mian, H.H.; Abid, M.; Topa, A.; Sheikh, M.Z.; Bilal, M. Experimental and numerical investigation of compressive behavior of lattice structures manufactured through projection micro stereolithography. *Mater. Today Commun.* **2020**, *25*, 101563. [CrossRef]
16. Xie, G.; Dong, Y.; Zhou, J.; Sheng, Z. Topology optimization design of hydraulic valve blocks for additive manufacturing. *Proc. Inst. Mech. Eng. Part C J. Mech. Eng. Sci.* **2020**, *234*, 1899–1912. [CrossRef]

17. Cheng, L.; To, A. Part-scale build orientation optimization for minimizing residual stress and support volume for metal additive manufacturing: Theory and experimental validation. *Comput. Aided Des.* **2019**, *113*, 1–23. [CrossRef]
18. Cheng, L.; Liang, X.; Bai, J.; Chen, Q.; To, J.L.A. On utilizing topology optimization to design support structure to prevent residual stress induced build failure in laser powder bed metal additive manufacturing. *Addit. Manuf.* **2019**, *27*, 290–304. [CrossRef]
19. Salmi, M. Additive Manufacturing Processes in Medical Applications. *Materials* **2021**, *14*, 191. [CrossRef] [PubMed]
20. Zhang, K.; Qu, H.; Guan, H.; Zhang, J.; Zhang, X.; Xie, X.; Yan, L.; Wang, C. Design and Fabrication Technology of Metal Mirrors Based on Additive Manufacturing: A Review. *Appl. Sci.* **2021**, *11*, 10630. [CrossRef]
21. Wang, X.; Wang, C.; Zhou, X.; Wang, D.; Zhang, M.; Gao, Y.; Wang, L.; Zhang, P. Evaluating Lattice Mechanical Properties for Lightweight Heat-Resistant Load-Bearing Structure Design. *Materials* **2020**, *13*, 4786. [CrossRef] [PubMed]
22. Ning, F.; Cong, W.; Jia, Z.; Wang, F.; Zhang, M. Additive manufacturing of CFRP composites using fused deposition modeling: Effects of process parameters 1989. In Proceedings of the ASME 2016 11th International Manufacturing Science and Engineering Conference, Blacksburg, Virginia, USA, 27 June – 1 July 2016; Volume 3, p. 003. [CrossRef]
23. Korshunova, N.; Alaimo, G.; Hosseini, S.B.; Carraturo, M.; Reali, A.; Niiranen, J.; Auricchio, F.; Rank, E.; Kollmannsberger, S. Bending behavior of octet-truss lattice structures: Modelling options, numerical characterization and experimental validation. *Mater. Des.* **2021**, *205*, 109693. [CrossRef]
24. Azzouz, L.; Chen, Y.; Zarrelli, M.; Pearce, J.M.; Mitchell, L.; Ren, G.; Grasso, M. Mechanical properties of 3-D printed truss-like lattice biopolymer non-stochastic structures for sandwich panels with natural fibre composite skins. *Compos. Struct.* **2019**, *213*, 220–230. [CrossRef]
25. Long, J.; Nand, A.; Ray, S. Application of Spectroscopy in Additive Manufacturing. *Materials* **2021**, *14*, 203. [CrossRef] [PubMed]
26. Obadimu, S.O.; Kourousis, K.I. Compressive Behaviour of Additively Manufactured Lattice Structures: A Review. *Aerospace* **2021**, *8*, 207. [CrossRef]
27. Ueno, A.; Guo, H.; Takezawa, A.; Moritoyo, R.; Kitamura, M. Temperature Distribution Design Based on Variable Lattice Density Optimization and Metal Additive Manufacturing. *Symmetry* **2021**, *13*, 1194. [CrossRef]
28. Huang, J.; Zhang, Q.; Scarpa, F.; Liu, Y.; Leng, J. Bending and benchmark of zero Poisson's ratio cellular structures. *Compos. Struct.* **2016**, *152*, 729–736. [CrossRef]
29. Chowdhury, S.; Mhapsekar, K.; Anand, S. Part Build Orientation Optimization and Neural Network-Based Geometry Compensation for Additive Manufacturing Process. *J. Manuf. Sci. Eng. Trans. ASME* **2018**, *140*, 031009. [CrossRef]
30. Choi, S.; Samavedam, S. Modelling and Optimisation of Rapid Prototyping. *Comput. Ind.* **2002**, *47*, 39–53. [CrossRef]
31. Stichel, T.; Laumer, T.; Linnenweber, T.; Amend, P.; Roth, S. Mass flow characterization of selective deposition of polymer powders with vibrating nozzles for laser beam melting of multi-material components. *Phys. Procedia* **2016**, *83*, 947–953. [CrossRef]
32. Chianrabutra, S.; Mellor, B.G.; Yang, S. A Dry Powder Material Delivery Device for Multiple Material Additive Manufacturing. In Proceedings of the 25th Annual International Solid Freeform Fabrication Symposium: An Additive Manufacturing Conference, Austin, TX, USA, 4–6 August 2014; Bourell, D.L., Ed.; University of Texas at Austin: Austin, TX, USA, 2014. ISBN 1053-2153.
33. Kruth, J.P.; Mercelis, P.; Van Vaerenbergh, J.; Froyen, L.; Rombouts, M. Binding mechanisms in selective laser sintering and selective laser melting. *Rapid Prototyp. J.* **2005**, *11*, 26–36. [CrossRef]
34. Babuska Tomas, F.; Krick Brandon, A.; Susan Donald, F.; Kustas Andrew, B. Comparison of powder bed fusion and directed energy deposition for tailoring mechanical properties of traditionally brittle alloys. *Manuf. Lett.* **2021**, *28*, 30–34. [CrossRef]
35. Grierson, D.; Rennie, A.E.W.; Quayle, S.D. Machine Learning for Additive Manufacturing. *Encyclopedia* **2021**, *1*, 576–588. [CrossRef]
36. Wang, C.; Tan, X.P.; Tor, S.B.; Lim, C.S. Machine learning in additive manufacturing: State-of-the-art and perspectives. *Addit. Manuf.* **2020**, *36*, 101538. [CrossRef]
37. Zhang, Y.; Dong, G.; Yang, S.; Zhao, Y.F. Machine learning assisted prediction of the manufacturability of laser-based powder bed fusion process. In Proceedings of the ASME Design Engineering Technical Conference; American Society of Mechanical Engineers (ASME), Anaheim, CA, USA, 18–21 August 2019; Volume 1. [CrossRef]
38. Zheng, Y.; Zhang, W.; Baca Lopez, D.M.; Ahmad, R. Scientometric Analysis and Systematic Review of Multi-Material Additive Manufacturing of Polymers. *Polymers* **2021**, *13*, 1957. [CrossRef]
39. García-León, R.A.; Gómez-Camperos, J.A.; Jaramillo, H.Y. Scientometric Review of Trends on the Mechanical Properties of Additive Manufacturing and 3D Printing. *J. Mater. Eng. Perform.* **2021**, *30*, 4724–4734. [CrossRef]
40. Jin, Y.; Ji, S.; Li, X.; Yu, J. A scientometric review of hotspots and emerging trends in additive manufacturing. *J. Manuf. Technol. Manag.* **2017**, *28*, 18–38. [CrossRef]
41. Osama, S.; Al-Ahmari Ameen, W.; Mian, S. 2019 Additive manufacturing: Challenge, Trends, and Applications. *Adv. Mech. Eng.* **2019**, *11*, 1687814018822880.
42. Chabaud, G.; Castro, M.; Denoual, C.; Le Duigou, A. Hygromechanical properties of 3D printed continuous carbon and glass fibre reinforced polyamide composite for outdoor structural applications *Addit. Manuf.* **2019**, *26*, 94–105. [CrossRef]
43. Liao, G.; Li, Z.; Cheng, Y.; Xu, D.; Zhu, D.; Jiang, S.; Guo, J.; Chen, X.; Xu, G.; Zhu, Y. Properties of oriented carbon fiber/polyamide 12 composite parts fabricated by fused deposition modeling. *Mater. Des.* **2018**, *139*, 283–292. [CrossRef]
44. Wauthle, R.; Vrancken, B.; Beynaerts, B.; Jorissen, K.; Schrooten, J.; Kruth, J.P.; Van Humbeeck, J. Effects of build orientation and heat treatment on the microstructure and mechanical properties of selective laser melted Ti6Al4V lattice structures. *Addit. Manuf.* **2015**, *5*, 77–84. [CrossRef]

45. Gorny, B.; Niendorf, T.; Lackmann, J.; Thoene, M.; Troester, T.; Maier, H.J. In situ characterization of the deformation and failure behaviour of non-stochastic porous structures processed by selective laser melting. *Mater. Sci. Eng. A* **2011**, *528*, 7962–7967. [CrossRef]
46. Mohsenizadeh, M.; Gasbarri, F.; Munther, M.; Beheshti, A.; Davami, K. Additively manufactured lightweight metamaterials for energy absorption. *Mater. Des.* **2018**, *1*, 39. [CrossRef]
47. Maconachie, T.; Leary, M.; Lozanovski, B.; Zhang, X.; Qian, M.; Faruque, O.; Brandt, M. 2019 SLM lattice structures: Properties, performance, applications, and challenges. *Mater. Des.* **2019**, *183*, 108137. [CrossRef]
48. Mark, C. Optimal lattice-structured materials. *J. Mech. Phys. Solids* **2016**, *96*, 162–183. [CrossRef]
49. Mona, M.; Methods for Modelling Lattice Structures. Kth Royal Institute of Technology School of Engineering Sciences. Master's Thesis, 2020. Available online: http://kth.diva-portal.org/smash/record.jsf?pid=diva2%3A1355716&dswid=2652 (accessed on 3 February 2022).
50. Pan, C.; Han, Y.; Lu, J. Design and optimization of lattice structures: A review. *Appl. Sci.* **2020**, *10*, 6374. [CrossRef]
51. Rosen, D.W.; Johnston, S.R.; Reed, M. Design of general lattice structures for lightweight and compliance applications. In Proceedings of the Rapid Manufacturing Conference, Loughborough, UK, 5–6 July 2006; pp. 1–14. Available online: https://www.researchgate.net/publication/43767416_Design_of_General_Lattice_Structures_for_Lightweight_and_Compliance_Applications (accessed on 3 February 2022).
52. Ashby, M.F. The properties of foams and lattices. *Philos. Trans. R. Soc. A Math. Phys. Eng. Sci.* **2006**, *364*, 15–30. [CrossRef] [PubMed]
53. Gibson, L.J.; Ashby, M.F. *Cellular Solids: Structure and Properties*; Cambridge University Press: Cambridge, UK, 1997.
54. Meza, L.R.; Das, S.; Greer, J.R. Strong, lightweight, and recoverable three-dimensional ceramic nanolattices. *Science* **2014**, *345*, 1322–1326. [CrossRef]
55. Wadley, H.N.; Fleck, N.A.; Evans, A.G. Fabrication and structural performance of periodic cellular metal sandwich structures. *Compos. Sci. Technol.* **2003**, *63*, 2331–2343. [CrossRef]
56. Alzahrani, M.; Choi, S.K.; Rosen, D.W. Design of truss-like cellular structures using relative density mapping method. *Mater. Des.* **2015**, *85*, 349–360. [CrossRef]
57. Tao, W.; Leu, M.C. Design of lattice structure for additive manufacturing. In Proceedings of the 2016 International Symposium on Flexible Automation (ISFA), Cleveland, OH, USA, 1–3 August 2016; pp. 325–332. [CrossRef]
58. Saleh Alghamdi, S.; John, S.; Roy Choudhury, N.; Dutta, N.K. Additive Manufacturing of Polymer Materials: Progress, Promise and Challenges. *Polymers* **2021**, *13*, 753. [CrossRef] [PubMed]
59. Giubilini, A.; Bondioli, F.; Messori, M.; Nyström, G.; Siqueira, G. Advantages of Additive Manufacturing for Biomedical Applications of Polyhydroxyalkanoates. *Bioengineering* **2021**, *8*, 29. [CrossRef]
60. Pryadko, A.; Surmeneva, M.A.; Surmenev, R.A. Review of Hybrid Materials Based on Polyhydroxyalkanoates for Tissue Engineering Applications. *Polymers* **2021**, *13*, 1738. [CrossRef]
61. Ruban, R.; Rajashekhar, V.S.; Nivedha, B.; Mohit, H.; Sanjay, M.R.; Siengchin, S. Role of Additive Manufacturing in Biomedical Engineering. In *Innovations in Additive Manufacturing. Springer Tracts in Additive Manufacturing*; Khan, M.A., Jappes, J.T.W., Eds.; Springer: Cham, Switzerland, 2022. [CrossRef]
62. Sheoran, A.J.; Kumar, H.; Arora, P.K. Moona, GBio-Medical applications of Additive Manufacturing: A Review. *Procedia Manuf.* **2020**, *51*, 663–670. [CrossRef]
63. Kumar, R.; Kumar, M.; Chohan, J. SThe role of additive manufacturing for biomedical applications: A critical review. *J. Manuf. Processes* **2021**, *64*, 828–850. [CrossRef]
64. Popov, V.V., Jr.; Muller-Kamskii, G.; Kovalevsky, A.; Dzhenzhera, G.; Strokin, E.; Kolomiets, A.; Ramon, J. Design and 3D-printing of titanium bone implants: Brief review of approach and clinical cases. *Biomed. Eng. Lett.* **2018**, *8*, 337–344. [CrossRef] [PubMed]
65. Shakibania, S.; Ghazanfari, L.; Raeeszadeh-Sarmazdeh, M.; Khakbiz, M. Medical application of biomimetic 4D printing. *Drug Dev. Ind. Pharm.* **2021**, *47*, 521–534. [CrossRef]
66. Javaid, M.; Haleem, A. 4D printing applications in medical field: A brief review. *Clin. Epidemiol. Glob. Health* **2019**, *7*, 317–321. [CrossRef]
67. Agarwal, T.; Hann, S.Y.; Chiesa, I.; Cui, H.; Celikkin, N.; Micalizzi, S.; Barbetta, A.; Costantini, M.; Esworthy, T.; Zhang, L.G.; et al. 4D printing in biomedical applications: Emerging trends and technologies. *J. Mater. Chem. B* **2021**, *9*, 7608–7632. [CrossRef]
68. Zhou, W.; Qiao, Z.; Zare, E.N.; Huang, J.; Zheng, X.; Sun, X.; Shao, M.; Wang, H.; Wang, X.; Chen, D.; et al. 4D-Printed Dynamic Materials in Biomedical Applications: Chemistry, Challenges, and Their Future Perspectives in the Clinical Sector. *J. Med. Chem.* **2020**, *63*, 15–8003. [CrossRef] [PubMed]
69. Zheng, X.; Smith, W.; Jackson, J.; Moran, B.; Cui, H.; Chen, D.; Ye, J.; Fang, N.; Rodriguez, N.; Weisgraber, T.; et al. Multiscale metallic metamaterials. *Nat. Mater.* **2016**, *15*, 1100–1106. [CrossRef] [PubMed]
70. Yang, L.; Cormier, D.; West, H.; Knowlson, K. NonStochastic Ti-6Al-4V Foam Structure That Shows Negative Poisson's Ratios. *Mater. Sci. Eng. A* **2012**, *558*, 579–585. [CrossRef]
71. Yang, L.; Harrysson, O.; Cormier, D.; West, H. Compressive Properties of Ti-6Al-4V Auxetic Mesh Structures Made by EBM Process. *Acta Mater.* **2012**, *60*, 3370–3379. [CrossRef]
72. Yang, L.; Harrysson, O.; Cormier, D.; West, H. Modeling of the Uniaxial Compression of a 3D Periodic ReEntrant Honeycomb Structure. *J. Mater. Sci.* **2012**, *48*, 1413–1422. [CrossRef]

73. Cuan-Urquizo, E.; Álvarez-Trejo, A.; Robles Gil, A.; Tejada-Ortigoza, V.; Camposeco-Negrete, C.; Uribe-Lam, E.; Treviño-Quintanilla, C.D. Effective Stiffness of Fused Deposition Modeling Infill Lattice Patterns Made of PLA-Wood Material. *Polymers* **2022**, *14*, 337. [CrossRef] [PubMed]
74. Kladovasilakis, N.; Charalampous, P.; Tsongas, K.; Kostavelis, I.; Tzetzis, D.; Tzovaras, D. Experimental and Computational Investigation of Lattice Sandwich Structures Constructed by Additive Manufacturing Technologies. *J. Manuf. Mater. Process.* **2021**, *5*, 95. [CrossRef]
75. RAENG. *Additive Manufacturing: Opportunities and Constraints*; Summary Report of Roundtable Forum Held on 23 May 2013; Royal Academy of Engineering (RAENG): London, UK, 2013; Available online: https://raeng.org.uk/media/ak3htcyo/additive_manufacturing.pdf (accessed on 22 July 2022).
76. Wohler, T. *Wohler's Report 2013-Additive Manufacturing and 3D Printing State of the Industry. Annual Worldwide Progress Report*; Wohler's Associates, Inc.: Washington, DC, USA, 2013; Available online: https://wohlersassociates.com/product/wohlers-report-2013/ (accessed on 22 July 2022).
77. Obi, M.U.; Pradel, P.; Sinclair, M.; Bibb, R. A bibliometric analysis of research in design for additive manufacturing. *Rapid Prototyp. J.* **2022**, *28*, 967–987. [CrossRef]
78. Caviggioli, F.; Ughetto, E. A bibliometric analysis of the research dealing with the impact of additive manufacturing on industry, business and society. *Int. J. Prod. Econ.* **2019**, *208*, 254–268. [CrossRef]
79. Patil, A.K.; Soni, G. A State of The Art Bibliometric Analysis For Additive Manufacturing. *Curr. Mater. Sci.* **2020**, *14*, 2021. [CrossRef]
80. Zhou, Y.; Tang, Y.; Hoff, T.; Garon, M.; Zhao, F.Y. The Verification of the Mechanical Properties of Binder Jetting Manufactured Parts by Instrumented Indentation Testing. *Procedia Manuf.* **2015**, *1*, 327–342. [CrossRef]
81. Forster, A.M. *Materials Testing Standards for Additive Manufacturing of Polymer Materials: State of the Art and Standards Applicability*; NIST Interagency/Internal Report (NISTIR), National Institute of Standards and Technology: Gaithersburg, MD, USA, 2015. [CrossRef]
82. Mao, M.; He, J.; Li, X.; Zhang, B.; Lei, Q.; Liu, Y.; Li, D. The Emerging Frontiers and Applications of High-Resolution 3D Printing. *Micromachines* **2017**, *8*, 113. [CrossRef]
83. Bezek, L.B.; Cauchi, M.P.; de Vita, R.; Foerst, J.R.; Williams, C.B. 3D Printing Tissue-Mimicking Materials for Realistic Transseptal Puncture Models. *J. Mech. Behav. Biomed. Mater.* **2020**, *110*, 103971. [CrossRef]
84. Miyanaji, H.; Ma, D.; Atwater, M.A.; Darling, K.A.; Hammond, V.H.; Williams, C.B. Binder Jetting Additive Manufacturing of Copper Foam Structures. *Addit. Manuf.* **2020**, *32*, 100960. [CrossRef]
85. Herzberger, J.; Sirrine, J.M.; Williams, C.B.; Long, T.E. Polymer Design for 3D Printing Elastomers: Recent Advances in Structure, Properties, and Printing. *Prog. Polym. Sci.* **2019**, *97*, 101144. [CrossRef]
86. Sturm, L.D.; Albakri, M.I.; Tarazaga, P.A.; Williams, C.B. In Situ Monitoring of Material Jetting Additive Manufacturing Process via Impedance Based Measurements. *Addit. Manuf.* **2019**, *28*, 456–463. [CrossRef]
87. Chatham, C.A.; Long, T.E.; Williams, C.B. A Review of the Process Physics and Material Screening Methods for Polymer Powder Bed Fusion Additive Manufacturing. *Prog. Polym. Sci.* **2019**, *93*, 68–95. [CrossRef]
88. Moher, D.; Shamseer, L.; Clarke, M.; Ghersi, D.; Liberati, A.; Petticrew, M.; Shekelle, P.; Stewart, L.A.; Group, P.-P.; Altman, D.G.; et al. Preferred reporting items for systematic review and meta-analysis protocols (PRISMA-P) 2015 statement. *Syst. Rev.* **2015**, *4*, 1–9. [CrossRef]
89. Martinez, P.; Al-Hussein, M.; Ahmad, R. A scientometric analysis and critical review of computer vision applications for construction. *Autom. Constr.* **2019**, *107*, 102947. [CrossRef]
90. Hood, W.W.; Wilson, C.S. The literature of bibliometrics, scientometrics, and informetrics. *Scientometrics* **2001**, *52*, 291–314. [CrossRef]
91. Jin, Y.; Li, X.; Campbell, R.I.; Ji, S. Visualizing the hotspots and emerging trends of 3D printing through scientometrics. *Rapid Prototyp. J.* **2018**, *24*, 801–812. [CrossRef]
92. Jemghili, R.; Taleb, A.A.; Mansouri, K. A bibliometric indicators analysis of additive manufacturing research trends from 2010 to 2020. *Rapid Prototyp. J.* **2021**, *27*, 1432–1454. [CrossRef]
93. Parvanda, R.; Kala, P.; Sharma, V. Bibliometric Analysis-Based Review of Fused Deposition Modeling 3D Printing Method (1994–2020). *3D Printing and Additive Manufacturing* **2021**. ahead of print. [CrossRef]
94. Dzogbewu, T.C.; Amoah, N.; Fianko, S.K.; Afrifa, S.; de Beer, D. Additive manufacturing towards product production: A bibliometric analysis. *Manuf. Rev.* **2022**, *9*, 1. [CrossRef]
95. Perianes-Rodriguez, A.; Waltman, L.; Van Eck, N.J. Constructing bibliometric networks: A comparison between full and fractional counting. *J. Informetr.* **2016**, *10*, 1178–1195. [CrossRef]
96. Granovsky, Y.V. Is it possible to measure science? V. V. Nalimov's research in scientometrics. *Scientometrics* **2001**, *52*, 127–150. [CrossRef]
97. Park, J.Y.; Nagy, Z. Data on the interaction between thermal comfort and building control research. *Data Brief* **2018**, *17*, 529–532. [CrossRef] [PubMed]
98. Wu, Z.; Yang, K.; Lai, X.; Antwi-Afari, M.F. A Scientometric Review of System Dynamics Applications in Construction Management Research. *Sustainability* **2020**, *12*, 7474. [CrossRef]

99. Soosaraei, M.; Khasseh, A.A.; Fakhar, M.; Hezarjaribi, H.Z. A decade bibliometric analysis of global research on leishmaniasis in Web of Science database. *Ann. Med. Surg.* **2018**, *26*, 30–37. [CrossRef]
100. Cash-Gibson, L.; Rojas-Gualdrón, D.F.; Pericàs, J.M.; Benach, J. Inequalities in global health inequalities research: A 50-year bibliometric analysis (1966–2015). *PLoS ONE* **2018**, *13*, e0191901. [CrossRef] [PubMed]
101. Sweileh, W.M.; Al-Jabi, S.W.; Zyoud, S.H.; Sawalha, A.F.; Abu-Taha, A.S. Global research output in antimicrobial resistance among uropathogens: A bibliometric analysis (2002–2016). *J. Glob. Antimicrob. Resist.* **2018**, *13*, 104–114. [CrossRef]
102. Krauskopf, E. A bibiliometric analysis of the Journal of Infection and Public Health: 2008–2016. *J. Infect. Public Health* **2018**, *11*, 224–229. [CrossRef]
103. Van Eck, N.J.; Waltman, L. *VOSviewer Manual: Manual for VOSviewer Version 1.6.18*; Universiteit Leiden: Leiden, The Netherlands, 24 January 2022; pp. 1–53. Available online: https://www.vosviewer.com/documentation/Manual_VOSviewer_1.6.18.pdf (accessed on 12 July 2022).
104. Van Eck, N.J.; Waltman, L. VOS: A new method for visualizing similarities between objects. In *Advances in Data Analysis: Proceedings of the 30th Annual Conference of the German Classification Society*; Lenz, H.-J., Decker, R., Eds.; Springer: Berlin/Heidelberg, Germany, 2007; pp. 299–306.
105. Van Eck, N.J.; Waltman, L. How to normalize cooccurrence data? An analysis of some well-known similarity measures. *J. Am. Soc. Inf. Sci. Technol.* **2009**, *60*, 1635–1651. [CrossRef]
106. Van Eck, N.J.; Waltman, L. Software survey: VOSviewer, a computer program for bibliometric mapping. *Scientometrics* **2010**, *84*, 523–538. [CrossRef]
107. Van Eck, N.J.; Waltman, L. Text mining and visualization using VOSviewer. *ISSI Newsl.* **2011**, *7*, 50–54.
108. Van Eck, N.J.; Waltman, L. Visualizing bibliometric networks. In *Measuring Scholarly Impact: Methods and Practice*; Ding, Y., Rousseau, R., Wolfram, D., Eds.; Springer: Berlin/Heidelberg, Germany, 2014; pp. 285–320.
109. Van Eck, N.J.; Waltman, L.; Dekker, R.; Van den Berg, J. A comparison of two techniques for bibliometric mapping: Multidimensional scaling and VOS. *J. Am. Soc. Inf. Sci. Technol.* **2010**, *61*, 2405–2416. [CrossRef]
110. Van Nunen, K.; Li, J.; Reniers, G.; Ponnet, K. Bibliometric analysis of safety culture research. *Saf. Sci.* **2018**, *108*, 248–258. [CrossRef]
111. Waltman, L.; Van Eck, N.J. A smart local moving algorithm for large-scale modularity-based community detection. *Eur. Phys. J. B* **2013**, *86*, 471. [CrossRef]
112. Waltman, L.; Van Eck, N.J.; Noyons, E.C.M. A unified approach to mapping and clustering of bibliometric networks. *J. Informetr.* **2010**, *4*, 629–635. [CrossRef]
113. Wood, J.; Khan, G.F. International trade negotiation analysis: Network and semantic knowledge infrastructure. *Scientometrics* **2015**, *105*, 537–556. [CrossRef]
114. Chandra, Y. Mapping the evolution of entrepreneurship as a field of research (1990–2013): A scientometric analysis. *PLoS ONE* **2018**, *13*, e0190228. [CrossRef] [PubMed]
115. Chen, C. Searching for intellectual turning points: Progressive Knowledge Domain Visualization. *Proc. Natl. Acad. Sci. USA* **2004**, *101*, 5303–5310. [CrossRef]
116. Chen, C. CiteSpace II: Detecting and visualizing emerging trends and transient patterns in scientific literature. *JASIST* **2006**, *57*, 359–377. [CrossRef]
117. Chen, C. System and Method for Automatically Generating Systematic Reviews of a Scientific Field. U.S. Patent US20110295903A1, 22 October 2010.
118. Chen, C. *CiteSpace: A Practical Guide for Mapping Scientific Literature*; Nova Science Publishers: New York, NY, USA, 2016.
119. Chen, C. Science mapping: A systematic review of the literature. *JDIS* **2017**, *2*, 1–40. [CrossRef]
120. Chen, C. A Glimpse of the First Eight Months of the COVID-19 Literature on Microsoft Academic Graph. *Front. Res. Metr. Anal.* **2020**, *5*, 607286. [CrossRef]
121. Chen, C.; Ibekwe-Sanjuan, F.; Hou, J. The structure and dynamics of co-citation clusters: A multiple-perspective co-citation analysis. *J. Am. Soc. Inf. Sci. Technol. (JASIST)* **2010**, *61*, 1386–1409. [CrossRef]
122. Chen, C.; Song, M. Visualizing a field of research: A methodology of systematic scientometric reviews. *PLoS ONE* **2019**, *14*, e0223994. [CrossRef] [PubMed]
123. Scimago. Scimago Journal & Country Rank. Scimango Lab. 2022. Available online: https://www.scimagojr.com/journalrank.php (accessed on 12 July 2022).
124. WoS. Web of Science Database. 2022. Available online: https://www-webofscience-com.ezproxy.lancs.ac.uk/wos/woscc/basic-search (accessed on 12 July 2022).
125. Scopus. SCOPUS Database. Elsevier B.V. 2022. Available online: https://www-scopus-com.ezproxy.lancs.ac.uk/search/form.uri?display=basic#basic (accessed on 12 July 2022).
126. Clarivate. Journal Citation Reports. 2022. Available online: https://jcr.clarivate.com/jcr/home (accessed on 12 July 2022).
127. Clarivate. Journal Citation Reports™ (JCR) Infographics: Make Better Informed, More Confident Decisions. 2022. Available online: https://clarivate.com/webofsciencegroup/web-of-science-journal-citation-reports-2021-infographic/ (accessed on 12 July 2022).
128. AlRyalat SA, S.; Malkawi, L.W.; Momani, S.M. Comparing Bibliometric Analysis Using PubMed, Scopus, and Web of Science Databases. *J. Vis. Exp.* **2019**, *152*, e58494. [CrossRef]

129. Zhang, Y.; Liu, H.; Kang, S.C.; Al-Hussein, M. Virtual reality applications for the built environment: Research trends and opportunities. *Autom. Constr.* **2020**, *118*, 103311. [CrossRef]
130. Kim, M.J.; Wang, X.; Love, P.E.D.; Li, H.; Kang, S.C. Virtual reality for the built environment: A critical review of recent advances. *J. Inf. Technol. Constr.* **2013**, *18*, 279–305. Available online: https://www.itcon.org/2013/14 (accessed on 22 July 2022).
131. Hirsch, J.E. An Index to Quantify an Individual's Scientific Research Output. *Proc. Natl. Acad. Sci. USA* **2005**, *102*, 16569–16572. [CrossRef]
132. Kamdem, J.P.; Duarte, A.E.; Lima, K.R.R.; Rocha, J.B.T.; Hassan, W.; Barros, L.M.; Roeder, T.; Tsopmo, A. Research Trends in Food Chemistry: A Bibliometric Review of its 40 Years Anniversary (1976–2016). *Food Chem.* **2019**, *294*, 448–457. [CrossRef] [PubMed]
133. Olawumi, T.O.; Chan, D.W. A scientometric review of global research on sustainability and sustainable development. *J. Clean. Prod.* **2018**, *183*, 231–250. [CrossRef]
134. Zheng, C.; Yuan, J.; Zhu, L.; Zhang, Y.; Shao, Q. From digital to sustainable: A scientometric review of smart city literature between 1990 and 2019. *J. Clean. Prod.* **2020**, *258*, 120689. [CrossRef]
135. Montoya, F.G.; Montoya, M.G.; Gómez, J.; Manzano-Agugliaro, F.; Alameda-Hernández, E. The research on energy in Spain: A scientometric approach. *Renew. Sustain. Energy Rev.* **2014**, *29*, 173–183. [CrossRef]
136. Pollack, J.; Adler, D. Emergent trends and passing fads in project management research: A scientometric analysis of changes in the field. *Int. J. Proj. Manag.* **2015**, *33*, 236–248. [CrossRef]
137. Chen, K.; Wang, J.; Yu, B.; Wu, H.; Zhang, J. Critical evaluation of construction and demolition waste and associated environmental impacts: A scientometric analysis. *J. Clean. Prod.* **2021**, *287*, 125071. [CrossRef]
138. Yin, X.; Liu, H.; Chen, Y.; Al-Hussein, M. Building information modelling for off-site construction: Review and future directions. *Autom. Constr.* **2019**, *101*, 72–91. [CrossRef]
139. Chen, D.; Bi, B.; Luo, Z.H.; Yang, Y.W.; Webber, M.; Finlayson, B. A scientometric review of water research on the Yangtze River. *Appl. Ecol. Environ. Res.* **2018**, *16*, 7969–7987. Available online: https://www.aloki.hu/pdf/1606_79697987.pdf (accessed on 12 July 2022). [CrossRef]
140. Tariq, S.; Hu, Z.; Zayed, T. Micro-electromechanical systems-based technologies for leak detection and localization in water supply networks: A bibliometric and systematic review. *J. Clean. Prod.* **2021**, *289*, 125751. [CrossRef]
141. Fang, J.; Pan, L.; Gu, Q.X.; Juengpanich, S.; Zheng, J.H.; Tong, C.H.; Wang, Z.Y.; Nan, J.J.; Wang, Y.F. Scientometric analysis of mTOR signaling pathway in liver disease. *Ann Transl. Med.* **2020**, *8*, 93. [CrossRef] [PubMed]
142. Oladinrin, O.; Gomis, K.; Jayantha, W.M.; Obi, L.; Rana, M.Q. Scientometric Analysis of Global Scientific Literature on Aging in Place. *Int. J. Environ. Res. Public Health* **2021**, *18*, 12468. [CrossRef]
143. Wininger, A.E.; Fischer, J.P.; Likine, E.F.; Gudeman, A.S.; Brinker, A.R.; Ryu, J.; Maupin, K.A.; Lunsford, S.; Whipple, E.C.; Loder, R.T.; et al. Bibliometric Analysis of Female Authorship Trends and Collaboration Dynamics Over JBMR's 30-Year History. *J. Bone Miner Res.* **2017**, *32*, 2405–2414. [CrossRef]
144. Palmblad, M.; van Eck, N.J. Bibliometric Analyses Reveal Patterns of Collaboration between ASMS Members. *J. Am. Soc. Mass Spectrom.* **2018**, *29*, 447–454. [CrossRef] [PubMed]
145. Yang, L.; Harrysson, O.; West, H.; Cormier, D. A Comparison of Bending Properties for Cellular Core Sandwich Panels. *Mater. Sci. Appl.* **2013**, *4*, 471–477. [CrossRef]
146. Ladani, L.; Romano, J.; Brindley, W.; Burlatsky, S. Effective liquid conductivity for improved simulation of thermal transport in laser beam melting powder bed technology. *Addit. Manuf.* **2017**, *14*, 13–23. [CrossRef]
147. Galati, M.; Iuliano, L.; Salmi, A.; Atzeni, E. Modelling energy source and powder properties for the development of a thermal FE model of the EBM additive manufacturing process. *Addit. Manuf.* **2017**, *14*, 49–59. [CrossRef]
148. Hwang, T.; Woo, Y.Y.; Han, S.W.; Moon, Y.H. Functionally graded properties in directed-energy-deposition titanium parts. *Opt. Laser Technol.* **2018**, *105*, 80–88. [CrossRef]
149. Bletzinger, K.U.; Ramm, E. Structural optimization and form-finding of lightweight structures. *Comput. Struct.* **2001**, *79*, 2053–2062. [CrossRef]
150. Mostafa, K.G.; Momesso, G.A.; Li, X.; Nobes, D.S.; Qureshi, A.J. Dual Graded Lattice Structures: Generation Framework and Mechanical Properties Characterization. *Polymers* **2021**, *13*, 1528. [CrossRef]
151. Mustafa, S.S.; Lazoglu, I. A new model and direct slicer for lattice structures. *Struct. Multidisc. Optim.* **2021**, *63*, 2211–2230. [CrossRef]
152. Guerra Silva, R.; Torres, M.J.; Zahr Viñuela, J.; Zamora, A.G. Manufacturing and Characterization of 3D Miniature Polymer Lattice Structures Using Fused Filament Fabrication. *Polymers* **2021**, *13*, 635. [CrossRef]
153. Al-Ketan, O.; Lee, D.; Al-Rub, R.K.A. Mechanical properties of additively-manufactured sheet-based gyroidal stochastic cellular materials. *Addit. Manuf.* **2021**, *48*, 102418. [CrossRef]
154. Maskery, I.; Parry, L.A.; Padrao, D.; Hague, R.J.M.; Ashcroft, I.A. FLatt Pack: A research-focussed lattice design program. *Addit. Manuf.* **2022**, *49*, 102510. [CrossRef]
155. Saremian, R.; Badrossamay, M.; Foroozmehr, E.; Kadkhodaei, M.; Foroghi, F. Experimental and numerical investigation on lattice structures fabricated by selective laser melting process under quasi-static and dynamic loadings. *Int. J. Adv. Manuf. Technol.* **2021**, *112*, 2815–2836. [CrossRef]
156. Habib, F.N.; Iovenitti, P.; Masood, S.H.; Nikzad, M. Fabrication of polymeric lattice structures for optimum energy absorption using Multi Jet Fusion technology. *Mater. Des.* **2018**, *155*, 86–98. [CrossRef]

157. Woodward, I.R.; Fromen, C.A. Scalable, process-oriented beam lattices: Generation, characterization, and compensation for open cellular structures. *Addit. Manuf.* **2021**, *48*, 102386. [CrossRef] [PubMed]
158. Hanks, B.; Frecker, M. 3D Additive Lattice Topology Optimization: A Unit Cell Design Approach. In Proceedings of the ASME 2020 International Design Engineering Technical Conferences and Computers and Information in Engineering Conference, 46th Design Automation Conference (DAC), Virtual, Online, 17–19 August 2020; Volume 11A. [CrossRef]
159. Hanks, B.; Frecker, M. Lattice Structure Design for Additive Manufacturing: Unit Cell Topology Optimization. In Proceedings of the ASME 2019 International Design Engineering Technical Conferences and Computers and Information in Engineering Conference, 45th Design Automation Conference, Anaheim, CA, USA, 18–21 August 2019; Volume 2A. [CrossRef]
160. Sienkiewicz, J.; Płatek, P.; Jiang, F.; Sun, X.; Rusinek, A. Investigations on the Mechanical Response of Gradient Lattice Structures Manufactured via SLM. *Metals* **2020**, *10*, 213. [CrossRef]
161. Abusabir, A.; Khan, M.A.; Asif, M.; Khan, K.A. Effect of Architected Structural Members on the Viscoelastic Response of 3D Printed Simple Cubic Lattice Structures. *Polymers* **2022**, *14*, 618. [CrossRef]
162. McConaha, M.; Anand, S. Design of Stochastic Lattice Structures for Additive Manufacturing. In Proceedings of the ASME 2020 15th International Manufacturing Science and Engineering Conference. Volume 1: Additive Manufacturing, Advanced Materials Manufacturing; Biomanufacturing, Life Cycle Engineering; Manufacturing Equipment and Automation, Virtual, Online, 3 September 2020. [CrossRef]
163. Al-Ketan, O.; Al-Rub, R.K.A. MSLattice: A free software for generating uniform and graded lattices based on triply periodic minimal surfaces. *Mater. Des. Processing Commun. (MDPC)* **2020**, *3*, e205. [CrossRef]
164. Riva, L.; Ginestra, P.S.; Ceretti, E. Mechanical characterization and properties of laser-based powder bed–fused lattice structures: A review. *Int. J. Adv. Manuf. Technol.* **2021**, *113*, 649–671. [CrossRef]
165. Nazir, A.; Abate, K.M.; Kumar, A.; Jeng, J.Y. A state-of-the-art review on types, design, optimization, and additive manufacturing of cellular structures. *Int. J. Adv. Manuf. Technol.* **2019**, *104*, 3489–3510. [CrossRef]
166. Kumar, A.; Collini, L.; Daurel, A.; Jeng, J. Design and additive manufacturing of closed cells from supportless lattice structure. *Addit. Manuf.* **2020**, *33*, 101168. [CrossRef]
167. Jamshidinia, M.; Kong, F.; Kovacevic, R. The Numerical Modeling of Fatigue Properties of a Bio-Compatible Dental Implant Produced by Electron Beam Melting® (EBM). Proceedings of the Conference: Twenty Forth Annual International Solid Freeform Fabrication Symposium. 2013. Available online: https://www.researchgate.net/publication/270822365_The_numerical_modeling_of_fatigue_properties_of_a_bio-compatible_dental_implant_produced_by_Electron_Beam_MeltingR_EBM (accessed on 12 July 2022).
168. Bacciaglia, A.; Ceruti, A.; Liverani, A. Proposal of a standard for 2D representation of bio-inspired lightweight lattice structures in drawings. *Proc. Inst. Mech. Eng. Part C J. Mech. Eng. Sci.* **2020**, 1–12. [CrossRef]
169. ISO/ASTM. International standard ISO/ASTM 52900 additive manufacturing—General principles—Terminology. *Int. Organ. Stand.* **2015**, *5*, 1–26. [CrossRef]
170. Rendeki, S.; Nagy, B.; Bene, M.; Pentek, A.; Toth, L.; Szanto, Z.; Told, R.; Maroti, P. An Overview on Personal Protective Equipment (PPE) Fabricated with Additive Manufacturing Technologies in the Era of COVID-19 Pandemic. *Polymers* **2020**, *12*, 2703. [CrossRef] [PubMed]
171. Tarfaoui, M.; Nachtane, M.; Goda, I.; Qureshi, Y.; Benyahia, H. Additive manufacturing in fighting against novel coronavirus COVID-19. *Int. J. Adv. Manuf. Technol.* **2020**, *110*, 2913–2927. [CrossRef] [PubMed]
172. Radfar, P.; Bazaz, S.R.; Mirakhorli, F.; Warkiani, M.E. The role of 3D printing in the fight against COVID-19 outbreak. *J. 3D Print. Med.* **2021**, *5*, 51–60. [CrossRef]
173. Agarwal, R. The personal protective equipment fabricated via 3D printing technology during COVID-19. *Ann. 3D Print. Med.* **2022**, *5*, 100042. [CrossRef]
174. Equbal, A.; Akhter, S.; Sood, A.K.; Equbal, I. The usefulness of additive manufacturing (AM) in COVID-19. *Ann. 3D Print. Med.* **2021**, *2*, 100013. [CrossRef]
175. Longhitano, G.A.; Nunes, G.B.; Candido, G.; da Silva, J.V.L. The role of 3D printing during COVID-19 pandemic: A review. *Prog. Addit. Manuf.* **2021**, *6*, 19–37. [CrossRef]
176. Zuniga, J.M.; Cortes, A. The role of additive manufacturing and antimicrobial polymers in the COVID-19 pandemic. *Expert Rev. Med. Devices* **2020**, *17*, 477–481. [CrossRef] [PubMed]
177. Kunkel, M.E.; Vasques, M.T.; Perfeito, J.A.J.; Zambrana, N.R.M.; Bina, T.D.S.; Passoni, L.H.D.M.; Ribeiro, T.V.; Rodrigues, S.M.S.; Castro, R.O.M.D.; Ota, N.H. Mass-production and distribution of medical face shields using additive manufacturing and injection molding process for healthcare system support during COVID-19 pandemic in Brazil. *Res. Sq.* **2020**. preprint. [CrossRef]
178. Swennen, G.R.J.; Pottel, L.; Haers, P.E. Custom-made 3D-printed face masks in case of pandemic crisis situations with a lack of commercially available FFP2/3 masks. *Int. J. Oral. Maxillofac. Surg.* **2020**, *49*, 673–677. [CrossRef]
179. Erickson, M.M.; Richardson, E.S.; Hernandez, N.M.; Bobbert II, D.W.; Gall, K.; Fearis, P. Helmet modification to PPE with 3D printing during the COVID-19 pandemic at Duke University Medical Center: A novel technique. *J. Arthroplast.* **2020**, *35*, S23–S27. [CrossRef] [PubMed]
180. Neijhoft, J.; Viertmann, T.; Meier, S.; Söhling, N.; Wicker, S.; Henrich, D.; Marzi, I. Manufacturing and supply of face shields in hospital operation in case of unclear and confirmed COVID-19 infection status of patients. *Eur. J. Trauma Emerg. Surg.* **2020**, *46*, 743–745. [CrossRef] [PubMed]

181. Belhouideg, S. Impact of 3D printed medical equipment on the management of the Covid19 pandemic. *Int. J. Health Plann. Manag.* **2020**, *35*, 1014–1022. [CrossRef]
182. Colorado, H.A.; Mendoza, D.E.; Lin, H.; Gutierrez-Velasquez, E. Additive manufacturing against the Covid-19 pandemic: A technological model for the adaptability and networking. *J. Mater. Res. Technol.* **2022**, *16*, 1150–1164. [CrossRef]
183. Amin, D.; Nguyen, N.; Roser, S.M.; Abramowicz, S. 3D printing of face shields during COVID-19 pandemic: A technical note. *J. Oral. Maxillofac. Surg.* **2020**, *78*, 1275–1278. [CrossRef] [PubMed]
184. Helman, S.N.; Soriano, R.M.; Tomov, M.L.; Serpooshan, V.; Levy, J.M.; Pradilla, G.; Solares, C.A. Ventilated upper airway endoscopic endonasal procedure mask: Surgical safety in the COVID-19 era. *Oper. Neurosurg.* **2020**, *19*, 271–280. [CrossRef]
185. Kalyaev, V.; Salimon, A.I.; Korsunsky, A.M. Fast mass-production of medical safety shields under COVID-19 quarantine: Optimizing the use of university fabrication facilities and volunteer labor. *Int. J. Environ. Res. Public Health* **2020**, *17*, 3418. [CrossRef]
186. Tino, R.; Moore, R.; Antoline, S.; Ravi, P.; Wake, N.; Ionita, C.N.; Morris, J.M.; Decker, S.J.; Sheikh, A.; Rybicki, F.J.; et al. COVID-19 and the role of 3D printing in medicine. *3D Print. Med.* **2020**, *6*, 1–8. [CrossRef]
187. Das, H.; Patowary, A. Uses of 3D Printing for Production of Ppe for Covid 19 like situations: Scope and future. *Am. J. Prev. Med. Public Health* **2020**, *6*, 76. [CrossRef]
188. Larrañeta, E.; Dominguez-Robles, J.; Lamprou, D.A. Additive Manufacturing can assist in the fight against COVID-19 and other pandemics and impact on the global supply chain. *3D Print. Addit. Manuf.* **2020**, *7*, 100–103. [CrossRef]
189. Novak, J.I.; Loy, J. A quantitative analysis of 3D printed face shields and masks during COVID-19. *Emerald Open Res.* **2020**, *2*, 42. [CrossRef]
190. Sinha, M.S.; Bourgeois, F.T.; Sorger, P.K. Personal protective equipment for COVID-19: Distributed fabrication and additive manufacturing. *Am. J. Public Health* **2020**, *110*, 1162–1164. [CrossRef] [PubMed]

Article

Hydroxyapatite from Natural Sources for Medical Applications

Laura Madalina Cursaru [1,*], Miruna Iota [1], Roxana Mioara Piticescu [1,*], Daniela Tarnita [2], Sorin Vasile Savu [3], Ionel Dănuț Savu [3], Gabriela Dumitrescu [4], Diana Popescu [4], Radu-Gabriel Hertzog [4] and Mihaela Calin [1,5]

1. National R&D Institute for Non-Ferrous and Rare Metals, INCDMNR-IMNR, 102 Biruintei Blvd, 077145 Pantelimon, Romania; iota.miruna@imnr.ro (M.I.); micalin@inoe.inoe.ro (M.C.)
2. Department of Applied Mechanics, Faculty of Mechanics, University of Craiova, 200585 Craiova, Romania; daniela.tarnita@edu.ucv.ro
3. Department of Engineering and Management of Technological Systems, Faculty of Mechanics, University of Craiova, 200585 Craiova, Romania; sorin.savu@edu.ucv.ro (S.V.S.); ionel.savu@edu.ucv.ro (I.D.S.)
4. "Cantacuzino" National Military Medical Institute for Research and Development, Splaiul Independenței nr. 103, Sector 5, 050096 Bucharest, Romania; dumitrescu.gabriela@cantacuzino.ro (G.D.); popescu.diana@cantacuzino.ro (D.P.); raduhg@yahoo.co.uk (R.-G.H.)
5. National Institute of Research and Development for Optoelectronics INOE 2000, 409 Atomistilor Street, 077125 Magurele, Romania
* Correspondence: mpopescu@imnr.ro (L.M.C.); roxana.piticescu@imnr.ro (R.M.P.); Tel.: +40-21-352-2046 (L.M.C. & R.M.P.)

Abstract: The aim of this work is to study the physical-chemical, mechanical, and biocompatible properties of hydroxyapatite obtained by hydrothermal synthesis, at relatively low temperatures and high pressures, starting from natural sources (Rapana whelk shells), knowing that these properties influence the behavior of nanostructured materials in cells or tissues. Thus, hydroxyapatite nanopowders were characterized by chemical analysis, Fourier-transform infrared spectroscopy (FT-IR), dynamic light scattering (DLS), scanning electron microscopy (SEM), and X-ray diffraction (XRD). In vitro studies on osteoblast cell lines (cytotoxicity and cell proliferation), as well as preliminary mechanical tests, have been performed. The results showed that the obtained powders have a crystallite size below 50 nm and particle size less than 100 nm, demonstrating that hydrothermal synthesis led to hydroxyapatite nanocrystalline powders, with a Ca:P ratio close to the stoichiometric ratio and a controlled morphology (spherical particle aggregates). The tensile strength of HAp samples sintered at 1100 °C/90 min varies between 37.6–39.1 N/mm^2. HAp samples sintered at 1300 °C/120 min provide better results for the investigated mechanical properties. The coefficient of friction has an appropriate value for biomechanical applications. The results of cell viability showed that the cytotoxic effect is low for all tested samples. Better cell proliferation is observed for osteoblasts grown on square samples.

Keywords: hydroxyapatite; nano-crystalline powders; hydrothermal synthesis; mechanical properties; cell viability; cell proliferation

Citation: Cursaru, L.M.; Iota, M.; Piticescu, R.M.; Tarnita, D.; Savu, S.V.; Savu, I.D.; Dumitrescu, G.; Popescu, D.; Hertzog, R.-G.; Calin, M. Hydroxyapatite from Natural Sources for Medical Applications. *Materials* 2022, *15*, 5091. https://doi.org/10.3390/ma15155091

Academic Editor: Irina Hussainova

Received: 30 June 2022
Accepted: 19 July 2022
Published: 22 July 2022

Publisher's Note: MDPI stays neutral with regard to jurisdictional claims in published maps and institutional affiliations.

Copyright: © 2022 by the authors. Licensee MDPI, Basel, Switzerland. This article is an open access article distributed under the terms and conditions of the Creative Commons Attribution (CC BY) license (https://creativecommons.org/licenses/by/4.0/).

1. Introduction

Hydroxyapatite (HAp) is a well-known calcium phosphate material, chemically identical to the mineral phase of the bone and the hard tissues of mammals. The most interesting property of this ceramic material is its ability to interact with living bone tissue, forming strong bonds with the bone, without causing toxicity or inflammatory response. It is commonly used for orthopedic, dental, and maxillofacial applications, either as a coating material for metal implants or as a bone filler. However, the material has some disadvantages. HAp is not thermally stable, with dehydroxylation starting at 800–1200 °C, depending on its stoichiometry [1,2]. It has poor mechanical properties (especially low fatigue strength), which means that it cannot be used in compact form for applications where the implant is subjected to heavy mechanical stresses (e.g., hip joint). The mechanical properties of HAp depend on porosity, density, sinterability, crystal size, and phase composition [3]. Nanoscale hydroxyapatite crystals show better mechanical properties and greater bioactivity than micron-sized crystals [4,5].

Hydroxyapatite in various forms, such as powder, porous blocks, or pearls, can be used to fill bone defects and free spaces in the bone [6–9]; these occur when parts of the bone have been removed due to a disease (bone cancer), or when bone extensions are needed (in the case of dental applications). The bone filling will form a skeleton and will facilitate the rapid filling of the pores by the growing natural bone tissue [2]. Hydroxyapatite as a filler is an alternative to autologous bone grafts, becoming part of the bone structure and reducing the time required to heal diseased tissue [10].

Recently, hydroxyapatite has been studied for other applications such as drug-delivery [11–13], collagen stimulation [14], skin regeneration [15,16], or sun protection in the pharmaceutical industry [15], as well as water purification [17–20], wastewater treatment [21–24], and in other chemical, optical, and electronics industries [1].

Therefore, various methods for synthesizing Hap, with tailored properties, have been investigated. These can be classified as dry methods (solid-state and mechanochemical), wet methods (chemical precipitation, hydrolysis, sol-gel, hydrothermal, emulsion, polymer-assisted routing, synthesis via biological tissue, ultrasonic spray freeze-drying, microwave irradiation, and sonochemical procedures), and high temperature processes (combustion and pyrolysis) [25–30].

Although many synthesis methods have been developed, the preparation of HAp with specific characteristics remains challenging because of the possibility of the formation of toxic intermediary products or impurities during the synthesis of HAp [5]. Thus, studies on new synthesis parameters of HAp are still in progress [27].

In recent years, many researchers combined the synthesis methods of HAp with the sustainable use of $CaCO_3$ natural resources, namely the processing of marine (seashells, fish bone, corals, algae) and agricultural wastes (eggshells, animal bones) for preparing calcium phosphates [31–34].

Compared to synthetic HAp, natural HAp is non-stoichiometric, containing traces of Na^+, Zn^{2+}, Mg^{2+}, K^+, Si^{2+}, Ba^{2+}, F^-, and $(CO_3)^{2-}$, which resembles the chemical composition of human bone [35,36].

In the present study, our goal is to study the physical-chemical, mechanical, and biocompatible properties of hydroxyapatite prepared by the hydrothermal method in different pressure conditions, starting from Rapana whelk shells from the Black Sea coast, knowing that these properties influence the behavior of nanostructured materials in cells or tissues. As a novelty, the influence of synthesis pressure on the physical-chemical properties of HAp is studied, aiming to prepare highly crystalline HAp with improved mechanical and biocompatible properties.

2. Materials and Methods

2.1. Materials

Rapana Thomasiana shells were collected from the Romanian Black Sea coast, cleaned of sand, and washed with water and detergent to remove algae and traces of visceral mass inside. The commercial materials used were $NH_4H_2PO_4$ p.a. (Lach-Ner, s.r.o., Neratovice, Czech Republic), HCl 37% p.a. (Cristal R Chim SRL, Bucharest, Romania), HNO_3 68% p.a. (Cristal R Chim SRL, Bucharest, Romania) and NH_3 25% p.a. (Cristal R Chim SRL, Bucharest, Romania).

2.2. Hydrothermal Synthesis

Prior to hydrothermal synthesis, Rapana Thomasiana shells were mechanically crushed, ground in a Retsch Vibratory Disc Mill RS 200 (Retsch GmbH, Haan, Germany), and dissolved in a mixture of $HCl:HNO_3$ = 2:1 (60 mL HCl 37% and 30 mL HNO_3 68%), resulting in a solution of 30–40% calcium (Ca precursor of HAp). Afterwards, this solution was magnetically mixed with $NH_4H_2PO_4$ as the phosphorus precursor, precipitated with NH_3 solution 25% until alkaline pH 10, and subjected to hydrothermal synthesis in a Teflon vessel placed in a closed system (Berghof reactor, Berghof Products + Instruments, GmbH, Eningen unter Achalm, Germany). The hydrothermal process was conducted at temperatures between 100–150 °C and various pressures (2, 6, and 10 MPa, respectively), followed by drying in a Memmert oven UFE 400 (MEMMERT GmbH + Co. KG, Schwabach, Germany) at 100 °C for 24 h. The pressure was created by bubbling Ar 5.0 gas (99.999% purity) over the aqueous solution in the Teflon vessel.

2.3. Characterization of Hydroxyapatite Powder

The characterization of the nanostructured HAp powder was performed by the following methods: flame atomic absorption spectrometry (FAAS), for the determination of Ca content; inductively coupled plasma optical emission spectrometry (ICP-OES), to determine P content; Fourier transform infrared spectroscopy (FT-IR), to highlight the vibrational modes; X-ray powder diffraction (XRD), for phase analysis and crystallite size determination; Dynamic light scattering (DLS), for particle size distribution in suspension; BET specific surface area determination; scanning electron microscopy (SEM), coupled with energy dispersive X-ray spectroscopy (EDX), for morphology analysis; and differential scanning calorimetry (DSC), coupled with thermogravimetry (TGA), for thermal stability.

2.3.1. Chemical Analysis

FAAS was performed using an Analytik Jena ZEEnit 700 P AAS Atomic Absorption Spectrometer (Jena, Germany). For ICP-OES analysis, an Agilent 725 ICP-OES system (Agilent Technologies, Santa Clara, CA, USA) was used.

2.3.2. Structural Analysis

The presence of functional groups characteristic of hydroxyapatite was identified by FT-IR, using an ABB MB 3000 FT-IR spectrometer (ABB Inc., Québec, QC, Canada), equipped with the EasiDiff device (PIKE Technologies, Inc., Madison, WI, USA) for working with powders. Measurements were conducted in the transmission mode, from 4000 to 550 cm^{-1}, with a scan resolution of 4 cm^{-1}. Experimental data were processed using the Horizon MBTM FTIR software version 3.4.0.3 (ABB Inc., Québec, QC, Canada).

In the case of X-ray diffraction, the data acquisition was performed on the BRUKER D8 ADVANCE diffractometer (Bruker AXS GmbH, Karlsruhe, Germany) using the DIFFRAC plus XRD Commander software, version 5.1.0.5 (32 Bit) Bruker AXS 2010-2019 (Bruker AXS GmbH, Karlsruhe, Germany), according to the Bragg-Brentano diffraction method, θ-θ coupling in vertical configuration, at a voltage of 40 kV and current of 40 mA, in the range $2\theta = 4 \div 74°$, and 2θ step of 0.03°. The phase identification was completed with the help of the DIFFRAC.EVA release 2019 program (Bruker AXS GmbH, Karlsruhe, Germany) from the DIFFRAC.SUITE.EVA software package and the ICDD PDF4 + 2022 database.

2.3.3. Particle Size Distribution

Particle size distribution was measured using a Zetasizer Nano ZS 90 particle analyzer, Malvern Instruments (Worcestershire, UK), in size range of 0.3–5.0 µm, temperature range of 2–90 °C, and endowed with Zetasizer software v8.01 (PSS0012-42, Malvern Instruments Ltd., Malvern Panalytical Ltd., Worcestershire, UK).

Sample preparation: a stable suspension was prepared by magnetic stirring and the sonication of nanostructured HAp powder with double distilled water, ethanol, and commercial dispersant Duramax^{TM} D3005 (Trademark of The Dow Chemical Company, Midland, MI, USA). The obtained suspension was filtered through a Millipore membrane (d = 0.22 µm) and transferred to the glass cuvette for particle size distribution measurement.

2.3.4. BET Specific Surface Area Measurements

The method used to determine specific surface area, pore volume or porosity, and pore shape and size is based on the physisorption of N_2 gas at 77 K (-196 °C), with an adsorption-desorption isotherm. Measurements were performed using a Micromeritics TriStar II Plus analyzer (Micromeritics Instrument Corporation, Norcross, GA, USA). The specific surface area was obtained by the Brunauer–Emmett–Teller (BET) method, while the pore volume and pore size distribution were determined by the Barrett–Joyner–Halenda (BJH) method. Prior to each determination, the powder samples were subjected to a heat treatment at 300 °C for several hours to remove traces of liquids and impurities using the VacPrep 061 degassing stations.

2.3.5. Morphological Analysis

To examine and correctly establish the morphology and size of the hydroxyapatite crystals obtained, the hydroxyapatite samples were studied in High Vacuum mode using a FEI Quanta 250 scanning electron microscope (FEI Company, Eindhoven, The Netherlands). The analyzed samples were metallized by coating with a 5 nm thick Au layer.

2.4. Hydroxyapatite Pellets Preparation

HAp nanopowders prepared by hydrothermal synthesis at 10 MPa were mechanically mixed with polyvinyl alcohol (PVA) 5% solution, dried in an oven at 100 °C for 24 h, and then uniaxially compacted at a pressure of 98 MPa into cylindrical pellets with a 16 mm diameter and a 12 mm height for mechanical evaluation, respectively, and round disks with a diameter of 9 mm and a height of 1.6 mm for in vitro testing. These pellets were further sintered in air atmosphere at different temperatures and sintering times (1100 °C/90 min; 1200 °C/90 min, 120 min, 180 min; 1300 °C/90 min, 120 min, 180 min; and 1400 °C/90 min, 120 min, 180 min, respectively) in an electric furnace at a heating and cooling rate of 1 °C/min and allowed to furnace cool. Sintering conditions were chosen based on literature data presented in [37,38]. HAp sintered specimens are presented in Table 1.

Table 1. Sintering conditions and type of sintered for HAp specimens.

Sintering Temperature, °C	Sintering Time, Min	Sample Type	Sample Codes	Sample Destination (Mechanical Test/In Vitro Test)
1100	90	Cylinder	5.2/5.3/5.5/5.6/5.7	Mechanical
1200	90	Cylinder	PE20, PE21, PE22	Mechanical
1200	120	Cylinder	PE17, PE18, PE19	Mechanical
1200	180	Cylinder	PE14, PE15, PE16	Mechanical
1200	180	Round disk	PE30, PE31, PE32, PE33, PE34, PE35	In vitro
1300	90	Cylinder	PE4, PE5, PE6	Mechanical
1300	120	Cylinder	PE1, PE2, PE3	Mechanical
1300	180	Cylinder	PE7, PE8, PE9	Mechanical
1400	90	Cylinder	PE10, PE11, PE12	Mechanical
1400	120	Cylinder	PE23, PE24, PE25	Mechanical
1400	180	Cylinder	PE13, PE27, PE29	Mechanical

2.5. Mechanical Properties Evaluation

To investigate the mechanical properties of the sintered specimens, all tests (compressive strength evaluation, micro-hardness measurement, and wear test) have been conducted in accordance with the international standards in force [39,40], but also based on our previous experience [41,42].

2.5.1. Compressive Strength Evaluation

The universal test machine LBG 100 (Maximum force: 100 kN) was used for the compression testing (performed according to our own procedure, based on EN 658-2:2002).

2.5.2. Micro-Hardness Measurement

The micro hardness (HV1) of the sintered specimens was determined according to ISO 14705:2000, via the Vickers indentation, with a NAMICON CV-400DM Microdurimeter produced by CV Instruments Europe BV (range: HV0,01-HV1, load 10–1000 g, resolution 0.03 µm, Vickers diamond indenter, 10×, 40× objectives, microscope with analog reading, automatic load force control, video image control). A total of 3 indentations were made and the resulting hardness values were averaged.

2.5.3. Wear Test

A CSM Instruments tribometer, with a maximum torque of 450 N.mm and a maximum load of 46 N., was used for the wear tests. The software used for analysis and graphical representation was InstrumX. The usable frequency is 1.6 Hz at a speed of ball movement in the range of 0.3–500 mm/s, the frequency of information acquisition being 10 Hz. The tests were performed according to our own developed procedure, based on ISO 22622:2019, with harder conditions (100 Hz instead 10 Hz).

2.6. Biocompatibility Assessment

2.6.1. Preparation of Cell Lines

To perform the cytotoxicity test of the HAp samples, a cell line of normal human osteoblasts (NHOst, Cat.No. CC-2538, Lonza, Germany) was used. The osteoblast cell line was cultured in a 25 cm^2 cell culture flask using osteoblast-specific growth medium, supplemented with 10% fetal bovine serum (FBS) and 50 µg/mL gentamicin. Cells from two flasks of 25 cm^2 cells were trypsinized with 0.025% trypsin-EDTA, centrifuged, and the pellet was resuspended in 5 mL of growth medium completely specific to the cell line. Initial cell counting was performed by staining with Tripan Blue 1:1 (50 µL cell suspension + 50 µL Tripan Blue). The osteoblast cell suspension has a concentration of 1.3×10^6 cells/mL.

2.6.2. Preparation of the Test Compound

A total of 12 HAp samples were assessed for cytotoxicity and cellular proliferation: 6 square samples, with dimensions of $15 \times 15 \times 5$ mm^3, fabricated using the 3D printing technique as described in [43], dried in oven at 100 °C, and not sintered; and 6 round samples (disks), with a diameter of 9 mm and a height of 1.6 mm, sintered at 1200 °C/180 min, as shown in Table 1.

The 12 tested samples were placed in a 12-well cell culture plate, with growth medium completely specific to the osteoblast cell line (OGM^{TM} Osteoblast Growth medium BulletKit^{TM}, Lonza, Germany), and incubated for 24 h at 37 °C, in a shaking incubator, 5% CO_2. The weight/volume ratio was 200 mg/mL, according to the recommendations of ISO 10993-12. After incubation, the stock solution was diluted in binary dilutions (1/2, 1/4, and 1/8).

2.6.3. Cytotoxicity Test

To perform the cytotoxicity test, the osteoblast cell suspension is cultured in 2 microplates with 96 wells and a flat bottom, 200 µL/well (2.6×10^5 cells/well), and incubated in a CO_2 incubator (5%) at 37 °C for 24 h.

The next day, the environment is changed, as follows:

- Only 200 µL of complete growth medium is added to the blank wells.
- In the cell control wells, the medium is removed, and 200 µL of complete, fresh growth medium is added.

- The medium is removed from the wells with the test compound, and 200 μL of fresh medium containing various dilutions of the compound (1/2, 1/4, and 1/8) are added. The blank, the cell control, and the test samples are distributed in duplicate for each dilution.

After 24 h, the 3-4,5-(dimethyl-2-thiazolyl)-2,5-diphenyl-2H-tetrazolium bromide (MTT) test is performed to measure the conversion of MTT to a stained product in living cells. The MTT-based cell growth assay kit (Sigma-Aldrich, Darmstadt, Germany) containing the MTT solution (5 mg/mL MTT in RPMI-1640 without phenol red) and the MTT solvent (0.1 N HCl in anhydrous isopropanol) were used for this assay. The microplates are removed from the incubator, 20 μL MTT/well (10% of the medium volume) is added, and the microplates are incubated for 4 h at 37 °C, in dark, under CO_2. After 4 h, the microplates are removed from the incubator, the medium is discarded, and 200 μL of MTT solution/well is added. The optical density (OD) is read at a wavelength of 570 nm, within a maximum of 1 hour from the time the solvent was added, using a multimodal reader (EnSight™ Multimode Microplate Reader, PerkinElmer, Akron, OH, USA).

Cell viability is calculated with the following formula:

$$\% \text{ Cell viability} = \frac{\text{OD positive control} - \text{OD blank}}{\text{OD negative control} - \text{OD blank}} \times 100$$

where: positive control = cells + compound + MTT + MTT solution; negative control = cells + MTT + solvent MTT; and blank = complete growth medium + MTT + MTT solution.

The negative control (cell control) and the samples (positive control) were run in 2 wells for each concentration of the compounds and at the end, the arithmetic mean of the optical density readings was made, and read at a wavelength of 570 nm.

2.6.4. Cell Proliferation

To highlight cell proliferation, the 12 samples to be tested were placed in the center of the wells of a 12-well plate and incubated with 400 μL growth medium in an incubator at 37 °C overnight. In addition, 200 μL of osteoblast cell suspension were added, the samples being thus cultured in a 5% CO_2 atmosphere at 37 °C, allowing the cells to be attached to the samples. A modified MTT test was used [1].

Briefly, the osteoblast cell line was seeded at a density of 2.6×10^4/mL, and the cell culture was for 24, 48, and 72 h. The growth medium was replaced daily during testing.

A total of 60 μL MTT solution (5 mg/mL) was added to the wells, followed by an incubation period at 37 °C for 4 h for the formation of MTT formazan. The supernatant was removed by aspiration, and the MTT solvent (600 μL DMSO) was added to dissolve the formazan crystals. Within a maximum of one hour after the addition of the solvent, 200 μL of each well were then transferred to a 96-well plate (in duplicate) to read the optical density at a wavelength of 570 nm, using the multimodal reader (EnSight™ Multimode Microplate Reader, PerkinElmer, Hopkinton, Massachusetts, USA). For the resumption of samples in 96-well plates, the same sample arrangement schemes, with the 2 adhered cell lines but at different contact times (at 24, 48, and 72 h), were used. The interpretation of the results is made using the previously mentioned formula to determine the cytotoxicity. The obtained values reflect cell proliferation in each well, for each sample [2].

3. Results

3.1. Hydrothermal Synthesis

Hydroxyapatite nanopowders prepared from natural sources were synthesized under hydrothermal conditions at temperatures, between 100–150 °C, and different pressures (2, 6, and 10 MPa). The resulting powders were denoted HAP-20, HAP-60, and HAP-100, with the numbers in the sample code signifying the working pressure in bar units.

The chemical reactions that lead to hydroxyapatite, starting with Rapana Thomasiana shells, are written below:

$$CaCO_3 + 2HCl = CaCl_2 + H_2O + CO_2\uparrow \quad (1)$$

$$CaCO_3 + 2HNO_3 = Ca(NO_3)_2 + H_2O + CO_2\uparrow \quad (2)$$

$$10CaCl_2 + 6NH_4 \cdot H_2PO_4 + 14NH_4OH = Ca_{10}(PO_4)_6OH_2 + 20NH_4Cl + 12H_2O \quad (3)$$

$$10Ca(NO_3)_2 + 6NH_4 \cdot H_2PO_4 + 14NH_4OH = Ca_{10}(PO_4)_6OH_2 + 20NH_4NO_3 + 12H_2O \quad (4)$$

Equations (1) and (2) describe chemical reactions which take place during the dissolving of Rapana shells in the HCl-HNO$_3$ mixture. Hydrothermal reactions are represented by Equations (3) and (4). Calcium solution, consisting of CaCl$_2$ and Ca(NO$_3$)$_2$ aqueous species, is mixed with NH$_4$·H$_2$PO$_4$ as the P precursor of HAp and precipitated with NH$_3$ 25% solution. During hydrothermal synthesis at a high temperature and pressure, crystalline HAp nanoparticles are formed.

It is well known that crystalline hydroxyapatite can be obtained using the hydrothermal method in a relatively wide temperature range (from 70 °C to 200 °C) [4,44–46]. An important advantage of using the hydrothermal process for HAp synthesis, besides controlled morphology and nanometer particle size, is that no hydroxyl defects are produced in the structure [44].

Hydrothermal synthesis takes place in a perfectly sealed reaction system, in aqueous solution, at a high temperature and pressure. Usually, pressure is created by the saturated vapor phase which forms above the solution, and the most varied hydrothermal synthesis parameters, according to literature data, are temperature, time, and pH [44,46]. Although preliminary results on hydrothermal synthesis of HAp obtained from Rapana shells in high pressure conditions (10 MPa) were reported in our previous paper [43], the effect of pressure on the physical-chemical properties of HAp was not investigated.

Numerous scientific papers [47–50] have developed various models for thermodynamic calculation of physicochemical processes that take place in aqueous or non-aqueous solutions during hydrothermal synthesis. Depending on the results obtained, not only can the appropriate solvent be selected, but also the pressure-temperature range that leads to the formation of the desired reaction products and allows for the control of the shape and size of the obtained particles. The behavior of the solvent under hydrothermal conditions, in terms of its structure in critical, super-critical, and sub-critical conditions, as well as the dielectric constant, pH variation, viscosity, coefficient of expansion, and density, must all be correlated with the temperature and working pressure.

In the case of high-pressure hydrothermal synthesis, an external pressure higher than the water vapor pressure at equilibrium is used. Under these conditions, remarkable results are obtained because the solubility of inorganic materials increases with increasing pressure. Thermodynamic stability also varies with pressure (at very high pressures, denser phases crystallize). Hydrothermal reactions are based on the equilibrium reactions of dissolution-reprecipitation and crystallization.

In the studied system, the pressure above the aqueous suspension in the autoclave vessel corresponds to the vapor pressure of the substances in the reaction system, but also to the external pressure of the inert gas introduced. The temperature is kept constant at values much lower than the critical water temperature (T = 100–150 °C << 374 °C) so that the liquid–gas equilibria will be neglected. In this way, due to the high pressure in the reaction system, the crystallization of hydroxyapatite under hydrothermal conditions takes place at temperatures much lower than the usual values for obtaining crystalline calcium phosphates. The influence of the pressure on the equilibrium constant for real systems is determined by the variation of the molar volumes of the participants in the reaction (reactants and reaction products), denoted $\Delta^r V$. In the case of hydrothermal reactions $\Delta^r V < 0$, the value of the equilibrium constant increases with increasing pressure, meaning the equilibrium shifts to the reaction products (HAp formation reaction is favored by the introduction of external pressure). The introduction of external pressure in the hydrothermal synthesis autoclave favors the obtaining of nanostructured crystalline HAp at relatively low temperatures.

The chemical analysis results (Ca:P ratio) are shown in Table 2.

Table 2. Chemical analysis results.

Sample Name	Ca, Weight %	P, Weight %	Ca:P Ratio
HAP-20	38.4	17.1	1.74
HAP-60	40.3	17.0	1.84
HAP-100	39.5	17.4	1.76

The Ca:P ratio is higher than theoretical value of Ca:P = 1.67, which is calculated from the HAp chemical formula $Ca_{10}(PO_4)_6(OH)_2$, which could be explained by the formation of non-stoichiometric hydroxyapatite, because of the natural Ca source used in the synthesis.

3.2. Structural Analysis

3.2.1. FT-IR Analysis

Figure 1 shows the superposed FT-IR spectra of the hydroxyapatite nanopowders prepared at different pressures (2, 6, and 10 MPa, respectively). The characterization of HAp nanostructured powders by FT-IR highlighted the presence of the following vibration bands for all the investigated samples: (i) the stretching vibration of the OH group (sharp band) from 3570 cm^{-1}; (ii) the stretching vibration of the H_2O molecule (broad band) from 3250–3500 cm^{-1} [25]; (iii) the deformation vibration of the OH group (1641–1651 cm^{-1}), associated with the stretching vibrations from 3250–3500 cm^{-1}; and (iv) the stretching vibrations of the $(PO_4)^{3-}$ group from 1095–1097 cm^{-1}, 1032–1038 cm^{-1}, and 962 cm^{-1}, respectively, characteristic of hydroxyapatite [51]. The medium intensity bands in the range of 1420–1489 cm^{-1} are due to the $(CO_3)^{2-}$ group [52].

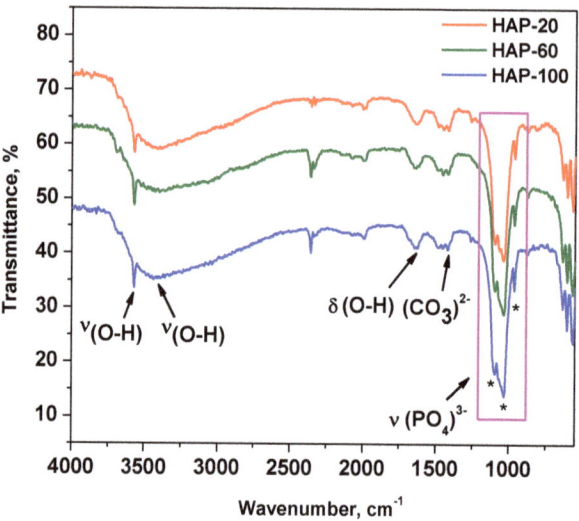

Figure 1. FT-IR spectra of HAp nanopowders.

Comparing the absorbance ratios A_{1097}/A_{1487} and A_{962}/A_{1421} of the peaks characteristic of $(PO_4)^{3-}$ (1097 and 962 cm^{-1}) to those characteristic of $(CO_3)^{2-}$ (1487 and 1421 cm^{-1}), at different pressures, provides information regarding the amount of hydroxyapatite in the nanopowder samples [53]. The values of the peak area ratios are displayed in Table 3.

Table 3. Values of absorbance ratios A_{1097}/A_{1487} and A_{962}/A_{1421} as obtained from the spectra of HAp nanopowders at different pressures.

Sample Name	A_{1097}/A_{1487}	A_{962}/A_{1421}
HAP-20	13.55	1
HAP-60	9	1
HAP-100	15.25	1.42

It can be observed that HAP-100 has the highest peak area ratios among the studied samples (15.25 and 1.42). These values represent the ratio between phosphate and carbonate. As a conclusion, nanopowder synthesized at 10 MPa has the highest content of HAp.

3.2.2. X-ray Diffraction Characterization

The XRD patterns of hydroxyapatite synthesized from natural sources are presented in Figure 2. For comparison, the XRD spectrum of Rapana Thomasiana shells is shown in Figure 2b. The main crystalline phases identified in these shells are calcite (chemical formula $CaCO_3$, PDF reference 01-083-3288), representing ~64.4% in weight, and aragonite (chemical formula $CaCO_3$, PDF reference 01-075-9982), representing ~35.6% in weight. The main crystalline phase identified in the powders obtained by the hydrothermal process is Hydroxylapatite, PDF reference 00-009-0432, with typical (h k l) Miller indices (002), (211), (300), (202), (310), (222), and (213). The (002) reflection peak from the XRD pattern was used to calculate the crystallite size of HAp nanopowders, using the Debye–Scherrer equation. The average particle size, measured using a BET analyzer, as well as the hydrodynamic diameter (d(H)), determined by the DLS method and the polydispersity index (PdI), are summarized in Table 4, along with the crystallite size.

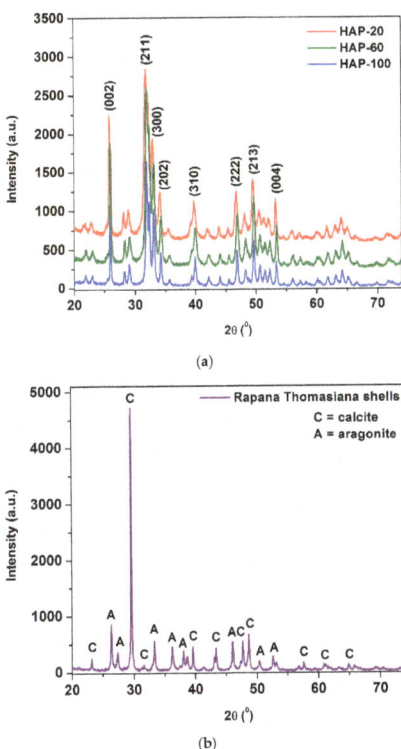

Figure 2. (a) XRD spectra of HAp nanopowders; (b) XRD spectrum of Rapana Thomasiana shells.

Table 4. Scherrer crystallite size, BET average particle size, and hydrodynamic diameter of HAP nanopowders.

Sample Name	Crystallite Size in (002) Direction, nm	CI, %	Average Particle Size (BET), nm	d (H), nm	PdI
HAP-20	28	71.1	18.7	76	0.009
HAP-60	33	77.8	21.7	84	0.087
HAP-100	38	79.4	37.3	97	0.085

The crystallinity index (CI), defined as the ratio between the total area of the narrow diffraction maxima, due to the crystalline phases, and the total diffracted area of the sample, after removing the background contribution [54,55], was determined for the three samples using the following equation:

$$\text{Crystallinity Index} = \frac{\text{Crystalline phase area}}{\text{Crystalline phase area} + \text{Amorphous phase area}}$$

It can be observed that the dimensions of HAp nanopowders increase with increasing pressure, regardless of the type and method used for determining them (Scherrer crystallite size, BET average particle size, hydrodynamic diameter). Pressure favors the crystallites growth during hydrothermal synthesis, leading to the formation of crystalline phases with a high degree of crystallinity. The crystallinity index also increases with pressure increase, from 71% to 79%. Based on these observations, correlated with peak area ratio calculated from FT-IR spectra, HAP-100 (sample synthesized at 10 MPa) was selected for further study of the mechanical properties.

3.3. Analysis of Particle Size Distribution by DLS Technique

Dynamic light scattering (DLS) is an established measurement technique for the characterization of particle sizes in suspension, based on the Brownian motion of particles. The smaller the particles, the faster they will move in a solution. The hydrodynamic diameter represents the particle size plus the dielectric layer, which adheres to its surface during movement through the liquid medium. The movement of the particles causes intensity fluctuations in the scattered light. From these fluctuations, the diffusion coefficient can be determined, and thus the hydrodynamic diameter of the particle is obtained from the Stokes–Einstein equation:

$$d(H) = \frac{kT}{3\pi\eta D}$$

where: d(H) = hydrodynamic diameter, k = Boltzmann's constant (1.38×10^{-23} NmK^{-1}), T = absolute temperature (K), η = solvent viscosity (N·s·m^{-2}), and D = diffusion coefficient (m^2·s^{-1}).

The results obtained for synthesized hydroxyapatite powders (hydrodynamic diameter and polydispersity index-PdI) are presented in Table 3 and Figure S1. The average particle size (hydrodynamic diameter in aqueous solutions) varies between 76 nm and 97 nm, with a monomodal size distribution. The low values of the polydispersity index suggest that the investigated samples are homogenous in size. The PdI is situated in the range of 0.009–0.087.

3.4. SEM Characterization

An SEM image of HAp nanopowder prepared at 10 MPa is presented in Figure 3.
All three analyzed samples are formed of irregularly shaped microcrystalline aggregates that have dimensions on the order of microns up to tens of microns. SEM images of HAP-20 and HAP-60 samples are presented in Figure S2. Microcrystalline aggregates are in turn formed of exceedingly small microcrystals (on the order of nanometers), whose shapes and sizes could not be highlighted. Moreover, morphological investigation revealed a porous structure of hydroxyapatite, regardless of the synthesis pressure. This porous

structure, with nano-sized pores (determined by BJH method), represents an advantage for medical applications (bone tissue reconstruction) [56]. Thus, the BJH adsorption average pore width is 2.75 nm for HAP-20, 2.74 nm for HAP-60, and 2.67 nm for HAP-100 powder samples.

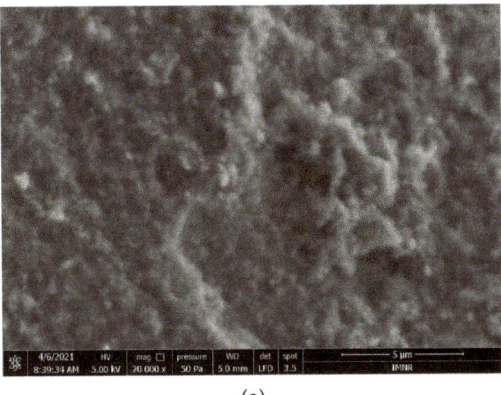
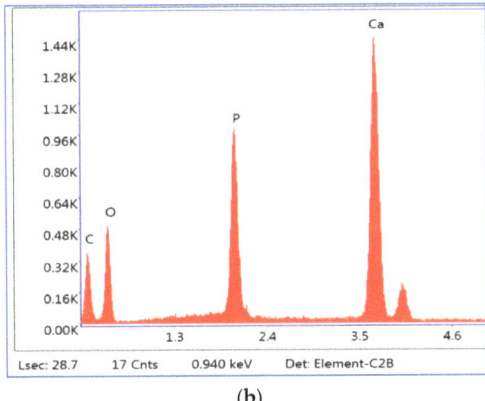

(a)   (b)

Figure 3. (a) SEM image at 5 μm scale bar, 5 kV voltage, and 20 kX magnification, and (b) EDS spectrum of HAP-100 nanopowder.

The EDS semiquantitative analysis results are presented in Table 5.

Table 5. Elemental compositions of HAp nanopowders.

Element	Weight %		
	HAP-20	HAP-60	HAP-100
Ca K	27.49	34.41	32.83
P K	12.26	14.83	14.44
O K	37.64	42.65	44.57
C K	17.68	8.11	3.89
Au K	4.92	-	4.26
Ca: P ratio (EDS analysis)	1.74	1.79	1.78
Ca:P ratio (chemical analysis)	1.74	1.84	1.76

The Ca:P ratios calculated based on EDS analysis agree with those obtained by chemical analysis.

Based on the results obtained from the physical-chemical characterization of the HAp samples, powders prepared at 10 MPa were further selected for mechanical and in vitro testing. The reason is that HAP-100 shows the highest crystallinity index, 79.4%, calculated from XRD measurements. In X-ray diffraction, it is revealed that the smaller the crystallite size, the more amorphous the material will be considered [57]. We also assume, based on our previous results in this field, that a higher pressure positively influences the biocompatible properties of the material [43,58].

3.5. Mechanical Properties Evaluation

3.5.1. Compressive Test Evaluation

The medical purpose application of the developed material requires appropriate compressive properties. Because of this, the compressive strengths of the tested specimens were measured, and these are presented in Table 6. Examples of strain/stress curves (for 90 min sintering time) are presented in Figure 4.

Table 6. The compressive strength of the specimens sintered at various sintering times.

Sintering Conditions	Compressive Strength, N/mm²		
	90 Min	120 Min	180 Min
1100 °C	37.65–39.13	-	-
1200 °C	49.08–102.24	49.79–76.28	64.54–127.15
1300 °C	20.21–76.93	58.81–86.12	26.72–73.81
1400 °C	37.00–65.81	75.55–120.23	51.17–206.40

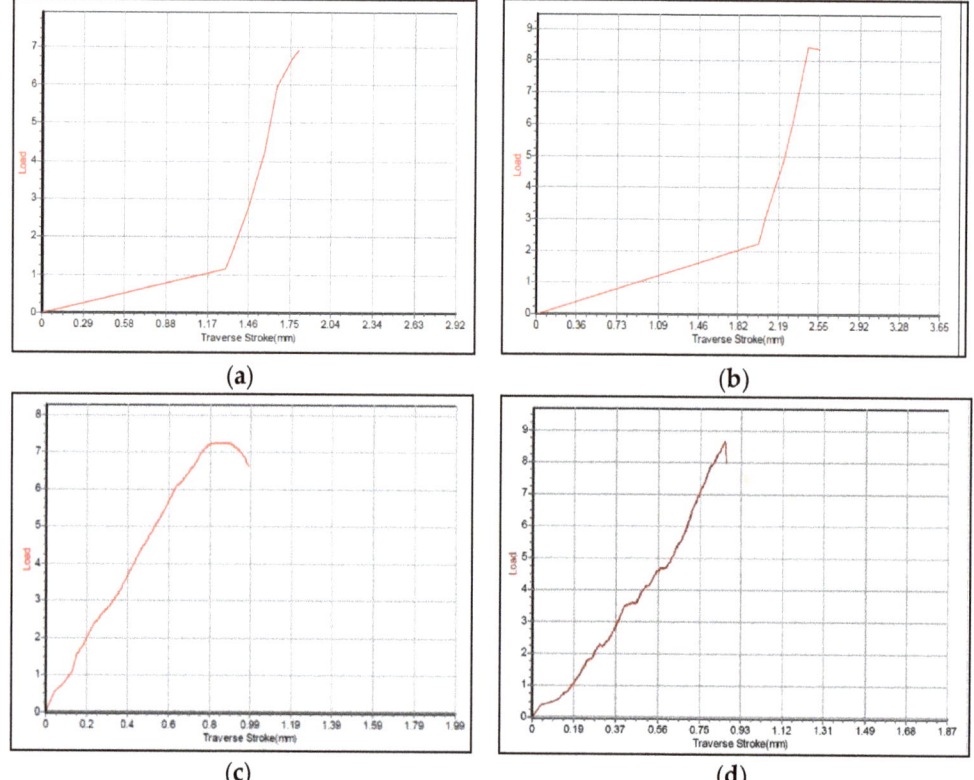

Figure 4. Examples of stress/strain curve for 90 min sintering: (**a**) 1100 °C/90 min; (**b**) 1200 °C/90 min; (**c**) 1300 °C/90 min; (**d**) 1400 °C/90 min.

Analyzing the ranges presented in Table 6, it can be concluded that appropriate/acceptable values are obtained in the case of HAp sintered at 1200 °C/180 min, as well as at 1300 °C and 1400 °C for 120 min.

3.5.2. Micro-Hardness Testing

For cylindrical specimens sintered at 1100 °C/90 min, 180 ± 5 HV1 were obtained. Experimental results for the specimens sintered at 1300 °C are presented in Figure 5.

As for the tensile strength measurements, it can be said that the set of samples sintered at 1300 °C/120 min provide better results compared to specimens sintered for 90 min and 180 min, respectively.

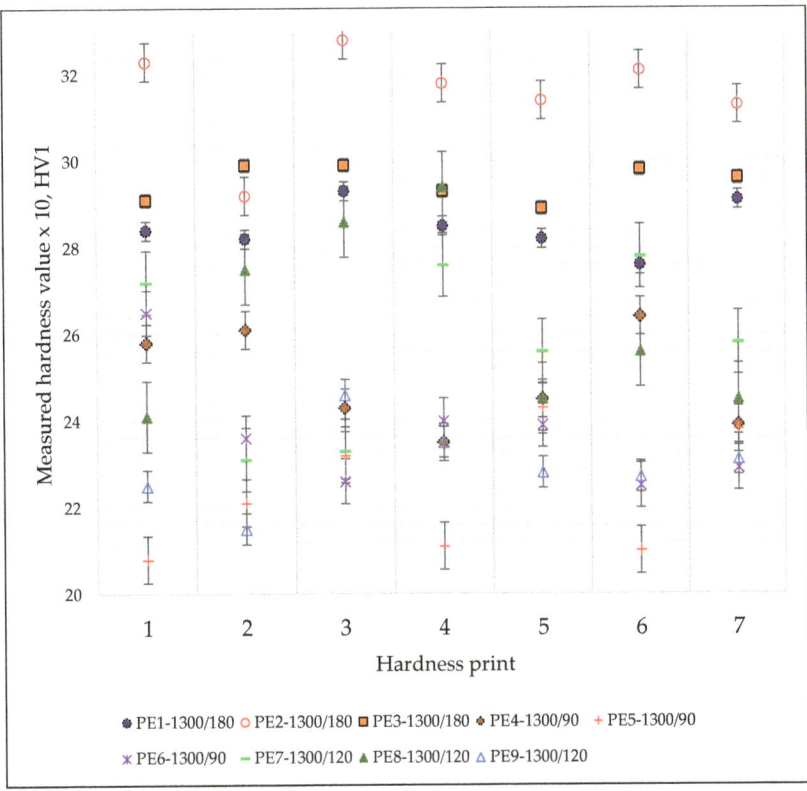

Figure 5. Microhardness values measured on the specimens sintered at 1300 °C.

3.5.3. Wear Test

Regarding the wear test, the material sintered at 1100 °C has a low wear resistance, the fingerprint depth being 0.88 mm. Each of the other specimens that were sintered above 1100 °C showed better wear resistance compared to the values recorded for the samples sintered at 1100 °C/90 min, the fingerprint depth being less than 0.8 mm. The coefficient of friction has an appropriate value for biomechanical applications for all the investigated samples (regardless of sintering temperatures). The measured values for the sintered specimens at the four sintering temperatures are presented in Table 7.

Table 7. Wear test-measured values for coefficient of friction and maximum penetration.

Sintering Temperature, °C	1100	1200			1300			1400		
Sintering Time, Minutes	90	90	120	180	90	120	180	90	120	180
Coefficient of Friction, μ	0.032	0.031	0.031	0.031	0.031	0.031	0.031	0.031	0.031	0.031
Maximum Depth, p, mm	0.88	0.76	0.72	0.72	0.75	0.69	0.72	0.71	0.68	0.69

3.5.4. SEM Characterization of Sintered Specimens

The sintered HAp pellets for which acceptable results were obtained from the mechanical properties evaluation were analyzed by scanning electron microscopy. Thus, SEM images of HAp specimens sintered at 1200 °C/180 min, 1300 °C/120 min, and 1400 °C/120 min, denoted as HAP-1200, HAP-1300 and HAP-1400, are shown in Figure 6.

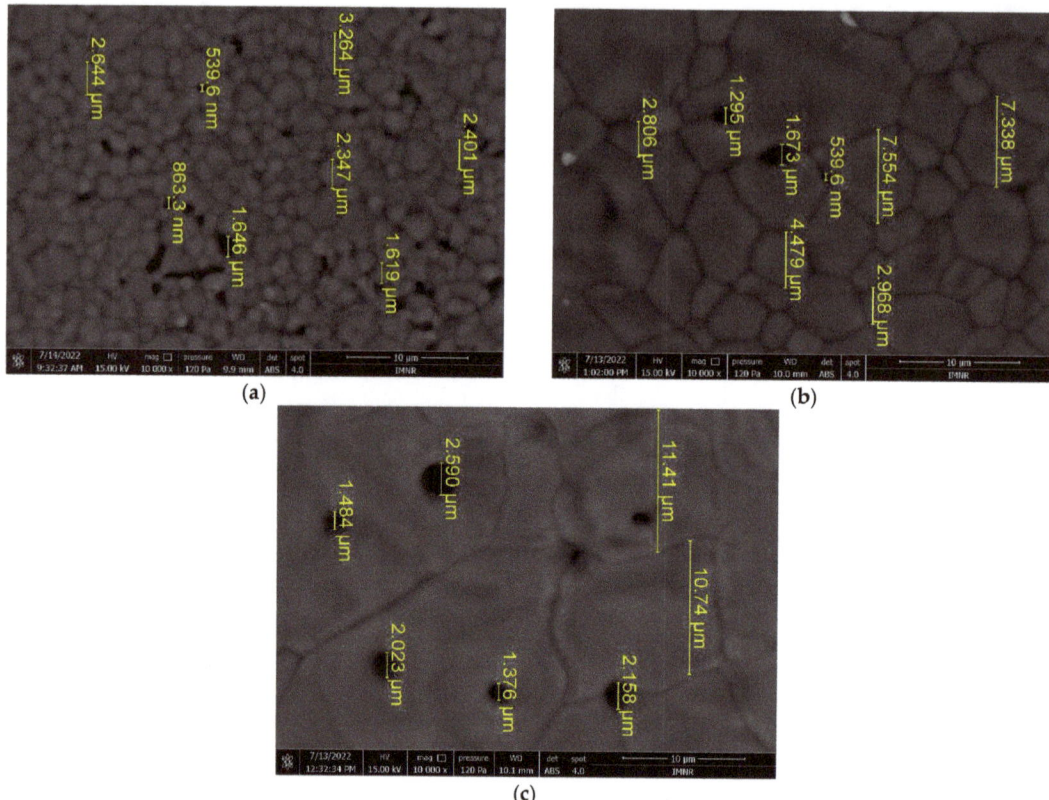

Figure 6. SEM images at 10 μm scale bar, 15 kV voltage, and 10 kX magnification for HAp specimens sintered at: (**a**) 1200 °C/180 min; (**b**) 1300 °C/120 min; (**c**) 1400 °C/120 min.

It can be seen that the grain size increases significantly with the sintering temperature, as expected. HAP-1200 has a grain size between 1.6–3.3 μm (Figure 6a) and open porosity (pore size in the range of 539 nm–1.64 μm). A much higher degree of densification can be noticed in the case of the HAP-1300 sample (Figure 6b), with a grain size of 2.8–7.5 μm and a few pores located at the grain boundaries which are similar in size to those in the HAP-1200 sample. HAP-1400 exhibits large grains, on the order of tens of microns (10.7–11.4 μm or more) and closed porosity, with pore sizes between 1.3–2.6 μm. When the sintering temperature was increased, the porosities of the HAp specimens were reduced due to the higher densification of the material [59].

3.6. Biocompatibility Assessment

Biocompatibility can be broadly defined as the physical, chemical, and biological compatibility between a biomaterial and body tissues and the optimal compatibility of a biomaterial with the mechanical behavior of the body.

The biocompatibility of any biomaterial (medical device) must be evaluated using in vitro and in vivo testing before use in patients. While animal experiments are expensive and require extended periods of experimentation, cell culture methods can be performed at lower cost, are faster and easier to perform, and can be easily reproduced. In recent years, a wide range of in vitro tests have been developed to evaluate the biocompatibility of different biomaterials (powders, solutions, hydrogels, medical devices). Such an in vitro assay uses MTT {3-(4,5-dimethithiazol-2-yl)-2,5-diphenyl tetrazolium bromide} and is a sensitive, quantitative, and reliable colorimetric assay that measures cell viability, proliferation, and activation. In living cells, water-soluble yellow MTT is reduced to a dark blue formazan product by the mitochondrial dehydrogenase enzyme. The amount of formazan produced is directly proportional to the number of viable cells present. Therefore, measuring the optical density will help to determine the amount of formazan produced and thus, the number of viable cells present.

3.6.1. Cytotoxicity Test

After reading the values of the optical densities for the osteoblast cell line used in testing the cytotoxicity of the 12 HAp samples, the arithmetic means of the values were calculated, and the viability calculation formula was applied. The results were plotted in Figure 7. Samples called HN-x are square specimens (3D printed samples with dimensions of $15 \times 15 \times 5$ mm^3), and they were studied for comparative reasons.

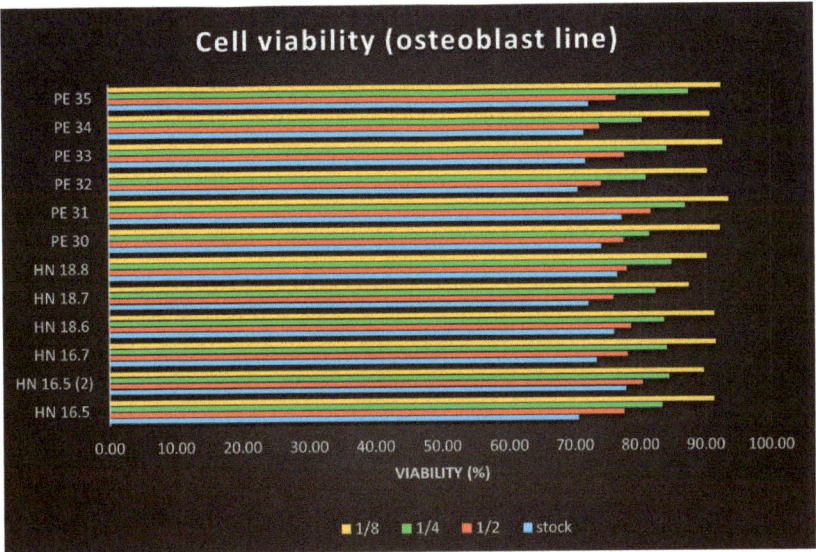

Figure 7. Cell viability of osteoblasts for the 12 HAp samples. The stock solution was diluted in binary dilutions (1/2, 1/4, and 1/8).

The cell viability results determined using the MTT test helped us to conclude the following:
- The cytotoxic effect of the 12 HAp samples tested is low, with very small differences depending on their size; square (printed) samples, dried at 100 °C showed better results.
- The cytotoxicity of the tested samples was dose-dependent; the lower the concentration of the tested product, the lower the cytotoxicity.
- The cell viability is lowest in culture wells with an extract stock concentration, and it increases in direct proportion to the increase in dilution.

3.6.2. Cell Proliferation Study

Osteoblast proliferation on the 12 HAp samples was analyzed at 24, 48, and 72 h of substrate–cell interaction. In the case of the cell proliferation test, after reading the absorbance values, calculating the average of the values read for each sample and applying the calculation formula, the results were synthesized as graphs (Figure 8).

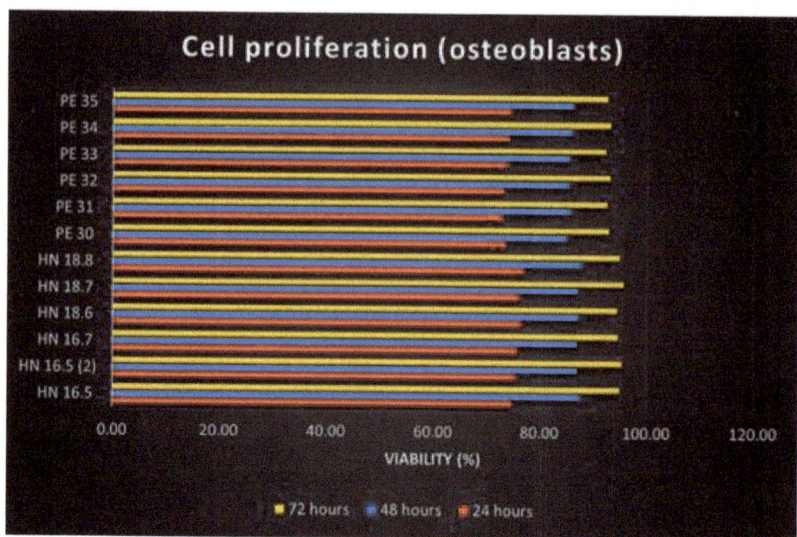

Figure 8. Cell proliferation of osteoblasts on HAp samples.

The increasing number of osteoblasts used for the proliferation tests of the 12 samples shows the proliferation production on all these samples. Comparing the 12 HAp samples, it is noticeable that the differences are insignificant. Better proliferation has been observed for osteoblasts grown on square (3D printed) samples. Regarding the time dynamics of cell proliferation, it increases in direct proportion to the increase in the substrate–cell contact period.

4. Conclusions

In this paper, the hydroxyapatite nanopowder was obtained from natural sources of Rapana Thomasiana using hydrothermal method at various synthesis pressures. The influence of applied pressure on the physical-chemical properties of HAp powders has been explored. It was found that the crystallite size and particle size of hydroxyapatite increase with a pressure increase. The biomedical potential of the obtained material was studied through mechanical evaluation (compressive test, microhardness test, wear test), as well as cytotoxicity and cell proliferation testing. HAp specimens sintered at 1300 °C/120 min present appropriate biomechanical properties. Hardness values up to around 200 HV1 were recorded for sintering temperatures above 1200 °C and sintering times above 120 min. The compressive properties of the material returned a large range of values, from about 50 N/mm^2 to around 120 N/mm^2. Both the hardness and the compressive properties are larger than the values specific to bones.

Due to its function as an implant to bones, the new material is subjected to wearing. Wear tests returned nearly the same value for the coefficient of friction, which is 0.031, with penetrations between 0.68 and 0.88. The lowest penetration depths were recorded for sintering above 1200 °C and sintering times above 120 min.

Cell viability and cell proliferation increase over time. These results are encouraging, demonstrating that natural sources can be successfully exploited for the synthesis of new materials.

Supplementary Materials: The following supporting information can be downloaded at: https://www.mdpi.com/article/10.3390/ma15155091/s1, Figure S1: Particle size distribution of: (a) HAP-20; (b) HAP-60 and (c) HAP-100 nanopowders; Figure S2: SEM images at 5 μm scale bar, 5 kV voltage and 20 kX magnification, and EDS spectra of HAp nanopowders: (a) and (b) HAP-20; (c) and (d) HAP-60.

Author Contributions: Conceptualization, L.M.C. and R.M.P.; methodology, G.D.; investigation, M.I., D.T., S.V.S., I.D.S. and D.P.; resources, R.-G.H.; writing—original draft preparation, L.M.C.; writing—review and editing, L.M.C., D.T. and M.C.; supervision, M.C.; project administration, L.M.C.; funding acquisition, R.M.P. All authors have read and agreed to the published version of the manuscript.

Funding: This work was supported by a grant from the Romanian Ministry of Education and Research, CCCDI-UEFISCDI, project number PN-III-P2-2.1-PED-2019-3090, contract no. 499PED/2020 (acronym 3D BIOPRO), within the PNCDI III and INOVADIT project of the Ministry of Research, Innovation, and Digitization through Program 1—Development of the national research-development system, Subprogram 1.2—Institutional performance projects for financing excellence in RDI, contract no. 9PFE/2021.

Institutional Review Board Statement: Not applicable. This study did not involve humans or animals.

Informed Consent Statement: Not applicable. This study did not use humans or animals as subjects.

Data Availability Statement: Not applicable. The results of the study were not published in other journals, conferences, public databases, etc.

Acknowledgments: The authors would like to thank to our colleagues from the Analysis Laboratory—IMNR for performing SEM/EDS analysis and the Laboratory of Advanced and Nanostructured Materials from IMNR for technical support.

Conflicts of Interest: The authors declare no conflict of interest.

References

1. Panda, S.; Biswas, C.K.; Paul, S. A comprehensive review on the preparation and application of calcium hydroxyapatite: A special focus on atomic doping methods for bone tissue engineering. *Ceram. Int.* **2021**, *47*, 28122–28144. [CrossRef]
2. Curran, D.J.; Fleming, T.J.; Towler, M.R.; Hampshire, S. Mechanical parameters of strontium doped hydroxyapatite sintered using microwave and conventional methods. *J. Mech. Behav. Biomed. Mater.* **2011**, *4*, 2063–2073. [CrossRef] [PubMed]
3. Kattimani, V.S.; Kondaka, S.; Lingamaneni, K.P. Hydroxyapatite—Past, Present, and Future in Bone Regeneration. *Bone Tissue Regen. Insights* **2016**, *7*, 9–19. [CrossRef]
4. Szterner, P.; Biernat, M. The Synthesis of Hydroxyapatite by Hydrothermal Process with Calcium Lactate Pentahydrate: The Effect of Reagent Concentrations, pH, Temperature, and Pressure. *Bioinorg. Chem. Appl.* **2022**, *2022*, 3481677. [CrossRef]
5. Sadat-Shojai, M.; Khorasani, M.-T.; Dinpanah-Khoshdargi, E.; Jamshidi, A. Synthesis methods for nanosized hydroxyapatite with diverse structures. *Acta Biomater.* **2013**, *9*, 7591–7621. [CrossRef]
6. Wei, S.; Dapeng, Z.; Peng, S.; Hemin, N.; Yuan, Z.; Jincheng, T. Strontium-doped Hydroxyapatite Coatings Deposited on Mg-4Zn Alloy: Physical-chemical Properties and in vitro Cell Response. *Rare Met. Mater. Eng.* **2018**, *47*, 2371–2380. [CrossRef]
7. Ma, K.; Huang, D.; Cai, J.; Cai, X.; Gong, L.; Huang, P.; Wang, Y.; Jiang, T. Surface functionalization with strontium-containing nanocomposite coatings via EPD. *Colloids Surf. B Biointerfaces* **2016**, *146*, 97–106. [CrossRef]
8. Gopi, D.; Ramya, S.; Rajeswari, D.; Kavitha, L. Corrosion protection performance of porous strontium hydroxyapatite coating on polypyrrole coated 316L stainless steel. *Colloids Surf. B Biointerfaces* **2013**, *107*, 130–136. [CrossRef]
9. Drevet, R.; Benhayoune, H. Pulsed electrodeposition for the synthesis of strontium-substituted calcium phosphate coatings with improved dissolution properties. *Mater. Sci. Eng. C* **2013**, *33*, 4260–4265. [CrossRef]
10. Sadiq, T.O.; Siti, N.; Idris, J. A Study of Strontium-Doped Calcium Phosphate Coated on Ti6Al4V Using Microwave Energy. *J. Bio-Tribo-Corros.* **2018**, *4*, 40. [CrossRef]
11. Tan, F.; Zhu, Y.; Ma, Z.; Al-Rubeai, M. Recent advances in the implant-based drug delivery in otorhinolaryngology. *Acta Biomater.* **2020**, *108*, 46–55. [CrossRef] [PubMed]
12. Soriano-Souza, C.; Valiense, H.; Mavropoulos, E.; Martinez-Zelaya, V.; Costa, A.M.; Alves, A.T.; Longuinho, M.; Resende, R.; Mourão, C.; Granjeiro, J.; et al. Doxycycline containing hydroxyapatite ceramic microspheres as a bone-targeting drug delivery system. *J. Biomed. Mater. Res. Part B Appl. Biomater.* **2020**, *108*, 1351–1362. [CrossRef] [PubMed]
13. Yang, P.; Quan, Z.; Li, C.; Kang, X.; Lian, H.; Lin, J. Bioactive, luminescent and mesoporous europium-doped hydroxyapatite as a drug carrier. *Biomaterials* **2008**, *29*, 4341–4347. [CrossRef] [PubMed]
14. Andrade, A.S.; Silva, G.F.; Camilleri, J.; Cerri, E.S.; Guerreiro-Tanomaru, J.M.; Cerri, P.S.; Tanomaru-Filho, M. Tissue Response and Immunoexpression of Interleukin 6 Promoted by Tricalcium Silicate–based Repair Materials after Subcutaneous Implantation in Rats. *J. Endod.* **2018**, *44*, 458–463. [CrossRef]
15. Pal, A.; Hadagalli, K.; Bhat, P.; Goel, V.; Mandal, S. Hydroxyapatite—A promising sunscreen filter. *J. Aust. Ceram. Soc.* **2020**, *56*, 345–351. [CrossRef]
16. Steffens, D.; Mathor, M.B.; Santi, B.T.; Luco, D.P.; Pranke, P. Development of a biomaterial associated with mesenchymal stem cells and keratinocytes for use as a skin substitute. *Regen. Med.* **2015**, *10*, 975–987. [CrossRef]

17. Saha, B.; Yadav, S.K.; Sengupta, S. Synthesis of nano-Hap prepared through green route and its application in oxidative desulfurisation. *Fuel* **2018**, *222*, 743–752. [CrossRef]
18. Narwade, V.N.; Mahabole, M.P.; Bogle, K.A.; Khairnar, R.S. Wastewater treatment by nanoceramics: Removal of lead particles. *Int. J. Eng. Sci. Innov. Technol.* **2014**, *3*, 324–329.
19. Vila, M.; Sánchez-Salcedo, S.; Vallet-Regí, M. Hydroxyapatite foams for the immobilization of heavy metals: From waters to the human body. *Inorg. Chim. Acta* **2012**, *393*, 24–35. [CrossRef]
20. Mtavangu, S.G.; Mahene, W.; Machunda, R.L.; van der Bruggen, B.; Njau, K.N. Cockle (*Anadara granosa*) shells-based hydroxyapatite and its potential for defluoridation of drinking water. *Results Eng.* **2022**, *13*, 100379. [CrossRef]
21. Ibrahim, M.; Labaki, M.; Giraudon, J.-M.; Lamonier, J.-F. Hydroxyapatite, a multifunctional material for air, water and soil pollution control: A review. *J. Hazard. Mater.* **2020**, *383*, 121139. [CrossRef] [PubMed]
22. Ideia, P.; Degli Esposti, L.; Miguel, C.C.; Adamiano, A.; Iafisco, M.; Castilho, P.C. Extraction and characterization of hydroxyapatite-based materials from grey triggerfish skin and black scabbardfish bones. *Int. J. Appl. Ceram. Technol.* **2020**, *18*, 235–243. [CrossRef]
23. Modolon, H.B.; Inocente, J.; Bernardin, A.M.; Klegues Montedo, O.R.; Arcaro, S. Nanostructured biological hydroxyapatite from Tilapia bone: A pathway to control crystallite size and crystallinity. *Ceram. Int.* **2021**, *47*, 27685–27693. [CrossRef]
24. Aziz, K.; Mamouni, R.; Azrrar, A.; Kjidaa, B.; Saffaj, N.; Aziz, F. Enhanced biosorption of bisphenol A from wastewater using hydroxyapatite elaborated from fish scales and camel bone meal: A RSM@BBD optimization approach. *Ceram. Int.* **2022**, *48*, 15811–15823. [CrossRef]
25. Akpan, E.S.; Dauda, M.; Kuburi, L.S.; Obada, D.O.; Bansod, N.D.; Dodoo-Arhin, D. Hydroxyapatite ceramics prepared from two natural sources by direct thermal conversion: From material processing to mechanical measurements. *Mater. Today Proc.* **2021**, *38*, 2291–2294. [CrossRef]
26. Hernández-Ruiz, K.L.; López-Cervantes, J.; Sánchez-Machado, D.I.; del Rosario Martínez-Macias, M.; Correa-Murrieta, M.A.; Sanches-Silva, A. Hydroxyapatite recovery from fish byproducts for biomedical applications. *Sustain. Chem. Pharm.* **2022**, *28*, 100726. [CrossRef]
27. Mohd Pu'ad, N.A.S.; Koshy, P.; Abdullah, H.Z.; Idris, M.I.; Lee, T.C. Syntheses of hydroxyapatite from natural sources. *Heliyon* **2019**, *5*, e01588. [CrossRef]
28. Sathiyavimal, S.; Vasantharaj, S.; Lewis Oscar, F.; Selvaraj, R.; Brindhadevi, K.; Pugazhendhi, A. Natural organic and inorganic–hydroxyapatite biopolymer composite for biomedical applications. *Prog. Org. Coat.* **2020**, *147*, 105858. [CrossRef]
29. Sharifianjazi, F.; Esmaeilkhanian, A.; Moradi, M.; Pakseresht, A.; Asl, M.S.; Karimi-Maleh, H.; Jang, H.W.; Shokouhimehr, M.; Varma, R.S. Biocompatibility and mechanical properties of pigeon bone waste extracted natural nano-hydroxyapatite for bone tissue engineering. *Mater. Sci. Eng. B* **2021**, *264*, 114950. [CrossRef]
30. Bee, S.-L.; Mariatti, M.; Ahmad, N.; Yahaya, B.H.; Abdul Hamid, Z.A. Effect of the calcination temperature on the properties of natural hydroxyapatite derived from chicken bone wastes. *Mater. Today Proc.* **2019**, *16*, 1876–1885. [CrossRef]
31. Mucalo, M. Animal-bone derived hydroxyapatite in biomedical applications. In *Hydroxyapatite (HAp) for Biomedical Applications*, 1st ed.; Mucalo, M.R., Ed.; Elsevier Science: Amsterdam, The Netherlands, 2015; Chapter 14; pp. 307–342.
32. Cree, D.; Rutter, A. Sustainable bio-inspired limestone eggshell powder for potential industrialized applications. *ACS Sustain. Chem. Eng.* **2015**, *3*, 941–949. [CrossRef]
33. Komur, B.; Lohse, T.; Can, H.M.; Khalilova, G.; Geçimli, Z.N.; Aydoğdu, M.O.; Kalkandelen, C.; Stan, G.E.; Sahin, Y.M.; Sengil, A.Z.; et al. Fabrication of naturel pumice/hydroxyapatite composite for biomedical engineering. *Biomed. Eng. Online* **2016**, *15*, 81. [CrossRef]
34. Miculescu, F.; Mocanu, A.-C.; Maidaniuc, A.; Dascălu, C.-A.; Miculescu, M.; Voicu, Ș.I.; Ciocoiu, R.-C. Biomimetic Calcium Phosphates Derived from Marine and Land Bioresources. In *Hydroxyapatite—Advances in Composite Nanomaterials, Biomedical Applications and Its Technological Facets*; Thirumalai, J., Ed.; IntechOpen: London, UK, 2018; Chapter 6; pp. 1–20.
35. Akram, M.; Ahmed, R.; Shakir, I.; Ibrahim, W.A.W.; Hussain, R. Extracting hydroxyapatite and its precursors from natural resources. *J. Mater. Sci.* **2014**, *49*, 1461–1475. [CrossRef]
36. Milovac, D.; Gallego Ferrer, G.; Ivankovic, M.; Ivankovic, H. PCL-coated hydroxyapatite scaffold derived from cuttlefish bone: Morphology, mechanical properties and bioactivity. *Mater. Sci. Eng. C* **2014**, *34*, 437–445. [CrossRef] [PubMed]
37. Dagmara Malina, D.; Biernat, K.; Sobczak-Kupiec, A. Studies on sintering process of synthetic hydroxyapatite. *Acta Biochim. Pol.* **2013**, *60*, 851–855.
38. Çalışkan, F.; Akça, S.G.; Tatlı, Z. Sintering Behaviour of Calcium Phosphate Based Powders prepared by Extraction Method. In Proceedings of the 6th International Symposium on Innovative Technologies in Engineering and Science, Antalya, Turkey, 9–11 November 2018; (ISITES2018). Volume 1, pp. 1273–1279.
39. ASTM C1424-04; Standard Test Method for Monotonic Compressive Strength of Advanced Ceramics at Ambient Temperature. ASTM Committee: West Conshohocken, PA, USA, 2013.
40. Hardness, A.B. *Standard Test Method for Microindentation Hardness of Materials*; ASTM Committee: West Conshohocken, PA, USA, 1999; Volume 384–399, pp. 1–24.
41. Savu, I.D.; Tarnita, D.; Savu, S.V.; Benga, G.C.; Cursaru, L.M.; Dragut, D.V.; Piticescu, R.M.; Tarnita, D.N. Composite Polymer for Hybrid Activity Protective Panel in Microwave Generation of Composite Polytetrafluoroethylene—*Rapana thomasiana*. *Polymers* **2021**, *13*, 2432. [CrossRef] [PubMed]

42. Savu, S.V.; Tarnita, D.; Benga, G.C.; Dumitru, I.; Stefan, I.; Craciunoiu, N.; Olei, A.B.; Savu, I.D. Microwave Technology Using Low Energy Concentrated Beam for Processing of Solid Waste Materials from *Rapana thomasiana* Seashells. *Energies* **2021**, *14*, 6780. [CrossRef]
43. Mocioiu, A.M.; Tutuianu, R.; Cursaru, L.M.; Piticescu, R.M.; Stanciu, P.; Vasile, B.S.; Trusca, R.; Sereanu, V.; Meghea, A. 3D structures of hydroxyapatite obtained from Rapana venosa shells using hydrothermal synthesis followed by 3D printing. *J. Mater. Sci.* **2019**, *54*, 13901–13913. [CrossRef]
44. Ebrahimi, S.; Sipaut, S.; Mohd Nasri, C.; Bin Arshad, S.E. Hydrothermal synthesis of hydroxyapatite powders using Response Surface Methodology (RSM). *PLoS ONE* **2021**, *16*, e0251009. [CrossRef]
45. Mohamad Razali, N.A.I.; Pramanik, S.; Abu Osman, N.A.; Radzi, Z.; Pingguan-Murphy, B. Conversion of calcite from cockle shells to bioactive nanorod hydroxyapatite for biomedical applications. *J. Ceram. Process. Res.* **2016**, *17*, 699–706.
46. Arokiasamy, P.; Al Bakri Abdullah, M.M.; Rahim, S.Z.A.; Luhar, S.; Sandu, A.V.; Jamil, N.H.; Nabiałek, M. Synthesis methods of hydroxyapatite from natural sources: A review. *Ceram. Int.* **2022**, *48*, 14959–14979. [CrossRef]
47. Yoshimura, M. Soft solution processing: Concept and realization of direct fabrication of shaped ceramics (nano-crystals, whiskers, films, and/or patterns) in solutions without post-firing. *J. Mater. Sci.* **2006**, *41*, 1299–1306. [CrossRef]
48. Schaf, O.; Ghobarkar, H.; Knauth, P. *Hydrothermal Synthesis of Nanomaterials in Nanostructured Materials. Selected Synthesis Methods, Properties and Applications*; Knauth, P., Schoonman, J., Eds.; Kluwer Academic Publisher: Amsterdam, The Netherlands, 2002; pp. 1–188.
49. Adschiri, T.; Hakuta, Y.; Sue, K.; Arai, K. Hydrothermal Synthesis of Metal Oxide Nanoparticles at Supercritical Conditions. *J. Nanopart. Res.* **2001**, *3*, 227–235. [CrossRef]
50. Lencka, M.M.; Riman, R.E. Intelligent Systems of Smart Ceramics. In *Encyclopedia of Smart Materials*; John Wiley & Sons: Hoboken, NJ, USA, 2002; Volume 1.
51. Naqshbandi, A.; Rahman, A. Sodium doped hydroxyapatite: Synthesis, characterization and zeta potential studies. *Mater. Lett.* **2022**, *312*, 131698. [CrossRef]
52. Osuchukwu, O.A.; Salihi, A.; Abdullahi, I.; Obada, D.O. Synthesis and characterization of sol–gel derived hydroxyapatite from a novel mix of two natural biowastes and their potentials for biomedical applications. *Mater. Today Proc.* **2022**, *62*, 4182–4187. [CrossRef]
53. Shaltout, A.A.; Allam, M.A.; Moharram, M.A. FTIR spectroscopic, thermal and XRD characterization of hydroxyapatite from new natural sources. *Spectrochim. Acta A Mol. Biomol. Spectrosc.* **2011**, *83*, 56–60. [CrossRef] [PubMed]
54. Poralan, G.M.; Gambe, J.E.; Alcantara, E.M.; Vequizo, R.M. X-ray diffraction and infrared spectroscopy analyses on the crystallinity of engineered biological hydroxyapatite for medical application. *IOP Conf. Ser. Mater. Sci. Eng.* **2015**, *79*, 012028. [CrossRef]
55. Park, S.; Baker, J.O.; Himmel, M.E.; Parilla, P.A.; Johnson, D.K. Cellulose crystallinity index: Measurement techniques and their impact on interpreting cellulase performance. *Biotechnol. Biofuels* **2010**, *3*, 10. [CrossRef]
56. Vinoth Kumar, K.C.; Jani Subha, T.; Ahila, K.G.; Ravindran, B.; Chang, S.W.; Mahmoud, A.H.; Mohammed, O.B.; Rathi, M.A. Spectral characterization of hydroxyapatite extracted from Black Sumatra and Fighting cock bone samples: A comparative analysis. *Saudi J. Biol. Sci.* **2021**, *28*, 840–846. [CrossRef]
57. Hassan, M.; Akmal, M.; Ryu, H.J. Cold sintering of as-dried nanostructured calcium hydroxyapatite without using additives. *J. Mater. Res. Technol.* **2021**, *11*, 811–822. [CrossRef]
58. Gradinaru, S.; Popescu, L.M.; Piticescu, R.M.; Zurac, S.; Ciuluvica, R.; Burlacu, A.; Tutuianu, R.; Valsan, S.-N.; Motoc, A.M.; Voinea, L.M. Repair of the Orbital Wall Fractures in Rabbit Animal Model Using Nanostructured Hydroxyapatite-Based Implant. *Nanomaterials* **2016**, *6*, 11. [CrossRef] [PubMed]
59. Islam, M.S.; Rahman, A.M.Z.; Sharif, M.H.; Khan, A.; Abdulla-Al-Mamun, M.; Todo, M. Effects of compressive ratio and sintering temperature on mechanical properties of biocompatible collagen/hydroxyapatite composite scaffolds fabricated for bone tissue engineering. *J. Asian Ceram. Soc.* **2019**, *7*, 183–198. [CrossRef]

Article

Comfort Analysis of Hafnium (Hf) Doped ZnO Coated Self-Cleaning Glazing for Energy-Efficient Fenestration Application

Srijita Nundy [1], Aritra Ghosh [1,*], Abdelhakim Mesloub [2], Emad Noaime [2] and Mabrouk Touahmia [3]

1 College of Engineering, Mathematics and Physical Sciences, Renewable Energy, University of Exeter, Penryn TR10 9FE, UK; s.nundy@exeter.ac.uk
2 Department of Architectural Engineering, Ha'il University, Ha'il 2440, Saudi Arabia; a.maslub@uoh.edu.sa (A.M.); e.noaime@uoh.edu.sa (E.N.)
3 Department of Civil Engineering, Ha'il University, Ha'il 2440, Saudi Arabia; m.touahmia@uoh.edu.sa
* Correspondence: a.ghosh@exeter.ac.uk

Abstract: To attain a comfortable building interior, building windows play a crucial role. Because of the transparent nature of the window, it allows heat loss and gain and daylight. Thus, they are one of the most crucial parts of the building envelope that have a significant contribution to the overall building energy consumption. The presence of dust particles on a window can change the entering light spectrum and creates viewing issues. Thus, self-cleaning glazing is now one of the most interesting research topics. However, aside from the self-cleaning properties, there are other properties that are nominated as glazing factors and are imperative for considering self-cleaning glazing materials. In this work, for the first time, Hf-doped ZnO was investigated as self-cleaning glazing and its glazing factors were evaluated. These outcomes show that the various percentages of ZnO doping with Hf improved the glazing factors, making it a suitable glazing candidate for the cold-dominated climate.

Keywords: glazing; Hf-ZnO; building; g-value; U-value; glare; thermal comfort; visual comfort; CCT; CRI

Citation: Nundy, S.; Ghosh, A.; Mesloub, A.; Noaime, E.; Touahmia, M. Comfort Analysis of Hafnium (Hf) Doped ZnO Coated Self-Cleaning Glazing for Energy-Efficient Fenestration Application. *Materials* 2022, 15, 4934. https://doi.org/10.3390/ma15144934

Academic Editor: Andrei Victor Sandu

Received: 22 June 2022
Accepted: 13 July 2022
Published: 15 July 2022

Publisher's Note: MDPI stays neutral with regard to jurisdictional claims in published maps and institutional affiliations.

Copyright: © 2022 by the authors. Licensee MDPI, Basel, Switzerland. This article is an open access article distributed under the terms and conditions of the Creative Commons Attribution (CC BY) license (https://creativecommons.org/licenses/by/4.0/).

1. Introduction

Currently, buildings consume 40% of energy globally, which is due to the heating, ventilation and air conditioning load. This consumption has an adverse impact on the environment [1]. According to United Nations, migration from rural to urban areas is alarming and increasing every day. This urban influx also increases modern buildings' energy consumption to maintain indoor comfort facilities [2–4]. Buildings generally consume high levels of energy due to their poorly thermally insulated envelopes [5]. Compared to other portions of envelopes, windows are critical, as they are the only parts of the building envelope that maintain the connection between the building's interior and the exterior and allow daylight to penetrate [6].

The glazing sector is predominantly controlled by antireflection, self-cleaning and energy-saving, which are the key three principal functions [7]. For a hot climate, reflecting the solar heat or more precisely reflecting the NIR and IR part of the solar spectrum is the most strategic decision, which in turn reduces the air conditioning load [8,9]. However, antireflection is not suitable for cold climates as it is essential for the reflecting solar spectrum to be transmitted through the window to enhance the room temperature [10,11]. Hence, there is a trend now of replacing the traditional single- and double-glazed windows with advanced technology such as smart switchable EC [12], SPD [13–16], PDLC [17], thermally activated PCM [18], hydrogel [19,20], aerogel [21] or vacuum [22,23] filled windows.

The self-cleaning type of window is another class or area that can be applied to any type of building window (e.g., traditional and smart). Atmospheric pollutants possess significant viewing challenges for window glazing. Dust includes emissions from agriculture and industry, bird droppings, pollen, mineral dust in a dry area, fibers, sand and clay [24]. Daylight transmission into buildings is affected by the deposition of atmospheric pollutants on glazing [25]. Even in a clean UK climate, building windows suffer from dust [26]. Thus, a cleaned window is indispensable for a sustainable building. Depending on the particle diameter, they either fall from the glass surface or stick on the surface. Even though the glass surface looks smooth, it has microscopically small pocks which enhance the attraction of dirt [27–29]. A self-cleaning glazing or window is a thin self-cleaning coating of film on the external surface of the glass, which protects it from dirt [30]. Generally, two types of self-cleaning technologies are available: hydrophobic and hydrophilic. Self-cleaning glazing is capable of cleaning its own surface. For self-cleaning coating, transparency is essential, as it should not create any obstacles to indoor viewing. In addition, long-term durability is crucial for cost-effectiveness.

In the past, several self-cleaning materials have been investigated particularly for photovoltaic applications [31,32] and window [33,34] applications. Zinc oxide (ZnO) is one of the most bio-friendly important semiconductors that have been investigated for self-cleaning applications. A superhydrophobic ZnO nanorods@cellulose membrane for efficient building radiative cooling was investigated [35]. ZnO-coated transparent wood was employed for building applications previously and showed 17% energy saving compared to a traditional window [36]. In another work, a ZnO nanoparticle enhanced paraffin-filled window was investigated for double glazing which showed improved efficiency [37]. To further enhance the ZnO properties, Dy_2WO_6-doped ZnO [38], Sm_{3+}-doped ZnO [39] and Hf-doped ZnO [40] have been investigated for self-cleaning. ZnO for self-cleaning is one of the most popular approaches [41].

Because of the similar ionic radii, Hf-doped ZnO has potential. Transition metal ion doping enhances the surface oxygen vacancies, which improves the self-cleaning behavior. The inclusion of lower-concentration hafnium increases oxygen vacancy defects and produces hydrophilic surfaces. Previously, hafnium oxide (HfO_2) was prepared by electron beam evaporation, and three layers of $HfO_2/Ag/HfO_2$ showed heat mirror properties for energy-efficient window application [42]. We previously developed morphologically varied ZnO for self-cleaning application [43] and synthesized high-quality Hf-ZnO thin films with various Hf contents [40]. However, the suitability of ZnO in terms of glazing for building window applications has not yet been investigated.

How a new material will behave as a building window can be understood by analyzing its thermal and visual comfort parameters [44,45]. The solar heat gain coefficient or solar factor is one of the major influential factors that determine the indoor room temperature and thus define the thermal comfort level [46]. Most often occupants prefer a 20 °C temperature in indoor conditions [47,48]. For a cold climate, a higher solar factor is essential as it increases the room temperature and maintains a comfortable level, whereas for a hot climate, the solar factor should be limited or rejected to limit the increase in room temperature [49,50]. Visual comfort includes both the illuminance level and the color properties. Bright ambient daylight is paramount for cognitive work [51]. However, this amount should not exceed a certain level, or else discomfort glare will dominate. External daylight transmitted through the window glazing attains wavelength changes, which can create discomfit for occupants [52–54]. Color property analysis tackles these challenges.

In this work, for the first time, Hf-doped ZnO was investigated for glazing application. Thus, to understand its suitability as a future self-cleaning fenestration, glazing factor and thermal and visual comfort analyses are essential. Employing the measured transmission spectrum of different Hf-doped ZnO, essential glazing factors such as solar and luminous transmission, solar material protection factor (SMPF) and solar skin protection factors (SSPFs) have been calculated. For thermal comfort analysis, the solar factor has been evaluated. Further, correlated color temperature (CCT), color rendering index (CRI) and glare have also been calculated to understand the visual comfort and suitability of this material for building fenestration application.

2. Experiments

2.1. Materials Fabrication for Glazing

The material for the self-cleaning glazing purpose was developed using hafnium IV chloride (HfCl$_4$), propanol (C$_2$H$_5$OH), triethanolamine (C$_6$H$_{15}$NO$_3$) and zinc acetate (Zn(CH$_3$COOH)2·2H$_2$O), which were purchased from Sigma Aldrich (St. Louis, MI, USA) and used without any further purification. Pure and Hf-doped ZnO were synthesized using the sol–gel synthesis method with Hf concentrations varying from 0 to 15%. Briefly, 2.2 g of Zn(CH$_3$COOH)2·2H$_2$O was made to dissolve completely in 10 mL of C$_2$H$_5$OH. Then, C$_6$H$_{15}$NO$_3$ was carefully poured into the above-prepared solution, where the molar ratio of triethanolamine:zinc acetate was kept at 3:5. The resultant mixture was maintained at room temperature for 5 min. Part of this sol was directly taken for preparation of pure ZnO, and the rest was separated into batches, wherein a particular amount of HfCl$_4$ was added, and stirred at 90 °C for 1 h, thereby forming the sol for Hf-doped ZnO. These as-prepared sols were taken for the thin-film coating on glass substrates, via spin coating (Ossila spin coater, Sheffield, UK), carried out at 500 rpm for 30 s. The as-deposited thin films were taken for annealing in a muffle furnace at 350 °C for 2 h. Finally, the pure and Hf-doped ZnO thin films on glass substrates were obtained after cooling down to room temperature and were further taken for characterization and application purposes. Figure 1 shows the schematic representation of different involved steps for the synthesis of the material.

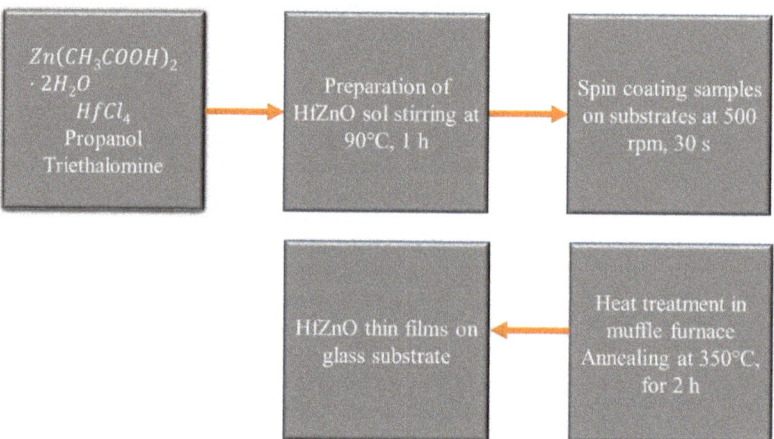

Figure 1. Schematic illustration of involved steps for synthesis of hafnium-doped ZnO.

2.2. Optical Characterization

For optical characterization of the developed glazing, a PerkinElmer Lambda 1050 spectrometer (Waltham, MA, USA) which could measure the visible and NIR transmission and reflection was employed. This system had a 150 nm diameter-based integrating sphere, and measurement was carried out at 10 nm intervals.

3. Methods

3.1. Glazing Factor Evaluation

Solar and luminous transmittance was evaluated by employing Equations (1) and (2), respectively. $T(\lambda)$ is the spectral transmission of glazing. The relative spectral distribution of the illuminant is D_{65}, $S(\lambda)$ is the relative spectral distribution of solar radiation, $V(\lambda)$ is the spectral luminous efficiency of a standard photopic observer, and wavelength interval is represented by $\Delta\lambda$.

Protection factors are crucial building window parameters that show the ability of a window to protect the building material and human skin (located behind the window) when they are exposed to solar radiation [55]. The solar material protection factor (SMPF) is associated with the protection of building material, and the solar skin protection factor (*SSPF*) is associated with the human skin [56]. *SMPF* and *SSPF* both vary between 0 and 1 [57]. Values close to 0 indicate a low protection level, whereas close to 1 indicate a high protection level. *SMPF* and *SSPF* are represented by Equations (3) and (4).

Solar transmission

$$\tau_s = \frac{\sum_{\lambda = 300 \text{ nm}}^{2500 \text{ nm}} S(\lambda) T(\lambda, \alpha) \Delta \lambda}{\sum_{\lambda = 300 \text{ nm}}^{2500 \text{ nm}} S(\lambda) \Delta \lambda} \tag{1}$$

Luminous transmission

$$\tau_v = \frac{\sum_{\lambda = 380 \text{ nm}}^{780 \text{ nm}} D_{65}(\lambda) T(\lambda, \alpha) V(\lambda) \Delta \lambda}{\sum_{\lambda = 380 \text{ nm}}^{780 \text{ nm}} D_{65}(\lambda) V(\lambda) \Delta \lambda} \tag{2}$$

Solar material protection factor (*SMRF*)

$$SMRF = 1 - \frac{\sum_{\lambda = 300 \text{ nm}}^{600 \text{ nm}} T(\lambda) C_\lambda S_\lambda \Delta \lambda}{\sum_{\lambda = 300 \text{ nm}}^{600 \text{ nm}} C_\lambda S_\lambda \Delta \lambda} \tag{3}$$

where $C_\lambda = e^{-0.012\lambda}$.

Solar skin protection factor (*SSPF*)

$$SSPF = 1 - \frac{\sum_{\lambda = 300 \text{ nm}}^{400 \text{ nm}} T(\lambda) E_\lambda S_\lambda \Delta \lambda}{\sum_{\lambda = 300 \text{ nm}}^{400 \text{ nm}} E_\lambda S_\lambda \Delta \lambda} \tag{4}$$

E_λ is the CIE erythemal effectiveness spectrum.

3.2. Thermal Comfort

The amount of solar energy transmitted through the transparent and semitransparent part of the window is represented by the solar heat gain coefficient or solar factor (g). This includes entering infrared radiation into a building's interior and solar transmittance [55,57].

$$\begin{aligned} g &= \tau_s + q_i = \tau_s + \alpha \frac{h_i}{h_i + h_e} \\ &= \tau_s + (1 - \tau_s - \rho_s) \frac{h_i}{h_i + h_e} \end{aligned} \tag{5}$$

where h_e and h_i are the external and internal heat transfer coefficients.

3.3. Visual Comfort

Quality and quantity of light in indoor conditions are essential to understanding and analyzing visual comfort. Correlated color temperature (*CCT*) and color rendering index (*CRI*) both indicate the quality of indoor daylight [58]. Compared to external daylight, *CRI* shows the rendering ability of the incoming daylight. *CCT* is measured in kelvin (K) and signifies a light source's "coolness" and "warmth". *CRI* over 80 is accepted for building window application, and *CRI* over 90 is outstanding [59–61]. For *CCT*, the range between 3000 K and 7500 K is desired for transmitted daylight.

CCT was calculated from McKamy's equation [62].

$$CCT = 449n^3 + 3525n^2 + 6823.3n + 5520.33 \tag{6}$$

where $n = \frac{(x - 0.3320)}{(0.1858 - y)}$ and x and y are chromaticity coordinates.

The color rendering index (*CRI*) is given by

$$CRI = \frac{1}{8}\sum_{i=1}^{8} R_i \qquad (7)$$

The total distortion ΔE_i is determined from

$$\Delta E_i = \sqrt{(U_{t,i}^* - U_{r,i}^*)^2 + (V_{t,i}^* - V_{r,i}^*)^2 + (W_{t,i}^* - W_{r,i}^*)^2} \qquad (8)$$

The special color rendering index R_i for each color sample is given by

$$R_i = 100 - 4.6\Delta E_i \qquad (9)$$

To understand the quality of the indoor light, daylight glare evaluation is essential; daylight glare was evaluated in this work by employing glare subjective rating (*SR*) (as shown in Equation (10)) [63]. Minimum engagement of photosensors makes this method widely available and useful because it saves time and cost [64]. Theoretically, glare control potential using this glazing was identified from measured outdoor illuminance on a vertical plane as shown in Figure 2. This *SR* index allows the estimation of discomfort glare experienced by subjects when working at a visual daylight task (VDT) placed against a window of high or non-uniform luminance.

$$SR = 0.1909 E_v^{0.31} \qquad (10)$$

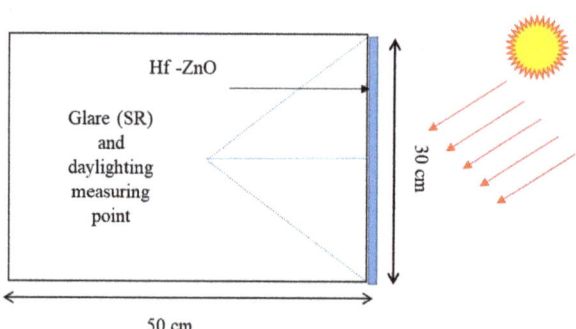

Figure 2. Schematic cross-section of a room with perovskite glazing mounted on vertical south facade.

SR for a typical sunny day in the cold-dominated climate of Penryn, UK (50.16° N, 5.10° W), was examined. Vertically south-facing Hf-doped ZnO glazing having dimensions of $30 \times 30 \times 0.5$ ($l \times w \times h$) cm in the scale model was considered, as shown in Figure 2. This large area resembles self-cleaning glazing as a large facade, while the internal surface was painted in white color with a reflectance of 0.8 [65]. Internal vertical illuminance (E_V) facing the window (worst case) was measured at the center of the room. Table 1 displays the criterion scale of *SR*. This method also allows the non-intrusive measuring equipment necessary for scale model daylighting assessments [66,67].

Table 1. Criterion scale of discomfort glare subjective rating (SR) [63].

Comfort Level Indicator	Glare Subjective Rating (SR)
Just intolerable	2.5
Just disturbing	1.5
Just noticeable/acceptable	0.5

4. Results

4.1. Optical Transmission

Figure 3a shows the spectral transmission of various Hf-doped ZnO for the wavelength range between 250 and 2500 nm. Transmission dropped for 15% Hf doping, while the highest transmission was observed for 6% Hf doping. The product of the spectral luminous efficiency for photopic vision $V(\lambda)$ and relative spectral distribution of illuminant $D65(\lambda)$ has been included for comparison; it varied from 400 nm to 700 nm, having its peak at 555 nm. Figure 3b shows the comparison of single value solar and visible transmission for pure and different Hf-doped ZnO. Pure ZnO showed 87% solar transmission, while 3%, 6%, 9%, 12% and 15% showed 87%, 99%, 88%, 86% and 73%, respectively. Extraordinary changes occurred while the Hf doping percentage was 6%. Visible transmissions for pure and different Hf-doped ZnO are 75% (pure), 88% (3%), 93% (6%), 69% (9%), 91% (12%) and 39% (15%).

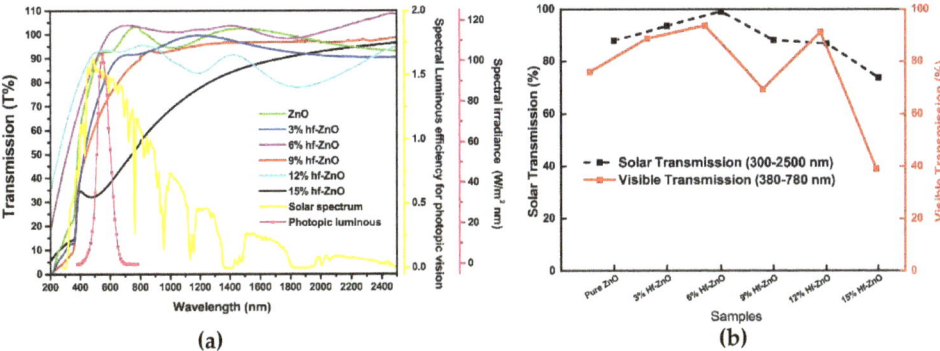

Figure 3. (a) Wavelength-dependent UV, visible and NIR transmission spectra of pure and Hf-doped ZnO thin films. (b) Relation between solar and visible transmission for ZnO with various levels of Hf doping.

Figure 4 illustrates the solar material protection factor and skin protection factor for pure and different Hf-doped ZnO. The material protection factor was higher for 15% Hf-doped ZnO, which was the reason for its lower transmission. Less solar transmission indicates lower degradation. The skin protection factor was lowest for the 12% Hf-doped ZnO, which was due to its highest transmission.

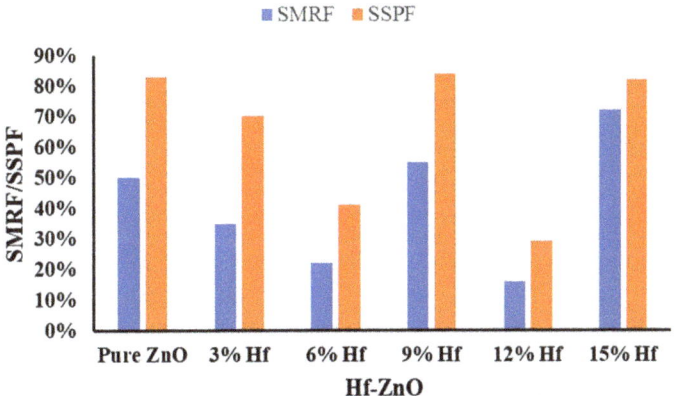

Figure 4. Solar material protection factor (*SMRF*) and solar skin protection factor (*SSPF*) for pure and different Hf-doped ZnO.

4.2. Comfort Analysis

Figure 5 illustrates the solar factor for self-cleaning glazing based on different Hf-doped ZnO. The solar factor is a crucial element for building glazing as its presence is highly recommended for a cold climate, whereas its rejection is essential for a hot climate. In this work, 6% Hf-doped ZnO showed the best solar factor for the cold-dominated climate. However, if this glazing is adopted in a heat-dominated climate, 15% Hf-doped ZnO should be selected. High values of solar factors indicate that the reflection of solar radiation from these glazings is minimal. This is also aesthetic as high reflection can cause issues for the other building users.

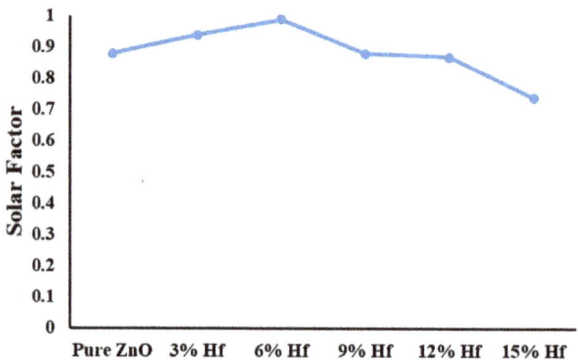

Figure 5. Solar factor for self-cleaning glazing based on different Hf-doped ZnO.

Color properties, including *CCT* and *CRI*, were calculated for Hf-ZnO glazing using Equations (6) (*CCT*) and (7) (*CRI*) and are shown in Figure 6. The 12% Hf-doped ZnO had the best CRI (>98) and CCT (>6200). Interestingly all the doped ZnO samples had higher *CRI* than the pure ZnO. These values satisfy the acceptance level for the comfort level criteria as prescribed in CIE *CIR* [68,69] and IES TM 30–15 [70]. In addition, it can be proposed that *CRI* and *CCTs* are not dependent on a single transmittance value, but their dependency relies on the overall spectrum range. A very similar outcome was previously demonstrated for other types of glazing [71,72].

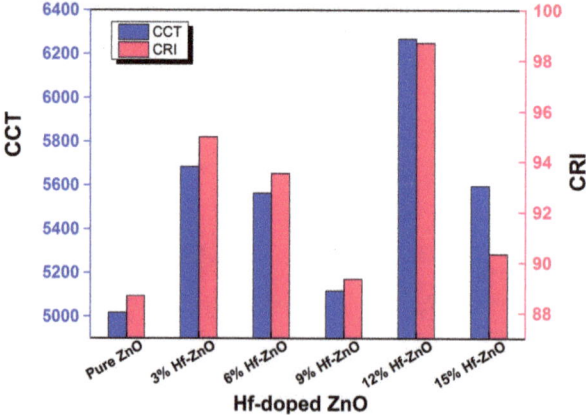

Figure 6. Color rendering index (*CRI*) and correlated color temperature (*CCT*) of pure and Hf-doped ZnO thin films.

Figure 7 shows the *SR* for different Hf-doped ZnO and pure ZnO-based glazing for a vertical south-facing large glazed facade located in cold-dominated climate of Penryn, in the southwest of the UK. A typical clear sunny day was considered for this analysis. The location of the subject is shown in Figure 2. It is clear from the figure that except for the 15% Hf-doped ZnO, others were not able to maintain the glare. This is definitely due to the high transmission rate for all the different Hf-doped and pure ZnO-based glazings. For a cold climatic country where the heating load is high, this penetration of higher solar light could be beneficial from a thermal comfort point of view, although visual comfort may be compromised. However, this argument is true for any type of window for which it is not possible to attain visual and thermal comfort concomitantly. The promising factor for this type of coating is a high transmission, which is key for any self-cleaning material. Transmission reduction on a double glass due to self-cleaning coating is not at all acceptable. Except for building windows, this analysis also strongly recommended the use of this material for self-cleaning coating for the PV system as no transmission reduction is attained and mostly very high transmission was achieved, particularly for the 3%, 6% and 12% Hf-doped ZnO.

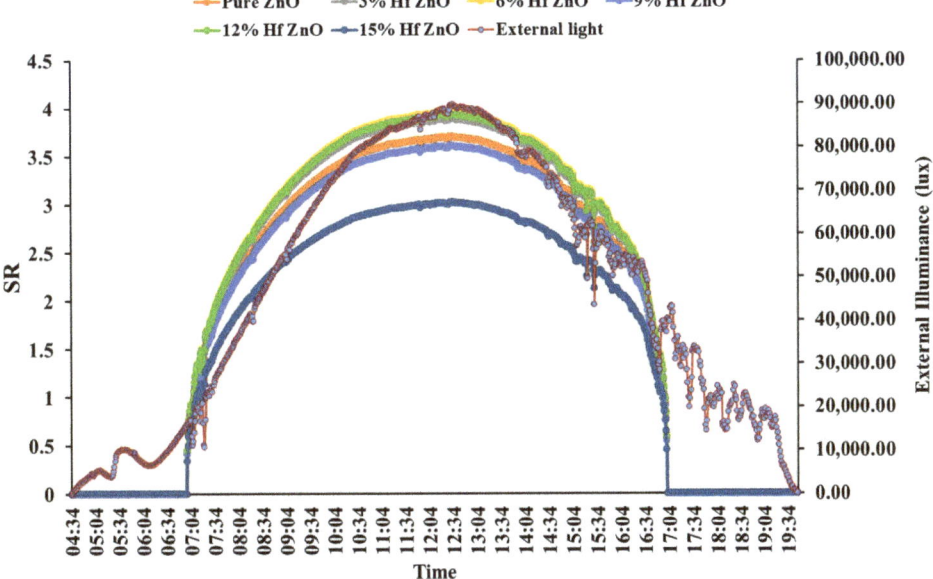

Figure 7. Glare (subjective rating) for pure and different Hf-doped ZnO.

5. Conclusions

In this work, glazing factors and thermal and visual comfort analyses of a self-cleaning coated glazing were examined. This particular self-cleaning coating was developed by the sol–gel method with the introduction of 3%, 6%, 9%, 12% and 15% Hf doping of ZnO. Results of these doped ZnO samples were also compared with pure ZnO. The visible transmission was always higher for the 6% doped ZnO. The protection factor had no trend with an increase in Hf doping. The lowest protection factor was observed at 12% Hf doping. CRI's threshold value of 80 was achieved for all the Hf-doped ZnO type glazings. A higher amount of solar factor also makes this glazing suitable for cold-dominated climates. This high solar factor also indicates that the glazing possesses lower reflection. The 15% doped ZnO showed an allowable SR limit compared to other doped ZnO samples. This was due to the lowest transmission level at the visible wavelength for 15% doped ZnO. This self-cleaning glazing can be a solution for future energy-efficient window applications.

Particularly for cold climate conditions, this self-cleaning can be a good candidate for building window application because of its high solar and visible transmission and high solar factor. In addition, because of lower reflection, it can also be applied on top of photovoltaic systems to diminish the soiling issues. In the future, further investigation is required to understand the reliability of this coating under real weather conditions after long-term outdoor exposure (following different Köppen climatic conditions).

Author Contributions: Conceptualization, S.N. and A.G.; methodology, S.N. and A.G.; software, S.N., A.G., A.M., E.N. and M.T., validation, S.N., A.G., A.M., E.N. and M.T.; formal analysis, S.N.; investigation, A.G.; resources, S.N., A.G., A.M., E.N. and M.T.; data curation, S.N. and A.G.; writing—original draft preparation, S.N. and A.G.; writing—review and editing, S.N., A.G., A.M., E.N. and M.T. visualization, S.N., A.G., A.M., E.N. and M.T. supervision, A.G.; project administration, A.G. and A.M.; funding acquisition, A.G. and A.M. All authors have read and agreed to the published version of the manuscript.

Funding: This research has been funded by the Scientific Research Deanship at the University of Ha'il, Saudi Arabia, through project number RG-21 029.

Institutional Review Board Statement: Not applicable.

Informed Consent Statement: Not applicable.

Data Availability Statement: Not applicable.

Conflicts of Interest: The authors declare no conflict of interest.

References

1. Nundy, S.; Ghosh, A. Thermal and visual comfort analysis of adaptive vacuum integrated switchable suspended particle device window for temperate climate. *Renew. Energy* **2020**, *156*, 1361–1372. [CrossRef]
2. Nematchoua Modeste Kameni; Sadeghi, M.; Reiter, S. Strategies and scenarios to reduce energy consumption and CO2 emission in the urban, rural and sustainable neighbourhoods. *Sustain. Cities Soc.* **2021**, *72*, 103053. [CrossRef]
3. Shi, G.; Lu, X.; Zhang, H.; Zheng, H.; Zhang, Z.; Chen, S.; Xing, J.; Wang, S. Environmental Science and Ecotechnology Air pollutant emissions induced by rural-to-urban migration during China's urbanization (2005–2015). *Environ. Sci. Ecotechnol.* **2022**, *10*, 100166. [CrossRef]
4. Yuan, R.; Rodrigues, J.F.D.; Wang, J.; Tukker, A.; Behrens, P. A global overview of developments of urban and rural household GHG footprints from 2005 to 2015. *Sci. Total Environ.* **2022**, *806*, 150695. [CrossRef] [PubMed]
5. Ghosh, A. Fenestration integrated BIPV (FIPV): A review. *Sol. Energy* **2022**, *237*, 213–230. [CrossRef]
6. Vasquez, N.G.; Rupp, R.F.; Andersen, R.K.; Toftum, J. Occupants' responses to window views, daylighting and lighting in buildings: A critical review. *Build. Environ.* **2022**, *219*, 109172. [CrossRef]
7. Garlisi, C.; Trepci, E.; Li, X.; Al, R.; Al-ali, K.; Pereira, R.; Zheng, L.; Azar, E.; Palmisano, G. Multilayer thin film structures for multifunctional glass: Self-cleaning, antireflective and energy-saving properties. *Appl. Energy* **2020**, *264*, 114697. [CrossRef]
8. Musa, A.; Hakim, M.L.; Alam, T.; Islam, M.T.; Alshammari, A.S.; Mat, K.; Salaheldeen, M.M.; Almalki, S.H.A. Polarization Independent Metamaterial Absorber with Anti-Reflection Coating Nanoarchitectonics for Visible and Infrared Window Applications. *Materials* **2022**, *15*, 3733. [CrossRef]
9. Jahid, A.; Wang, J.; Zhang, E.; Duan, Q.; Feng, Y. Energy savings potential of reversible photothermal windows with near infrared-selective plasmonic nanofilms. *Energy Convers. Manag.* **2022**, *263*, 115705. [CrossRef]
10. Mesloub, A.; Ghosh, A. Daylighting performance of light shelf photovoltaics (LSPV) for office buildings in hot desert-like regions. *Appl. Sci.* **2020**, *10*, 7959. [CrossRef]
11. Mesloub, A.; Ghosh, A.; Touahmia, M. Performance Analysis of Photovoltaic Integrated Shading Devices (PVSDs) and Semi-Transparent Photovoltaic (STPV) Devices Retrofitted to a Prototype Office Building in a Hot Desert Climate. *Sustainability* **2020**, *12*, 10145. [CrossRef]
12. Chidubem Iluyemi, D.; Nundy, S.; Shaik, S.; Tahir, A.; Ghosh, A. Building energy analysis using EC and PDLC based smart switchable window in Oman. *Sol. Energy* **2022**, *237*, 301–312. [CrossRef]
13. Mesloub, A.; Ghosh, A.; Touahmia, M.; Abdullah, G.; Alsolami, B.M.; Ahriz, A. Assessment of the overall energy performance of an SPD smart window in a hot desert climate The International Commission on Illumination. *Energy* **2022**, *252*, 124073. [CrossRef]
14. Ghosh, A.; Norton, B. Durability of switching behaviour after outdoor exposure for a suspended particle device switchable glazing. *Sol. Energy Mater. Sol. Cells* **2017**, *163*, 178–184. [CrossRef]
15. Ghosh, A.; Norton, B.; Duffy, A. Measured overall heat transfer coefficient of a suspended particle device switchable glazing. *Appl. Energy* **2015**, *159*, 362–369. [CrossRef]
16. Ghosh, A.; Norton, B.; Duffy, A. First outdoor characterisation of a PV powered suspended particle device switchable glazing. *Sol. Energy Mater. Sol. Cells* **2016**, *157*, 1–9. [CrossRef]

17. Shaik, S.; Nundy, S.; Ramana, V.; Ghosh, A.; Afzal, A. Polymer dispersed liquid crystal retrofitted smart switchable glazing: Energy saving, diurnal illumination, and CO_2 mitigation prospective. *J. Clean. Prod.* **2022**, *350*, 131444. [CrossRef]
18. Roy, A.; Ullah, H.; Ghosh, A.; Baig, H.; Sundaram, S.; Tahir, A.A.; Mallick, T.K. Understanding the Semi-Switchable Thermochromic Behavior of Mixed Halide Hybrid Perovskite Nanorods. *J. Phys. Chem. C* **2021**, *125*, 18058–18070. [CrossRef]
19. Hemaida, A.; Ghosh, A.; Sundaram, S.; Mallick, T.K. Evaluation of thermal performance for a smart switchable adaptive polymer dispersed liquid crystal (PDLC) glazing. *Sol. Energy* **2020**, *195*, 185–193. [CrossRef]
20. Hemaida, A.; Ghosh, A.; Sundaram, S.; Mallick, T.K. Simulation study for a switchable adaptive polymer dispersed liquid crystal smart window for two climate zones (Riyadh and London). *Energy Build.* **2021**, *251*, 111381. [CrossRef]
21. Ghosh, A.; Norton, B. Advances in switchable and highly insulating autonomous (self-powered) glazing systems for adaptive low energy buildings. *Renew. Energy* **2018**, *126*, 1003–1031. [CrossRef]
22. Ghosh, A.; Norton, B.; Duffy, A. Effect of atmospheric transmittance on performance of adaptive SPD-vacuum switchable glazing. *Sol. Energy Mater. Sol. Cells* **2017**, *161*, 424–431. [CrossRef]
23. Ghosh, A.; Norton, B.; Duffy, A. Effect of sky clearness index on transmission of evacuated (vacuum) glazing. *Renew. Energy* **2017**, *105*, 160–166. [CrossRef]
24. Ghosh, A. Soiling Losses: A Barrier for India's Energy Security Dependency from Photovoltaic Power. *Challenges* **2020**, *11*, 9. [CrossRef]
25. Ullah, M.B.; Kurniawan, J.T.; Poh, L.K.; Wai, T.K.; Tregenza, P.R. Attenuation of diffuse daylight due to dust deposition on glazing in a tropical urban environment. *Light. Res. Technol.* **2003**, *35*, 19–29. [CrossRef]
26. Sharples, S.; Stewart, L.; Tregenza, P.R. Glazing daylight transmittances: A field survey of windows in urban areas. *Build. Environ.* **2001**, *36*, 503–509. [CrossRef]
27. Chanchangi, Y.N.; Ghosh, A.; Baig, H.; Sundaram, S.; Mallick, T.K. Soiling on PV performance influenced by weather parameters in Northern Nigeria. *Renew. Energy* **2021**, *180*, 874–892. [CrossRef]
28. Chanchangi, Y.N.; Ghosh, A.; Sundaram, S.; Mallick, T.K. An analytical indoor experimental study on the effect of soiling on PV, focusing on dust properties and PV surface material. *Sol. Energy* **2020**, *203*, 46–68. [CrossRef]
29. Chanchangi, Y.N.; Ghosh, A.; Sundaram, S.; Mallick, T.K. Angular dependencies of soiling loss on photovoltaic performance in Nigeria. *Sol. Energy* **2021**, *225*, 108–121. [CrossRef]
30. Midtdal, K.; Jelle, B.P. Self-cleaning glazing products: A state-of-the-art review and future research pathways. *Sol. Energy Mater. Sol. Cells* **2013**, *109*, 126–141. [CrossRef]
31. Syafiq, A.; Balakrishnan, V.; Ali, M.S.; Dhoble, S.J.; Rahim, N.A.; Omar, A.; Halim, A.; Bakar, A. Application of transparent self-cleaning coating for photovoltaic panel: A review. *Curr. Opin. Chem. Eng.* **2022**, *36*, 100801. [CrossRef]
32. Adak, D.; Bhattacharyya, R.; Barshilia, H.C. A state-of-the-art review on the multifunctional self-cleaning nanostructured coatings for PV panels, CSP mirrors and related solar devices. *Renew. Sustain. Energy Rev.* **2022**, *159*, 112145. [CrossRef]
33. Roy, A.; Ghosh, A.; Mallick, T.K.; Tahir, A.A. Smart glazing thermal comfort improvement through near-infrared shielding paraffin incorporated SnO_2-Al_2O_3 composite. *Constr. Build. Mater.* **2022**, *331*, 127319. [CrossRef]
34. Ghunem, R.; Cherney, E.A.; Farzaneh, M.; Momen, G.; Illian, H.A.; Mier, G.; Peesapati, V.; Yin, F. Development and Application of Superhydrophobic Outdoor Insulation: A Review. *IEEE Trans. Dielectr. Electr. Insul.* **2022**. [CrossRef]
35. Zhao, B.; Yue, X.; Tian, Q.; Qiu, F.; Zhang, T. Controllable fabrication of ZnO nanorods @ cellulose membrane with self-cleaning and passive radiative cooling properties for building energy-saving applications. *Cellulose* **2022**, *29*, 1981–1992. [CrossRef]
36. Hu, X.; Zhang, Y.; Zhang, J.; Yang, H.; Wang, F.; Fei, B.; Noor, N. Sonochemically-coated transparent wood with ZnO: Passive radiative cooling materials for energy saving applications. *Renew. Energy* **2022**, *193*, 398–406. [CrossRef]
37. Ma, M.; Xie, M.; Ai, Q. Study on photothermal properties of Zn-ZnO/paraffin binary nanofluids as a filler for double glazing unit. *Int. J. Heat Mass Transf.* **2022**, *183*, 122173. [CrossRef]
38. Thirumalai, K.; Shanthi, M.; Swaminathan, M. Hydrothermal fabrication of natural sun light active Dy_2WO_6 doped ZnO and its enhanced photo- electrocatalytic activity and self-cleaning properties. *RSC Adv.* **2017**, *7*, 7509–7518. [CrossRef]
39. Saif, M.; Hafez, H.; Nabeel, A.I. Chemosphere Photo-induced self-cleaning and sterilizing activity of Sm^{3+} doped ZnO nanomaterials. *Chemosphere* **2013**, *90*, 840–847. [CrossRef]
40. Nundy, S.; Ghosh, A.; Tahir, A.; Mallick, T.K. Role of Hafnium Doping on Wetting Transition Tuning the Wettability Properties of ZnO and Doped Thin Films: Self-Cleaning Coating for Solar Application. *ACS Appl. Mater. Interfaces* **2021**, *13*, 25540–25552. [CrossRef]
41. El-Hossary, F.M.; Mohamed, S.H.; Noureldein, E.A.; Abo EL-Kassem, M. ZnO thin films prepared by RF plasma chemical vapour transport for self-cleaning and transparent conducting coatings. *Bull. Mater. Sci.* **2021**, *44*, 82. [CrossRef]
42. Al-Kuhaili, M.F. Optical properties of hafnium oxide thin films and their application in energy-efficient windows. *Opt. Mater.* **2004**, *27*, 383–387. [CrossRef]
43. Nundy, S.; Ghosh, A.; Mallick, T.K. Hydrophilic and Superhydrophilic Self-Cleaning Coatings by Morphologically Varying ZnO Microstructures for Photovoltaic and Glazing Applications. *ACS Omega* **2020**, *5*, 1033–1039. [CrossRef]
44. Alrashidi, H.; Ghosh, A.; Issa, W.; Sellami, N.; Mallick, T.K.; Sundaram, S. Evaluation of solar factor using spectral analysis for CdTe photovoltaic glazing. *Mater. Lett.* **2019**, *237*, 332–335. [CrossRef]
45. Ghosh, A.; Sarmah, N.; Sundaram, S.; Mallick, T.K. Numerical studies of thermal comfort for semi-transparent building integrated photovoltaic (BIPV)-vacuum glazing system. *Sol. Energy* **2019**, *190*, 608–616. [CrossRef]

46. Selvaraj, P.; Ghosh, A.; Mallick, T.K.; Sundaram, S. Investigation of semi-transparent dye-sensitized solar cells for fenestration integration. *Renew. Energy* **2019**, *141*, 516–525. [CrossRef]
47. Ghosh, A.; Norton, B.; Duffy, A. Behaviour of a SPD switchable glazing in an outdoor test cell with heat removal under varying weather conditions. *Appl. Energy* **2016**, *180*, 695–706. [CrossRef]
48. Ghosh, A. Potential of building integrated and attached/applied photovoltaic (BIPV/BAPV) for adaptive less energy-hungry building's skin: A comprehensive Review. *J. Clean. Prod.* **2020**, *276*, 123343. [CrossRef]
49. Nundy, S.; Ghosh, A.; Mesloub, A.; Abdullah, G.; Mashary, M. Impact of COVID-19 pandemic on socio-economic, energy-environment and transport sector globally and sustainable development goal (SDG). *J. Clean. Prod.* **2021**, *312*, 127705. [CrossRef]
50. Nundy, S.; Mesloub, A.; Alsolami, B.M.; Ghosh, A. Electrically actuated visible and near-infrared regulating switchable smart window for energy positive building: A review. *J. Clean. Prod.* **2021**, *301*, 126854. [CrossRef]
51. Ghosh, A.; Selvaraj, P.; Sundaram, S.; Mallick, T.K. The colour rendering index and correlated colour temperature of dye-sensitized solar cell for adaptive glazing application. *Sol. Energy* **2018**, *163*, 537–544. [CrossRef]
52. Ghosh, A.; Norton, B.; Duffy, A. Measured thermal & daylight performance of an evacuated glazing using an outdoor test cell. *Appl. Energy* **2016**, *177*, 196–203. [CrossRef]
53. Ghosh, A.; Norton, B.; Duffy, A. Measured thermal performance of a combined suspended particle switchable device evacuated glazing. *Appl. Energy* **2016**, *169*, 469–480. [CrossRef]
54. Ghosh, A.; Norton, B.; Duffy, A. Daylighting performance and glare calculation of a suspended particle device switchable glazing. *Sol. Energy* **2016**, *132*, 114–128. [CrossRef]
55. Jelle, B.P.; Gustavsen, A.; Nilsen, T.N.; Jacobsen, T. Solar material protection factor (SMPF) and solar skin protection factor (SSPF) for window panes and other glass structures in buildings. *Sol. Energy Mater. Sol. Cells* **2007**, *91*, 342–354. [CrossRef]
56. Grosjean, A.; Le Baron, E. Longtime solar performance estimations of low-E glass depending on local atmospheric conditions. *Sol. Energy Mater. Sol. Cells* **2022**, *240*, 111730. [CrossRef]
57. Ghosh, A.; Mallick, T.K. Evaluation of optical properties and protection factors of a PDLC switchable glazing for low energy building integration. *Sol. Energy Mater. Sol. Cells* **2017**, *176*, 391–396. [CrossRef]
58. Bhandari, S.; Ghosh, A.; Roy, A.; Kumar, T.; Sundaram, S. Compelling temperature behaviour of carbon-perovskite solar cell for fenestration at various climates. *Chem. Eng. J. Adv.* **2022**, *10*, 100267. [CrossRef]
59. Ghosh, A.; Mesloub, A.; Touahmia, M.; Ajmi, M. Visual Comfort Analysis of Semi-Transparent Perovskite Based Building Integrated Photovoltaic Window for Hot Desert. *Energies* **2021**, *14*, 1043. [CrossRef]
60. Ghosh, A.; Norton, B. Optimization of PV powered SPD switchable glazing to minimise probability of loss of power supply. *Renew. Energy* **2019**, *131*, 993–1001. [CrossRef]
61. Ghosh, A.; Sundaram, S.; Mallick, T.K. Colour properties and glazing factors evaluation of multicrystalline based semi-transparent Photovoltaic-vacuum glazing for BIPV application. *Renew. Energy* **2019**, *131*, 730–736. [CrossRef]
62. McCamy, C.S. Correlated color temperature as an explicit function of chromaticity coordinates. *Color Res. Appl.* **1992**, *17*, 142–144. [CrossRef]
63. Lee, E.S.; DiBartolomeo, D.L. Application issues for large-area electrochromic windows in commercial buildings. *Sol. Energy Mater. Sol. Cells* **2002**, *71*, 465–491. [CrossRef]
64. Ghosh, A.; Bhandari, S.; Sundaram, S.; Mallick, T.K. Carbon counter electrode mesoscopic ambient processed & characterised perovskite for adaptive BIPV fenestration. *Renew. Energy* **2020**, *145*, 2151–2158. [CrossRef]
65. Ghosh, A.; Norton, B.; Mallick, T.K. Daylight characteristics of a polymer dispersed liquid crystal switchable glazing. *Sol. Energy Mater. Sol. Cells* **2018**, *174*, 572–576. [CrossRef]
66. Sudan, M.; Tiwari, G.N. Daylighting and energy performance of a building for composite climate: An experimental study. *Alex. Eng. J.* **2016**, *55*, 3091–3100. [CrossRef]
67. Thanachareonkit, A.; Scartezzini, J.L.; Andersen, M. Comparing daylighting performance assessment of buildings in scale models and test modules. *Sol. Energy* **2005**, *79*, 168–182. [CrossRef]
68. CIE Publication. *Spectral Luminous Efficiency Functions Based upon Brightness Matching for Monochromatic Point Sources with 2° and 10° Fields*; CIE Publication: Vienna, Austria, 1988; p. 75. ISBN 3900734119.
69. CIE. *CIE 1988 2° Spectral Luminous Efficiency Function for Photopic Vision*; CIE: Vienna, Austria, 1990; Volume 2.
70. *Illuminanting Engineering Society of North America 2015*.
71. Ghosh, A.; Norton, B. Interior colour rendering of daylight transmitted through a suspended particle device switchable glazing. *Sol. Energy Mater. Sol. Cells* **2017**, *163*, 218–223. [CrossRef]
72. Piccolo, A.; Pennisi, A.; Simone, F. Daylighting performance of an electrochromic window in a small scale test-cell. *Sol. Energy* **2009**, *83*, 832–844. [CrossRef]

MDPI
St. Alban-Anlage 66
4052 Basel
Switzerland
www.mdpi.com

Materials Editorial Office
E-mail: materials@mdpi.com
www.mdpi.com/journal/materials

Disclaimer/Publisher's Note: The statements, opinions and data contained in all publications are solely those of the individual author(s) and contributor(s) and not of MDPI and/or the editor(s). MDPI and/or the editor(s) disclaim responsibility for any injury to people or property resulting from any ideas, methods, instructions or products referred to in the content.

www.ingramcontent.com/pod-product-compliance
Lightning Source LLC
LaVergne TN
LVHW070210100526
838202LV00015B/2027